Petroleum Technology and the Environment

Petroleum Technology and the Environment

Editor: Jane Urry

www.callistoreference.com

Callisto Reference,
118-35 Queens Blvd., Suite 400,
Forest Hills, NY 11375, USA

Visit us on the World Wide Web at:
www.callistoreference.com

ISBN: 978-1-64116-082-7 (Hardback)

Trademark Notice: Registered trademark of products or corporate names are used only for explanation and identification without intent to infringe.

Cataloging-in-Publication Data

Petroleum technology and the environment / edited by Jane Urry.
    p. cm.
Includes bibliographical references and index.
ISBN 978-1-64116-082-7
1. Petroleum engineering. 2. Petroleum--Prospecting. 3. Petroleum--Prospecting--Environmental aspects.
I. Urry, Jane.
TN870 .P48 2019
665.5--dc23

Table of Contents

Preface

Petroleum is a naturally occurring fluid which is refined to extract varied types of fuels. It is primarily obtained by the process of drilling. Kerosene and gasoline are the primary forms of fuels derived from petroleum through chemical processes of distillation. The by-products of petroleum extraction include asphalt, tar, paraffin wax, etc. Petroleum is a non-renewable form of energy and its reservoirs are depleting at a rapid pace. This book delves into the diverse aspects of petroleum extraction, refining, production, etc. and the impact they have on the environment. The various aspects of this field along with technological progress that have future implications have also been glanced at. For all readers who are interested in petroleum technology, the case studies included in this book will serve as an excellent guide to develop a comprehensive understanding.

Various studies have approached the subject by analyzing it with a single perspective, but the present book provides diverse methodologies and techniques to address this field. This book contains theories and applications needed for understanding the subject from different perspectives. The aim is to keep the readers informed about the progresses in the field; therefore, the contributions were carefully examined to compile novel researches by specialists from across the globe.

Indeed, the job of the editor is the most crucial and challenging in compiling all chapters into a single book. In the end, I would extend my sincere thanks to the chapter authors for their profound work. I am also thankful for the support provided by my family and colleagues during the compilation of this book.

Editor

Petroleum Hydrocarbon Spills in the Environment and Abundance of Microbial Community Capable of Biosurfactant Production

Chirwa EMN* and Bezza FA

Water Utilisation and Environmental Engineering Division, Department of Chemical Engineering, University of Pretoria, Pretoria 0002, Republic of South Africa

Abstract

Petroleum industries generate wastes, some of which may be considered hazardous because of the presence of toxic organics and heavy metals. The wastes can bioaccumulate in food chains where they disrupt biochemical or physiological activities of many organisms, thus causing carcinogenesis of some organs, mutagenesis in the genetic material, impairment in reproductive capacity. Bioremediation constitutes the primary mechanism for the elimination of hydrocarbons from contaminated sites by natural existing populations of microorganisms. In this work, microbial community composition and metabolic potential have been explored in petroleum-hydrocarbon contaminated wood treatment plant soil. A collection of strains, adapted to grow on minimal medium supplemented with coal tar creosote, was obtained and the diversity of the bacterial collection was analysed by 16S rRNA gene-based 454 pyrosequencing. Sequencing of the bands revealed a high proportion of Proteobacteria represented by the Alpha, Beta and Gamma subclasses, suggesting that Proteobacteria and especially Gamma subclass is a dominant group in coal tar creosote contaminated soils. The biotechnological potential of the operational taxonomic units (OUTs) revealed a significant degradation of creosote PAHs and production of biosurfactant with important emulsification activities during the bioremediation process.

Keywords: Biosurfactant; Bioremediation; Hydrocarbonoclastic Bacteria; Petroleum Hydrocarbons

Introduction

A considerable amount of oily sludge can be generated from the petroleum industry during its crude oil exploration, production, transportation, storage, and refining processes [1]. In particular, the sludge generated during the petroleum refining process has received increasing attention in recent years. It contains a high concentration of petroleum hydrocarbons (PHCs) and other recalcitrant components. As being recognized as a hazardous waste in many countries, the improper disposal or insufficient treatment of oily sludge can pose serious threats to the environment and human health [2]. Short et al. [3] reported that thirteen years after the *Exxon Valdez* oil spill in Prince William Sound, the toxic effects are still being felt due to the remaining bulk of the less-weathered subsurface oil. A random sampling of underground fuel storage tanks conducted by U.S. Environmental Protection Agency (USEPA) in the United States revealed about 35% leaks in these tanks [4,5].

Due to a lot of factors, oil pollution has continued to be of great concern to the entire world. One of the main issues faced by refineries and petrochemical industries is related to the safe disposal of this residue, since its destination and/or inappropriate treatment can cause serious impact to the environment and potential risk to human health [6]. Petroleum hydrocarbons are organic pollutants of major concern due to their wide distribution, persistence, complex composition, and toxicity. They can bioaccumulate in food chains where they disrupt biochemical or physiological activities of many organisms, thus causing carcinogenesis of some organs, mutagenesis in the genetic material, impairment in reproductive capacity [5]. The most common petroleum hydrocarbons include aliphatic, branched, and cycloaliphatic alkanes, as well as monocyclic and polycyclic aromatic hydrocarbons (PAHs). PAHs include naphthalene, fluorene, phenanthrene, anthracene, fluoranthene, pyrene, benzo[a]anthracene, benzo[a]pyrene. Combined cycloaliphatic–aromatic structures can also be found in crude oil. Each petroleum fraction is usually composed of hundreds of different hydrocarbon molecules rather than a defined composition. Thus, fractions are dissimilar in terms of volatility, bioavailability, toxicity, degradability, and persistence. This complex array of compounds depicts the tremendous challenge for designing effective bioremediation strategies, which can be illustrated by the effects of major contamination events in the past [7,8].

Various physicochemical methods are available for the treatment of petroleum waste, although many of these technologies are costly, energy intensive, inefficient and not eco-friendly. Bioremediation technology on the other hand, which is based on natural microbial population of contaminated sites has been recognized as a sustainable, economic, environmentally friendly and versatile alternative clean-up strategy [9]. The success of bioremediation technologies applied to hydrocarbon-polluted environments highly depends on the biodegrading capabilities of native microbial populations or exogenous microorganisms used as inoculants [10]. The communities which were exposed to hydrocarbons become adapted, exhibiting selective enrichment and genetic changes [11]. The adapted microbial communities can respond to the presence of hydrocarbon pollutants within hours [11] and exhibit higher biodegradation rates than communities with no history of hydrocarbon contamination [12].

The major issue hindering rapid degradation of the petroleum sludge in biological method is the poor availability of hydrocarbons to the microorganisms due to their complexity and water insoluble nature.

---

*Corresponding author: Chirwa EMN, Water Utilisation and Environmental Engineering Division, Department of Chemical Engineering, University of Pretoria, Pretoria 0002, Republic of South Africa, E-mail: Evans.Chirwa@up.ac.za

Hence, the greatest challenge for microbiologists and bioengineers in the area of petroleum sludge bioremediation is developing the technology that will ensure the interaction of microorganisms with the complex hydrocarbon overcoming the pollutant complexity and insolubility [13,14].

Biosurfactants play a pivotal role in the biodegradation of hydrophobic aromatic compounds contained in petroleum and heavy oil , but the extent to which various petroleum-degrading microorganisms produce these substances is still unknown [15]. Since long term exposure to petroleum hydrocarbons would be expected to select for the development of biosurfactant-producing bacteria via horizontal gene transfer and metabolic switching [16], chronically contaminated sites should contain bacteria that produce effective surfactants that can be used by many different petroleum-degrading species that are indigenous to petroleum-dominated habitats [15].

In our previous study of creosote contaminated soil bioremediation (Bezza and Chirwa, 2015, in press) microbial consortia was enriched from the wood treatment plant soil using Creosote as a source of carbon and energy. The soil sample was collected from the top surface layer (15 cm) at a wood impregnation plant in the outskirts of Pretoria (Gauteng, South Africa). The sampling site is contaminated by a variety of industrial waste products related to Creosote processing that have been released over the last twenty years. The soil contained about 21 percent by weight TOC and 1.5% of PAH (such as naphthalene, fluorene, phenanthrene, anthracene, fluoranthene, pyrene, benzo[a] anthracene, benzo[a]pyrene). The bioremediation technique proved efficient in the removal of PAHs in the contaminated soil through supplying oxygen and nutrients. An examination of the waste residue after 45 days of treatment revealed a 79% PAH removals in the nutrient amended bioslurry reactors. PAH percent removals varied from 52.0 percent for 5 and 6 ring PAHs to 82.0 percent for 2 and 3 ring PAHs in nutrient amended samples where the community dynamics was studied. The evaluated process was aerobic thus it supported an array of aerobic bacteria. The objective of this work was to assess the microbial community composition and abundance of *Hydrocarbonoclastic* and biosurfactant producing bacteria from the creosote contaminated soil using high-resolution 16S rRNA tag pyrosequencing technique as described by Albers et al. [17]. The abundance of different *Hydrocarbonoclastic* bacteria was assessed in the creosote contaminated soil samples before the inoculation and after the bioremediation of the soil samples.

Materials and Methods

Sample collection

Creosote degrading microbial cultures were enriched from soil obtained from wood treatment plant in Pretoria West (Pretoria, South Africa). Due to the creosote contamination at the site, high levels of PAHs, PCBs and other petrochemical organic contaminants were detected in the soil. Microbial inocula were prepared by shaking 5 g (wet weight) of the inoculum source in 100 ml of MSM [18] containing 5% (v/v) creosote as a source of carbon and energy. Microbial cultures for the pyrolysis analysis were sampled by withdrawing 5 ml liquid culture from the shake flask before the enrichment. The enrichment culture was grown for 7 days under continuous shaking at 120 rpm in a Labcon SPL-MP 15 Orbital Shaker (Labcon Laboratory Services, South Africa). Aliquots of enrichment cultures (5 ml) were aseptically transferred to fresh 5% (v/v) creosote containing medium and incubation continued. This enrichment procedure was repeated for five successive transfers.

Biosliurrty reactors and sampling procedure

The 2L bioslurry reactors sampled were set up with the contaminated soil to give 30 percent soil-water slurry. The reactors were seeded with the enriched microbial cultures and sufficient nutrients N and P were added to obtain a C:N:P ratio of 100:10:1 [19] to satisfy the demands of the microbes degrading the petroleum hydrocarbon contaminants. The reactors were vigorously mixed using overhead mechanical mixers and run in triplicates at 37°C for 45 days. After the 45 days of bioremediation samples were collected for microbial dynamics study. The reactors were sampled by withdrawing slurry material from each reactor. The samples for the microbial analysis were collected in sterile 15 ml plastic bottles. During transportation to the laboratory, the 15-ml samples were frozen immediately in a cooler with dry ice. The samples were stored in the laboratory at 20°C.

DNA extractions

To extract DNA from samples, frozen sample filters were removed from the freezer and immediately crushed into small pieces in the tube by using a sterile spatula. The frozen filter pieces were added to a tube containing a bead-beating matrix and buffers according to the standard protocol for the Fast DNA spin kit for soil (MP Biomedicals, Solon, OH). DNA extractions were further carried out according to the manufacturers' instructions. The amount and quality of the extracted DNA were estimated on 1% agarose gels and using Nano Drop spectrophotometer readings (Thermo Scientific, Wilmington, DE). Extracted DNA was stored at -20°C for downstream processing.

PCR amplification and pyrosequencing

Samples were prepared for 454 pyrosequencing using two-step PCR [20]. Initial PCR Mastermix was $1 \times$ Phusion HF buffer (with $MgCl_2$; Finnzymes Oy, Espoo, Finland), 0.2 mM of dNTP mixture, 0.5 U Phusion Hot Start DNA polymerase (Finnzymes), 0.5 μM of each primer, 1 μl of template, and sterile Milli-Q water to a final volume of 25 μl. Primers were MPRK341F (5′CCTAYGGGRBGCASCAG-3′) and MPRK806R (5′GGACTACNNGGGTATCTAAT-3′), slightly modified from Yu et al. [21]. The primer set targets the 16S rRNA genes flanking the V3 and V4 regions with an overall coverage of 85% and 80% for bacteria and archaea, respectively. PCR conditions were an initial denaturation step of 98°C for 30s, followed by 30 cycles of denaturation at 98°C for 5 s, annealing at 5°C for 20 s, elongation at 72°C for 20 s, and a final extension step of 72°C for 5 min. Immediately before running on 1% agarose gels with ethidium bromide for UV visualization, PCR products were incubated at 7°C for 3 min and then transferred to ice. The bands of PCR products were cut from the gels and purified using Montage DNA Gel Extraction kits (Millipore, Bedford, MA).

To add adaptor and tags to the PCR products, we performed a second round of PCR using DNA fragments from the purified bands as templates. The second PCR amplification was performed as described above, except that we used the primers MPRK341F and MPRK806R with adaptors and 22 barcodes of 10 nucleotides length (on the forward primer). Further, the number of cycles for denaturation, annealing, and elongation was reduced to 20. The PCR products were processed, run on agarose gels, and purified as described above

Amplified fragments with adapters and tags were quantified using a Qubit fluorometer (Invitrogen, Life technologies, Carlsbad, CA) and mixed in equal concentrations (10^8 copies/μl) to ensure equal representation of each sample. A 454 sequencing run was performed on a 70_75 GS Pico Titer Plate using a GS FLX pyrosequencing system according to the manufacturer's instructions (Roche, Mannheim,

Petroleum Hydrocarbon Spills in the Environment and Abundance of Microbial Community Capable...

3

Germany). Sorting and trimming of sequences>150 bp was done by the Pipeline Initial Process (http://rdp.cme.msu.edu) as previously described [17,22].

Bioinformatic analyses

Sequence processing was performed in Mothur [23]. In Mothur, poor quality sequences were set as sequences with a length less than 550 bases, contained ambiguous bases and homopolymers greater than 6 bases or did not have a barcode and a primer sequence. Multiple sequence alignments were performed using the program MAFFT, version 6.925 [24], with the E-INS-i strategy assuming multiple conserved regions and long gaps. After subsequent preclustering to an alignment of known 16S bacterial sequences, chimeras check with UCHIME [25], the sequences were aligned and clustered into operational taxonomic units (OTUs) using the furthest neighbour algorithm with 97% similarity threshold. Bacterial data were summarized at phylum, class, order, family and genus and species levels (Figure 1). These OTUs were taxonomically identified by the RDP-II Naïve Bayesian Classifier [26] using an 80% confidence threshold.

Results and Discussion

Distribution and abundance of bacterial groups

Using a similarity threshold of 97% to cluster sequences within the same operational taxonomic units (OTUs), a total of 320 OUTs were found from creosote contaminated soil before the bioremediation was conducted (ENG A) and 53 OUTs were obtained in the sample after the bioremediation treatment (ENGB), (Figures 1 and 2a). The overall distribution of the main prokaryotic groups (phyla or classes) showed a dominance of sequences within, *Gammaproteobacteria*, *Alphaproteobacteria*, *Bacilli*, *Betaprotobacteria* and, *Acidobacteria* (Supplementary Figure 1). At the phylum level, the foremost populations in the ENGA were *Proteobacreia* whereas and ENGB reactors included *Proteobacreia*, *Firmicutes*, *Actinobacteria*, *Acidobacteria* at 1% cut off. *Proteobacteria* was the dominant bacterial phylum, representing 99% of the 16S rDNA reads from the 320 most abundant bacterial OTUs in sample A at 1% cut off (Supplementary Figure 1; Relative abandance). *Pseudomonadaceae* (94%) was the most dominant family from the sample whereas *Enterobacteriaceae* (0.018%) and *Xanthomonadaceae* (0.0026%) were observed at 1% cut off level (Figure 1; Relative abandance). Proteobacteria represented 88.8% of the phyla in the ENGB whereas *Firmicutes* (2.22%), *Actinobacteria* (1.6%),

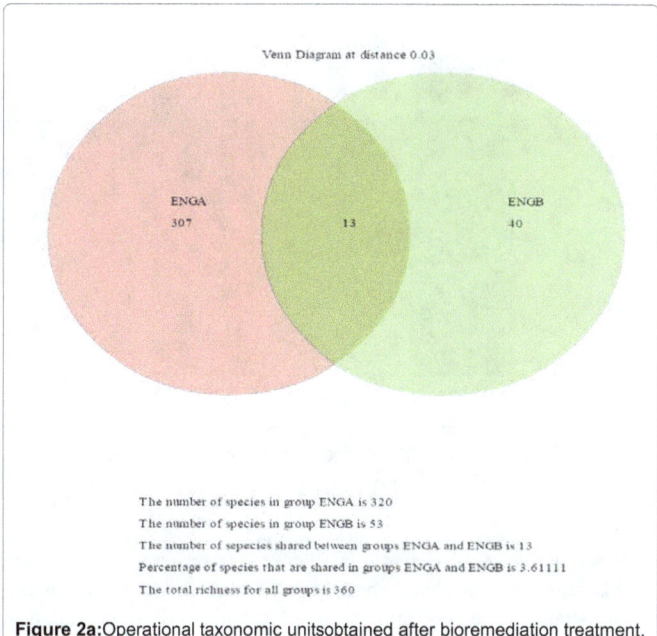

The number of species in group ENGA is 320
The number of species in group ENGB is 53
The number of sepecies shared between groups ENGA and ENGB is 13
Percentage of species that are shared in groups ENGA and ENGB is 3.61111
The total richness for all groups is 360

Figure 2a: Operational taxonomic unitsobtained after bioremediation treatment.

Acidobacteria (1.6%). *Proteobacteria* have been identified in many studies as the predominant phylum in soil samples [27,28] playing an integral role in nutrient cycling [28].The *Proteobacteria* encompass enormous morphological, physiological and metabolic diversity, and are of great importance to global carbon, nitrogen and sulphur cycles [27].

Bacteria in the phyla *Proteobacteria*, *Actinobacteria*, *Firmicutes*, *Bacteroidetes* and *Chlamydiae* have been reported as hydrocarbon degraders [29]. The current study showed that besides the above mentioned phyla *Verrucomicrobia*; *Chloroflexi*; *Planctomycetes*; *WPS-2*; *Chloroflexi*; *Armatimonadetes*; *Gemmatimonadetes*; *WPS-2*; *TM7* bacterial phyla were observed in the creosote contaminated soil. Our results are in agreement with most studies showing the importance of the *Proteobacteria*, especially the *Gamma* division, in hydrocarbon-polluted soil microbial communities or natural asphalts [27,30]. Shared families in both samples were *Enterobacteriaceae*; *Pseudomonadaceae*; *Xanthomonadaceae*; *Micrococcaceae*; *Brucellaceae*; *Bradyrhizobiaceae*; *Sphingomonadaceae*; *Bacillaceae*; *Comamonadaceae*; *Methylobacteriaceae*; *Oxalobacteraceae* and *Weeksellaceae*. Among the 13 bacterial OTUs shared between A and B (Figure 1; Relative abundance) *Pseudomonadaceae* is the dominant family. Strains affiliated with the *gamma-proteobacteria* group are associated with members of the genera *Pseudomonas*, *Stenotrophomonas*, which have been reported as being hydrocarbonoclastic strains [16,31]. The term "hydrocarbonoclastic" has been used to describe hydrocarbon utilizing microorganisms. This specifically relates to microbes that are capable of degrading hydrocarbons, and all of which share some characteristics like having a capable and efficient hydrocarbon uptake system, have receptor sites for binding hydrocarbons and are capable of producing surfactants [32].

Most of the shared OTUs belong to the genera *Pseudomonas*, *Stenotrophomonas*, *Ochrobactrum*, *Achromobacter*, *Enterococcus*, *Cellulosimicrobium*, *Lactobacillus*, *Brevibacillus*, *Ornithinibacillus*, *Arthrobacter*, *Paenibacillus* (Figure 1; Shared OTUs). Species from these genera are reported as efficient biosurfactant producers [33-38]. The community in sample ENGB was dominated by members

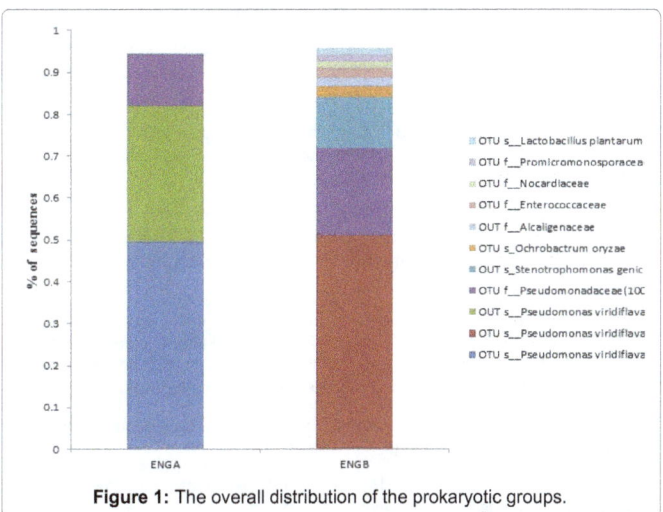

Figure 1: The overall distribution of the prokaryotic groups.

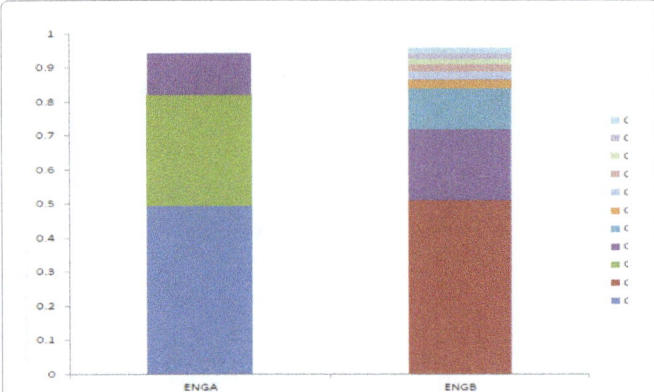

Figure 2b: Relative abundance of the microbial community before the inoculation of the enriched consortium (ENGA) and after 40 days of bioremediation of the creosote contaminated soil (ENGB).

of the *Enterobacteriaceae family* (Figure 2a) representative genera in this family like *Klebsiella* and *Serratia* were reported as efficient biosurfactant producers and capable of hydrocarbon degradation [39,40]. *Pseudomonas spp* which are dominant in sample A (Figure 2b) are able to endure and metabolize contaminants that are considered very toxic to other bacteria. *Pseudomonads* have a huge potential for bioremediation, several studies have proved that *Pseudomonas* sp. can utilize a vast range of contaminants either naturally present or xenobiotic [41]. *Pseudomonas sp.* is a prolific producer of a number of biosurfactants and extra cellular enzymes (like lipase). Importantly, these organisms are able to form stable surface associated microbial communities (biofilms), which have been described as potent alternative to planktonic cells regarding their application as biocatalyst, especially when solvents or otherwise toxic compounds are involved in the processes [42]. The term 'biofilm' refers to surface-associated microbial communities as well as microbes forming flocks and aggregates. Biofilm organisms develop on all kinds of interfaces (e.g., oil/water/air) and are embedded in self-produced extracellular polymeric substances (EPS) in which they live in a coordinate fashion, thereby benefitting from ecological niches formed within the biofilm [43]. Biofilm-growing organisms are self-regenerating, spatially and metabolically well organized, and are in general less affected by toxic substrates and/or products [43]. Accordingly the *Pseudomonas sp.* can be considered as exceptional biocatalysts and can accelerate bioremediation when other species are not fit. (Figure 1) Venn diagram showing unique and shared OTUs (97%) in each sample ENGA and ENGB and the total richness for all groups by pyrosequencing. Figure Relative abundance of the microbial community before the inoculation of the enriched consortium (ENGA) and after 40 days of bioremediation of the creosote contaminated soil (ENG B).

Identification of biosurfactant-producing isolates

Biosurfactant-producing microorganisms are ubiquitous, inhabiting both water (sea, fresh water, groundwater) and land (soil, sediment, sludge) as well as extreme environments (e.g. Oil reservoirs), and thriving at a wide range of temperatures, pH values and salinity. They can also be isolated from undisturbed environments where they have physiological roles not involving the solubilisation of hydrophobic pollutants e.g., antimicrobial activity, biofilm formation or processes of motility and colonization of surfaces [44]. However, it is among the hydrocarbon-degrading microbial communities that the capability to produce biosurfactants is most widespread [45]. Microorganisms have

adopted different strategies to enhance the bioavailability and gain access to hydrophobic compounds, such as hydrocarbons, including (1) biosurfactant mediated solubilisation, (2) direct access of oil drops and (3) biofilm-mediated access [46]. The production of biosurfactants and bioemulsifiers is generally involved, although to different degrees, in all the above strategies [45]. Biosurfactant structural uniqueness resides in the coexistence of a hydrophilic (a sugar or peptide) and a hydrophobic (fatty acid chain) domain in the same molecule, which allows them to occupy the interface of mixed phase systems (e.g., oil/water, air/water, oil/solid/water) and consequently to alter the forces governing the equilibrium conditions. This is the prerequisite for a broad range of surface activities to take place including emulsification, dispersion, dissolution, solubilisation, wetting and foaming [45,47].

The current study suggests that the predominant OTUs identified are ubiquitous in petroleum hydrocarbon contaminated soil and were proved efficient Creosote degrading with efficient biosurfactant production capabilities. The present work shows the potential use of these microorganisms and the total consortium for the bioremediation of crude oil and petroleum polluted environmental media.

Acknowledgement

This research was funded by the National Research Foundation (NRF) of South Africa through the Incentive Funding for Rated Researchers Grant No. IFR2010042900080awarded to Prof. Evans M.N.Chirwa of the University of Pretoria.We thank Prof Fanus Venter of the Department of Microbiology and Plant Pathology, University of Pretoria, for his assistance with the DNA sequencing and characterisation of bacteria from the soil.

References

1. Xu N, Wang W, Han P, Lu X (2009) Effects of ultrasound on oily sludge deoiling. Journal of hazardous materials 171: 914-917.

2. Hu G, Li J, Zeng G (2013) Recent development in the treatment of oily sludge from petroleum industry: A review. Journal of hazardous materials 261: 470-490.

3. Short JW, Lindeberg MR, Harris PM, Maslko J, Rice SD (2002) Vertical oil distribution within intertidal zone 12 years after the Exxon Valdez oil spill in Prince William Sound, Alaska. Proceedings of the 25th Arctic and Marine Oilspill Program (AMOP) Technical seminar, Calgary, Alberta pp: 57-72.

4. United Press International (1986) Gasoline Reported threatening water. Boston Globe.

5. Onwurah INE, Ogugua VN, Onyike NB, Ochonogor AE, Otitoju OF (2007) Crude oil spills in the environment, Effects and Some Innovative Clean-up Biotechnologies. International Journal of Environmental Research 1: 307-320.

6. Cerqueira VS, Maria do Carmo RP, Camargo FA, Bento FM (2014) Comparison of bioremediation strategies for soil impacted with petrochemical oily sludge. International Biodeterioration and Biodegradation 95: 338-345.

7. Atlas RM, Hazen TC (2011) Oil Biodegradation and Bioremediation: A Tale of the Two Worst Spills in U.S. History. Environmental science and technology 45: 6709-6715.

8. Fuentes S, Méndez V, Aguila P, Seeger M (2014) Bioremediation of petroleum hydrocarbons: catabolic genes, microbial communities, and applications. Applied microbiology and biotechnology 98: 4781-4794.

9. Das R, Kazy SK (2014) Microbial diversity, community composition and metabolic potential in hydrocarbon contaminated oily sludge: prospects for in situ bioremediation. Environmental Science and Pollution Research 21: 7369-7389.

10. Diaz-Ramirez IJ, Escalante-Espinosa E, Favela-Torres E, Gutiérrez-Rojas M, Ramírez-Saad H (2008) Design of bacterial defined mixed cultures for biodegradation of specific crude oil fractions, using population dynamics analysis by DGGE. International Biodeterioration & Biodegradation 62: 21-30.

11. Atlas RM, Bartha R (1998) Microbial Ecology Fundamentals and Applications (4thedn.) Benjamin Cummings.

12. Al-Wasify RS, Hamed SR (2014) Bacterial Biodegradation of Crude Oil using Local Isolates. International Journal of Bacteriology.

13. Mohan SV, Reddy BP, Sarma PN (2009) Ex situ slurry phase bioremediation of chrysene contaminated soil with the function of metabolic function: Process evaluation by data enveloping analysis (DEA) and Taguchi design of experimental methodology (DOE). Bioresource technology 100: 164-172.

14. Mohan SV, Chandrasekhar K (2011) Self-induced bio-potential and graphite electron accepting conditions enhances petroleum sludge degradation in bio-electrochemical system with simultaneous power generation. Bioresource Technology 102: 9532-9541.

15. Belcher RW, Huynh KV, Hoang TV, Crowley DE (2012) Isolation of biosurfactant-producing bacteria from the Rancho La Brea Tar Pits. World Journal of Microbiology and Biotechnology 28: 3261-3267.

16. Brito EM, Guyoneaud R, Goñi-Urriza M, Ranchou-Peyruse A, Verbaere A, et al. (2006) Characterization of hydrocarbonoclastic bacterial communities from mangrove sediments in Guanabara Bay, Brazil. Research in microbiology 157: 752-762.

17. Albers CN, Ellegaard-Jensen L, Harder CB, Rosendahl S, Knudsen BE, et al. (2014) Groundwater chemistry determines the prokaryotic community structure of waterworks sand filters. Environmental Science and Technology 49: 839-846.

18. Trummler K, Effenberger F, Syldatk C (2003) An integrated microbial/enzymatic process for production of rhamnolipids and L-(+)-rhamnose from rapeseed oil with Pseudomonas sp. DSM 2874. European journal of lipid science and technology 105: 563-571.

19. Cookson JT (1995) Bioremediation engineering: design and application. Food and Agricultural Organization of the United Nations.

20. Sundberg C, Al-Soud WA, Larsson M, Alm E, Yekta SS, et al. (2013) 454 pyrosequencing analyses of bacterial and archaeal richness in 21 full-scale biogas digesters. FEMS microbiology ecology 85: 612-626.

21. Yu Y, Lee C, Kim J, Hwang S (2005) Group-specific primer and probe sets to detect methanogenic communities using quantitative real-time polymerase chain reaction. Biotechnology and Bioengineering 89: 670-679.

22. Cole JR, Wang Q, Cardenas E, Fish J, Chai B, et al. (2009) The Ribosomal Database Project: improved alignments and new tools for rRNA analysis. Nucleic acids research 37: D141-D145.

23. Schloss PD, Westcott SL, Ryabin T, Hall JR, Hartmann M, et al. (2009) Introducing mothur: open-source, platform-independent, community-supported software for describing and comparing microbial communities. Applied and environmental microbiology 75: 7537-7541.

24. Katoh K, Asimenos G, Toh H (2009) Multiple alignment of DNA sequences with MAFFT. Methods Mol Biol 537: 39-64.

25. Edgar RC, Haas BJ, Clemente JC, Quince C, Knight R (2011) UCHIME improves sensitivity and speed of chimera detection. Bioinformatics 27: 2194-2200.

26. Wang Q, Garrity GM, Tiedje JM, Cole JR (2007) Naive Bayesian classifier for rapid assignment of rRNA sequences into the new bacterial taxonomy. Appl Environ Microbiol 73: 5261-5267.

27. Militon C, Boucher D, Vachelard C, Perchet G, Barra V, et al. (2010) Bacterial community changes during bioremediation of aliphatic hydrocarbon-contaminated soil. FEMS Microbiology Ecology 74: 669-681.

28. Sutton NB, Maphosa F, Morillo JA, Al-Soud WA, Langenhoff AA, et al. (2013) Impact of long-term diesel contamination on soil microbial community structure. Applied and Environmental Microbiology 79: 619-630.

29. Prince RC, Gramain A, McGenity TJ (2010) Prokaryotic hydrocarbon degraders. Handbook of Hydrocarbon and Lipid Microbiology 19: 1669-1692.

30. Kim JS, Crowley DE (2007) Microbial diversity in natural asphalts of the Rancho La Brea Tar Pits. Applied and environmental microbiology 73: 4579-4591.

31. Mahjoubi M, Jaouani A, Guesmi A, Amor SB, Jouini A, et al. (2013). Hydrocarbonoclastic bacteria isolated from petroleum contaminated sites in Tunisia: isolation, identification and characterization of the biotechnological potential. New Biotechnology 30: 723-733.

32. Philp JC, Bamforth SM, Singleton I, Atlas RM (2005) Environmental pollution and restoration: a role for Bioremediation. Bioremediation: Applied Microbial Solutions for Real-World Environmental Cleanup 1-48.

33. Bodour AA, Drees KP, Maier RM (2003) Distribution of biosurfactant-producing bacteria in undisturbed and contaminated arid Southwestern soils. Applied and Environmental Microbiology 69: 3280-3287.

34. Rodrigues L, Moldes A, Teixeira J, Oliveira R (2006) Kinetic study of fermentative biosurfactant production by Lactobacillus strains. Biochemical Engineering Journal 28: 109-116.

35. Maciel BM, Dias JC, Dos Santos AC, Filho RC, Fontana R, et al. (2007) Microbial surfactant activities from a petrochemical landfarm in a humid tropical region of Brazil. Canadian journal of microbiology 53: 937-943.

36. Najafi AR, Rahimpour MR, Jahanmiri AH, Roostaazad R, Arabian D, et al. (2011) Interactive optimization of biosurfactant production by Paenibacillus alvei ARN63 isolated from an Iranian oil well. Colloids and Surfaces B: Biointerfaces 82: 33-39.

37. Patil SN, Aglave BA, Pethkar AV, Gaikwad VB (2012) Stenotrophomonas koreensis a novel biosurfactant producer for abatement of heavy metals from the environment. Afr J Microbiol Res 6: 5173-5178.

38. Amer R, Fattah YRA (2014) Hydrocarbonclastic marine bacteria in Mediterranean Sea, El-Max, Egypt: isolation, identification and site characterization. Jokull journal 64: 223-249.

39. Vieira PA, Vieira RB, De França FP, Cardoso VL (2007) Biodegradation of effluent contaminated with diesel fuel and gasoline. Journal of Hazardous Materials 140: 52-59.

40. Chamkha M, Trabelsi Y, Mnif S, Sayadi S (2011) Isolation and characterization of Klebsiella oxytoca strain degrading crude oil from a Tunisian off-shore oil field. Journal of basic microbiology 51: 580-589.

41. Nikolopoulou M (2013) Oil spills bioremediation in marine environment: biofilm characterization around oil droplets.

42. Lang K, Zierow J, Buehler K, Schmid A (2014) Metabolic engineering of Pseudomonas sp. strain VLB120 as platform biocatalyst for the production of isobutyric acid and other secondary metabolites. Microbial cell factories 13.

43. Halan B, Buehler K, Schmid A (2012) Biofilms as living catalysts in continuous chemical syntheses. Trends in biotechnology 30: 453-465.

44. Van Hamme JD, Singh A, Ward OP (2006) Physiological aspects: Part 1 in a series of papers devoted to surfactants in microbiology and biotechnology. Biotechnology Advances 24: 604-620.

45. Marchant R, Smyth APTJP, Banat IM (2010) 47 Production and Roles of Biosurfactants and Bioemulsifiers in Accessing Hydrophobic Substrates, pp.1502-1510.

46. Hommel RK (1990) Formation and physiological role of biosurfactants produced by hydrocarbon-utilizing microorganisms Biosurfactants in hydrocarbon utilization. Biodegradation 1(2-3): 107-119.

47. Desai JD, Banat IM (1997) Microbial production of surfactants and their commercial potential. Microbiology and Molecular biology reviews 61: 47-64.

2

Evaluation and Prospect Identification in the Olive Field, Niger Delta Basin, Nigeria

Emina R, Obiadi II* and Obiadi CM

Department of Geological Sciences, Nnamdi Azikiwe University, Awka, Nigeria

Abstract

The study area lies in the Greater Ughelli Depobelt of Niger Delta. Recently, a major focus within the Niger Delta is the rejuvenation of older fields (also called brown fields) and the identification of new prospects from these old fields. This study is focused on the evaluation of the Olive Field and the identification of new prospects within the field. Data used were; 3D seismic cube, four composite well logs and check shot. 3 dimension seismic, well log and structural interpretation were done to evaluate the petroleum potentials of the reservoirs using the petrel 2010 and the interactive petrophysic v.36 softwares. Well data were used in the identification of reservoirs and determination of petrophysical parameters and hydrocarbon presence. Four horizons that corresponded to selected well tops were mapped after well to seismic tie. Time and depth structural maps were created from the mapped horizons. Four Hydrocarbon bearing reservoirs within the depth range of 6743 ft – 9045 ft, having volume of shale (Vsh) ranging from 15.32% - 29.06% were interpreted. The total porosity of the reservoirs ranges from 24.63% - 34.01%, while the effective porosity ranges from 17.26% - 31.71%, indicating the reservoirs have very good porosities. The ratio of the Net to Gross Thickness of the reservoirs ranges from 0.720 – 0.980 while the water saturation values ranges from 19.87% - 29.07%. From the water saturation deductions, the hydrocarbon saturation ranges from 70.93% - 78.86% of gas in the given reservoirs. Amplitude attribute extraction and analysis of the horizon maps was used in the identification of areas with hydrocarbon accumulations which are conformable with structures. The use of structural and attribute maps has aided the identification of prospects in the Olive Field. Therefore, it is recommended that wells be drilled to target the new prospects which will improve the hydrocarbon recovery in Olive Field.

Keywords: Olive; Delta; Reservoirs; Petroleum hydrocarbons

Introduction

The Olive Field is located within the Greater Ughelli Depobelt, Onshore Niger Delta, Nigeria (Figure 1). Recently, major focus within the Niger Delta has been directed to the rejuvenation of older fields and the identification of new prospects from these old fields. Studies on the Niger Delta, has shown that reservoirs in this Basin exhibit range of complexities in the sedimentological and petrophysical characteristics due to differences in hydrodynamic conditions prevalent in their depositional setting.

Reservoir characterization technology requires the integration of all available subsurface data such as well logs, check shot, core data, biostratigraphic data, and seismic data. These data are the result of measurement carried out by sophisticated instrumentations and processed using highly developed software [1]. The parameters characterized in a reservoir include structure (fold, fault, and fractures), internal architecture (homogeneity), petrophysical properties (porosity and permeability) and hydrocarbon properties (thermodynamics) [2].

Considering the increasing global demand for energy, the high cost of hydrocarbon production and the associated risk, more robust ways of evaluating and characterizing reservoirs is required to enhance the identification of new prospect and improve yield / rejuvenate older fields as its much cheaper and less risky than exploring in frontier / unknown areas [3]. Aizebeokhai and Olayinka established that reservoir heterogeneity and formation evaluation problems can make it difficult to characterize fluid distribution, estimate hydrocarbon in place and determine permeability. They suggested that the approach used in characterizing a reservoir should involve a combination of analysis of geological framework of the reservoir, hydrocarbon trapping components (stratigraphic and structural), formation evaluation and the calculation of volumetric hydrocarbon in place. Obiadi et al. generated horizon and structural maps of the subsurface geology of

parts of the Niger Delta by interpreting seismic data [4]; and on the bases of structural closures identified leads in deeper horizons of producing fields [5]. Adetaye and Enikanselu extracted reflection attributes from subsurface maps and used it to map lateral boundaries of reservoirs [6]. Omoboriowo et al. used suit of geophysical well-logs for detailed petrophysical and well log sequence stratigraphic analysis in the Niger Delta. Their findings show that reservoirs vary in petrophysical properties due to differences in their environment of deposition. This paper is aimed at analysing and integrating seismic and well log data to aid reservoir characterization and prospect identification in the Olive Field, an old field in the Niger delta.

Geology and tectonic setting of the Niger delta basin

The tectonic framework of the continental margin along the West Coast of equatorial Africa is controlled by Cretaceous fracture zones which subdivide the margin into individual basins, and, in Nigeria, form the boundary faults of the Cretaceous Benue-Abakaliki trough. The trough represents a failed arm of a rift triple junction associated with the opening of the South Atlantic. In this region, rifting started in the Late Jurassic and persisted into the Middle Cretaceous [7]. In the region of the Niger Delta, rifting diminished altogether in the Late Cretaceous. After rifting ceased, gravity tectonics became the primary deformational process. Shale mobility induced internal deformation

*Corresponding author: Obiadi II, Department of Geological Sciences, Nnamdi Azikiwe University, Awka, Nigeria, E-mail: izuchukwuig@yahoo.com

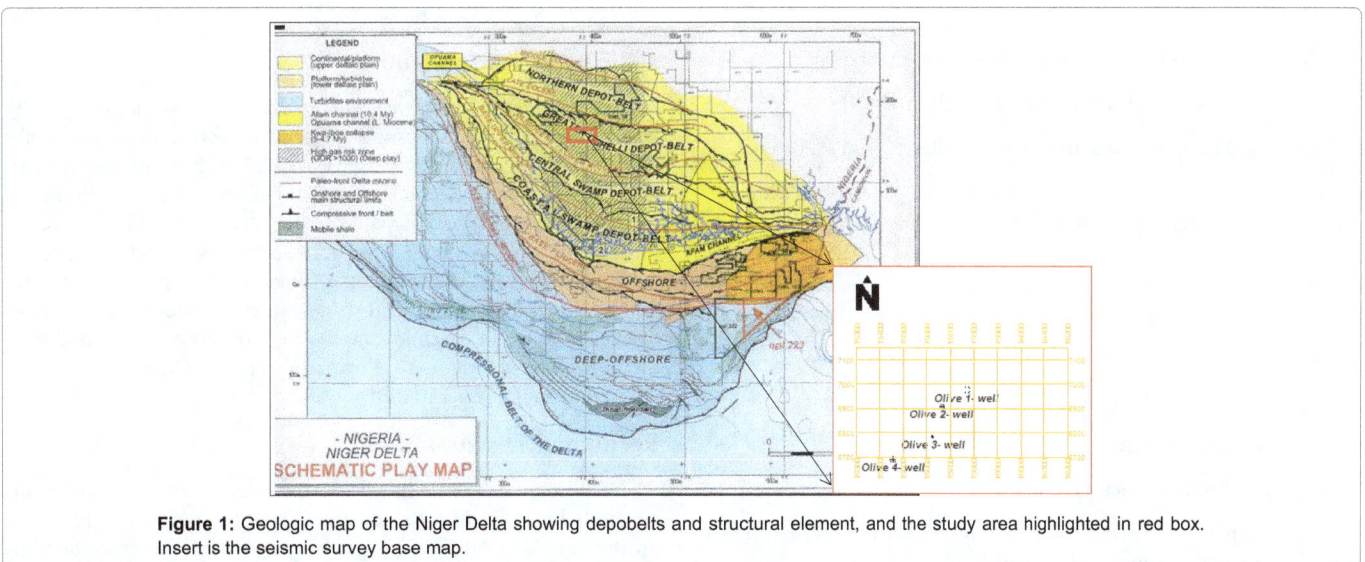

Figure 1: Geologic map of the Niger Delta showing depobelts and structural element, and the study area highlighted in red box. Insert is the seismic survey base map.

and occurred in response to two processes [8]. Firstly, shale diapirs formed from loading of poorly compacted, over-pressured, prodelta and delta-slope clays (Akata Formation) by the higher density delta-front sands (Agbada Formation.); and secondly, slope instability occurred due to a lack of lateral, basinward support for the under-compacted delta-slope clays. For any given depobelt, gravity tectonics were completed before deposition of the Benin Formation and are expressed in complex structures, including shale diapirs, roll-over anticlines, collapsed growth fault crests, back-to-back features, and steeply dipping, closely spaced flank faults [9]. These faults mostly offset different parts of the Agbada Formation and flatten into detachment planes near the top of the Akata Formation. Three lithostratigraphic units are distinguished in the Tertiary Niger Delta. The basal Akata Formation which is predominantly marine prodelta shale is overlain by the paralic sand/shale sequence of the Agbada Formation. The topmost section is the continental upper deltaic plain sands – the Benin Formation. Virtually all the hydrocarbon accumulations in the Niger Delta occur in the sands and sandstones of Agbada Formation where they are trapped by rollover anticlines related to growth fault development [10,11].

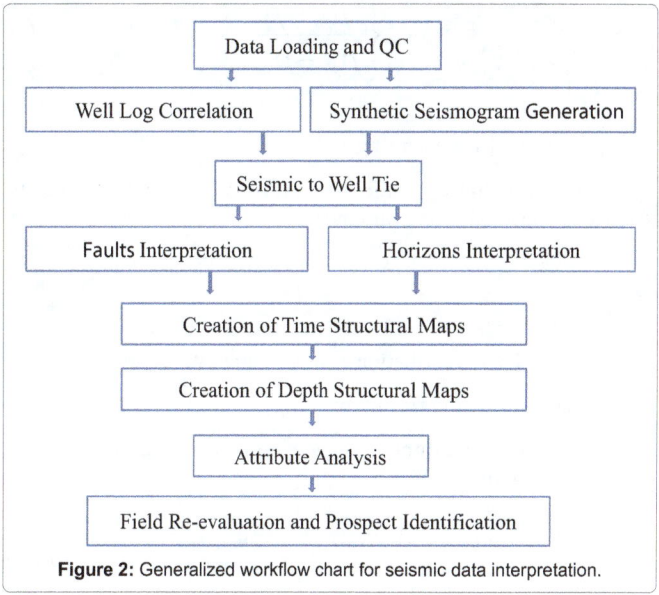

Figure 2: Generalized workflow chart for seismic data interpretation.

Well log analysis/formation evaluation

Reservoirs were identified and correlated across the wells using Gamma and Resistivity logs. The gamma ray log was used in identifying the lithology penetrated by the wells. A shale base line was established, and deflection of the log signature to the right of the shale base line was interpreted as shale (non-reservoir lithology) while deflection to the left of the shale base line was interpreted as sandstone (reservoir lithology). For the resistivity log, deflections to the left were interpreted as low resistivity (or high conductivity). Saline water bearing formations are characterised by low resistivity while hydrocarbon bearing intervals (reservoirs) have high resistivity. Having identified the reservoirs, the Net-to-Gross ratio was calculated.

Hydrocarbon typing was done using the sonic / resistivity logs comparison method. It is best applied as an overlay technique and provides a quick look identification of hydrocarbon bearing zones and type. In water bearing intervals, the lateral deflections (log signatures) due to porosity changes on a resistivity and sonic log are very similar in magnitude, when displayed on a standard scale. If the logs are adjusted to overlay in water bearing zones, hydrocarbon bearing zones will be evident by a deflection of the resistivity log to the right. Gas bearing zones can be identified by a tendency for the sonic to shift to the left due to the slowing of the compressional wave. Thus in gas bearing intervals, the separation between the sonic and resistivity will be greater than in oil bearing intervals.

Volume of Shale V_{Sh} was estimated from the Gamma Ray Log using the equation:

$$V_{Sh} = \frac{GR_{log} - GR_{clean}}{GR_{Shale} - GR_{Clean}} \qquad (1)$$

Where

GR_{log} = Gamma Ray log reading for formation interval

GR_{Shale} = Maximum Gamma Ray log reading (shale)

GR_{Clean} = Minimum Gamma Ray log reading (Clean sand)

Porosity Φ was estimation in the mapped reservoirs using the Density Log reading and the equation:

$$\Phi = \frac{\rho_{ma} - \rho_b}{\rho_{ma} - \rho_{fluid}} \quad (2)$$

Where:

ρ_{ma} = Rock matrix density

ρ_b = Measured density

ρ_{fluid} = Flushed zone measured density

Effective porosity (interconnected pore spaces) was also estimated from Density Log readings and the equation:

$$\Phi_e = \frac{\rho_{ma} - \rho_b}{\rho_{ma} - \rho_{fluid}} - V_{Sh}\left\{\frac{\rho_{ma} - \rho_{Sh}}{\rho_{ma} - \rho_{fluid}}\right\} \quad (3)$$

Where:

ρ_{sh} = measured shale density

Water saturation S_w of the un-invaded was estimated using the Resistivity Log reading and the equation:

$$S_w^2 = \frac{R_0}{R_T} \quad (4)$$

Where:

R_0 = Resistivity of formation at 100% water saturation

R_T = True formation resistivity

The hydrocarbon saturation S_{hy} was estimated from the calculated water saturation thus:

$$S_{hy} = 1 - S_w \text{ or } S_{hy}\% = 100 - S_w\% \quad (5)$$

Seismic data interpretation

Synthetic seismogram was generated using the well logs (Figure 3), horizons picked at the top of the mapped reservoirs and tied to the seismic data (Figure 4). A good tie was obtained between the synthetic seismogram and the seismic data. Faults defined by abrupt termination and dislocation of reflection pattern were identified and interpreted all through the seismic volume. The picked horizons were also mapped all through the seismic volume. Horizon/structural maps were generated for the horizons mapped, and using the check-shot data, converted from time to depth. Seismic (amplitude) attribute was extracted from the horizon/structural maps. Prospects were then identified from the generated maps and seismic attributes.

Result and Discussion

Well log analysis resulted in the identification of reservoirs penetrated by wells in the study area. These reservoirs were identified on the basis of relatively low gamma ray values and corresponding high resistivity reading. The reservoirs, designated Reservoir 1-4, were correlated across the well from the most distal (landwards) to the most proximal (seaward) well (Figure 5).

Reservoir 1 was penetrated by all four well. It is characterized by relatively high resistivity values and signature of the sonic log within this reservoir interval shows its hydrocarbon type to be gas (Figure 6). Estimates of petrophysical parameters computed form different logs from the well are summarised in Table 1. Reservoir 2 also was penetrated by all four wells in the Olive field. The thickness of this reservoir increases basinwards suggesting incision from a lowstand. The well log signature shows it to be hydrocarbon bearing and the hydrocarbon type to be gas. Reservoir 3 was identified in wells 1 and 2 but did now appear in wells 3 and 4. Hydrocarbon type is interpreted to be gas also. Reservoir 4 was penetrated by wells 1, 2 and 4, and the hydrocarbon type is gas.

Results of the well log / petrophysical analysis show that the reservoir quality decreases in the basinward (distal) direction. The net-

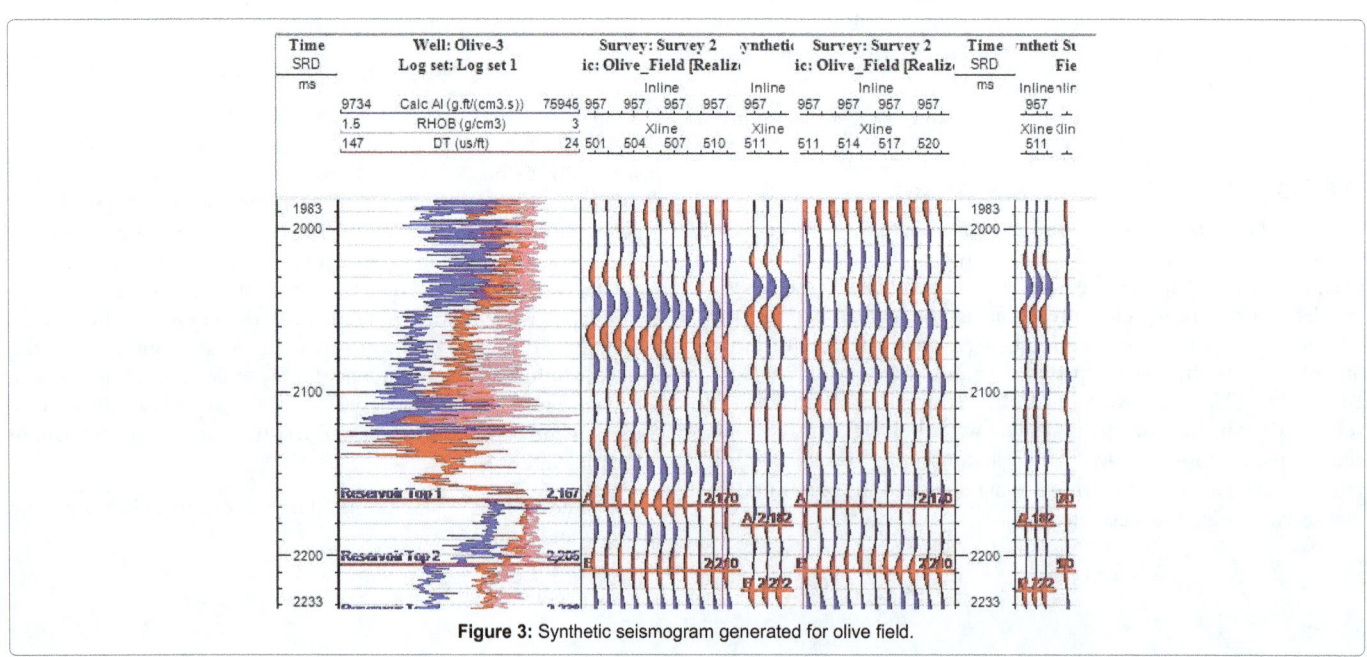

Figure 3: Synthetic seismogram generated for olive field.

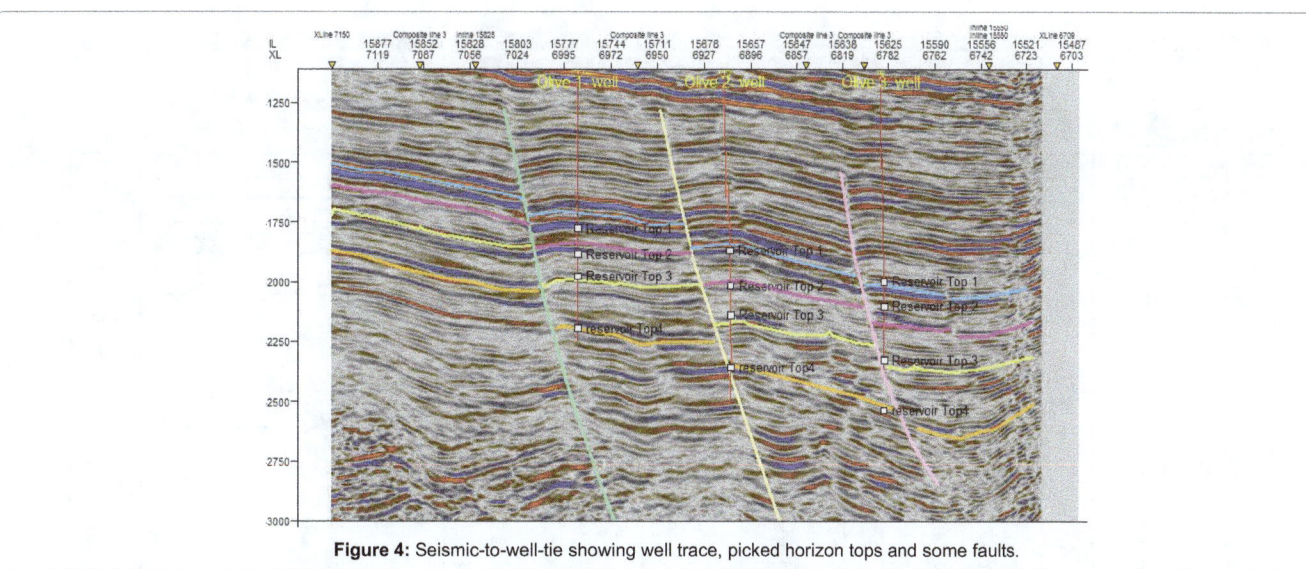

Figure 4: Seismic-to-well-tie showing well trace, picked horizon tops and some faults.

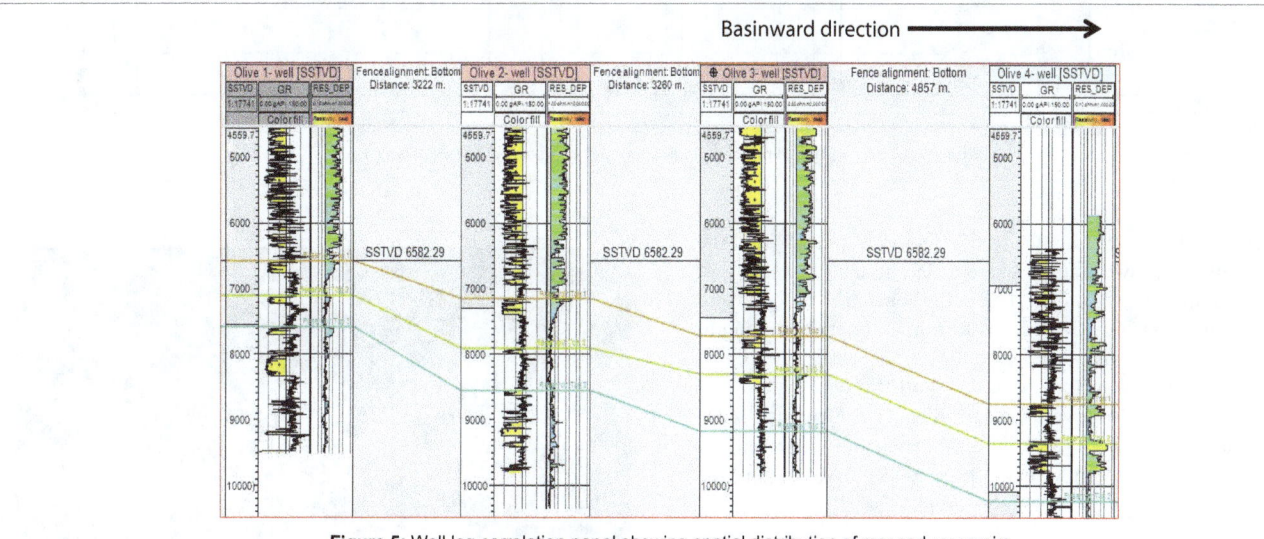

Figure 5: Well log correlation panel showing spatial distribution of mapped reservoirs.

Petro-physical Parameter	Estimated Values			
	Olive Well 1	**Olive Well 2**	**Olive Well 3**	**Olive Well 4**
Reservoir 1 interval	6743-6915 ft	7310-7549 ft	8454-8599 ft	8900-9045 ft
Hydrocarbon type	Gas	Gas	Gas	Gas
Gross Thickness	172 ft	239 ft	144.50 ft	155.5 ft
Net thickness	168 ft	213 ft	104 ft	12 5 ft
Net/Gross	0.980	0.89	0.72	0.81
Volume of shale/clay	0.153 or 15.33%	0.17 or 17.36%	0.29 or 29.06%	0.23 or 22.67%
Total porosity	0.34 or 34.01%	0.31 or 30.79%	0.26 or 25.62%	0.25 or 24.63%
Effective porosity	0.32 or 32.00%	0.28 or 28.18%	0.21 or 21.26%	0.17 or 17.26%
Water saturation	0.19 or 19.8%	0.22 or 21.58%	0.29 or 29.07%	0.21 or 21.14%
Hydrocarbon saturation	0.80 or 80.13%	0.78 or 78.42%	0.71 or 70.93%	0.79 or 78.86%

Table 1: Summary of some estimated petro-physical parameters.

to-gross ratio decreased from 0.98 in the most proximal well (Olive well 1) to 0.81 in the most distal well (Olive well 4). This is also seen in the Volume of Shall which increased from 15.33% in Olive Well 1 to 22.67% in Olive Well 4. A very important property of reservoir is effective porosity, and this too is estimated to have decreased from a value of 32% for Olive Well 1 to 17.26% in Olive Well 4. Hydrocarbon saturation has also followed similar trend with a value of 80.13% in the most proximal well to 78.86 in the most distal well. Corresponding reservoir intervals in all the wells are seen to be penetrated at greater depths from Olive Well 1 to Olive Well 4. This is typical of the Niger Delta Basin and suggestive of progradation. Hydrocarbon typing showed that all the reservoirs contain gas.

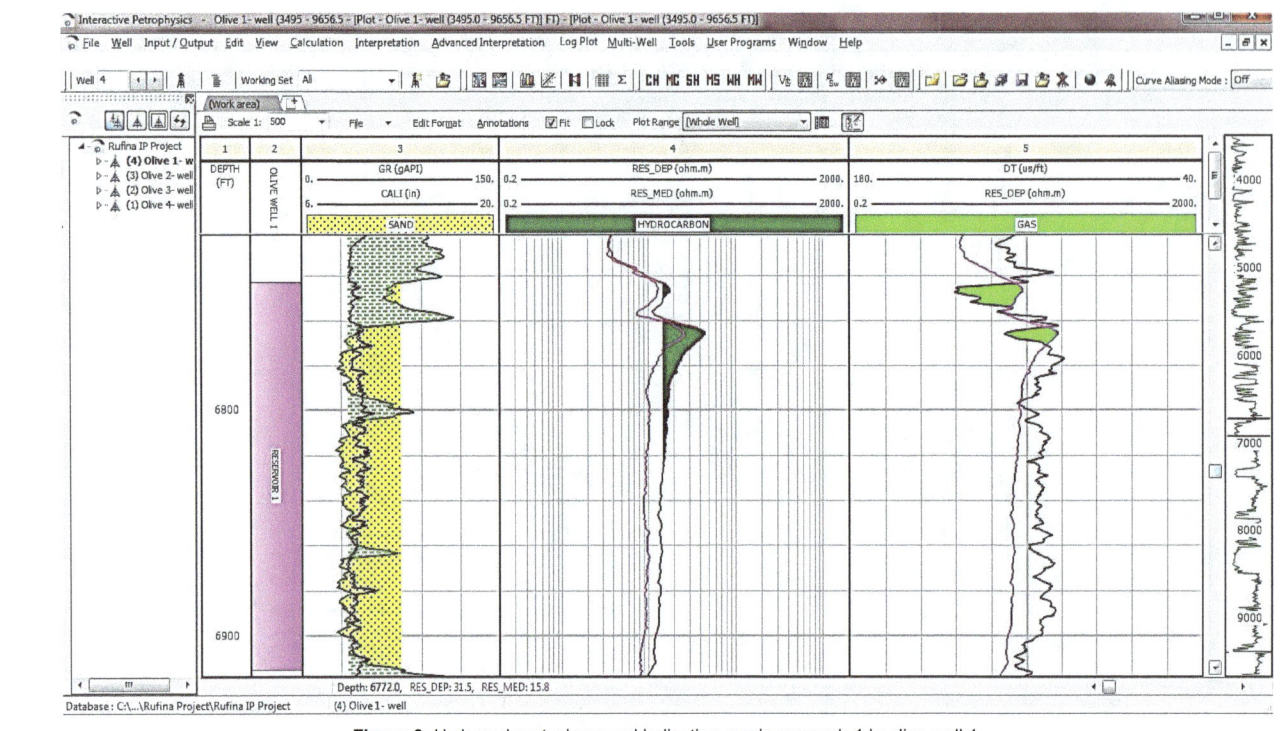

Figure 6: Hydrocarbon typing panel indicating gas in reservoir 1 in olive well 1.

Several faults were picked in the study area (Figure 7). Most of the faults are synthetic growth faults characterised by fault surfaces that are concave in the basinward direction, while some are antithetic and terminate against the synthetic faults. Four horizons corresponding to the tops of the reservoirs mapped during the well log analysis were tied to the seismic volume with the aid of the synthetic seismogram and picked all through the seismic volume to produce Time Structural Maps for all the mapped horizons (Figure 8). With the aid of the velocity data, the Time Structural Map was converted to Depth Structural Map (Figure 9).

The structural maps displayed in time and depths are very similar suggesting good horizon interpretation. The structural maps showed that the field is compartmentalized into several blocks by faults, and these faults and fault blocks formed closures (traps) for hydrocarbon accumulation. The four wells drilled in the field targeted these traps (Figures 7 and 8).

Seismic amplitude attribute map generated from the four horizon maps showed that the fault closures are characterised by high amplitude (Figure 9). Since the horizons mark the top of reservoirs, the high amplitude recorded on the horizons indicates the presence of hydrocarbon within the mapped reservoirs and supports the results of the well log analysis.

Prospect Identification

The Akata Shale source rock, which are thick marine shales deposited at the base of the delta in an oxygen deficient environment is present in large volume beneath the Agbada Formation (main reservoir unit) and is at least volumetrically sufficient to generate enough hydrocarbon for a world class province like the Niger Delta [4,11]. Total Organic Carbon TOC values for the Akata source rock ranges from 0.4% -14.4% [12]. The sandstone reservoir units of the overlying

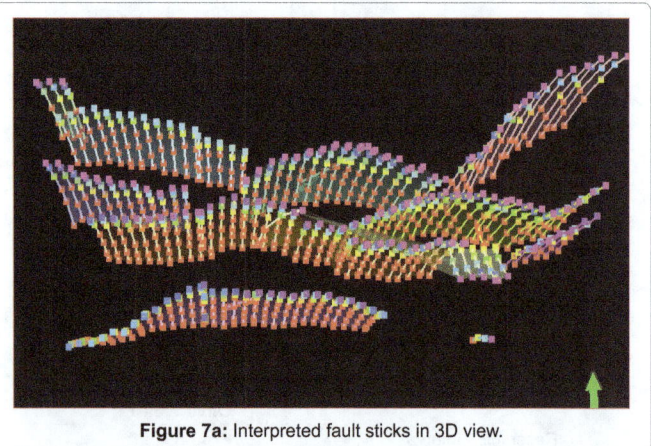

Figure 7a: Interpreted fault sticks in 3D view.

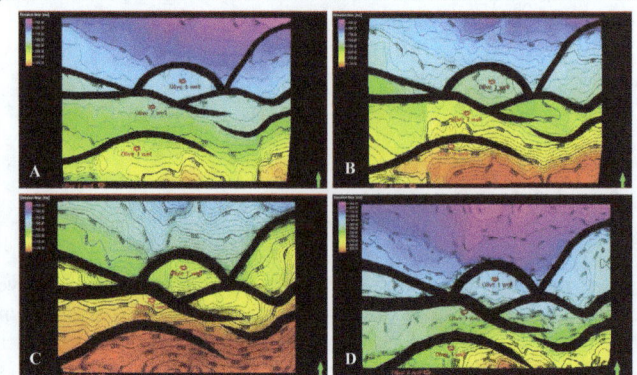

Figure 7b: Time Structural Maps of Reservoir Top 1 (A), Reservoir Top 2 (B), Reservoir Top 3 (C) and Reservoir Top 4 (D) with well positions.

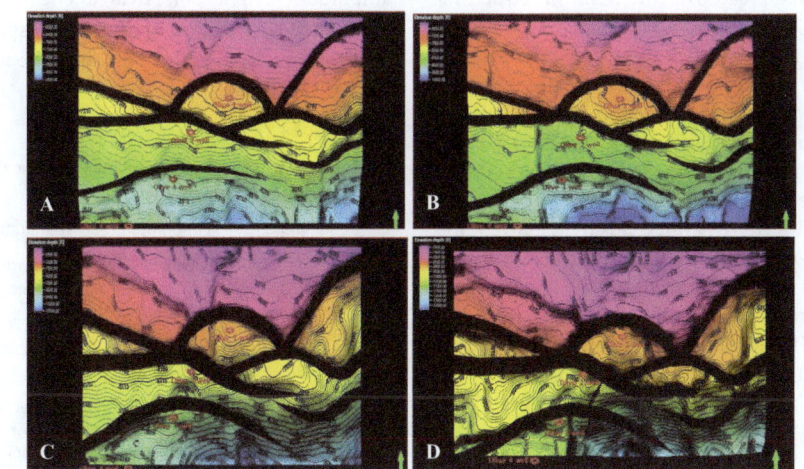

Figure 8: Depth structural maps of Reservoir Top 1 (A), Reservoir Top 2 (B), Reservoir Top 3 (C) and Reservoir Top 4 (D) with well positions.

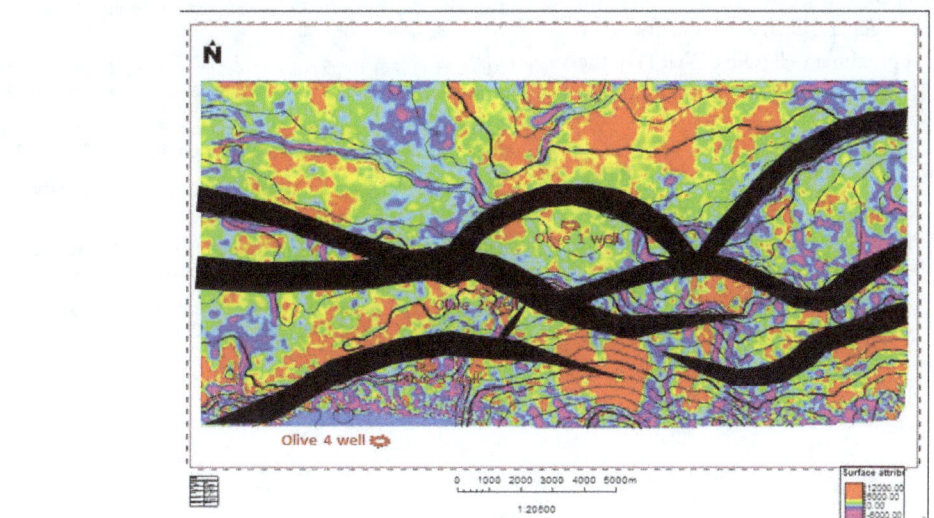

Figure 9: Amplitude attribute map of Reservoir Top 4 showing faults closures characterised by high amplitudes and well drilled to target them.

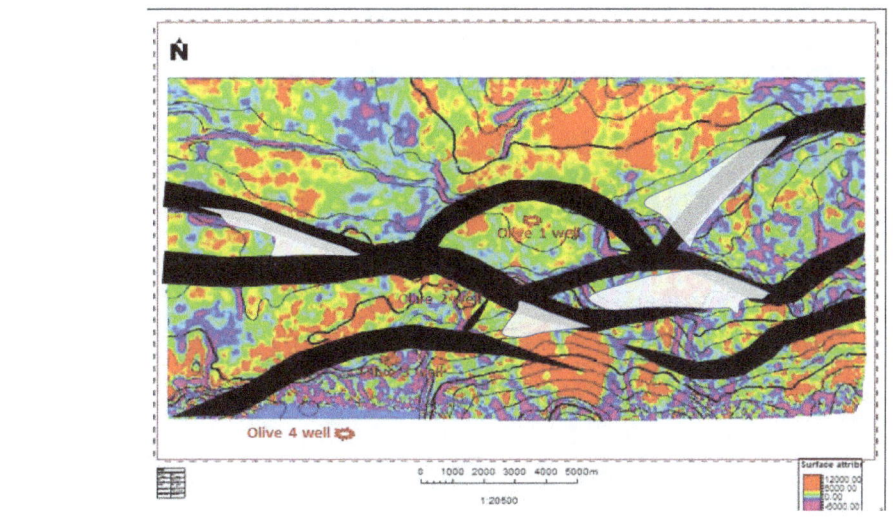

Figure 10: Structural map of Reservoir Top 4 with overlay of amplitude attribute showing four identified prospect (highlighted in white).

Agbada Formation are also widespread over the basin in. Structural interpretation shows that the Olive Field is highly faulted with several fault block forming closure and potential traps for hydrocarbon entrapment and accumulation. The four well drilled in the field are all vertical wells and targeted some of these structural traps. However, from the structural maps produced from the structural interpretation of the seismic volume data showed the presence of other structural closures/traps which has not been drilled into. The amplitude attribute map extracted over these untested structural closures showed the occurrence of high amplitude over them, leading to the identification of four (4) prospects in the Olive Field (Figure 10). These prospects are characterised by closures against two branching faults and relatively high amplitude (bright spot). It is recommended that these prospects be tested as this will improve the viability of the Olive Field.

Conclusion

Evaluation of the Olive Field, Niger Delta basin was done using well logs and seismic data volume. Well log / petrophysical analysis was carried out using log suites from four well. Four reservoirs at different intervals in the well logs were mapped in the field. The petrophysical parameters of the reservoirs showed a general decrease in quality in the basin ward direction from the most proximal well (Olive Well 1) to the most distal well (Olive Well 4). Hydrocarbon typing showed that the reservoirs all contain gas.

Seismic interpretation showed that the field is highly faulted with the faults forming structures for hydrocarbon entrapment and accumulation. Four of these trapping structures have been targeted and exploited by the four wells drilled in the field. Seismic amplitude attribute maps extracted from the tops of the mapped reservoirs showed that the reservoirs are characterised by relatively high amplitudes (bright spots) in areas enclosed by the structural traps. This led to the identification of four (4) prospects within the Olive Field. It is recommended that these prospects be tested to improve the viability of the field.

References

1. Jarvis K (2006) Integrating well and seismic data for reservoir characterization: Risks and Rewards. AESG 1: 1-4.

2. Cosentino L (2001) Integrated reservoir studies. Technip.

3. Aizebeokhai AP, Olayinka I (2011) Structural and stratigraphic mapping of Emi field, offshore Niger Delta. J Geol Mini Rese 3: 25-38.

4. Obiadi II, Ozumba BM, PL Osterloff (2012) Sequence stratigraphy and hydrocarbon distribution in parts of the eastern central swamp depobelt, Niger Delta, Nigeria. NAPE Bulletin 24: 43-51.

5. Adetoye TO, Enikanselu PA (2009) Hydrocarbon reservoir mapping and volumetric analysis using seismic and bolehole data over extreme field, South Western Niger Delta. Ocean Journal of Applied Sciences 2: 429-431.

6. Omoboriowo AO, Chiaghanam OI, Chiadikobi KC, Oluwajana OA, Soronnadi-Ononiwu et al. (2012) Reservoir characterization of Konga Field, Onshore Niger Delta, Southern Nigeria. Int. J Sci. Emerging Tech 3: 19-31.

7. Lehner P, De Ruiter PAC (1977) Structural history of Atlantic Margin of Africa. AAPG Bulletin 61: 961-981.

8. Kulke H (1994) Regional Petroleum Geology of the World. Part II: Africa, America, Australia and Antarctica: Berlin Borntraeger G 21-22: 143-172.

9. Evamy BD, Haremboure J, Kamerling P, Molloy FA, Rowland PH (1978) Hydrocarbon habitat of Tertiary Niger Delta. AAPG Bulletin 63: 1-39.

10. Ekweozor CM, Daukoru ED (1994) Northern delta depobelt portion of the Akata-Agbada (!) petroleum system, Niger Delta, Nigeria pp: 599-613.

11. Michele LW, Ronald RC, Michael EB (1999) The Niger Delta Petroleum System: Niger Delta Province, Nigeria Cameroon, and Equatorial Guinea, Africa.

12. Ekweozor CM, Okoye NV (1980) Petroleum source-bed evaluation of Tertiary Niger Delta. AAPG Bulletin 64: 1251-1259.

3

Analytical Study of Viscosity Effects on Waterflooding Performance to Predict Oil Recovery in a Linear System

Abbas Mamudu[1]*, Olafuyi Olalekan[2] and Giegbefumwen Peter Uyi[3]

University of Benin, Benin City, Nigeria

Abstract

Waterflood displacement efficiency is affected by the viscosity ratio of the displaced to the displacing fluid. Therefore, the oil recovered in a water flooding process is largely determined by the viscosity ratio.

This paper presents a quantitative analysis of the viscosity effects on oil recovery in a linear system using Buckley-Leverett equation and other related mathematical models to simulate the effects on two stages: Case one, when the viscosity of the displaced fluid was varied from $5cp$ to $300cp$ and that of the displacing fluid remained constant at $1cp$. And case two, when the viscosity of the displaced fluid was at $2cp$ and that of the displacing fluid varied from $2cp$ to $10cp$ with the assumption of miscibility between the viscous water and the interstitial water or previously injected water. With the aid of the fractional flow curves, the value for the average water saturations, $\overline{S}_w$ behind the shock front associated with each change in the viscosity ratio was obtained and the corresponding recoveries were predicted.

The results show appreciable recovery at a viscosity ratio as high as 100, however, the S-shape of the fractional flow curve diminishes with increasing viscosity ratio. At $200cp$ and above, the S-shape totally disappears. Viscous fluid appreciably improves oil recovery particularly in reservoirs containing viscous oil. The difference between S_{wf} and $\overline{S}_w$ is constant at various viscosity ratios till the disappearance of the S-shape of the fractional flow curve. Recovery increases with decreasing viscosity ratio and decreases with increasing viscosity ratio. At a very low viscosity ratio, $\dfrac{\mu_o}{\mu_w}$ of 0.4, $\overline{S}_w$ equals the end point water saturation, and this gives the highest possible oil recovery (the optimum). The oil produced, N_p and the average water saturation, $\overline{S}_w$ in an immiscible displacement system are linearly related.

Keywords: Average water saturation; Buckley-Leverett equation; Fractional flow curve; Fractional flow equation; Oil Initially-in-place equation; Oil produced; Optimum recovery; S-shape; Viscosity ratio and OIP- Oil in place

Introduction

During waterflooding, the objective is to displace oil successfully. However, achieving this seems difficult without proper analysis. Frontal advance theory provides the answer to this in 1-D. It's been observed over time that the oil recovered in an immiscible displacement system is largely a function of the viscosity ratio. This is due to the fact that the waterflood displacement efficiency is affected by the viscosity ratio of the displaced to the displacing fluid [1]. Therefore, an analytical study of viscosity ratio alteration to avoid unfavorable viscosity ratios and to predict recoveries correspondingly becomes of paramount importance. It should also be noted that an important aspect of any EOR process is the effectiveness of the process fluids in removing oil from the rock pores at the microscopic scale [2]. Microscopic displacement efficiency, E_D largely determines the success or failure of a process. For crude oil, E_D is reflected in the magnitude of S_{or} (residual oil saturation) [2-4] whose mobilization is the primary aim of waterflooding. The volume of oil displaced during a viscous waterflood is determined by computing the average water saturation in the system at various points in time as done for waterflooding calculations [5-10]. However, the assumption that the viscous water is miscible with the interstitial water or previously injected water was upheld throughout this study.

In 1942, Buckely and Leverett presented what is recognized as the basic equation for describing immiscible displacement in one dimension [11,12]. For water displacing oil, the equation determines the velocity of a plane of a constant water saturation Travelling through a linear system. The equation is derived based on developing a material balance for the displacing fluid as it flows through any element in the given media [11-13]. This well-established theory called frontal displacement theory is very useful in finding solutions to problems when is shown that E_D will continually increase with increasing water saturation in the reservoir by developing an approach for determining the increase in average water saturation in the swept area as a function of the cumulative water injected (or injection time) or when there is increase in oil produced at different increases in viscosity ratio by developing an approach for determining the increase in average water saturation in the swept area as a function of the viscosity ratio alteration. The later is the area of interest in this paper. This classic theory consists of two equations: Fractional flow equation and Frontal advance equation.

The development of the fractional flow equation is attributed to leverett. For two immiscible fluids, oil and water, the fractional flow of water, f_w (or any immiscible displacing fluid) is defined as the water flow rate divided by the total flow rate (Figure 1).

Thomas, Mahoney and Winter pointed out that in determining the suitability of a candidate reservoir for waterflooding The following

*Corresponding author: Abbas Mamudu, Research Assistant, University of Benin, Benin City, Nigeria, E-mail: abbasagim@yahoo.com

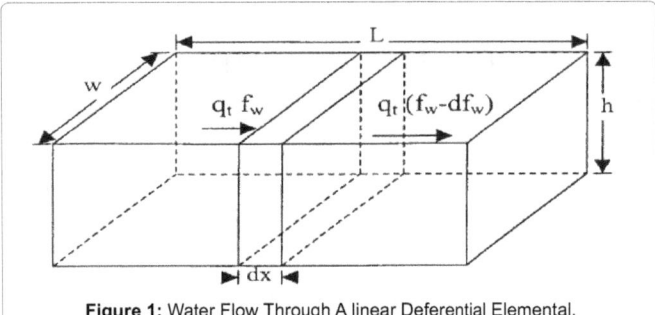

Figure 1: Water Flow Through A linear Deferential Elemental.

characteristics must be considered:

- Reservoir geometry
- Fluid properties
- Reservoir depth
- Lithology and rock properties
- Fluid saturation.

This analytical study was done assuming that all the above characteristics are favorable and using selected relative permeability data. The data must be available to construct the fractional flow curves. It should be noted that relative permeabilities are measured in the laboratory under the diffuse flow condition. This normally results from displacing one fluid by another in thin core plugs at high flow rates [13]. As such, the laboratory or rock relative permeabilities must be regarded as point relative permeabilities which are functions of the point water saturation in the reservoir. During the analysis, with the aid of the relative permeability data, several fractional flow curves were constructed at different viscosity ratios. The corresponding values of the average water saturation behind the shock front, $\overline{S}_w$ were obtained to predict the corresponding oil recoveries. All the required reservoir parameters were assumed with the shock front at breakthrough. The models used include Buckley-Leverett equation, fractional flow equation and oil in place equation.

In (2011), Ghosh and Alshalabi did a similar research termed "Solvent Induced Oil Viscosity Reduction and Its Effect On Waterflood Recovery Efficiency". While Fried in 1955, worked on "Effect of Oil Viscosity On The Recovery Of Oil by Water Flooding".

The main objectives of this study are to analyze the effects of viscosity on immiscible displacement systems to know the most favorable viscosity ratios at which optimum recoveries could be predicted, and to know the effects of its alterations on oil recovery, average water saturation behind the shock front, shock front saturation and the fraction flow curve itself.

Frontal Advance and Related Equations

Frontal advance theory provides the answers to many challenges in immiscible displacement systems. The equations derived from this theory and that of OIIP were used in this present study based on the following assumptions:

- Rate is constant
- The system is linear
- The water in the rock is initially at interstitial water saturation
- The rock is a uniform horizontal reservoir

- Porosity is constant
- Permeability is constant
- The injected viscous water is miscible with the interstitial water or previously injected water
- The displacement process is at breakthrough
- Gravity forces are negligible
- Capillary forces are negligible
- The fluids are incompressible
- The reservoir is rectangular

Buckley -Leverett equation

This equation, called the frontal advance or Buckley-Leverett equation was derived by Willhite as [8].

$$\frac{dS_{S_w}}{dt} = \frac{q_T}{A\varnothing}\left(\frac{\partial f_w}{\partial S_w}\right)_{S=S_w} \tag{1}$$

Where x_{S_w} = location of water saturation, S_w measured from x = 0, A = cross sectional area, $\varnothing$ = porosity, q_T = injection rate, f_w = fractional flow of water and t =time from the beginning of injection.

Eqn (1) could be integrated and expressed in terms of the distance travelled as

$$x\big|_{S_w} = \frac{q_T t}{A\varnothing}\frac{df_w}{dS_w}\bigg|_{S_w} \tag{2}$$

And finally, in terms of average water saturation as

$$\overline{S}_w = S_{wf} + \left(1 - f_w\big|_{S_{wf}}\right)\left(\frac{1}{\dfrac{df_w}{dS_w}\bigg|_{S_{wf}}}\right) \tag{3}$$

Fractional flow equation

As developed by Leverett, for two immiscible fluids, oil and water, the water flow rate is given as

$$f_w = \frac{q_w}{q_T} = \frac{q_w}{q_w + q_o} \tag{4}$$

where f_w = fraction of water in the flowing stream, i,e., water cut, q_T = total flow rate, q_w = water flow rate and q_o = oil flow rate.

This effort of Leverett assists in determining the water cut at any given point in time.

Oil In Place Equation (OIP)

The volume of oil displaced in a water flooding project is determined by computing the average water saturation in the swept zone or behind the shock front. When the initial oil saturation is $1-S_{iw}$, the oil displaced is given as

$$N_p = \frac{A\varnothing L\left(\overline{S}_w - S_{iw}\right)}{B_o} \tag{5}$$

Where N_p = oil produced, A = reservoir cross sectional area, L = reservoir length, $\varnothing$ = reservoir porosity, S_w = average water saturation in the swept zone and S_{iw} = interstitial water saturation.

The effects of viscosity ratio on oil recovery were analyzed using the above mathematical equations at breakthrough, $x_f = L$.

At breakthrough, Eqn (3) becomes

$$\overline{S}_{w_{bt}} = S_{w_{bt}} + \left(1 - f_w\big|_{S_{w_{bt}}}\right)\left(\dfrac{1}{\dfrac{df_w}{dS_w}\big|_{S_{w_{bt}}}}\right) \tag{6}$$

and Eqn (5) becomes

$$N_{p_{bt}} = \dfrac{A\varnothing L\left(\overline{S}_w - S_{iw}\right)}{B_o} \tag{7}$$

The following rock and fluid parameters were assumed for this analysis.

$h = 40\,ft, \varnothing = 0.18, B_o = 1.3\,\dfrac{RB}{STB}, S_{iw} = 0.20, L = 2000\,ft,$ and $A = 25000\,ft^2$

To simulate the effects of the viscosity ratio, we considered two cases. Case one: Where we considered oil with increasing viscosity, and case two: Where we considered viscous water (viscous fluid) (Figure 2).

Case 1: With the viscosity of water at $1cp$, we varied that of oil from $5cp$ to $300cp$. Between $5cp$ and $50cp$, we took a step of $5cp$ in the viscosity of oil to initiate viscosity ratio alteration which was enough to provide us with fractional flow curves with distinct features. However, above $50cp$, the changes in the step became so minimal that they were initiated based on the information required from the curves. Precisely, $50cp$ and $100cp$ were the changes initiated. For each change made in the viscosity ratio, the corresponding value of the average water saturation was determined using eqn (6) with the aid of the fractional flow curve accordingly constructed. But it could still be read directly from the curve. This value is then imputed in eqn (2.7) to predict the resulting recovery. The constructed fractional flow curves are depicted in Figure 3a-n. Visual Basin. Net was the program used to program the equations for the simulation process.

Case 2: Here, we considered viscous water. The viscosity of the oil was kept constant at $2cp$ and that of the viscous water (fluid) varied from $2cp$ to $10cp$. The step here is $2cp$ and the same calculations as done in case 1 were repeated. The constructed graphs were presented Figure 3a-c respectively.

Results and Discussion

Figure 2 shows the results of the predicted oil recoveries at different viscosity ratios. It shows the oil recovery as a function of the viscosity ratio. It reveals that the recovery decreases with increasing viscosity ratio. A very swift decline occurs in the recovery between 1 and 100, indicating the extent of the effect the viscosity ratio has on the average water saturation in the swept zone within this range of viscosity ratio. As it increases further, the corresponding change in the recovery becomes relatively less such that maintaining a viscosity ratio of 100, 200 and 300 gives a loss of recovery of $263\times10^4 STB$, $245\times10^4 STB$ and $102\times10^3 STB$ respectively. At a viscosity ratio of 300, it gives a recovery of $693\times10^3 STB$.

Figure 3a-l show the fractional flow profiles with the constructed tangent for each of the value of the viscosity ratio considered. Figure 3a shows the profile when the viscosity ratio is 1. The s-shape of the fractional flow is very distinct, and at $f_{w1} = 1$, $\overline{S}_w = 0.73$. It shows that

the values of the fractional flow at the shock front, $f_w\big|_{S_{wf}}$ and the shock front saturation, S_{wf} are 0.88 and 0.66 respectively.

Figure 2 shows the graph of oil recovery versus increasing viscosity ratio.

Figure 3b shows the fractional flow profile when the viscosity ratio is 5. At $f_{w1} = 1$, $\overline{S}_w = 0.61$ and it also shows that the values of the fractional flow at the shock front, $f_w\big|_{S_{wf}}$ and the shock front saturation, S_{wf} are 0.8 and 0.53 respectively. This shows that the average water saturation decreases with increasing viscosity ratio. The respective values of the viscosity ratio, average water saturation, fractional flow at the displacement front and shock front saturation for all the constructed fractional flow curves are presented in Table 1. The graphs from which these values are picked are in Figures 3a-l. They show that the s-s shape of the fractional flow curve gets distorted with increasing viscosity ratio, which greatly affect recovery.

Figures 3l-n, show the effect of high viscosity ratio on the nature of the fractional flow curve and recovery. Figure 3m shows that at 100, the s-shape of the fractional flow curve is almost entirely distorted; however, oil can still be produced. Beyond this, as shown in Figure 3m-n, the s-shape is completely distorted. The values of the average water saturation are 0.32 and 0.3 respectively. The recoveries are presented in Figure 2.

Figure 4a-c shows the fractional flow profile of viscous water. Though viscosity ratio of 0.2, 0.22, 0.25, 0.29, 0.33, 0.40, 0.50, and 0.67 were used to construct the fractional flow curves shown in Figure 2.2, However, Figure 3.3c shows that at a viscosity ratio of less 0.40 under the assumed conditions, no practical significance would be shown on the fractional flow curve because it shows that the average water saturation, $\overline{S}_w$ at $f_w = 1$ is 0.8, that is, it's equal to the end point water saturation and this gives the highest possible oil recovery (It means that all the movable oil has been produced) as depicted in Figure 3.1. Table 2 shows the selected water saturation and relative permeability data used in this analysis.

Figure 5 shows average water saturation profile. Here, the water saturation is a function of the viscosity ratio. It shows that as the viscosity ratio increases the average water saturation decreases. And it also suffices to say that the response of the average water saturation, $\overline{S}_w$ to the viscosity ratio is the same as that of the recovery to it. This is tenably justified in Figure 6 which shows that a linear relationship exists between the two parameters.

Figure 7 shows comparative responses of the average water saturation and the water saturation at the shock front. It shows that

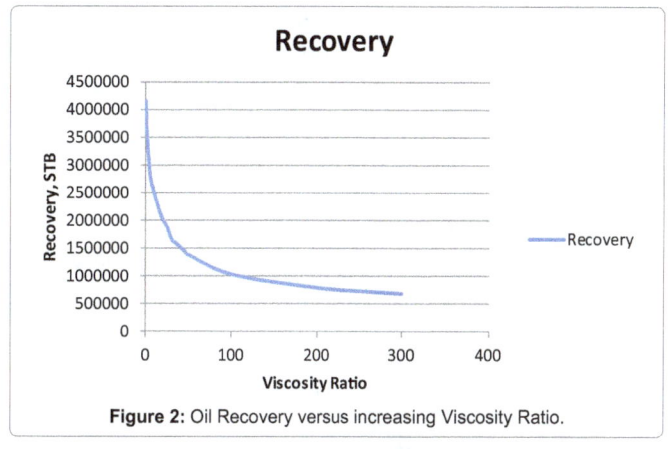

Figure 2: Oil Recovery versus increasing Viscosity Ratio.

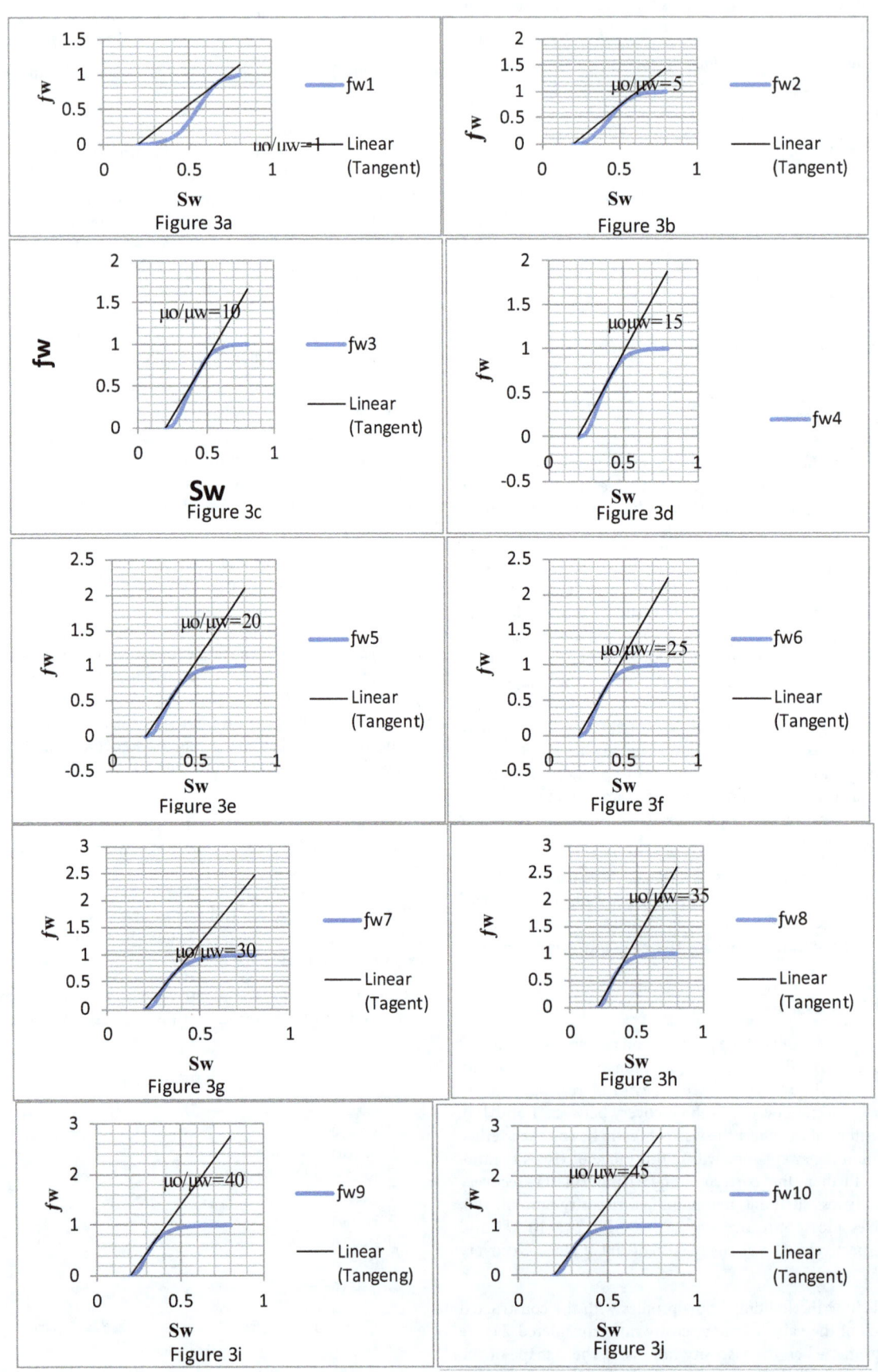

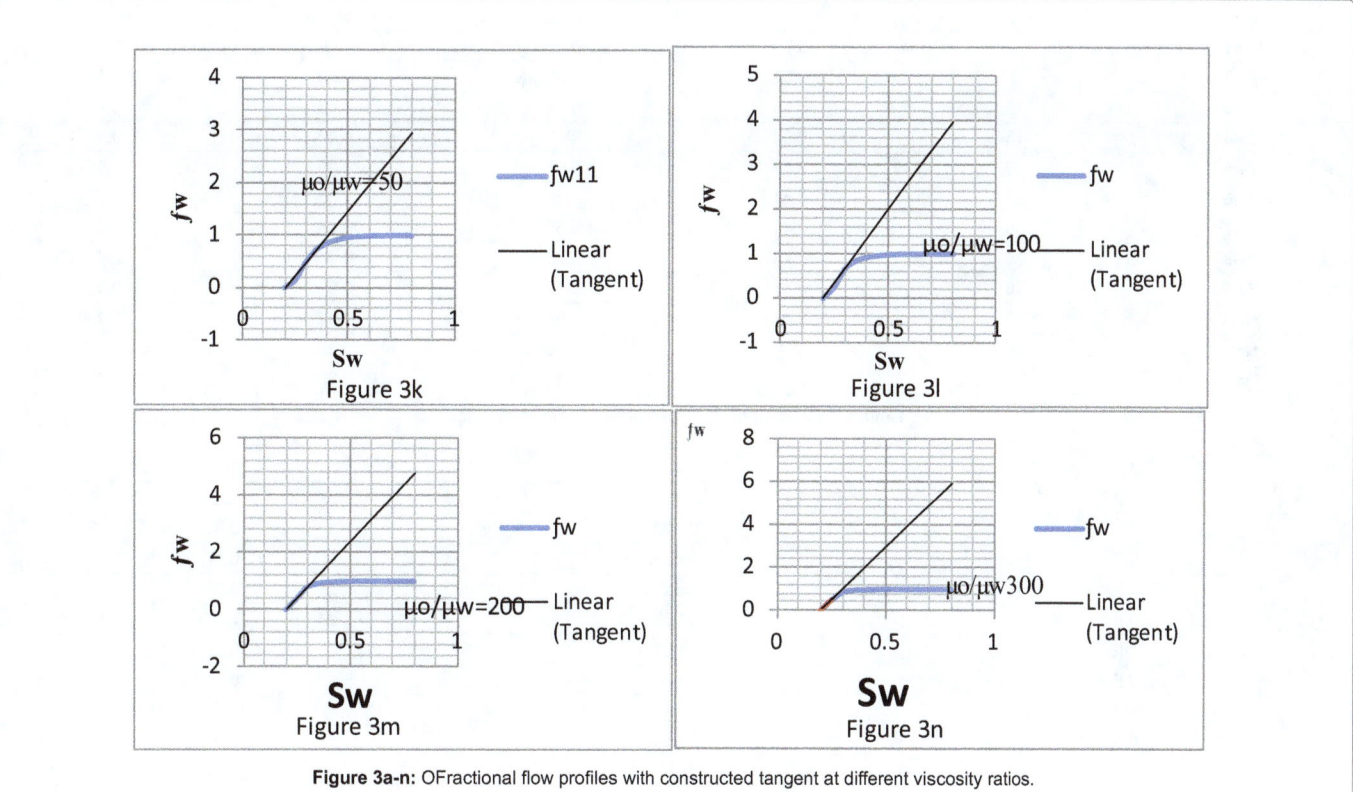

Figure 3a-n: OFractional flow profiles with constructed tangent at different viscosity ratios.

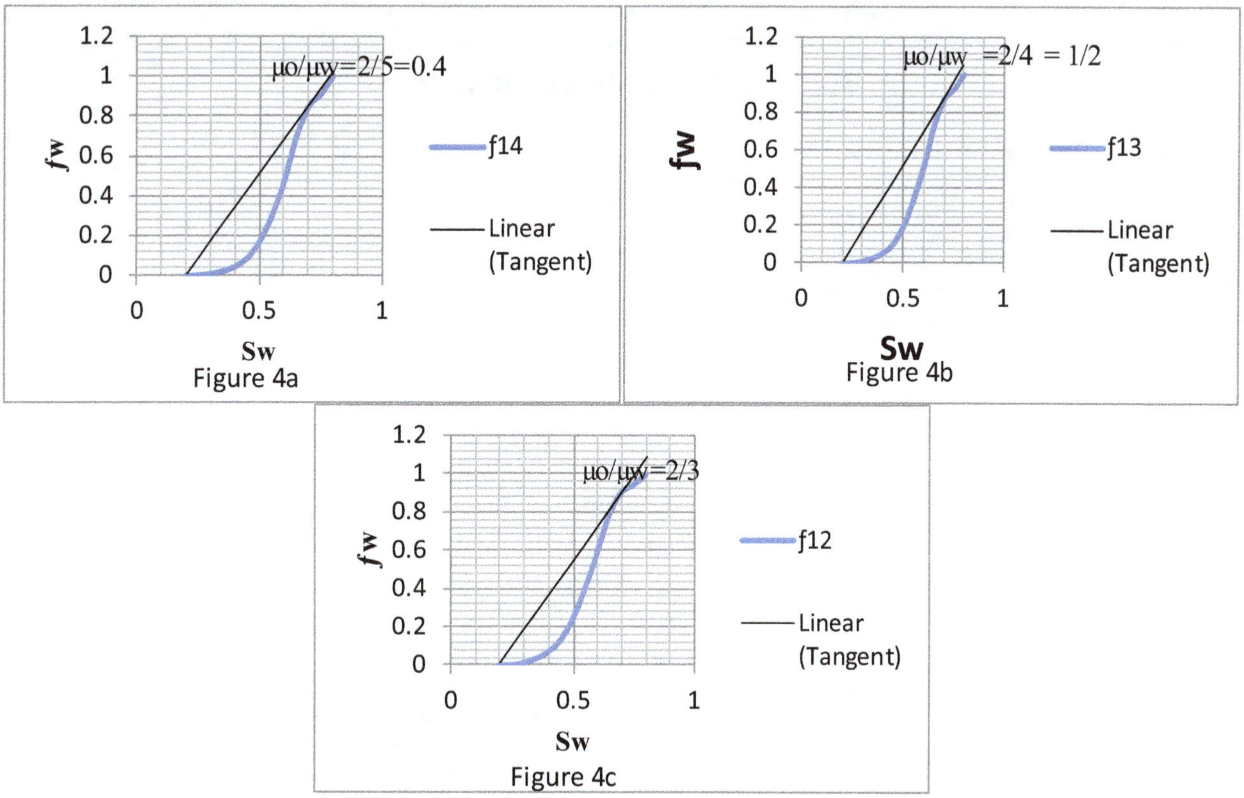

Figure 4a-c: Fractional flow profiles with constructed tangents at various viscosity ratios of less than 1.

Fw	μo/μw	Ŝw	fw|swf	Swf	Ŝw - Swf
fw1	1	0.73	0.88	0.66	0.1
fw2	5	0.61	0.8	0.53	0.1
fw3	10	0.56	0.72	0.46	0.1
fw4	15	0.52	0.7	0.42	0.1
fw5	20	0.49	0.68	0.39	0.1
fw6	25	0.47	0.6	0.36	0.1
fw7	30	0.44	0.58	0.35	0.1
fw8	35	0.43	0.58	0.34	0.1
fw9	40	0.42	0.58	0.33	0.1
fw10	45	0.41	0.58	0.32	0.1
fw11	50	0.4	0.58	0.32	0.1
f'w12	100	0.35	0.58	0.29	0.1

Table 1: Viscosity ratio and fractional flow characteristics.

S_w	k_{rw}	k_{ro}
0.2	0	0.8
0.25	0.002	0.61
0.3	0.009	0.47
0.35	0.02	0.37
0.4	0.033	0.285
0.45	0.051	0.22
0.5	0.075	0.15
0.55	0.1	0.095
0.6	0.132	0.06
0.65	0.17	0.03
0.7	0.208	0.015
0.75	0.251	0.01
0.8	0.3	0

Table 2: Selected relative permeability and water saturation data.

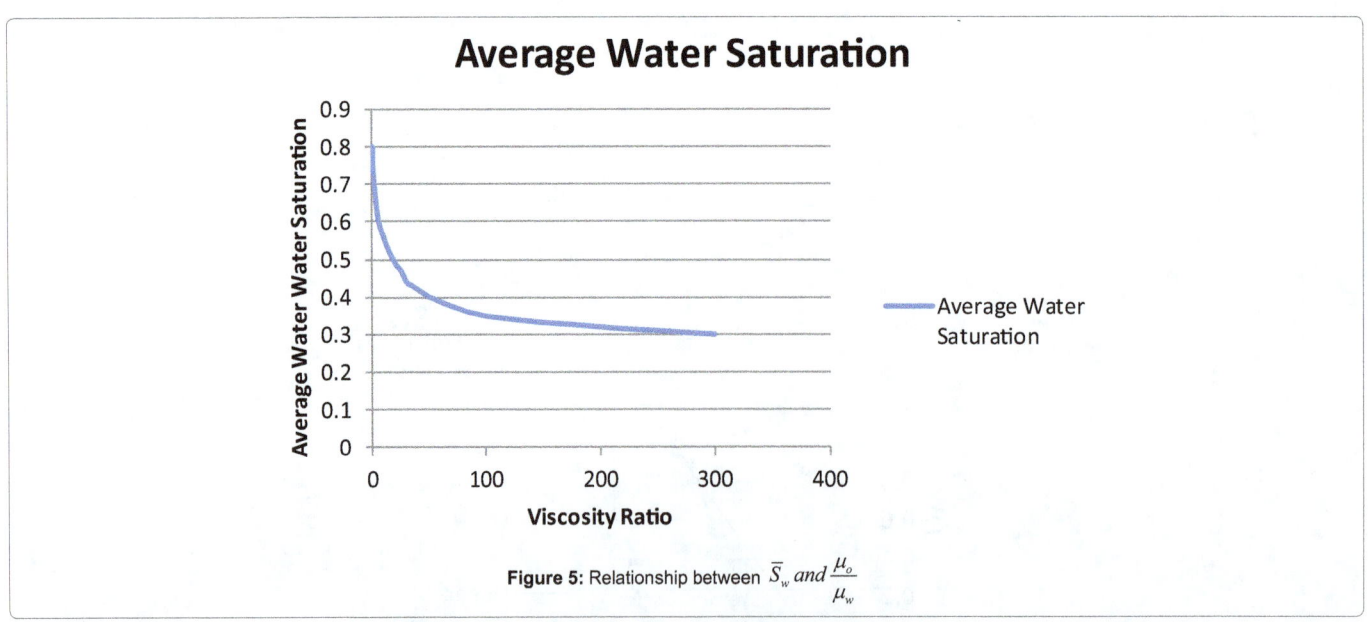

Figure 5: Relationship between $\overline{S}_w$ and $\dfrac{\mu_o}{\mu_w}$

the difference between the values of both at a specific viscosity ratio remains constant at all increases in the viscosity ratio. This fact is presented in Figure 8.

Conclusion

By the theoretical analysis of viscosity effects on water flooding performance, we come to the following conclusions. Optimum recovery is achieved at a viscosity ratio of 0.4 where the average water saturation equals the end point water saturation and any viscosity ratio less than this shows no significance as revealed on the fractional flow curve because all the movable oil has been produced. Viscous fluid appreciably improves oil recovery in reservoirs containing viscous oil.

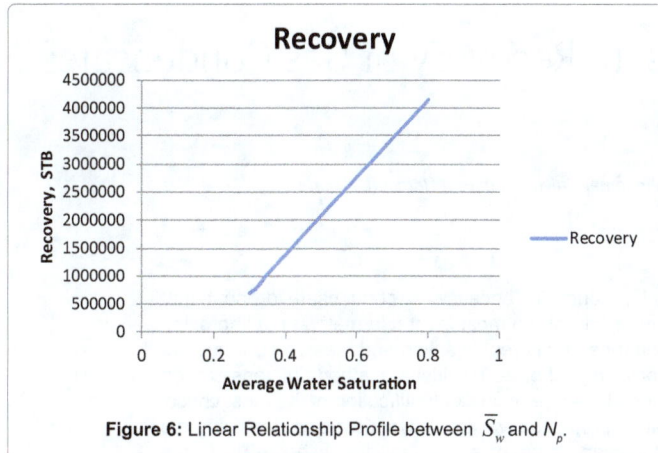

Figure 6: Linear Relationship Profile between $\overline{S}_w$ and N_p.

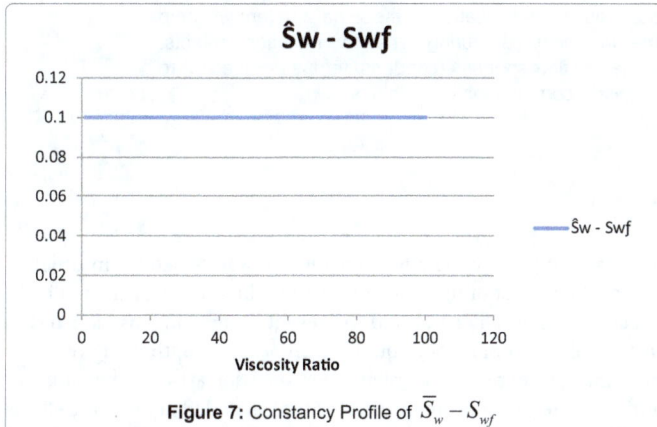

Figure 7: Constancy Profile of $\overline{S}_w - S_{wf}$

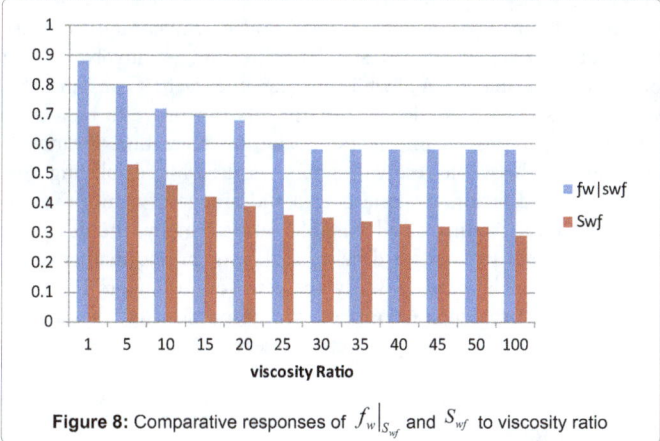

Figure 8: Comparative responses of $f_w|_{S_{wf}}$ and S_{wf} to viscosity ratio

References

1. Don WG, Paul WG (1998) Enhanced Oil Recovery. Richardson, Texas, USA.

2. Willhits GP (1986) Waterflooding Textbook Series. SPE, Richardson, TX, USA.

3. Amyx JW, Bass DM, Whiting RL (1986) Petroleum Reservoir Engineering. Mc Graw-Hill Book Co, New York City, USA.

4. Craig FF (1991) The Reservoir Engineering Aspects of Water Flooding, Monograph Series, SPE, Richardson, TX, USA.

5. Claridge EL, Bonder PL (1974) A Graphical Method for Calculating linear displacement with mass transfer and continuously Changing Mobility 14.

6. Patton JT, Coast KH, Colegrove GT (1971) Prediction of Polymer Flood Performance. SPEJ 72-84, 11.

7. Pope GA (1980) The Application of Fractional Flow Theory to Enhanced Oil Recovery 20.

8. Welge HJA (1952) Simplified Method for Computing Oil Recoveries by Gas or Water Drive. Trans, AIME 195: 91-98.

9. Koval EJ (1963) A Method for Predicting the Performance of unstable miscible displacement in Heterogeneous media. SPEJ 145-54; Trans., AIME, 228.

10. Taber JJ (1969) Dynamic and Static Forces Required to Remove a Discontinuous Oil Phase from Porous Media containing both Oil and Water. SPEJ 3-12.

11. Dake LP (1998) Fundamentals of Reservoir Engineering.

12. Buckley SE, Leverett MC (1942) Mechanism of Fluid Displacement in Sand. trans., AIME: 146-116.

13. Tarek A, Pual DM (2005) Advanced Reservoir Engineering.

Oil can still be recovered even at a viscosity ratio as high as 100. The nature of the s-shape of the fractional flow curve is directly dependent on the viscosity ratio. Efforts made to ensure that viscous water (fluid) is considered to maintain a viscosity ratio of 1 or less would be fruitful. The oil produced,(N_p) and the average water saturation behind the shock front, $(\overline{S}_w)$ in an immiscible displacement system are linearly related. Recovery increases with decreasing viscosity ratio and decreases with increasing viscosity ratio.

Acknowledgement

This research was supervised and supported by the head of department of petroleum engineering, University of Benin, Benin City, Nigeria.

4

Parametric Study of Enhanced Condensate Recovery of Gas Condensate Reservoirs using Design of Experiment

Nkemakolam Izuwa and Basil C Ogbunude*

Department of Petroleum Engineering, Federal University of Technology, Owerri, Peace Wokoma, University of Port Harcourt, Nigeria

Abstract

Gas condensate reservoirs usually exhibit reduced well productivity because of condensate dropout that occurs below the dew point pressure. Gas recycling has become one of the most favorable methods of improving recovery of condensed liquid. However, understanding the influence of different injection and reservoir parameters on productivity is of great importance when planning a gas recycling scheme. Traditional methods of sensitization during reservoir simulation for gas condensate fields creates the challenge of quick identification of the most critical properties for sensitization, and hence delay of overall simulation project delivery. This work aims at identifying the key variables that influence productivity of a gas condensate reservoir under a gas recycling scheme using the design of experiment approach (DOE). DOE represents a more effective method for computer-enhanced, systematic approach to experimentation, considering all the factors simultaneously. Identification of these parameters will help simulators achieve best optimization targets and also save time and resources during dynamic simulation projects. Furthermore, it will be shown that experimental design can be used to fit responses (condensate/gas production) to mathematical models that will be able to predict outputs for any given combination of variables.

Keywords: Gas condensate; Gas recycling; Design of experiments

Introduction

Rich gas or retrograde condensate gas reservoir is a common type of hydrocarbon reservoir around the world. Much of the 6,183 trillion cubic feet of worldwide gas reserves can be found in gas condensate reservoirs [1-3]. Hence, gas condensate reservoirs are important to today's energy demand/supply challenges. On the other hand, gas condensate systems have been recognized as the reservoir type with the most complex flow behavior and thermodynamic characteristics [4]. The gas condensate systems exist as a single-phase fluid (gas) at original reservoir conditions, but unlike a wet or dry gas reservoir, it separates into two phases, a gas and a liquid (condensate) at pressures below the saturation pressure of the reservoir [5]. The main problems associated with gas condensate systems are the formation damage effects leading to a reduced relative permeability of gas because of liquid condensate dropout, and permanent loss of valuable liquid due to the trapping capillary effects in the reservoir [6] (Figure 1).

Historically, there are three main methods for gas condensate recovery: natural pressure depletion to the abandonment pressure, full pressure maintenances by gas cycling and partial pressure maintenance by means of gas cycling after previous natural depletion. In order to reduce the impact of the condensate accumulations near the wellbore, gas cycling is usually employed to prevent liquid condensation and to also vaporize dropped out liquid [7]. In properly optimizing recovery from this type of reservoir system, a key question arises to the timing of initiating the gas injection project, as well as understanding the effects of different parameters on the recovery potential of the injection. Though gas-recycling will always improve recovery, there is a need to identify the set of parameters that will lead to a maximum recovery when optimized. Traditional simulation techniques involve testing one factor at a time (OFAT) while holding other factors constant. This work shows how the design of experiments can prove to be a cost-effective way to provide information about the interaction of variables and the way the whole reservoir system works while displaying how interconnected factors respond over a wide range of values without requiring direct testing of all possible values. Finally, the design of experiment will be used to develop a system-specific mathematical model that can be used to study the reservoir behaviors based on optimal statistical interactions of the responses (condensate/gas production) and variables (production/reservoir/injection properties).

Methodology

Generally, injection of gas into the reservoir results in an increase in production [8]. However, to obtain optimum productivity, different production and injection conditions are required to be sensitized.

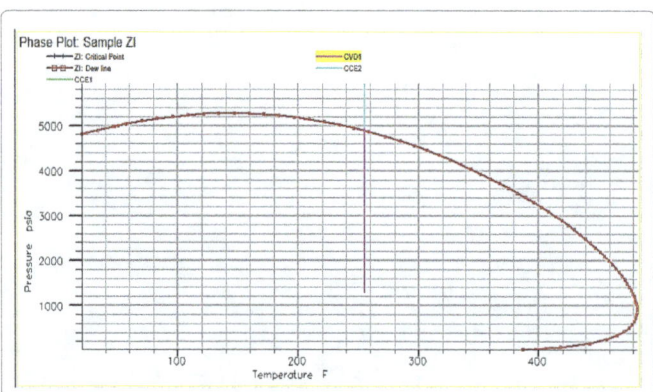

Figure 1: Phase envelope for the gas condensate sample; Tr=255°F, Pi=4953psia.

*Corresponding author: Basil C Ogbunude, Department of Petroleum Engineering, Federal University of Technology, Owerri, Peace Wokoma, University of Port Harcourt, Nigeria, E-mail: basilogbunude@yahoo.com

Such conditions include the injection pressure, injection rate, and the various reservoir and fluid properties. For the purpose of this study, two reservoir models were used create a dynamic simulation model which was used (together with the reservoir, injection and production variables) as input for the design of experiment.

One of the models is the fluid model which was designed using a set of real fluid data obtained from a Niger Delta retrograde gas field. The other model comprises the bulk reservoir, including its petro-physical properties which were hypothetically designed within the confines of Niger Delta reservoir characteristics.

Fluid characterization and generation of compositional PVT tables

The fluid properties including the phase behavior are greatly dependent on the properties of each component or pseudo-component and composition [9]. The Peng-Robinson (PR) equation of state (EOS) was applied to design the fluid behavioral patterns at different reservoir temperatures and pressures. The results of this design were compared to the laboratory generated results gotten through various routine tests like constant composition expansion (CCE) and constant volume depletion (CVD). Discrepancies in the two models were adjusted by applying heptane-plus characterization techniques and EOS tuning methods. The heavier components (heptane-plus) have various isomers for the same carbon number components and hence they have different characteristics by the presence of different isomers [10]. The heptane-plus characterization involved splitting into three fractions; C7+, C14+ and C25+ before lumping into groups of all pseudo-components according to their molecular weights. The first pseudo-component GRP1 is composed of carbon dioxide only as the only significant non-hydrocarbon. The second pseudo-gas contains nitrogen, methane, and ethane. The amount of nitrogen is not significant; hence, it is assumed that this pseudo-component contains only methane and ethane. The third pseudo-component contains the gasolines; propane, butanes, pentanes, and hexanes. The fourth group is C7 to C13, while the fifth is C14 to C24. The final group is the heaviest, C25+ components (Table 1).

The EOS tuning method applied was the 3-Parameter PR model which involved multiple non-linear regression techniques. After several regressions, the fluid was able to be matched. The parameters used to validate the match are shown in Figures 2-6.

Reservoir Model and Experimental Design

A simple five-spot model was designed using hypothetical grid blocks, rock properties and initialization properties. The synthetic model has Cartesian coordinates with block-centered geometry having length of 328 ft. in the X and Y directions having 10x10x7 grids. The reservoir which was at a depth of 9560 ft. below seas level has an initial reservoir pressure of 4953 psia (Figure 7).

Sensitivity analyses are common during reservoir simulations. To understand the prevailing factors that are most contributory to the final

Components	Mol %	Weight fraction, %
GRP1	3.35	6.7572
GRP2	90.69	70.112
GRP3	3.69	9.1893
GRP4	1.9992	11.142
GRP5	0.26079	2.6309
GRP6	0.010017	0.16876

Table 1: Composition of pseudo-components.

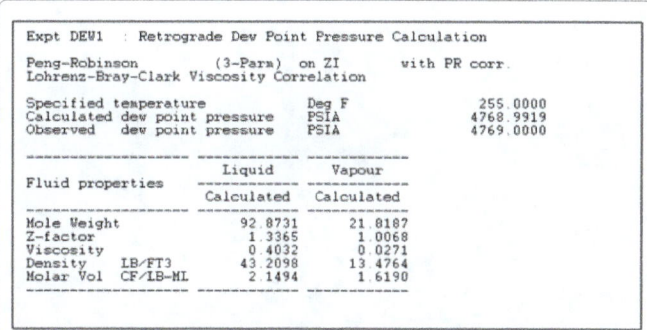

Figure 2: Result of saturation pressure EOS tuning showing the matched dew point pressure.

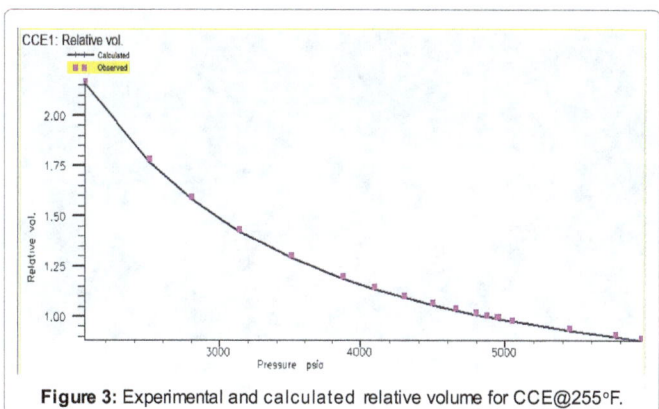

Figure 3: Experimental and calculated relative volume for CCE@255°F.

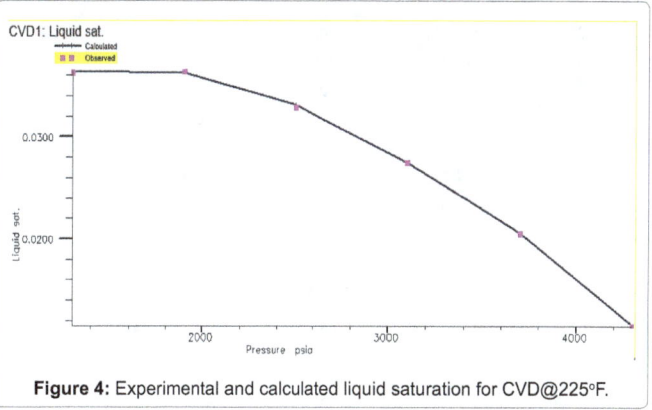

Figure 4: Experimental and calculated liquid saturation for CVD@225°F.

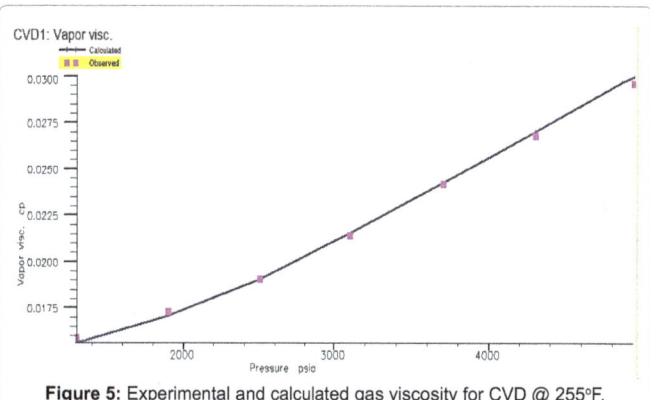

Figure 5: Experimental and calculated gas viscosity for CVD @ 255°F.

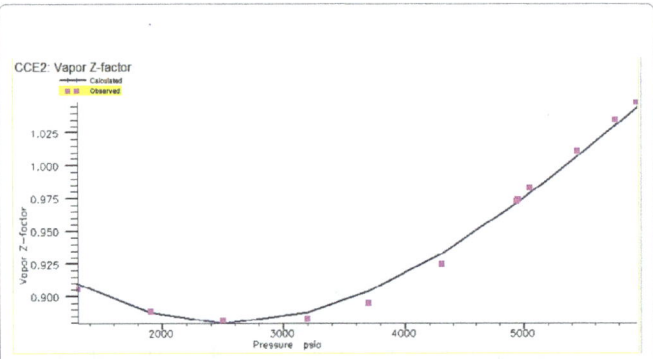

Figure 6: Experimental and calculated gas compressibility factor data for CVD @255°F.

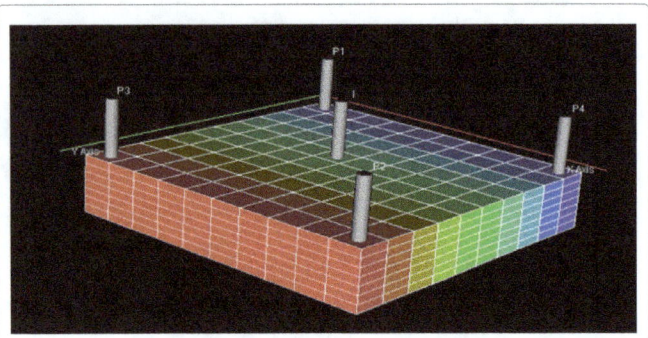

Figure 7: 3D simulation model of reservoir.

responses in the dynamic modeling of a gas recycling project in a gas condensate reservoir, the DOE technique was applied. In this method, eleven properties expected to influence gas and condensate production are taken as factors to be used in the experimental design procedure and thus determine the statistical effects of these different parameters on gas and condensate recovery. Responses are the condensate and gas production, generated for each combination of parameters. DOE provides information about the interactions of the factors and responses and how interconnected factors respond over a wide range of values, without the need to test all possible values directly. The Plackett-Burman DOE Design for selection of significant parameters was used for the eleven factors (parameters), where each factor was varied over two levels (low and high) based on regional petrophysical and operational characteristics (Table 2).

This generates a set of saturated screening designs based on Plackett-Burman structures, the number of factors being one less than the number of required runs. These runs are a mixture of the different levels of the factors as shown in Table 3.

Results

With the results of the design, the various levels of significance of the eleven parameters on gas and condensate production responses were observed using the normal probability plot, the half-normal probability plot and the Pareto chart. Interpretation of the charts gave rise to identification of seven factors that showed the most significant impact on the production responses.

These parameters are

- Porosity
- Net-to-Gross ratio
- Kv/Kh
- Injection Rate
- Injection Pressure
- Thickness
- Reservoir Pressure

Development of proxy

As an extension to the work, a mathematical model was developed using D-optimal Response Surface Method (RSM) to study the effects

Factor	Name	Unit	Low	High
A	PORO	fraction	0.1	0.38
B	PERM	mD	100	1000
C	NTG	fraction	0.4	0.9
D	Kv/Kh	fraction	0.01	0.1
E	Scc	fraction	0.1	0.4
F	CGR	stb/scf	50	240
G	Qinj	scf/day	2480	24800
H	Pinj	psia	1400	7000
J	H	ft.	40	200
K	Pr	psia	3000	7000
L	Krg	fraction	0.2	0.85

Table 2: Plackett-Burman Design showing the factors and levels (Low and High).

	Factor 1	Factor 2	Factor 3	Factor 4	Factor 5	Factor 6	Factor 7	Factor 8	Factor 9	Factor 10	Factor 11	Response 1	Response 2
Run	A:PORO	B:PERM	C:NTG	D:Kv/Kh	E:Scc	F:CGR	G:Qinj	H:Pinj	J:H	K:Pr	L:Krg	Gas Prod	Cond Prod
	fraction	mD	fraction	fraction	fraction	stb/scf	scf/day	psia	ft.	psia	fraction	Mscf	bbls
1	0.38	100	0.9	0.1	0.1	240	24800	7000	40	3000	0.2	1.36E+08	1113184
2	0.1	1000	0.4	0.1	0.4	50	24800	7000	200	3000	0.2	1.36E+08	747511.7
3	0.1	100	0.4	0.01	0.1	50	2480	1400	40	3000	0.2	2242553	51224.29
4	0.38	100	0.9	0.1	0.4	50	2480	1400	200	3000	0.85	95895064	2225146
5	0.1	100	0.9	0.01	0.4	240	2480	7000	200	7000	0.2	52022840	1503648
6	0.38	1000	0.9	0.01	0.1	50	24800	1400	200	7000	0.2	1.36E+08	4400179
7	0.1	1000	0.9	0.01	0.4	240	24800	1400	40	3000	0.85	1.36E+08	212694.7
8	0.1	1000	0.9	0.1	0.1	50	2480	7000	40	7000	0.85	9727697	276220.8
9	0.38	1000	0.4	0.1	0.4	240	2480	1400	40	7000	0.2	15023919	425837.5
10	0.38	100	0.4	0.01	0.4	50	24800	7000	40	7000	0.85	1.36E+08	1034213
11	0.1	100	0.4	0.1	0.1	240	24800	1400	200	7000	0.85	85265984	196071.7
12	0.38	1000	0.4	0.01	0.1	240	2480	7000	200	3000	0.85	4651961	101806.9

Table 3: Experimental Design Table showing the factors and responses used for the design.

of these factors on gas and condensate recovery. RSM designs help to quantify the relationships between one or more measured responses and the vital input factors or parameters. The D-optimal criteria is one of the optimalities that selects design points in a way that minimizes the variance associated with the estimates of specified model coefficients. The aim is to generate a model that represents the responses using quadratic interactions of the factors. Using the quadratic model, an overall candidate point set was created, after which fifty –five specific design points (the experimental runs that would be done) were chosen after which the proxy was generated.

This proxy was tested using statistical indicators to ascertain its degree of error as shown in Tables 4 and 5 for gas and condensate production respectively.

At the end of the experimental design, the following equations were generated for the gas and condensate production;

$$
\begin{aligned}
G_p, C_p = {} & A_1 + A_2(\varnothing) + A_3(NTG) + A_4\left(\frac{K_V}{K_H}\right) + A_5(Q_{inj}) + A_6(H) + A_7(P_r) + A_8(P_{inj}) \\
& + A_9(\varnothing * NTG) + A_{10}\left(\varnothing * \frac{K_V}{K_H}\right) + A_{11}(\varnothing * Q_{inj}) + A_{12}(\varnothing * H) + A_{13}(\varnothing * P_r) \\
& + A_{14}(\varnothing * P_{inj}) + A_{15}\left(NTG * \frac{K_V}{K_H}\right) + A_{16}(NTG * Q_{inj}) + A_{17}(NTG * H) + A_{18}(NTG * P_r) \\
& + A_{19}(NTG * P_{inj}) + A_{20}\left(\frac{K_V}{K_H} * Q_{inj}\right) + A_{21}\left(\frac{K_V}{K_H} * H\right) + A_{22}\left(\frac{K_V}{K_H} * P_r\right) + A_{23}\left(\frac{K_V}{K_H} * P_{inj}\right) \\
& + A_{24}(Q_{inj} * H) + A_{25}(Q_{inj} * P_r) + A_{26}(Q_{inj} * P_{inj}) + A_{27}(H * P_r) + A_{28}(H * P_{inj}) + A_{29}(P_r * P_{inj}) \\
& + A_{3029}(\varnothing^2) + A_{31}(NTG^2) + A_{32}\left[\left(\frac{K_V}{K_H}\right)^2\right] + A_{33}(Q_{inj}^2) + A_{34}(H^2) + A_{35}(P_r^2) + A_{36}(P_{inj}^2)
\end{aligned}
$$

The values of the coefficients for gas and condensate equations are represented in Table 6.

The mathematical model was validated by comparing them to results generated from an independent dynamic simulator. For the gas production model, the relative error when compared to simulation results was found to be 3.8%, while that for condensate production prediction model was 3.6%.

Finally, a sensitivity analysis was carried out on these parameters using the mathematical models to help understand how these factors influence production in a gas condensate reservoir.

Effects of injection rate and pressure on gas and condensate production

Five injection rates were chosen for the injection process ranging from 19,800 Mscf/day to 5,900 Mscf/day, and the effects of each rate on gas and condensate production was analyzed.

It can be seen from the graphs above that the maximum gas production occurs at the maximum injection rate. This also coincides with the maximum injection pressure. However, the lowest injection

Indicator	Value	Indicator	Value
Std. Dev.	-	R-Squared	0.987586
Mean	81063030	Adj R-Squared	0.964719
C.V. %	12.75986	Pred R-Squared	0.822069
PRESS	2.91E+16	Adeq Precision	20.91767

Table 4: Statistical summary for gas prediction model.

Indicator	Value	Indicator	Value
Std. Dev.	-	R-Squared	0.994997
Mean	1352464	Adj R-Squared	0.98578
C.V. %	10.06817	Pred R-Squared	0.934002
PRESS	4.65E+12	Adeq Precision	40.23579

Table 5: Statistical summary for condensate prediction model.

Constants	Coefficients	
	Gas Production	**Condensate Production**
A_1	$4.60738 * 10^7$	$1.04572 * 10^6$
A_2	$-5.30696 * 10^7$	$-4.41239 * 10^6$
A_3	$-1.35533 * 10^8$	$-3.74945 * 10^6$
A_4	$-7.34830 * 10^7$	$2.43062 * 10^6$
A_5	1176.33385	51.19948
A_6	$4.49593 * 10^5$	-4815.17296
A_7	-7244.68379	-15.93230
A_8	-9111.02877	-25.52929
A_9	$1.56111 * 10^8$	$4.7435 * 10^6$
A_{10}	$5.30660 * 10^8$	$1.5562 * 10^6$
A_{11}	-6257.17754	11.78361
A_{12}	$7.33963 * 10^5$	33198.09102
A_{13}	-2228.53231	630.74498
A_{14}	13414.38501	12.50100
A_{15}	$-2.22378 * 10^8$	$-3.20724 * 10^6$
A_{16}	-2319.93384	-7.00348
A_{17}	$3.29370 * 10^5$	10852.73623
A_{18}	4522.62490	143.36173
A_{19}	-208.50250	-22.02281
A_{20}	5071.25296	40.98690
A_{21}	-78705.33842	3950.85908
A_{22}	7658.12602	-72.92741
A_{23}	33312.15701	335.85908
A_{24}	-14.36637	0.048639
A_{25}	-0.25318	$-2.87403 * 10^{-3}$
A_{26}	$3.33435 * 10^{-3}$	0.000000
A_{27}	42.58720	1.43811
A_{28}	29.45186	0.074765
A_{29}	1.22107	$7.49886 * 10^{-3}$
A_{30}	$8.03938 * 10^6$	$-3.59965 * 10^6$
A_{31}	$7.70745 * 10^7$	$1.84886 * 10^6$
A_{32}	$7.13256 * 10^8$	$-1.52469 * 10^7$
A_{33}	0.31962	$-7.03805 * 10^{-4}$
A_{34}	-3031.83758	-35.35163
A_{35}	0.13735	-0.017308
A_{36}	-0.80404	$-1.94740 * 10^{-3}$

Table 6: Coefficients of gas and condensate equation.

rate, 5,900 Mscf/day does not give the lowest cumulative gas production. Generally, the optimum injection rate will always depend on the prevailing economic conditions of the operating environment (Figures 8-11).

Effects of permeability ratio on gas and condensate production

For this parameter, the sensitivity was done at different injection rates. This was aimed at studying the possible existing of interaction between the two parameters for both gas and condensate production and to confirm if the little changes in condensate production observed with increasing injection pressure observed in Figure 12 was particular to injection rates only.

The permeability ratio does not have a lot of variation on gas production, especially at very low injection rate. However, the effect of permeability ratio on condensate production is very pronounced when correlated with injection pressure, as seen in Figures 13. At very high

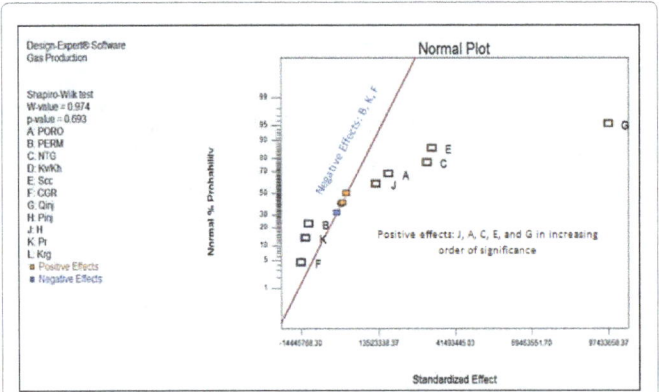

Figure 8: Graph of Normal % Probability vs. Standardized Effects for Gas Production.

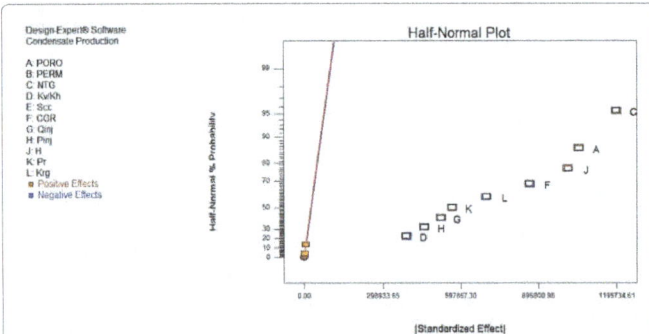

Figure 9: Half-Normal % Probability vs. Absolute Standardized Effects for Cond. Production.

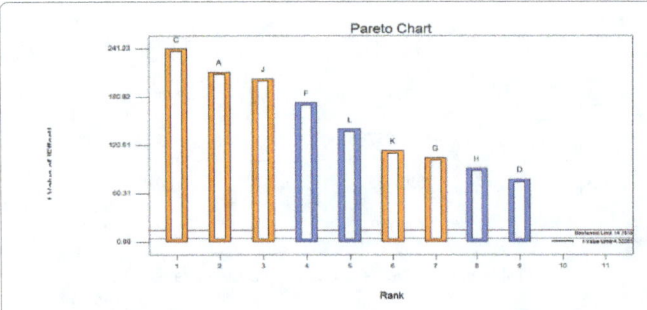

Figure 10: Pareto Chart for Cond Prod., showing factor Casthemost significant parameter.

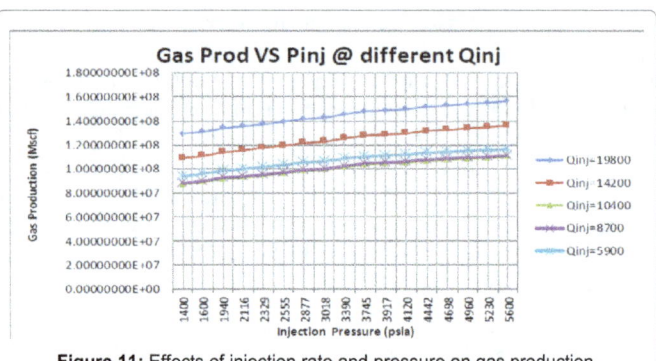

Figure 11: Effects of injection rate and pressure on gas production.

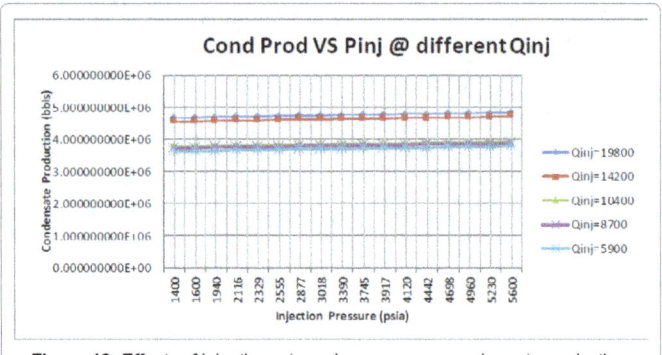

Figure 12: Effects of injection rate and pressure on condensate production.

injection rates and injection pressures, the highest permeability ratio (Kv/Kh=0.1) gives the maximum condensate production while at very low injection rates and injection pressures.

Effects of net-to-gross ratio on gas and condensate production

Both gas and condensate production showed similar effects with NTG sensitivity (Figures 14 and 15). As expected, higher values of NTG gave lower responses of productivity.

Effects of porosity on gas and condensate production

For this sensitivity, the porosity was correlated with different injection rates to study their effects on gas and condensate production (Figures 16 and 17). As expected, the least production occurs in the least porous system. However, the least gas production for each given porosity system does not coincide with the lowest injection rate. A similar observation was also made when studying the effects of injection rate at different injection pressures (Figure 11). Again, this shows that economic conditions could influence the nature of the outcome of the sensitivity involving injection rates. Similar observations were made in the condensate analysis.

Effects of gross thickness on gas and condensate production at varying porosity

Using an NTG of 0.96, the thickness was sensitized on at different porosity. At high porosity, it is observes that the maximum production coincides with the highest thickness. However, as the porosity decreases, this fails to hold. At the lowest porosity system of 0.11, it is observed that the highest gas production does not coincide with the highest thickness of 200 ft. All the above hold true for the condensate

Figures 13: Effects of different permeability ratios on gas and cond. production at different injection rates.

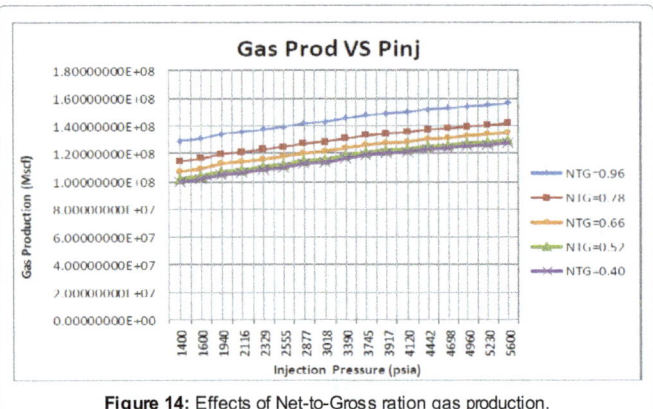

Figure 14: Effects of Net-to-Gross ration gas production.

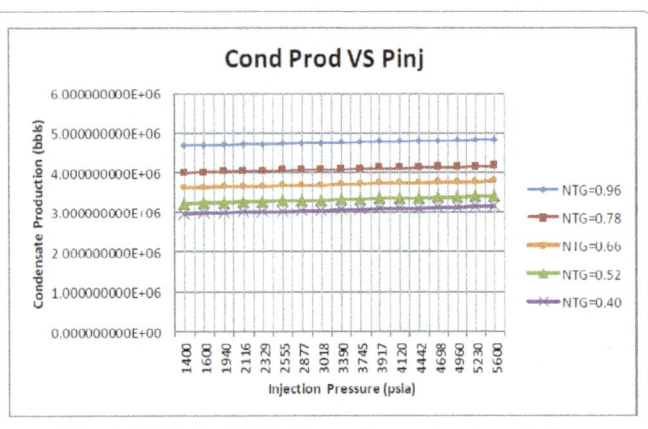

Figure 15: Effects of Net-to-Gross ration condensate production.

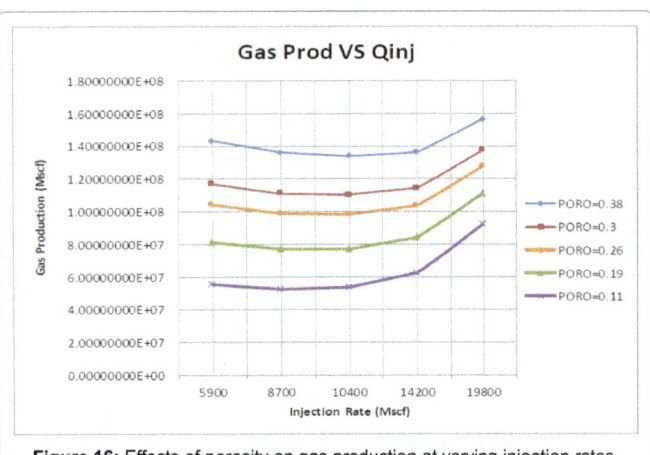

Figure 16: Effects of porosity on gas production at varying injection rates.

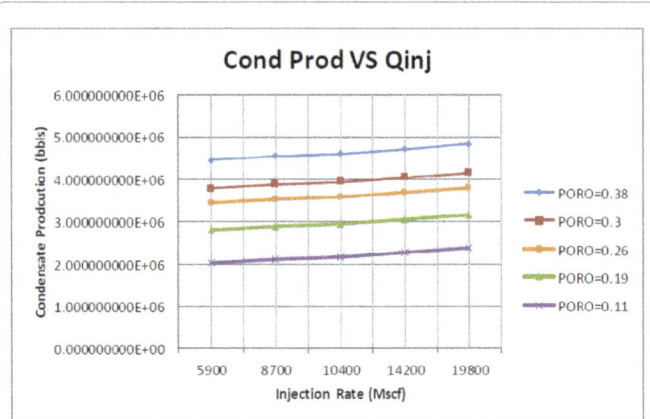

Figure 17: Effects of porosity on condensate production at varying injection rates.

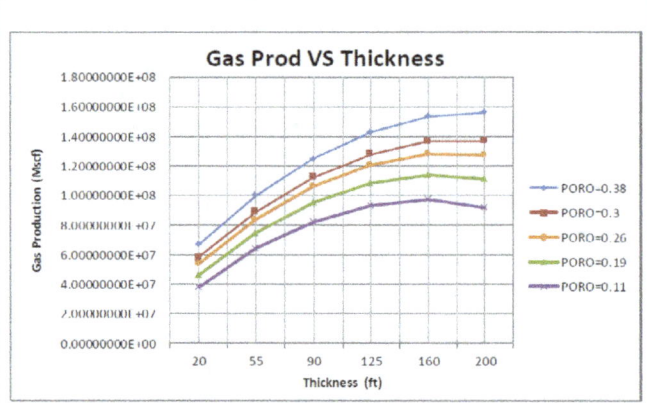

Figure 18: Effects of thickness on gas production at varying porosity values.

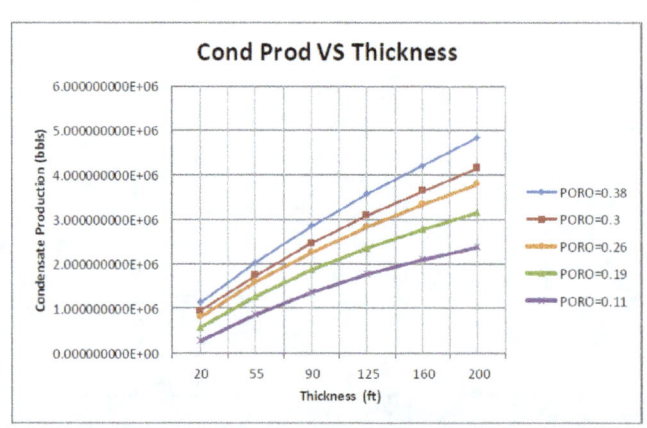

Figure 19: Effects of thickness on condensate production at varying porosity values.

production, except that even at low porosity systems, the maximum production still coincides with the maximum thickness of 200 ft (Figures 18 and 19).

Conclusion

Gas condensate reservoirs are known to be very valuable because of the condensate's high API value. Producing this fluid however has been met with several challenges over the years. This is abated by injection of produced gas into the formation to evaporate the condensed fluid.

Due to the sensitive nature of this kind of reservoir, it is very important to understand the parameters that influence production, and know how these parameters influence production. This work proposed a hypothetical model that was used to study the effects of different parameters on gas and condensate production through statistical optimization. It was discovered that several parameters did affect production of reservoir fluids under varying conditions more than others.

References

1. Ahmed T, Evans J, Kwan R, Vivian T (1998) "Wellbore liquid blockage in gas condensate reservoirs". Paper SPE 51050, Presented at the 1998 SPE Eastern Regional meeting, Pittsburgh.

2. Anderson M (1997) "Design of Experiments", The Industrial Physicist, American Institute of Physics.

3. Nathan M, Hadi N, Ding Z (2010) "Application of Horizontal Wells to Reduce Condensate Blockage in Gas Condensate Reservoirs". Paper SPE 130996 presented at CPS/SPE International Oil and Gas Conference and Exhibition, Beijing, China.

4. Chunmei S (2009) "Flow Behavior of Gas Condensate Wells". A Dissertation Submitted to the Department of Energy Resources Engineering, Stanford University, California, USA.

5. Fevang O, Whitson CH (1996) "Modeling Gas Condensate Well Deliverability". Paper SPE 30714 presented at the SPE Annual Technical conference and Exhibition, Dallas.

6. Barnum RS, Brinkman FP, Richardson TW, Spillette AG (1995) "Gas Condensate Reservoir Behavior: Productivity and Recovery Reduction due to Condensation". Dallas, Texas, USA.

7. Kenyon DE, Behie GA (1987) "Third SPE Comparative Solution Project: Gas Cycling of Retrograde Condensate Reservoirs". Paper SPE 122 Chunmei 78, Journal of Petroleum Technology.

8. Bourbiaux B (1994) "Parametric study of Gas condensate Reservoir Behavior during Depletion: A Guide for Development planning", Paper SPE 28848, presented at the 1994 EUROPEC, London, UK.

9. Henderson GD, Danesh A, Trehran DH, Peden JM (1993) "An Investigation Governing the Flow and Recovery in Different Flow Regimes Present in Gas Condensate Reservoirs", Paper SPE 26661, presented at the 1993 Annual Technical Conference and Exhibition, Houston, Texas.

10. Whitson CH, Torp SB (1983) "Evaluating constant-volume depletion data". J Pet Tech 35: 610-20.

5

Modelling Gelation Time of Organically Cross-linked Water-shutoff Systems for Oil Wells

Marfo SA[1], Appah D[2], Joel OF[3] and Ofori-Sarpong G[4]

[1]World Bank African Centre of Excellence in Oilfield Chemicals Research, IPS, Uniport, Nigeria : Petroleum Engineering Department, University of Mines and Technology, Tarkwa, Ghana
[2]Department of Gas Engineering, Uniport, Nigeria
[3]Centre for Petroleum Research and Training, IPS, Uniport, Nigeria
[4]Minerals Engineering Department, University of Mines and Technology, Tarkwa, Ghana

Abstract

Water production is one of the major challenges in the petroleum industry, especially brown fields and water drive reservoirs. Water production can weaken the cementation of sand grains, thereby rendering formations partially or completely unconsolidated. This in turn initiates fines migration and aggravates safety concerns. The gelation time is an important characteristic of water-shutoff systems and it is influenced by different parameters. The gelation time gives an indication of the time required for a gel to transit from free flowing fluid to solid or semi-solid gel making it difficult to pump. Organically cross-linked gels have numerous applications in the industry such as shutoff systems. They are used to control water and gas production in oil wells. The effect of cross-linker concentration, temperature and brine concentration on the gelation time of an organically crosslinked system was studied. Based on the experimental results a mathematical model was developed for predicting the gelation time of the water-shutoff system. The results showed that temperature had the highest impact on the gelation time of the water-shutoff system with an effect estimate of (-2.292). The brine concentration of the mix water recorded the lowest impact with an effect estimate of 0.2083 and the interaction between cross-linker concentration and brine concentration of mix water had neutral impact. A predictable and effective water-shutoff system has been developed with an excellent initial viscosity which can easily be pumped and applied to solve water production and its associated problems. The gelation time of organically crosslinked water-shutoff system can be optimised in water control operations using this model and the effect estimates of these parameters.

Keywords: Water-shutoff system; Gelation time; Polymer; Cross linker; Brine; Modelling

Introduction

During petroleum production, an acceptable amount of water production is expected and can even be beneficial in the initial stages of the life of a well. This water however can be problematic when it is in excess [1]. Excessive water production is one of the major challenges facing the oil and gas industry [2-4]. Water production affects the economics of producing wells as it has to be treated and disposed of in an environmentally friendly manner at an additional cost [3,5]. Produced water cost the petroleum industry about $45 billion in 2002 [5,6], and this could be on the increase with development of additional wells. High water cut comes with its associated problems such as corrosion, sand production, scale formation and loss of productivity [1,7-9]. This occurrence is common with mature fields [10].

Excessive water production can be the result of the natural depletion of a reservoir where either naturally or artificially active water drive has swept away most of the oil that the reservoir can produce. Causes of water production can either be completion-or reservoir-related. Many researchers [1,11,12] identified the completion-related issues as casing leaks, channel behind casing, completion into water; and the reservoir-related mechanisms as coning or cusping, fractures or faults, and stimulation out of zone, among others.

Excessive water production can be controlled either by mechanical or chemical means [6,13]. The mechanical methods include using hardware or cement as mechanical seals or isolations to control water production, coning control through draw-down reduction and co-production and downhole separation. These mechanical methods such as casing and tubing patches, bridge plugs and cement squeeze can work but not for all types of water production mechanisms. This results in searching for other means which are more efficient and can penetrate the formations to either partially or completely block water producing zones.

The search for efficient chemical means of water control over the decades has evolved with different chemicals; both organic and inorganic being developed. Researchers [1,6,13-16] have identified polymer gels for near wellbore area (organically or inorganically crosslinked) and microgels for deep profile modification (Bright water and Methylene bisacrylamide Aggregates) as the broad categories of chemicals for water sealants in wells producing excessive and unwanted water. According to Watters, Kabir, Sydansk, these broad categories include; inorganic gels, resins or elastomers, monomer based systems, polymer gels, viscous systems, bio-polymers and foam gel among others [1,13,17,18].

The gelation time of water-shutoff systems is a function of the concentrations of polymer (base gel), cross-linker concentration, temperature, salinity of mixing water and pH of the fluid system among others [6]. Temperature has an important effect on the rheological

---

*Corresponding author: Marfo SA, World Bank African Centre of Excellence in Oilfield Chemicals Research, IPS, Uniport, Nigeria; Petroleum Engineering Department, University of Mines and Technology, Tarkwa, Ghana, E-mail: smarfo@umat.edu.gh

properties of polymeric melts, just as it does on polymeric solids. The gelation point of polymers is temperature and time dependent at a determined favourable cross-linker concentration. There is a gradual rise in viscosity followed by an asymptotic increase to infinity as the gel point is approached Marfo [19]. Al-Muntasheri [14] developed correlation for gelation time as a function of temperature for water-shutoff systems. These equations showed that increasing the temperature shortened the gelation time of organically crosslinked gels. Reddy [9] developed correlations for gelation time as a function of degree of salinity of mixing fluid for KCl and NaCl. Their findings indicated a linear relation between gelation time and the salt concentration. It was concluded that increasing degree of salinity increased the gelation time of organically crosslinked water-shutoff systems. El-Karsani, Reddy BR, Al-Muntasheri and Das concluded that increasing the polymer and cross-linker concentrations resulted in a decrease in the gelation time of water-shutoff systems [4,6,9,14,20].

The research findings reported so far examined one parameter at a time in developing correlations as a function of gelation time. This paper, however, explains how the interaction between three parameters; cross-linker concentration, temperature and brine concentration impact on the gelation time of organically crosslinked water-shutoff systems. The parameter effect estimates and the interaction between these parameters with their corresponding impacts on gelation time were studied using factorial design and are presented in this paper. Efficient model with excellent gelation time prediction and gelation time equation as a function of temperature, mix water salinity and cross-linker concentration are developed and presented.

Materials and Methods

Materials

Organically crosslinked system comprising of acrylamide/acrylate copolymer crosslinked with polyamine was used in this experiment and these were in aqueous form. Different formulations were designed using 2% and 5% potassium chloride (KCl) brine as the mix water. The concentrations of the polyamine (cross-linker) used were 6 wt%, 9 wt% and 12 wt% corresponding to 60 gal/1000 gal, 90 gal/1000 gal and 120 gal/1000 gal of the base polymer respectively. These were formulated at a constant base polymer of 350 gal/1000 gal representing 35 wt%. All the chemicals used are American Chemical Standard (ACS) grade. The pH of the solutions was measured using HI991003 pH meter. The gelation times of the various formulations were determined at preset temperatures using Fann 35 Viscometer.

Procedure

The mix fluid and the required amount of polymer were stirred thoroughly in a blender. The pH of the solution was then measured. The required amount of cross-linker was added to the solution and stirred thoroughly, and the pH of the solution was measured. The apparent viscosity of the sample was monitored and poured into 8 oz glass jar and placed in water preset to the test temperature. The apparent viscosity was monitored at regular interval (30 minutes) with a Fann 35 viscometer, equipped with F1 spring, B1 bob and R1 rotor @ shear rate of 511 s⁻¹.

Results and Discussion

The water-shutoff system was observed to be alkaline as the pH of the system was in the range of 9.91 to 10.52. Addition of the polymer to the different mix fluid concentrations resulted in pH in the range of 5 to 8 indicating a slightly acidic medium. Upon adding the cross-linker,

the system's pH changed from 9.91 to 10.52, indicating that cross-linker is a strong alkaline. The water-shutoff system developed has similar pH (approximately 10) with other conformance sealants [5,10]. Maintaining the pH of the system in slightly alkaline medium enhances the hydration process of polymers. This indicates that adjusting the pH to acidic medium will change the gelation time as confirmed by Boye [5]. It is thus advised to minimise the exposure of this system to acidic medium and other contaminants as the pH of fluids control cross-linker function and polymer hydration [1].

The viscosity build-up of the different formulations is both time and temperature dependent. To study the viscosity build-up of the system with temperature and time, different formulations were prepared and the viscosity measured at regular intervals (30 minutes) until the system gelled. When water-shutoff system was placed in the water bath at the present temperature, the apparent viscosity of the system decreased slightly from what was obtained at room temperature before it started building-up, and this is due to the thinning effect of the polymer and cross-linker. The viscosity build-up with time for water-shutoff design systems at different brine concentration and formulations are shown in Figures 1 and 2.

Temperature and cross-linker effect

The temperature and cross-linker effect on water-shutoff design for brine (2% and 5% KCl) are shown in Figures 3 and 4 respectively. The effect of cross-linker concentration on gelation time of the system was determined by varying the cross-linker concentrations (6, 9 and 12 wt%) with all other parameters kept constant. The effect of cross-linker concentrations was determined at temperatures 140 °F and 190 °F. Cross-linker concentrations, salinity of mix water and temperature have effect on the gelation time of water-shutoff systems. Increasing

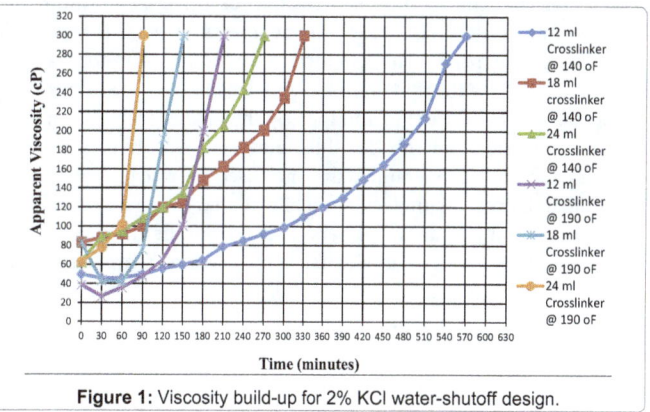

Figure 1: Viscosity build-up for 2% KCl water-shutoff design.

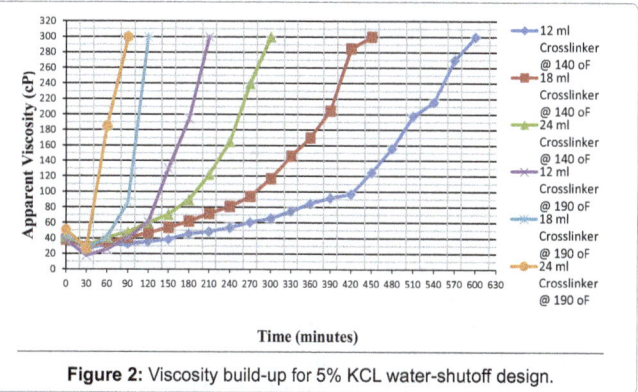

Figure 2: Viscosity build-up for 5% KCL water-shutoff design.

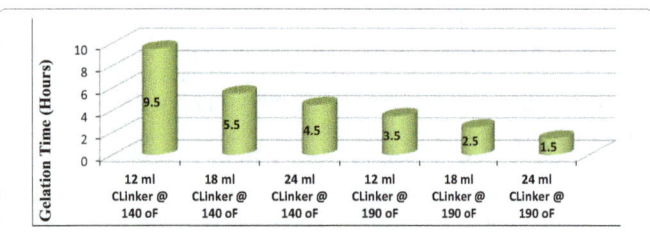

Figure 3: Temperature and crosslinker effect on gel time of water-shutoff for 2% KCl.

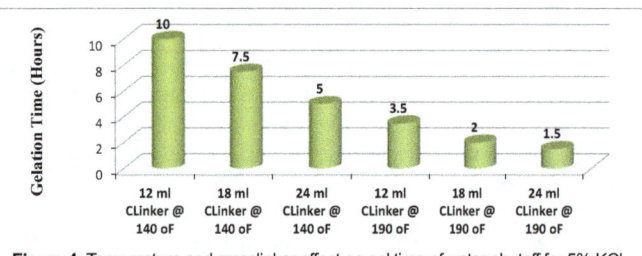

Figure 4: Temperature and crosslinker effect on gel time of water-shutoff for 5% KCl.

the cross-linker concentration shortens the gelation time and this occurs in all the formulations, at different brine concentrations and test temperatures. This is as a result of creation of more sites for crosslinking [1,10], leading to hydration and enhancing rate of reaction. So the more concentrated the cross-linker, the faster the rate of reaction and the shorter the gelation time as shown in the Figures 3 and 4.

Temperature had the greatest impact on the gelation time, and this was determined by measuring the gelation time at different temperatures (140 °F and 190 °F). Irrespective of the cross-linker and brine concentration, an increase in temperature drastically decreased the gelation time with the greatest impact occurring at lower cross-linker concentrations. This is predictive of gel systems indicating endothermic reaction as increase in temperature decreases gelation time [14]. An increase in temperature is a good platform for viscosity of gel systems to build-up as crosslinking density is increased resulting in increased rate of reaction and thereby shortening the gelation time. Other explanations for this could be increased hydrolysis of the base polymer at higher temperatures that increased the rate of crosslinking, resulting in increase in molecular mobility and creating more crosslink sites for reaction [10] among others. Therefore in the gelation process, the rate of crosslinking is accelerated with increased temperature, and the gelation time is decreased.

Brine concentration effect

Increasing brine concentration increased the gelation time of the system, and this impact was greater at lower temperatures. At 140 °F, 5% brine gave a gelation time of 7.5 hours whereas decreasing brine concentration to 2% shortened the time to 5.5 hours for the same formulation. At a higher temperature of 190 °F, the effect of brine concentration on gelation time diminished as different concentrations gave the same gelation time value for the same formulation. The effect of brine concentration (2% KCl and 5% KCl) on gelation time of water-shutoff design at temperatures 140 °F and 190 °F are shown in Figures 5 and 6 respectively. From Figure 5, it is observed that increasing brine concentration from 2% to 5% KCl at 140 °F increased the gelation time for the system and this was noticed in all the formulations. This phenomenon can be attributed to the effect of brine on the hydrolysis of polymer. Brine is known to cause shrinkage of polymer hydration thereby reducing the crosslinking sites available for reaction leading to slow reaction rate, and thus the elongation in the gelation time.

This explains the trend of increase in gelation time when the brine concentration was increased from 2% to 5% KCl mixed fluid. However, at 190 °F the effect of brine concentration was not significant as it did not affect the gelation time of the designs. The possible explanation though not conclusive could be the interactive effect of high temperature and brine concentration. The only exception was when 18 ml of cross-linker was used. This did not follow the trend seen at 140 °F where increasing brine concentration increased the gelation time (Figures 5 and 6).

This is in line with the findings of researchers [1,6,10,20] as the gelation time of the developed system is influenced by factors such as temperature, salinity of mix water, pH, cross-linker and base polymer concentration, among others. The effect of cross-linker concentration on gelation time for 2% and 5% KCl with predictive equations for temperatures at 140 °F and 190 °F are shown in Figures 7 and 8 respectively. These equations can be used to predict the gelation time at the given temperatures and at constant base polymer concentration. The equations make it easier to determine the required cross-linker concentration to achieve the desired gelation time at a given temperature and constant polymer base concentration.

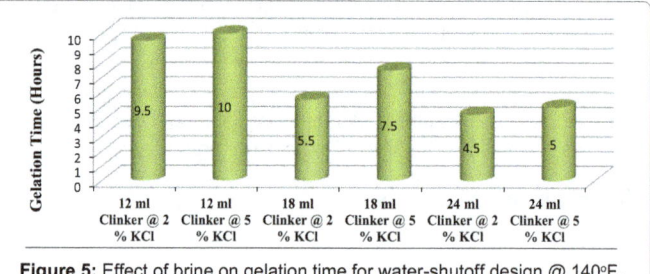

Figure 5: Effect of brine on gelation time for water-shutoff design @ 140ºF.

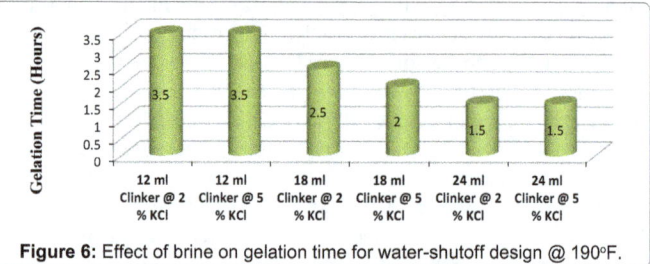

Figure 6: Effect of brine on gelation time for water-shutoff design @ 190ºF.

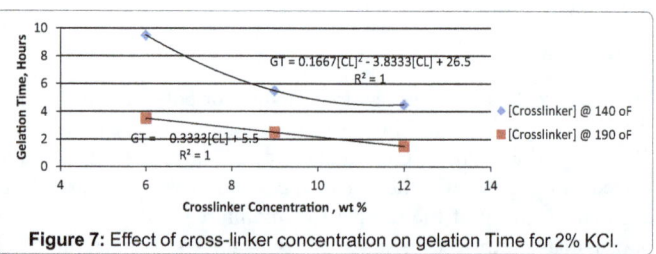

Figure 7: Effect of cross-linker concentration on gelation Time for 2% KCl.

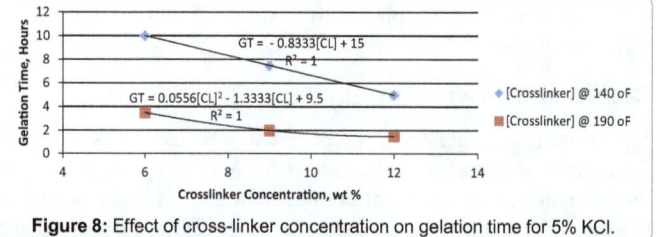

Figure 8: Effect of cross-linker concentration on gelation time for 5% KCl.

To study the interactive effect of the design parameters; temperature, salinity of mix water, and cross-linker concentration on the gelation time of the water-shutoff system, a factorial design was performed. The results obtained from the experiment were modelled using JMP; a statistical analysis software. The summary of fit and the analysis of variance (ANOVA) generated for the water-shutoff model developed are shown in Tables 1 and 2 respectively. The R-Square which is the ratio of the sum of squares of the model to the sum of squares of the total obtained from the model, measures how well a model will predict new data. The R-Square obtained for the water-shutoff system is 0.982 meaning the model is a good predictor and expected to explain about 98% of the variability of the gelation time of the water-shutoff system in a new data designed using this system. The corresponding R-square adjusted for this model is about 96%. The R-Square adjusted (R² adj) is a statistic adjusted for the size of the model which takes care of the number of factors considered in a model.

The parameter estimates for the model are shown in Table 3. The highest impact for the model is from the temperature parameter (-2.292) with a negative value indicating an inverse relationship between it and the response variable. The mix water salinity (0.20833) gave the least impact on the model, with the interaction between cross-linker concentration and mix water salinity having neutral impact on the model.

The residual by predicted gelation time is shown in Figure 9. There are residuals for this model and this confirms the 98% predictability for the model.

The interaction between the design parameters and the effect on the gelation time for the model is shown in the cube plot in Figure 10. The cube plot can be used in determining which parameters to control to achieve a desired gelation time. Additionally, it can be used as an optimisation tool. The gelation times are superimposed on the eight corners of the box and these values correspond to the assigned values for the designed parameters. It can be seen that the longest gelation time (10 hours) for the system occurred when the temperature and cross-linker concentration parameters are at the lowest level and the

R-Square	0.982498359
R-Square Adj	0.96149639
Root Mean Square Error	0.577350269
Mean of Response	4.708333333
Observations (or Sum Weights)	12

Table 1: Summary of fit for water-shutoff model.

Source	DF	Sum of Squares	Mean Square	F Ratio	Prob > F
Model	6	93.5625	15.5938	46.7813	0.0003
Error	5	1.666667	0.3333		
C. Total	11	95.229167			

Table 2: Analysis of variance for water-shutoff model.

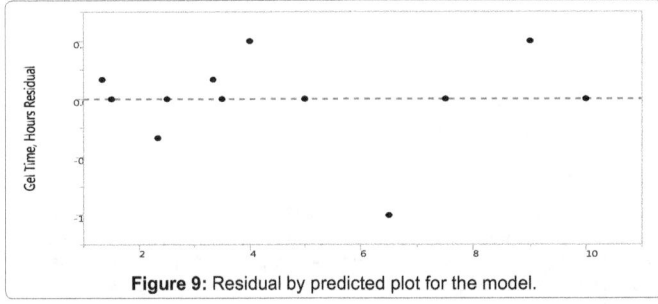

Figure 9: Residual by predicted plot for the model.

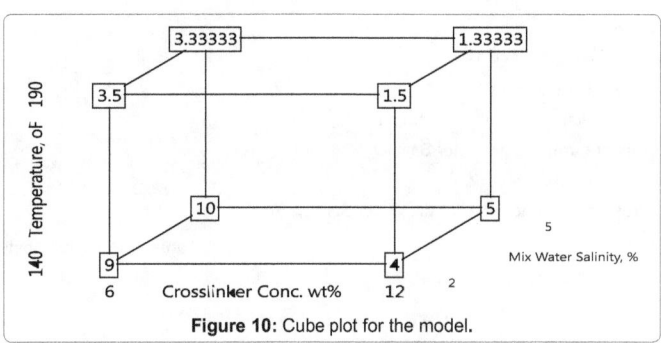

Figure 10: Cube plot for the model.

mix water salinity is at the highest level. For the shortest gelation time (1.333 hours) it occurred at the highest point for all the design factors; temperature, cross-linker concentration and mix water salinity.

The two dimensional contour plots for temperature vs cross-linker concentration, temperature vs mix water salinity and cross-linker concentration vs mix water salinity are shown in Figures 11-13 respectively. These plots predict the behaviour of the gelation time in the various regions created by the design parameters in the model developed. For Figure 11, the maximum gelation time is reached around 140 °F and 6 wt% of cross-linker concentration whereas the minimum is attained around 190 °F and 12 wt% of cross-linker. The longest gelation time for Figure 12 occured around 140 °F and 5% mix water salinity and the shortest gelation time took place around 185 °F irrespective of the degree of salinity of mix water. For Figure 13, the maximum gelation time is recorded around 5% mix water salinity and 6 wt% of cross-linker whereas the minimum took place around 11 wt% of cross-linker and 2% of mix water salinity.

A predictive equation for gelation time GT (hours) was developed from the model (Equation 1). The water-shutoff model's equation has an R-Square of 98% and the design parameters are temperature T (°F), salinity of mix water S (%) and cross-linker concentration C (wt%).

$$GT = 38.4333 - \frac{13}{75}T + \frac{19}{30}S - \frac{67}{30}C + \frac{1}{100}T*C - \frac{1}{300}T*S \quad [1]$$

Conclusions

From the research, the following conclusions could be made:

a. Effective and efficient predictable water-shutoff system using an organically crosslinked polymer with an excellent initial viscosity has been designed. This is indicative of a system that can easily be pumped into the formation matrix.

b. The performance in 2% and 5% KCl brine mix fluid and in different formation temperatures has been tested, and the results indicate that the system can be designed to achieve its desired function in different formations.

c. Temperature, cross-linker concentration and mix fluid salinity and the interaction between these parameters have effect on the gelation time of water-shutoff systems. The research revealed that the longest gelation time (10 hours) for the model occurred when temperature and cross-linker concentration were at the lowest levels and the mix water salinity was at the highest level. The shortest gelation time (1.333 hours) for the system occurred at the highest levels for all the design parameters.

d. Temperature recorded the highest impact on the gelation time with an effect estimate of (-2.292). This was followed by

Term	Estimate	Std Error	t Ratio	Prob>|t|
Temperature, °F(140,190)	-2.291666667	0.166666667	-13.75	<0.0001
Crosslinker Conc. wt%	-0.583333333	0.068041382	-8.57	0.0004
(Crosslinker Conc. wt%-9)*Temperature, °F	0.25	0.068041382	3.67	0.0144
Temperature, °F*Mix Water Salinity, %	-0.291666667	0.166666667	-1.75	0.1405
Mix Water Salinity, %(2.5)	0.208333333	0.166666667	1.25	0.2666
(Crosslinker Conc. wt%-9)*Mix Water Salinity, %	0	0.068041382	0	1

Table 3: Parameter estimate for water-shutoff model.

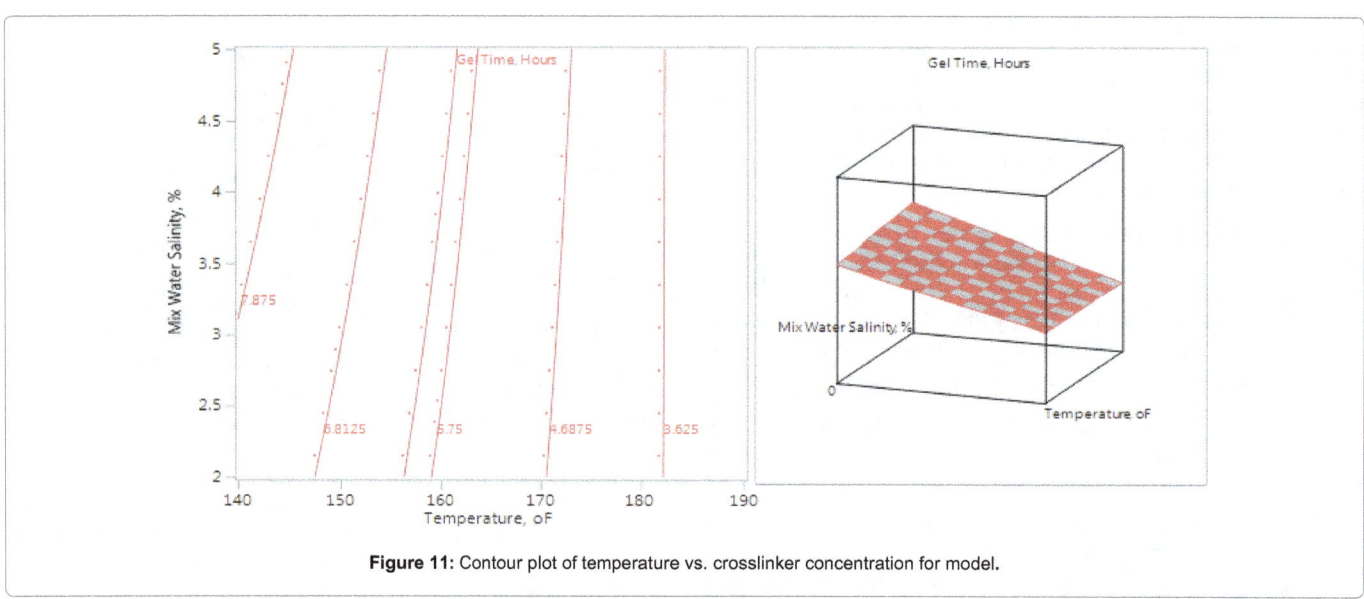

Figure 11: Contour plot of temperature vs. crosslinker concentration for model.

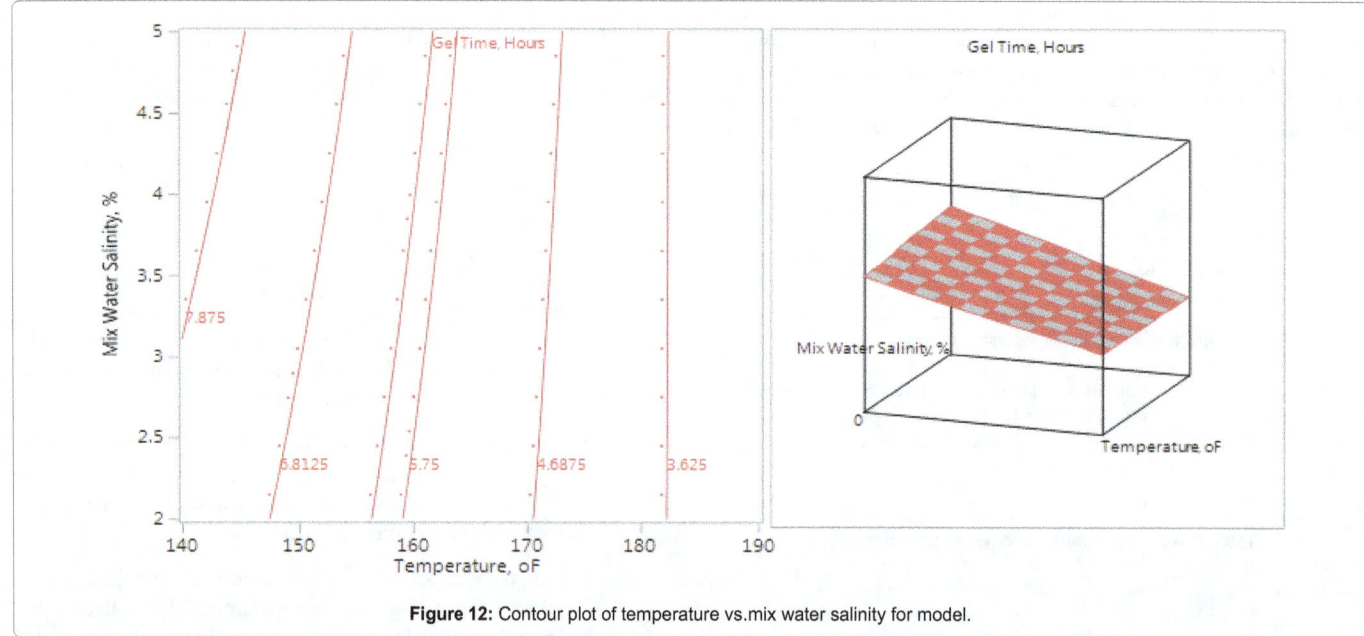

Figure 12: Contour plot of temperature vs.mix water salinity for model.

cross-linker concentration (-0.5833), temperature – mix water salinity interaction (-0.2917), cross-linker concentration – temperature interaction (0.25) and mix water salinity (0.2083). The interaction between the cross-linker concentration and mix water salinity had neutral impact on the gelation time.

e. An efficient and good predictor model with an R-square of about 98% and equations expressing the gelation time as a function of temperature, cross-linker concentration and mix water salinity has been developed.

Acknowledgements

The authors wish to express their thanks to the World Bank for the offer of PhD scholarship at the World Bank African Centre of Excellence, Institute of Petroleum Studies, University of Port Harcourt, Nigeria and University of Mines and Technology, Tarkwa, Ghana for financially supporting this research work.

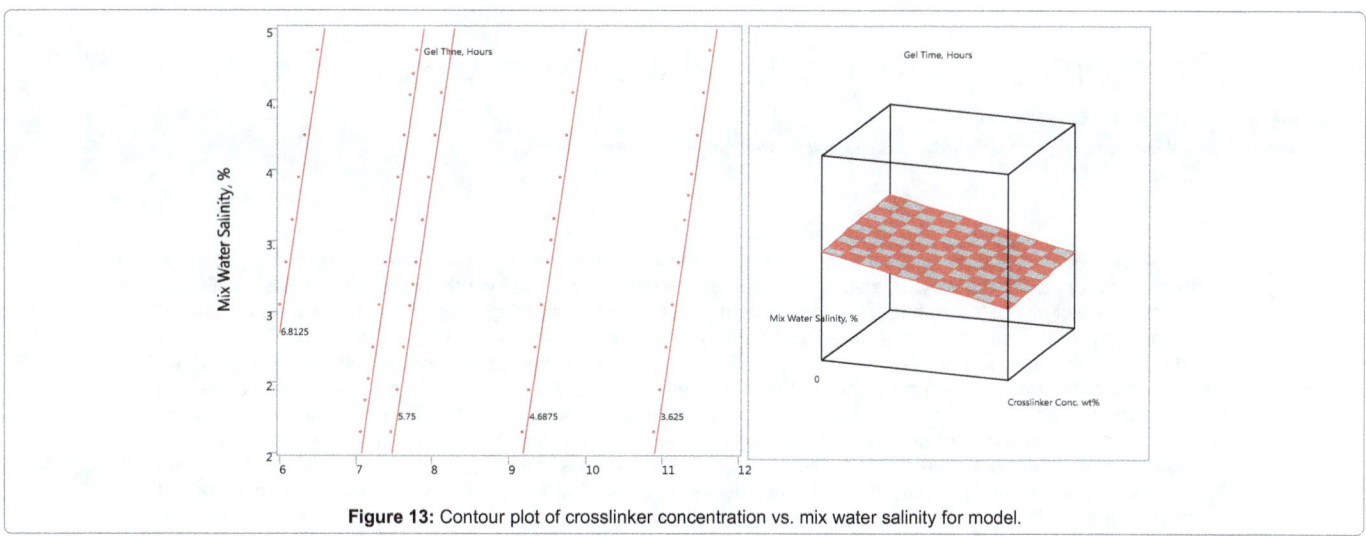

Figure 13: Contour plot of crosslinker concentration vs. mix water salinity for model.

References

1. Economides MJ, Watters L (1997) Petroleum well construction. John Wiley and Sons.

2. Lei G, Li L, Nasr-El-Din HA (2010) New gel aggregates for water shut-off treatments. Proceedings of the SPE Improved Oil Recovery Symposium, Tulsa, Oklahoma, USA.

3. Salgaonkar L, Das P (2012) Laboratory evaluation of organically crosslinked polymer for water shutoff in high-temperature well applications. Proceedings of the SPE International Petroleum Conference and Exhibition, Kuwait.

4. Vasquez JE, Tuck D (2015) Environmentally acceptable porosity-fill sealant systems for water and gas control applications. Proceedings of the SPE HSE and Sustainability Conference, Colombia.

5. Boye B, Rygg A, Jodal C, Klungland I (2011) Development and evaluation of a new environmentally acceptable conformance sealant. Proceedings of the SPE European Formation Damage Conference, Noordwijk, The Netherlands.

6. El-Karsani KS, Al-Muntasheri GA, Hussein IA (2014) Polymer systems for water shutoff and profile modification: A review over the last decade.

7. Van der Hoek JE, Botermans W, Zitha PLJ (2001) Full blocking mechanism of polymer gels for water control. Proceedings of the SPE European Formation Damage Conference, The Hague, Netherlands.

8. You Q, Tang Y, Wang B, Zhao F (2011) Enhanced oil recovery and corrosion inhibition through a combined technology of gel treatment for water shutoff and corrosion inhibitor huff & puff in oil well. Procedia Engineering 18: 7–12.

9. Reddy BR, Crespo F, Eoff L (2012) Water shutoff at ultralow temperature using organically crosslinked polymer gels. Proceedings of the SPE Improved Oil Recovery Symposium, Tulsa, Oklahoma, USA.

10. El-Karsani KS, Al-Muntasheri GA, Sultan AS, Hussein IA (2014) Gelation of a water-shutoff gel at high pressure and high temperature: Rheological investigation. SPE Journal.

11. Barrufet MA, Burnett D, Macauley J (1998) Screening and evaluation of modified starches as water shutoff agents in fractures and in matrix flow configurations. Proceedings of the SPE Improved Oil Recovery Symposium, Tulsa, Oklahoma.

12. Clegg JD (2007) Production operations engineering: Well production problems - Water control. Petroleum engineering handbook. Lake LW, Richardson, Texas SPE 4: 381-384.

13. Kabir AH (2001) Chemical water & gas shutoff technology - An overview. Proceedings of the SPE Asia Pacific Improved Oil Recovery Conference, Kuala Lumpur, Malaysia.

14. Al-Muntasheri GA, Nasr-El-Din HA, Zitha PLJ (2008) Gelation kinetics and performance evaluation of an organically crosslinked gel at high temperature and pressure. SPE Journal.

15. Deolarte C, Vasquez J, Sorian E, Santillan A (2009) Successful combination of an organically crosslinked polymer system and a rigid-setting material for conformance control in Mexico. SPE Production & Operation: 522-529.

16. Reddy BR, Eoff L, Crespo F, Lewis C, (2013) Recent advances in organically crosslinked conformance polymer systems. Proceedings of the SPE International Symposium on Oilfield Chemistry, The Woodlands, Texas, USA.

17. Sydansk RD, Southwell P (2000) More than 12 years of experience with a successful conformance-control polymer gel technology. Proceedings of the SPE AAPG Western Regional Meeting.

18. Sydansk RD, Seright RS (2007) When and where relative permeability modification water-shutoff treatments can be successfully applied. Proceedings of the SPE Symposium Improved Oil Recovery, USA.

19. Marfo SA, Appah D, Joel OF, Ofori-Sarpong G (2015) Sand consolidation operations, challenges and remedy. Proceedings of the SPE Nigeria Annual International Conference and Exhibition, Nigeria.

20. Das P, Patil P, Agashe S (2013) Application of relative permeability modifier for sealant diversion. Proceedings of the SPE North Africa Technical Conference and Exhibition, Cairo, Egypt.

6

Solvent Dewaxing of Heavy Crude Oil with Methyl Ethyl Ketone

As'ad AM, Yeneneh AM and Obanijesu EO*

Department of Chemical Engineering, Curtin University, Bentley Campus, Perth, W.A. 6102, Australia

Abstract

Transportation of waxy crude oil along horizontal pipeline usually requires extra energy that costs additional billions of dollars to the industry. This study investigated the feasibility of Methyl Ethyl Ketone (MEK) as a selective solvent to dewax an Australian heavy crude oil and the possible optimum conditions. Experiments were conducted on three solvent to crude oil ratios (10 : 1, 15 : 1 and 20 : 1), three mixing temperatures (40°C, 50°C, and 60°C) and three cooling temperatures (-10°C, -15°C and -20°C). Each crude oil sample was weighed out and mixed with MEK at a predetermined mass ratio; the mixture was then heated in a hot water bath and stirred until a thermal equilibrium was achieved. The mixture was then placed in an ethylene glycol bath which had been cooled to the desired temperature using dry ice until the target temperature was achieved. The crystallised wax which forms in the mixture was then vacuum filtered, dried, and weighed. Three samples were prepared for each unique parametric variation, and the average result recorded. The results indicated that MEK dewaxing performance improved at higher mixing temperatures. This could be explained by the disruption of dispersion forces which exist between the molecules in the crude oil, allowing new intermolecular bonds to form between MEK and oil molecules in greater preference than with the wax molecules. It was also discovered that the use of a higher solvent to oil ratio resulted in a greater wax yield that is attributed to a greater oil solubility, considering MEK's greater affinity for oil than wax, as well as a greater number of unbounded MEK molecules for dispersion forces to form when a high solvent to oil ratio is used. In contrast, it was found that a lower cooling temperature resulted in a greater extraction of wax from the mixture. This can be associated with the fact that the decrease in temperature encourages the crystallisation of the wax, as well as providing the system with a preferential condition in which an exothermic process, such as the formation of solute to solvent interactions to take place. Finally, the greatest wax yield (27.9 wt%) was achieved at a solvent to oil ratio of 15:01, a mixing temperature of 50°C and a cooling temperature of -20°C. Similar results of approximately 27.6 wt% wax yield was obtained at a cooling temperature of -15°C, which leads us to consider whether the additional energy required to achieve a lower cooling temperature is worth the increased revenue which may be obtained at the marginally greater wax yield when considering a large scale solvent dewaxing application.

Keywords: Horizontal pipelines; Methyl ethyl ketone; Solvent dewaxing; Unconventional hydrocarbon; Waxy crude

Introduction

The global oil demand is expected to increase to about 40 billion barrel of oil (Gbo) per year by 2020 as a result of an increasing population and industrialization [1]. This indicated 60% increase from the present global demand will assert pressure on the conventional oil supply with current proven global reserves of 1,238 billion barrels. Out of a range of alternative resources, only heavy oils and oil sands which are unconventional hydrocarbon resources are sizable enough to supplement and possibly substitute for the shortage in the conventional oil. Consequently, the increasing difficulty in exploration of the depleting conventional reservoirs leading to more expensive production cost will favour the shift to the unconventional resources [2]. Roughly 75% of the world's heavy oil reserves can be found in the Orionoco Belt in Venezuela, and in Northern Alberta and Saskatchewan provinces of Canada. Both heavy oils and oil sands have higher viscosity and lower gravity of less than 20° API which makes them extremely difficult to produce from subsurface reservoirs and subject the reservoirs to thermal stimulation [3,4]. Wax deposition, which is crystallization resulting from phase separation of paraffinic solids from crude oil due to temperature drop also occurs during their transmission in pipelines. Accumulation of these solids could cause severe flow assurance problems that ultimately lead to pipe leakage, rupture and explosion [5,6]. An increased wax deposition is often accompanied by high pigging frequency as curative treatment which costs extra millions of dollars in deferred revenue [7]. Therefore, wax deposition must be properly managed in order to reduce the associated problems as well as increase the heavy oil flowability for an increased market values and ease of processing in refineries.

An effective treatment method for wax deposition as developed by Hamilton and Herman [8] used passive energy to stabilize micelle structures that promote the deposition of paraffin's, asphaltenes and mineral scale particularly in heavy oil pipelines. Solvent dewaxing is another effective treatment method which employs recovery of microcrystalline (or paraffin waxes) from the heavy crude before it is processed [9]. This study investigates the range of optimum performance conditions for solvent dewaxing using methyl ethyl ketone (MEK). MEK is selected for this study due to its rapid evaporation rate, higher solvency, lower viscosity, good miscibility with most hydrocarbons without impact on their characteristic as well as the favourable volume/mass ratio due to its low density [10,11]. Nimer et al., [12] conducted a similar study by using a mixture of toluene and MEK as carrier solvents; however, pure MEK is used in this study due to its established lower selectivity for paraffinic compounds compared to toluene. Though,

*Corresponding author: Obanijesu EO, Department of Chemical Engineering, Curtin University, Bentley Campus, Perth, W.A. 6102, Australia
E-mail: e.obanijesu@curtin.edu.au

both are good solvents for oil, presence of toluene could negatively impact the solvent dewaxing performance [12]. The effects of changes in solvent to oil ratio, heating temperature and cooling temperature on the amount of wax yielded are studied in this work.

Methodology and Analysis

Chemicals

The crude oil with THE composition shown in Table 1 was obtained from Cliff Head oilfield of Roc Oil Pty Ltd, Australia. The MEK, ethylene glycol, and ethanol used for this study were of 100% purity and supplied by Chem-Supply Pty Ltd, Australia. The pure dry ice used to cool the mixture to desired temperatures was obtained from BOC Limited Pty Ltd. A weighing balance of ± 0.5 mg accuracy was used for all weight measurements. The studied crude oil sample was stabilised with 10 ppm hydrogen sulphide and studies were conducted on the mixing and heating, chilling, filtration, and drying stages.

Experimentation

Heating and mixing: For each unique combination in solvent to oil ratio, heating temperature, and cooling temperature, three identical samples were prepared and the experiments were conducted simultaneously such that an average of the results can be obtained for consistency. To do this, three 150 ml beakers were prepared, into each of which, a 5 g quantity of crude oil sample was added using a 30 ml syringe. Using a 150 ml measuring cylinder, the specified quantity of solvent was weighed out. The first set of experiments used 50 g of MEK, the second set of experiments used 75 g of MEK, and the third set of experiments used 100 g of MEK, corresponding to solvent ratios of 10:1, 15:1, and 20:1 respectively.

A magnetic stirrer was added to each beaker, before it was placed in the hot water bath and covered with a watch glass. The temperature of the water bath was kept constant at 40°C, 50°C or 60°C using the hotplate which also served to stir the mixture at a constant speed of 180 RPM. An alcohol thermometer with an accuracy of ± 0.5°C was used to measure the temperature of the mixture. The heating and mixing stage was concluded when the desired heating temperature was achieved in the mixture. This stage is immediately followed by chilling and the beakers are immediately placed in the ethylene glycol cooling bath.

Cooling and filtration of mixture: Following the heating and mixing stage, the mixture was immediately placed in the ethylene glycol and dry ice bath to allow the wax to crystallize. An alcohol thermometer with an accuracy of ± 0.5°C was used to monitor the temperature of the cooling bath to ensure that the temperature is maintained at -10°C, -15°C, or -20°C. To maintain the cooling temperature, dry ice was added periodically as the temperature of the cooling bath tends to increase as the hot mixture is submerged in the ethylene glycol. An identical alcohol thermometer was also used to monitor the temperature and to stir the mixture to ensure that the wax does not deposit on the walls of the beaker and remains suspended in the liquid mixture. The cooling process was completed after the desired temperature was achieved in the mixture.

Following the cooling stage, the mixture was stirred to ensure that no crystallised wax adhered to the walls of the beaker. The magnetic stirrer was then removed from the mixture using a pair of tweezers and the mixture was vacuum-filtered using 45 μm Wathman filter paper. The filter paper containing the wax was then removed and placed on a watch glass in the fume hood for a period of 20 minutes to allow the remaining volatile solvent to evaporate. The filter paper containing

the residue was weighed using the weighed. The results were recorded and expressed in yield as a weight fraction of the overall crude oil sample. The filtration procedure was carried out on all three mixtures individually.

Determination of wax yield: The extracted wax was expressed as a percentage yield of the total weight of the original oil sample and calculated using equation 1:

$$Y = \frac{W_e}{W_o} \times 100 \tag{1}$$

Where Y is the yield as a percentage of the total weight of the sample; W_e is the weight of the extract and W_o is the weight of the crude oil sample itself.

The total wax content of the crude oil sample used in this study was 29.9%, or 1.5 g per 5 g of crude oil sample. After determining the optimum physical conditions of the solvent extraction process, further analysis of the sample was conducted to determine the overall effectiveness of the process.

Results and Discussion

Effect of mixing temperature

Figure 1A shows the wax yield for the varying mixing temperatures. Higher mixing temperature corresponds to a greater yield of wax from the mixture. For a 10:1 solvent ratio, there is a pronounced difference in wax yield between the different mixing temperatures. However, when a higher solvent ratio is used, the yield obtained by the mixtures which were heated to 50°C and 60°C only differ by up to 0.7 wt%, but marginally outperformed the mixture which was only heated to 40°C by up to 1.2 wt% .

Similarly, the cooler chilling temperatures of -15°C and below show a small difference in wax yield of approximately 0.2 wt% between the mixtures which are heated to 50°C and 60°C, however, the yield is up to 0.7 wt% more than that which was obtained by the mixture which was only heated to 40°C as shown in Figure 1B.

Figure 1B also clearly shows that the dewaxing performance is improved when the mixture is heated to a higher temperature of 50°C to 60°C. This is due to the fact that the higher temperature serves to break apart the dispersion forces between the molecules in the crude oil sample, allowing the MEK to form such forces with free oil molecules in greater preference than that to the paraffinic compounds, resulting in a greater filtration and overall separation performance.

Since the overall wax yields of the mixtures heated to 50°C and 60°C are only separated by no more than 0.7 wt%, the more economically viable option must be applied if the process was to be applied on a greater scale. It is better to heat the mixture to a temperature no higher than 50°C as the additional utility requirements to heat the mixture to 60°C is not economically feasible based on the marginal increase in the overall wax yield, as well as the degree of cooling which is required later on in the process. However, it is advisable to conduct a thorough economic analysis and optimization study that accounts for additional

Ingredient	Formula	CAS No.	Content
Hydrogen Sulphide	H_2S	7783-06-4	< 0.005%
Aliphatic Hydrocarbon(s)	N/A	N/A	> 60%
Aromatic Hydrocarbon(s)	N/A	N/A	< 1%

Table 1: Properties of the crude oil used for the study

utility requirements on one hand and the revenue from a slightly higher wax yield on the other hand.

Effect of solvent to oil ratio

Figure 2 shows that there is a distinct trend between the mixing and chilling temperatures against the overall wax yield. More specifically, it appears that when a high solvent to oil ratio is used, a higher yield is achieved. Similar yield was achieved from the samples which used a 15:1 and 20:1 solvent to oil ratios.

It can be said that although the extraction performance was similar between 15:1 and 20:1 solvent ratios, the maximum wax extraction was achieved at a solvent to oil ratio of 15:1 or greater. The results can be justified by the fact that the high solvent ratio results in a greater reduction of the mixture's viscosity, and the solvent itself enhances the formation and growth of wax crystals, which in turn, resulted in an improved filtration performance. More so, the greater ratio of solvent to oil molecules which have a greater solubility preference for the oil than that of the wax serves to improve the dewaxing performance. This is reflected by the high yield obtained from mixtures with 15:1 and 20:1 solvent ratios.

Although a high performance can be achieved when a high fraction of solvent is used, it is important to consider the implications a high volumetric flow rate of mixture may have on the economics of a large scale process. Compared to the wax yield of mixtures which used a 10:1 solvent to oil ratio, mixtures using a 15:1 ratio were able to achieve up to 2.1 wt% greater yield. However, no significant performance differences separate the mixtures using a 20:1 and a 15:1 ratio.

Higher volumetric flow rate per unit of wax yield results in a significantly greater pumping power requirement, and a greater duty in the units which serve to recover the solvent. Additionally, a higher raw material cost will be incurred as a result of the greater requirement of solvent. In turn, applying a 20:1 solvent ratio in a commercial scale process may not be economically justifiable, considering the fact that a similar performance can be achieved by using a 15:1 solvent to oil ratio. An economic study which takes into account the additional utilities required for higher volumetric flow rates of mixture can be conducted for further confirmation.

Effect of chilling temperature

The results presented in Figure 3 shows that a lower chilling temperature results in a greater wax yield, and that the greatest average yield was obtained for solvent to oil ratio of 15:1. Although the wax yield differs by up to 3% between chilling temperatures of -10°C and -15°C, there is no more than 1% difference in yield when chilling

temperatures of -15°C and -20°C are applied to the mixture.

This can be explained by the fact that due to a greater temperature difference, a longer amount of time is required for the mixture to reach thermal equilibrium. This translates to a greater crystallisation time, allowing a greater amount of wax crystals to form [13]. In addition to this, the greater temperature gradient which the mixture is exposed to further encourages the formation of new crystals or nucleation, which enables a greater rate of formation of the wax crystals [13].

Although a greater performance is evident when a chilling temperature of -20°C is applied, the small increase in wax yield may not make it preferable than chilling the mixture to -15°C. The optimum chilling temperature can be determined by studying the economic trade-off between a greater cooling duty and a higher wax yield.

Effect of solvent to oil ratio on cake formation

In the vacuum filtration process, various filter cake characteristics were observed. The appearance of wax crystal clusters varied in size and formation according to the variation in parameters. It was generalised that as the solvent to oil ratio was increased, smaller clusters of wax were formed on the filter paper post filtration. The change in chilling temperature also changed the appearance of the filter cake such that a lower chilling temperature produced larger wax clusters as seen in Figure 4.

A possible explanation for such various filter cake characteristics is the variance in concentration and cooling residence time. A lower solvent ratio will result in a greater wax concentration, which in turn will result in the formation of large wax crystal clusters. In addition to this, the greater period of cooling for the lower chilling temperatures will result in an increased formation of crystals, which translates to a larger cluster size seen in the filter bed.

Optimum yield point

At the completion of the experimental study, it was found that the maximum possible wax extraction was obtained when the solvent was mixed with the crude sample at a ratio of 15:1, then heated to a temperature of 50°C in a hot bath with simultaneous stirring, before immediately chilling it to a temperature of -20°C. Three experiments carried out under these conditions produced an average dry wax mass of 1.39 g from a 5 g crude oil sample, corresponding to a yield of 27.9 wt%.

Conclusion

This study has successfully established the effects of parametric variations on wax extraction from an Australian heavy crude oil using

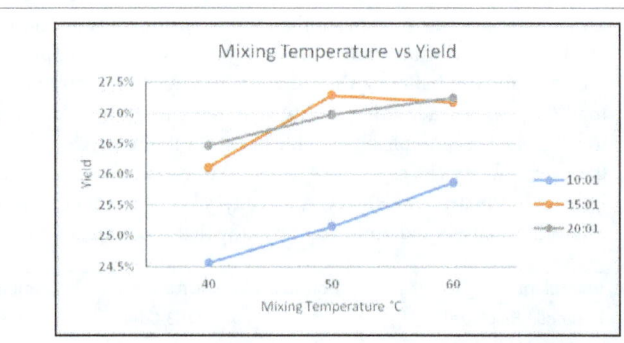

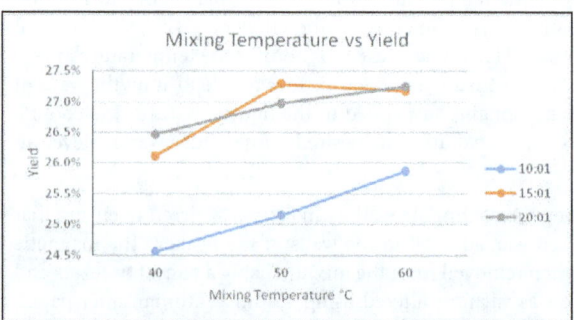

A: Solvent Ratio Comparison   B: Chilling Temperature Comparison

Figure 1: The results of mixing Temperature against Yield.

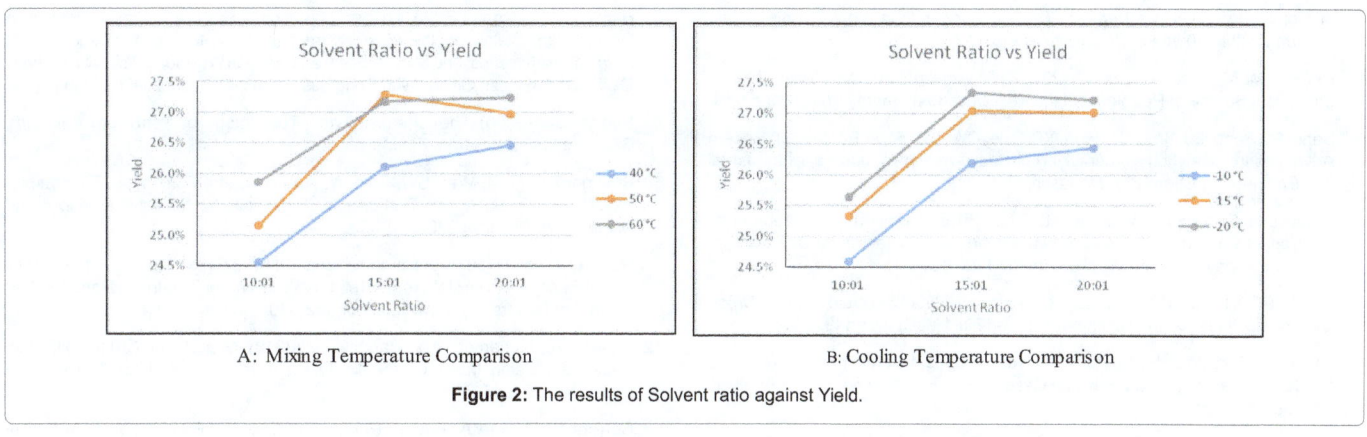

A: Mixing Temperature Comparison

B: Cooling Temperature Comparison

Figure 2: The results of Solvent ratio against Yield.

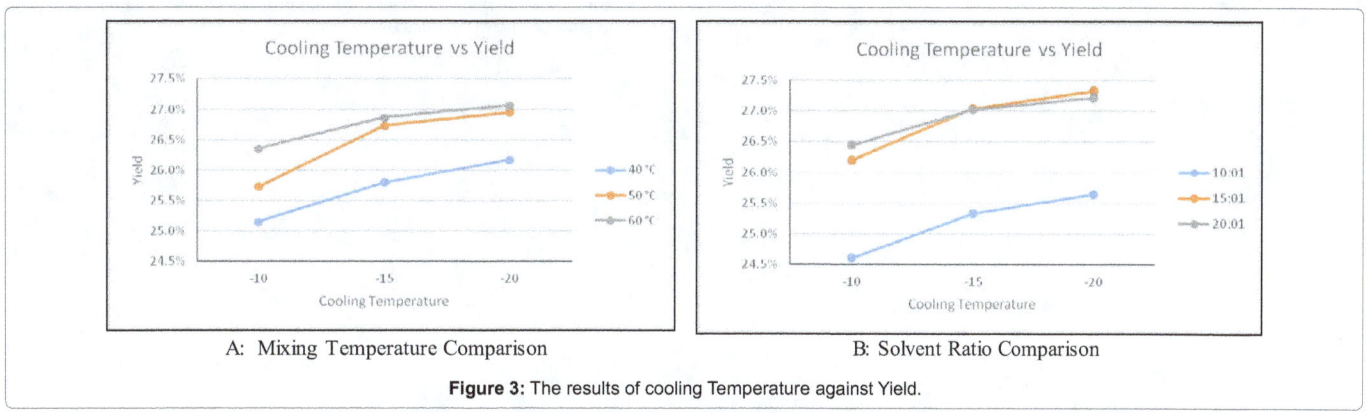

A: Mixing Temperature Comparison

B: Solvent Ratio Comparison

Figure 3: The results of cooling Temperature against Yield.

(a)   20:1   (b)   15:1   (c)   10:1

Figure 4: Structure and formation of different cake clusters after vacuum filteration.

a pure Methyl Ethyl Ketone solvent. It was observed that the wax yield increases with increasing mixing temperature and solvent to oil ratio, as well as a decrease in chilling temperature. Based on the experimental results it was concluded that the optimum wax yield of 27.9 wt% was obtained using a 15:1 solvent to crude oil ratio, at a mixing temperature of 50°C, and a chilling temperature of -20°C.

High wax yields of 27.6 wt% were also observed at a solvent to oil ratio of 15:1 under a 50°C mixing temperature and a chilling temperature of -15°C, as well as at a solvent to oil ratio of 20:1, mixed at 50°C and chilled at -20°C. With no more than 0.3 wt% yield separating the top three results, which have been obtained under a relatively small margin of operating conditions, more experimental work should be conducted to obtain a better optimum condition for solvent dewaxing over a wider range of operational parameters.

In addition to the experimental optimization, economic optimization and feasibility analysis should be conducted to observe the possibilities for commercial scale implementation. This may include a study on the trade-off between the additional revenue which may be achieved through a higher yield of wax and the additional cost of utilities and raw materials required to achieve such high yields, taking into account other parameters such as processing time, solvent recovery, capital and operating costs which may affect the operation as a whole.

Acknowledgement

Dr. Emmanuel OBANIJESU wishes to acknowledge ROC Oil Pty, Perth, Australia for providing the waxy crude oil that was used for the study.

References

1. Campbell CJ, Laherrere JH (1998) The End of Cheap Oil: Global Production

of Conventional Oil will Begin to Decline Sooner than most People Think, Probably within 10 years, Scientific American.

2. DeSena MFM, Rosa LP, Szklo A (2013), Will Venezuelan Extra-Heavy Oil be a Significant Source of Petroleum in the Next Decades? Energy Policy 61: 51-59.

3. Sahu R, Song BJ, Im JS, Jeon YP, Lee CW (2015) A Review of Recent Advances in Catalytic Hydrocracking of Heavy Residues. Journal of Industrial and Engineering Chemistry, Manuscript In Press.

4. Giacchetta G, Leporini M, Marchetti B (2015) Economic and Environmental Analysis of a Steam Assisted Gravity Drainage (SAGD) Facility for Oil Recovery from Canadian Oil Sands. Applied Energy 142: 1-9.

5. Visintin RFG, Lockhart TP, Lapasin R, D'Antona P (2008) Structure of Waxy Crude Oil Emulsion Gels. Journal of Non-Newtonian Fluid Mechanics 149: 34-39.

6. Liu H, He W, Guo J, Huang Q (2015) Risk Propagation Mechanism: Qingdao Crude Oil Leaking and Explosion Case Study . Engineering Failure Analysis ,Manuscript In Press.

7. Lu Y, Huang Z, Hoffmann R, Amundsen L, Fogler HS (2012) Counterintuitive Effects of the Oil Flow Rate on Wax Deposition. Energy & Fuels 26: 4091-4097.

8. Hamilton DS, Herman B (2011) The Application of Passive Energy to Production Optimization; Stabilizing the Micelle Structure in Oil to Prevent Deposition of Paraffin, Asphaltenes, and Mineral Scale and Reduce Well-head Viscosity in Heavy Oil. South American Oil and Gas Congress, Maracaibo, Venezuela, October 18.

9. Speight JG (2006) The Chemistry and Technology of Petroleum. (5thedn), CRC Press, Boca Raton, Florida

10. Beringer LT, Xu X, Wan SW, Shih W, Habas R, et al. (2015) An Electrospun PVDF-TrFe Fiber Sensor Platform for Biological Applications. Sensors and Actuators A: Physical 222: 293-300.

11. Hu G, Li J, Hou H (2015) A Combination of Solvent Extraction and Freeze Thaw for Oil Recovery from Petroleum Refinery Wastewater Treatment Pond Sludge, Journal of Hazardous Materials 283: 832-840.

12. Nimer AA, Mohamed AA, Rabah AA (2010) Nile Blend Crude Oil: Wax Separation Using Mek Toluene Mixtures. Arabian Journal for Science and Engineering 35: 17-24.

13. Correra S, Fasano A, Fusi L, Primicerio M, Rosso F (2007) Wax Diffusivity under given Thermal Gradient: A Mathematical Model. Journal of Applied Mathematics and Mechanics 87: 24-36.

7

Electro-Coagulation Treatment and De-oiling of Wastewaters Arising from Petroleum Industries

Sellami MH[1]*, Loudiyi K[2], Bellemharbet K[1] and Djabbour N[1]

[1]Process Engineering Department, Laboratory of Process Engineering, Ouargla University 30000 Algeria
[2]Renewable Energies Laboratory (REL) Al Akhawayne University, Ifrane, Morocco

Abstract

Petroleum region of Haoud Berkaoui (HBK) is one of the first areas to provide great efforts in the field of environmental protection; this region has a resort of de-oiling, a recovery unit flaring gas, a potable water station and will soon have a wastewater treatment plant. Generally, wastewaters discharged by petroleum industries may contain hydrocarbons and suspended solids. The charge of wastewaters derived from petroleum industry exists in various forms: solutions, colloids and particles. It is important to treat this wastewater at low cost by electro-coagulation compared with traditional chemical treatments as conventional coagulation-flocculation process instead of reusing or re-injecting it directly into oil wells. The treatment by electro-coagulation was applied. This treatment provides two separate phases by settling or flotation. The electro-coagulation produces a similar separation as coagulation -flocculation using reagents such as $FeCl_3$ or $Al_2(SO4)_3$ but in this case, the coagulant is derived from the anodic dissolution of Aluminum. In this experiment we have studied the effect of several parameters, namely: the solution pH, current intensity and clarification time on the process efficiency. The results showed that the Oil and Wastes Removal Efficiency (OWRE) was 79.61% for both current intensities of 0.3 and 0.65 A in a medium basic solution pH (9-11) and respectively after 100 and 80 min of clarification time. To see the impact of the treated water on vegetations, some irrigation tests have been conducted regarding two types of plants (date palm and shaft apocalyptic) for 13 months. The tests showed that the thick layer of 5 cm and fine particles diameter of dune sand removes most of the remaining oil. The layer that fills the basin surrounding the shaft is removed and replaced every 03 months. So, fine dune sand plays the role of biological filter. The little garden plants appear and grow normally.

Keywords: De-oiling; Wastewater; Electro-coagulation; Environment; Colloids; Emulsion; Dune sand

Abbreviations: HBK: Petroleum region of Haoud Berkaoui (Ouargla/Algeria); OWRE: Oil and Wastes Removal Efficiency (%).

Nomenclature

Turbidity $_{in}$= Wastewater turbidity before treatment.

Turbidity $_{out}$ = Wastewater turbidity after treatment.

Introduction

The discharge of petroleum products into the wild leads to the proliferation of microorganisms able to grow on hydrocarbons and their degradation products. Their number is chronically much higher in the polluted areas and increases after intake of oil in sites without contamination.

Industrial wastewaters discharged by petroleum industries contains: oil, heavy metals and chemicals used in the process of oil separation and treatment. These wastewaters are a source of soil, water and air pollution, and lead a mortal danger to the ecosystem.

In petroleum plants, the removal of oil and suspended solids is primary and relatively performed by purely physical separation methods such as CPI (density difference), decantation, filtration, centrifugation etc; nevertheless, the fine particles behave as a remaining colloidal suspension and require separation by chemical flocculation. The latter consists in neutralizing the colloidal suspension by the addition of an electrolyte causing particles agglomeration and consequently their flocculation.

The biodegradation is also one of the primary mechanisms leading to the elimination of oil from the environment; however, wastewaters are usually difficult to degrade and have high turbidity and toxicity. Thus, physical separation step is important to ensure that the wastewater being treated before reuse or release to wild [1]. Coagulation and flocculation is one of the methods used to treat industrial wastewaters. Colloidal particles in nature normally carry charges on their surface which lead to the stabilization of suspension. The attractive force between particles, exist in case of colloidal particles in suspension, but the electrostatic repulsion of surface charges opposes the particles to come together and form agglomerates. The principal mechanism controlling the stability of both hydrophobic and hydrophilic particles is the electrostatic repulsion [2].

Addition of chemicals can change the surface property of such colloidal particles or lead to precipitate dissolved material so as to facilitate the separation of solids by filtration or settling. The conversion of stable state dispersion to the unstable state is expressed as coagulation and flocculation [3,4].

The coagulation, is the process whereby destabilization of a given colloidal suspension or solution is taking place. The role of coagulation is to overcome the factors that promote the stability of a given system.

*Corresponding author: Sellami MH, Process Engineering Department, Laboratory of Process Engineering, Ouargla University 30000 Algeria, E-mail: sellami2000dz@gmail.com

It is realized by the use of appropriate chemicals, usually Aluminum or iron salts called coagulant agents. The flocculation refers to the induction of destabilized particles in order to come closer, to coalesce and thereby, to form large agglomerates, which can be generally separated easier by gravity settling [5].

The agglomerates formed by coagulation are compact and loosely tied, whereas the flocs are of larger size, strongly attached and porous in case of flocculation. In mineral processing industries, the range of application of flocculants is much greater than the coagulants.

The load of wastewaters discharged from petroleum industries exists in various forms: solutions, colloids and particles. It is important to treat this wastewater at low cost instead of reusing or injecting it directly into oil wells. In order to improve the physicochemical processes of destabilization of emulsions, the treatment by electro-coagulation was used. Typically, wastewater is treated by coagulation-flocculation using reagents such as $FeCl_3$ or $Al_2(SO4)_3$. This treatment provides two separate phases by settling or flotation [6].

The electro-coagulation produces a similar separation but, in this case the coagulant is derived from the anodic dissolution of Aluminum. In addition, the treatment lends itself to automation, conservation issues reagents are negligible because the anodes can last long and the response time to a change in flow rate or concentration of the effluent load remains very fast [7].

For good knowledge of pollution level of the area, M. Valipour et al. [8-10] proposed an Environmental flow diagram (EFD) based on energy reference system (RES) and process flow diagram (PFD) for each industrial company because without the exact information about quantity and quality of pollution sources, decrease or eliminate industrial pollutions are difficult. Environmental flow diagram was carried out by authors to find and locate sources of pollutants, and explaining impact of solutions to the energy optimization and reduces environmental pollutants.

Some investigators on wastewater treatments in Ouargla University (southern Algeria) use a new promising technique for wastewater treatment; this technique is the use of dune sand as a biological filter. The use of this local material as a filtration support for domestic wastewater treatment demonstrated its efficiency at pilot scale. Using two layers of dune sand in various column heights, they carried out measurements of porosities, suspended matter and fouling time-course for a week. Nevertheless, dune sand filter receives considerably amounts of suspended matter, causing biological clogging, decreasing therefore the filter porosity [11,12].

During electrostatic oil desalting process in treatment units, about 5% of the volume of the crude oil is added as fresh water (wash water) to reach and extract emulsified salts. Haoud Berkaoui region (HBK) is one of the most important areas for the production of crude oil. So, the amount of wastewater discharged from this region's refinery is significant; its spill in nature without treatment or its reinjection in wells to increase the reservoir pressure is polluting our ecosystem. The treatment of this water becomes an obligation and the choice of a soft and cheap technique remains one of the solutions we have for the moment to decrease the risk that this wastewater exerts on the environment.

In this experiment we have optimized the conditions of destabilizing emulsions by studying several parameters, namely: the solution pH, intensity of applied current and clarification time. The Oil and Wastes Removal Efficiency (OWRE) was calculated using:

$$OWRE(\%) = \left(\frac{Turbidity_{in} - Turbidity_{out}}{Turbidity_{in}} \right) x100$$

Finally, irrigation tests of treated water using fine particles of dune sand as biological filter in private garden plants during 13 months have been carried out.

Materials and Methods

Two experiments were performed for removal of wastes and hydrocarbons by using electrochemical cell with Aluminum electrodes supplied by an electric generator (Figure 1). In the first experiment, the current intensity was varied and the solution pH was maintained as possible at a constant value (without the addition of any acid or base). In the second experiment the current intensity was fixed and the solution pH was varied. Finally the Oil and Wastes Removal Efficiency (OWRE) was calculated.

Sampling

Sampling operations must be performed quickly to avoid any change in wastewater quality. The water was taken either in clean glass bottles (rinsed and dried), or in polyethylene bottles. To avoid any change in the characteristics of the wastewater between sampling and analysis, the samples were stored at standard temperature in the dark and were sent to the laboratory within 24 hours after collection. Different samples were taken from the same wastewater and various parameters were analyzed and determined, such as solution pH, conductivity, turbidity, different ions concentrations, different pollutants elements and heavy metals. Because of analytical samples were taken during the week, the primary test (pH, turbidity etc ...) were not exactly the same; so we have used for each experiment a different wastewater. We were interested on clarification time and Oil and wastes removal efficiency only. The Table 1 below given as an example, shows the mean values of key parameters of the wastewater (N°1) used in the first experiment compared with standard values authorized by the official gazette of Algeria shown in the same table [13].

First experiment (Effect of current intensity)

Firstly the initial pH and turbidity were measured (Table N°1). A sample of one liter of oily wastewater was placed in the electrochemical cell. A generator for supplying the cell was used, the current intensity was adjusted to I = 0.25 A with a voltmeter.

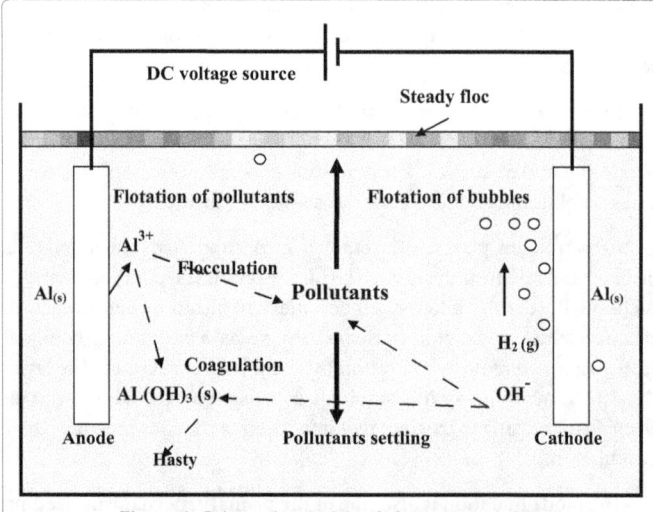

Figure 1: Schematic principle of electro-coagulation.

Parameter	Concentration (mg/l)	Maximal value authorized [13]
(Ca^{2+}) (mg/l)	10832	----
(Mg^{2+}) (mg/l)	2733	----
(Fe^{2+}) (mg/l)	601	3
(Cl^-) (mg/l)	69393	----
(SO_4^{2-}) (mg/l)	411	----
pH	6.05	6.5 à 8.5
Conductivity (µs/cm)	1131	----
Total salinity expressed in (NaCl) (mg/l)	115000	----
Turbidity NTU	209	----
SM (mg/l)	169	40
(BOD_5) (mg/l)	3.91	35
(COD) (mg/l)	948	120
(NO_3^-) (mg/l)	2.10	----
(PO_4^{3-}) (mg/l)	3.49	10
(Pb) (mg/l)	4.01	0.5
(Cu) (mg/l)	0.09	0.5
(Cd) (mg/l)	0.34	0.2
(Zn) (mg/l)	22	3
(Mn) (mg/l)	115	1

Table 1: Key parameters values of wastewater (N°1) and standard values authorized.

5 g of Na_2SO_4 was weighed, and then added to improve the solution conductivity. The duration of clarification time is two hours (120 min). The solution pH and the turbidity were measured every 20 min, respectively by pH-meter and spectrophotometer. The experiment was repeated for each current (I = 0.3 A - 0.35 A - 0.40 A - 0.45 A - 0.50 A - 0.55 A - 0.60 A - 0.65 A - 0.7 A).

Second experiment (Effect of solution pH)

For acid solution (pH is between 4 and 6), sulfuric acid (H_2SO_4) was used. For a bit basic solution (pH between 6 and 9), sodium hydroxide (NaOH) was used because the solution pH at the entrance is usually close to 6.

For medium basic solution (pH between 9 and 11), also sodium hydroxide (NaOH) was used.

After the first experiment we have chosen two best current intensities which given minimum of turbidity and maximum of efficiency for acidic and basic solution. The clarification time was measured for each of the best intensities.

Finally the better of the two current intensities was chosen to give: the lower turbidity, the maximum oil and wastes removal efficiency and the lower clarification time.

Results and Discussion

To avoid graphical clutter during the study of the current intensities effect (first experiment), each series of 10 curves has been divided in two parts of 5 curves.

Effect of current intensity

Figures 2a and 2b present the effect of the current intensity on the solution pH. As shown, generally the solution pH increased slowly from 6 (weakly acid) to reach neutral or weakly basic solution after clarification time (120 min). For the lower intensity (I = 0.25 A) the variation of solution pH was insignificant because the solution remained weakly acid.

As displayed in Figures 3a and 3b, the turbidity decreases with processing time from 209 NTU to less than 125 NTU. The lower turbidity observed was 21 NTU for I = 0.65 A obtained after 40 min, succeeded by 28 NTU for I=0.3 A obtained at the end of experiment (120 min). Generally the increase of turbidity observed after 40 min of experiment was due to the solution pH which becomes weakly acid and allows the dissolution of Aluminum. This was remarkably observed for the current intensity of 0.25 A.

Figures 4a and 4b show the effect of the current intensity on the efficiency of removing oil and wastes vs. time for different current

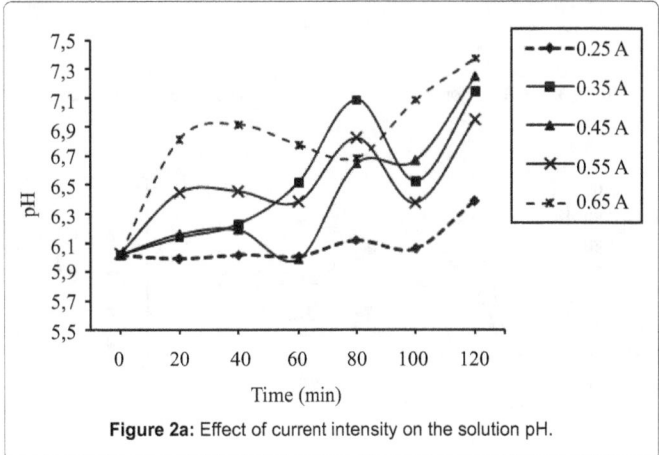

Figure 2a: Effect of current intensity on the solution pH.

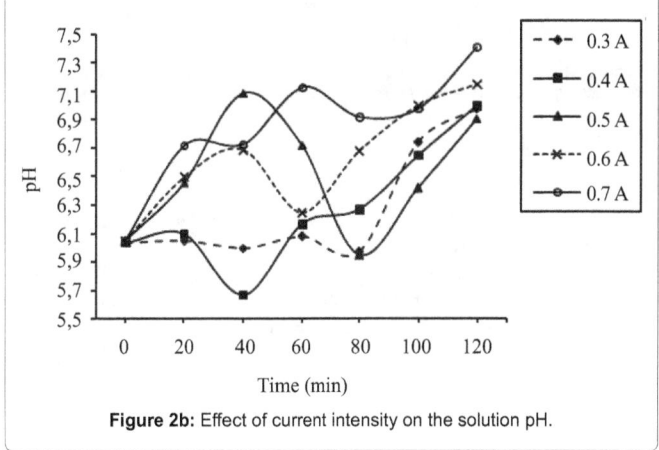

Figure 2b: Effect of current intensity on the solution pH.

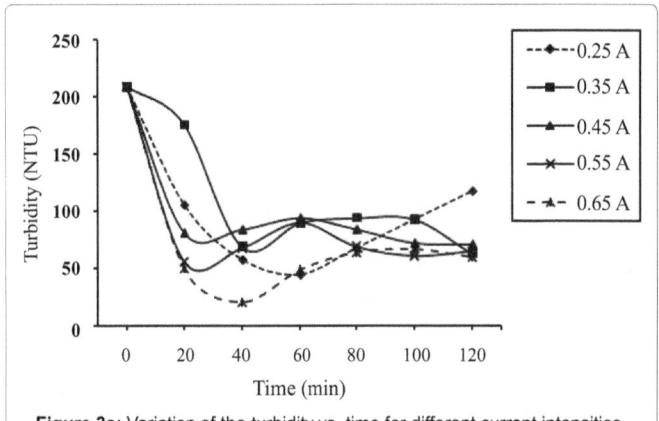

Figure 3a: Variation of the turbidity vs. time for different current intensities.

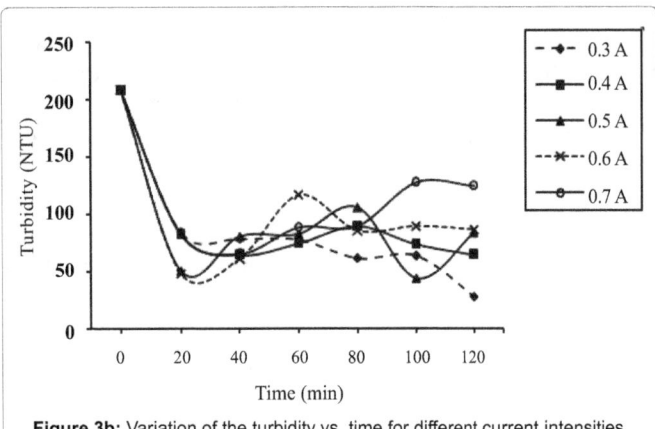

Figure 3b: Variation of the turbidity vs. time for different current intensities.

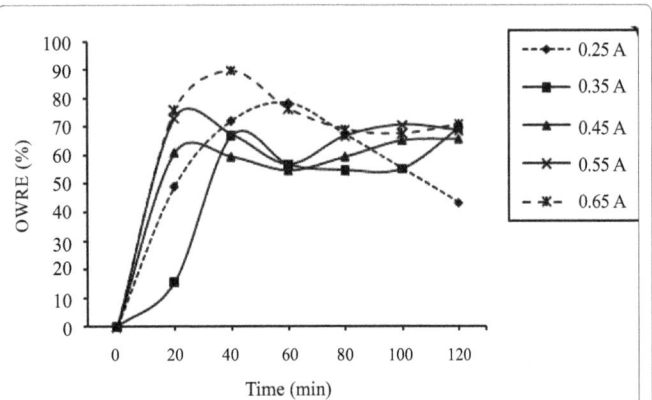

Figure 4a: Variation of oil and wastes removal efficiency (OWRE) vs. processing time for different current intensities.

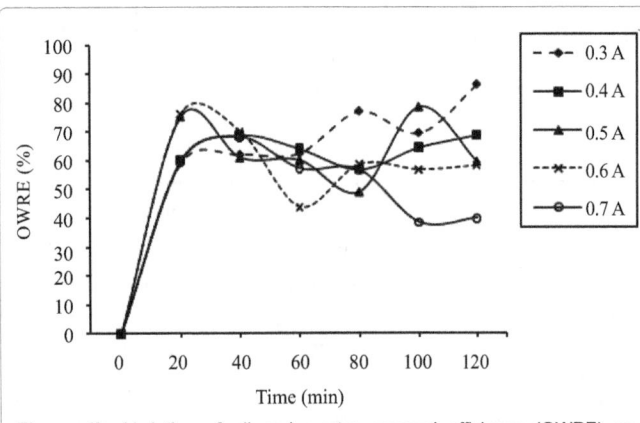

Figure 4b: Variation of oil and wastes removal efficiency (OWRE) vs. processing time for different current intensities.

intensities. As shown, and from (OWRE) equation, the current intensities which give the lower turbidity must inevitably give the high efficiency. The high values recorded for the (OWRE) were: 89.95 and 86.6% respectively for I = 0.65 A after 40 min and I = 0.3 A after 120 min. So from Figures 3 and 4, we concluded that the best current intensities recorded were 0.65 A and 0.3 A.

Effect of solution pH

In this experiment the wastewater (N°2) used has 157 NTU of initial turbidity and pH= 5.98.

During this second experiment, and for choosing the better current intensity between 0.65 and 0.3 A, we changed the solution pH by adding sulfuric acid to obtain acidic solution and by adding sodium hydroxide to obtain basic solution. For each pH condition, we have measured the turbidity and calculated the (OWRE) for the two intensities cited.

pH range (4-6): Figure 5 displays the measured turbidity according to processing time in medium acidic solution (pH range: 4-6) for the two chosen intensities (0.3 and 0.65 A). The initial turbidity (157 NTU) decreased for the both current intensities to reach 59 and 34 NTU respectively for 0.3 and 0.65 A after 120 min. Between 60 and 100 min, applying 0.65 A of current intensity, increased the turbidity slightly from 67 to 103 NTU, this was due to the dissolution of Aluminum in this acidic environment.

Figure 6 presents the oil and wastes removal efficiency variation vs. processing time for the two chosen intensities for the same pH solution range (4-6). The maximum (OWRE) obtained were 78.34% and 62.42% respectively for 0.65 and 0.3 A, that it was observed at the end of experiment (120 min).

pH range (6-9): For a neutral or slightly basic environment, it is clear from Figure 7 that the turbidity decreased all of a sudden from 157 to 45 NTU for the current intensity of 0.3 A only after 20 min, but increased again between 40 and 100 min to decrease again at the end of experiment reaching its lower value (35 NTU).

On the contrary, and for the current intensity of 0.65 A, the

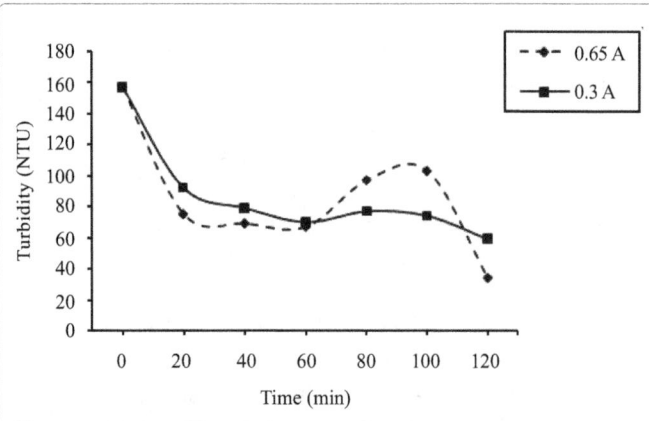

Figure 5: Variation of the turbidity vs. time for the two current intensities (0.65 and 0.3A) - Solution pH (4–6).

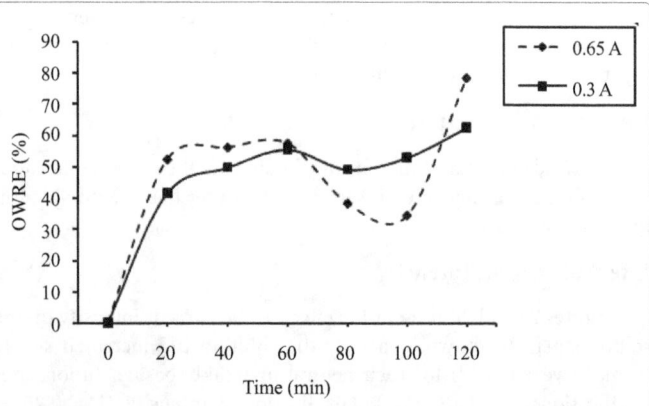

Figure 6: Variation of the OWRE vs. processing time for the two current intensities (0.65 and 0.3A) - Solution pH (4–6).

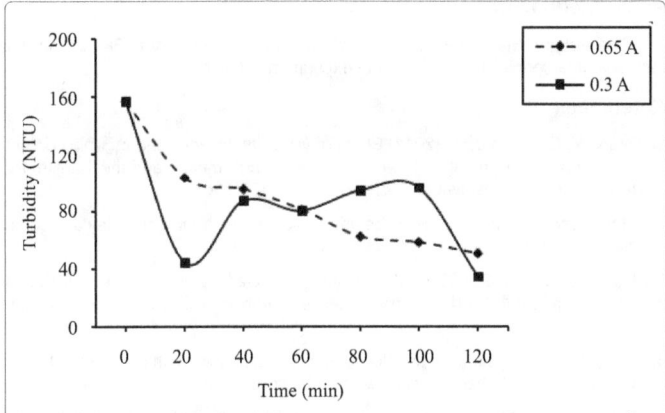

Figure 7: Variation of the turbidity vs. time for the two current intensities (0.65 and 0.3A) - Solution pH (6–9).

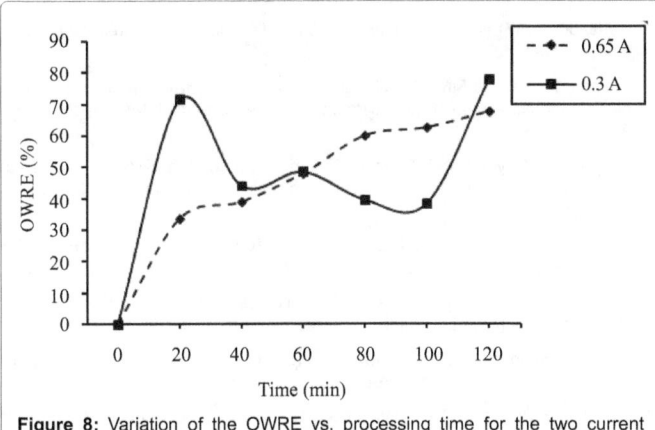

Figure 8: Variation of the OWRE vs. processing time for the two current intensities (0.65 and 0.3A) - Solution pH (6-9).

turbidity decreased in a regular manner throughout the experiment until 51 NTU.

Figure 8 illustrates the variation of oil and wastes removal efficiency vs. processing time for the two chosen intensities for a neutral or slightly basic environment. In these environment conditions, it is clear that the current intensity of 0.3 A has given the best (OWRE). So, after 20 min of treatment time, the latter reached 71.33% and decreased again, finally it reached 77.7% after clarification time (120 min). On the contrary, the (OWRE) calculated for the current intensity of 0.65 A after clarification time (120 min) was only 67.51%.

pH range (9-11): In the absence of any production of turbidity due to dissolution of Aluminum in acidic solution, it can be seen from Figure 9 that the turbidity decreased almost regularly for the both of current intensities. The lower turbidity value recorded for the both intensities was 32 NTU, after 80 min and 100 min respectively for 0.65 and 0.3 A of current intensity. At the end of treatment time, the turbidities recorded for the two intensities were not interesting.

Figure 10 further clarifies the Figure 9; the best (OWRE) calculated were after 80 and 100 min of processing time respectively for 0.65 and 0.3 A of current intensity. Obviously and respectively, 79.62 and 79.61% of oil and wastes removal efficiency has been calculated.

Vegetation Tests

In aim to see the impact and the behavior of the treated water

towards plants, daily irrigation tests have been conducted in a little private garden on two types of plants (shaft apocalyptic and date palm) for 13 months. The tests showed that the layer of 5 cm thickness and 0.08 mm of average particles diameter of dune sand removes most of remaining oil and impurities. The sand layer that fills the basin surrounding the shaft is removed and replaced every 03 months. So, we can deduce that fine dune sand plays the role of natural filter. After 13 months of control, garden plants appear and grow normally.

Finally, we suggest the filtration of wastewaters through fine particles of dune sand primary to any chemical treatment. Encouraging trials are in progress in this axis, as well as biological and mineral tests on the eventual remaining heavy metals and hydrocarbons contained in plants.

Conclusion

In order to remove the emulsified oil and wastes from wastewater arising out of petroleum industries, a treatment by electro-coagulation-flotation using Aluminum electrodes with a voltage of 12 V and a contact time of 120 min was realized and gave good results. It allowed the reduction of:

a) 78.34% of oil and wastes using 0.65 A of current intensity in an environment of solution pH (4-6) after 120 min of clarification time.

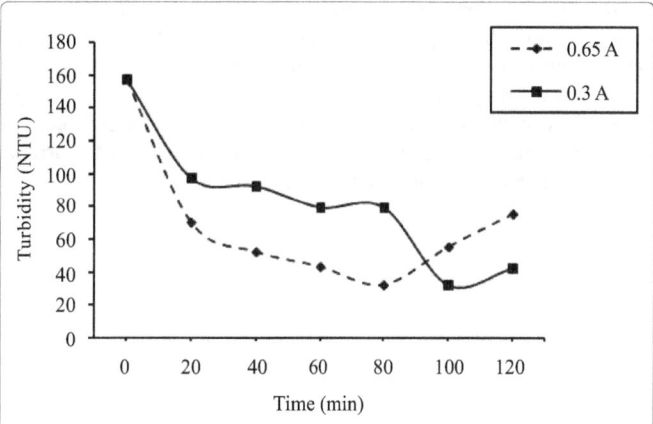

Figure 9: Variation of the turbidity vs. time for the two current intensities (0.65 and 0.3A) - Solution pH (9-11).

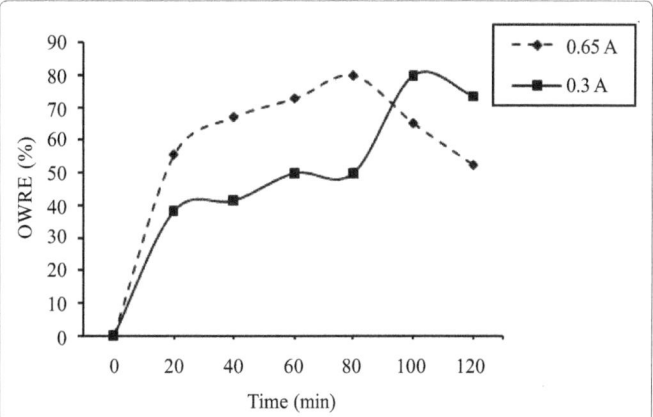

Figure 10: Variation of the OWRE vs. processing time for the two current intensities - Solution pH (9-11).

b) 77.70% of oil and wastes using 0.3 A of current intensity in an environment of solution pH (6-9) after 120 min of clarification time.

c) 79.62% of oil and wastes using 0.65 A of current intensity in an environment of solution pH (9-11) after 80 min of clarification time.

d) 79.61% of oil and wastes using 0.3 A of current intensity in an environment of solution pH (9-11) after 100 min of clarification time.

The four results are close in turbidity term, but are different in term of clarification time and solution pH. The last two seem to be the best results; their Oil and Wastes Removal Efficiencies are almost equal; nevertheless, they require an economic discussion: either save time or gain electrical energy. To save electrical energy and lose 20 min of clarification time implies the use of 0.3 A of current intensity in medium basic environment. A simple observation shows that the use of 0.65 A instead of 0.3 A of current intensity leads to excessive consumption of electrical energy despite the gain of 20 min of processing time.

In aim to see the impact of the treated water on plants, daily irrigation tests were conducted within a little private garden on two types of plants (date palm and the shaft apocalyptic) for more than a year. The tests showed that the layer of about 5 cm thickness and 0.08 mm of particles diameter of fine dune sand removes most of the remaining oil and impurities. The sand layer that fills the basin surrounding the shaft is sometimes removed and replaced by new dune sand four times per year. After 13 months of control, garden plants appear and grow normally. Thus, we suggest the filtration of wastewaters through dune sand primary to any chemical treatment because of encouraging trials recorded in this axis; as well as biological and mineral tests on the probable remaining traces of hydrocarbons and heavy metals contained within plants.

Acknowledgment

The authors thank the personnel of Sonatrach Laboratory /HBK/Algeria, for their valuable cooperation in doing the experimental tests.

References

1. Wang Y, Gao B, Yue Q (2011) Effect of viscosity, basicity and organic content of composite flocculants on the discoloration performance and mechanism for reactive dyeing wastewater. J EnvironSci 23: 1626-1633.

2. Montgomery JM (1985) Principles and design water treatment. Wiley J & Sons, New York p: 116.

3. Hughes MA (1990) Coagulation and flocculation, Part-1, in solid-liquid separation, (3rdedn). L. Svarosky, (ed.). Butterworth & Co (Publishers) Ltd. p: 74.

4. Gregory J (1993) Stability and flocculation of suspensions, in Proc. Solid-Liquid Dispersions. P. Ayazi Shamlou (ed.) Butterworth-Heinemann Ltd. - Ch. 3.

5. Bratby J (2006) Coagulation and flocculation in water and wastewater treatment. (2ndedn) IWA Publishing, London. pp: 50-68.

6. Hassen Sellami M (2015) Treatment and reuse of wastewaters discharged by petroleum industries (HMD/Algeria). Interna J waste res 6: 1-6.

7. Cansado (1997) Water treatment by electro-flocculation. Montréal-University-Quebec, Canada.

8. Valipour M, Seyyed MM, Reza V, Ehsan R (2012) Air, water, and soil pollution study in industrial units using environmental flow diagram. J Basic Appl Sci Res 2: 12365-12372.

9. Valipour M, Seyyed MM, Reza V, Ehsan R (2013) Deal with environmental challenges in civil and energy engineering projects using a new technology. J Civ & Enviro Eng 3: 1-6.

10. Valipour M, Seyyed MM, Reza V, Ehsan R (2013) A new approach for environmental crises and its solutions by computer modeling. Civilica 3: 1-6.

11. Ammour F (2008) Use of dune sands as a bio-filter for the wastewater treatment in Ouargla City (Algeria). Public knowledge project pp: 1-7.

12. Touil Y, Yamina G, Rachid I, Abdeltif A (2014) Biological filtration on sand of dunes filters fouling. Energy Procedia 50: 471-478.

13. Gezette (2006) Official Gazette of the Democratic and Popular Republic of Algeria N° 26.

8

Application of 3D Reservoir Modeling on Zao 21 Oil Block of Zilaitun Oil Field

Peprah Agyare Godwill* and Jackson Waburoko

China University of Geosciences, Wuhan, Hubei, China

Abstract

Reservoir modeling is an effective technique that assists in reservoir management as decisions concerning development and depletion of hydrocarbon reserves must be taken considering the uncertainties of the formation involved. The paper focuses on the use of Petrel Software to construct a 3D-dimensional reservoir model that characterizes and evaluate Zao 21 block reservoir located in Dagang Oil field zilaitun area in Hebei Province of China which has an oil bearing area of 0.9 km^2.

The approach is based on integration of data from seismic, well logs of 41 wells obtained from geology, geophysics, petrophysics, to characterize and provide an accurate description of the internal architecture and visualization of reservoir heterogeneity. These data are used to build the lithofacies, porosity, permeability and oil saturation model which are the parameters that describe the reservoir and provide information on effective evaluation of the need to develop the potential of the remaining oil in the reservoir. The lithological facies architecture is simulated using Sequential Indicator Simulation to guide the distribution of petro-physical properties of the reservoir since they are intimately related. In addition, the petro-physical parameters are simulated using Sequential Gaussian Simulation. The reservoir structural model shows system of different oriented faults which divided the model into two segments, the major and minor segments.

Statistical analysis of the Porosity model, permeability model for Zao 21 block showed that porosity is mainly concentrated between 12.5% to 22.5% with an average porosity of 15.5%; and permeability mainly between 40 mD~ 110 mD, with a mean permeability of 81 mD; overall good reservoir properties. The estimation of these values was used to quantify the geological reserve of Zao 21 reservoir block oil deposit.

This study has shown the effectiveness of 3D reservoir modeling technology as a tool for adequate understanding of the spatial distribution of petro-physical properties and in addition framework for future performance and production behavior of Zao 21 block reservoir.

The reservoir model reveals that the reservoir properties of the north-eastern part of the oil field are very promising and wells should be drilled to investigate and exploit the oil.

Keywords: 3D reservoir modeling; Characterize; Heterogeneity; Porosity; Permeability; Oil saturation

Introduction

The demand for oil product has placed huge effort on the search for oil with development in Technology to assess the certainty of hydrocarbon, reducing the risk associated with hydrocarbon. Many Government of oil producing countries rely on money generated from oil and their products. It is essential to model the reservoir as accurately as possible in other to calculate the reserves and to determine the most effective way of recovering as much of the petroleum economically as possible. In addition it enables for 3D visualization of the subsurface, which improves understanding of reservoir heterogeneity and helps to enhance oil recovery. To build this model an understanding of the data integrity was done as well as the reservoir with its host rock. In other to drill to the target, 3D seismic data interpretation and well data were used to build a 3D reservoir model that would make the data more reliable [1-5].

Geological location

The Zao 21 Block is located on Zilaitun area of Dagang oil field, Hebei Province Huanghua Depression area which belongs to a tectonic unit of Bohai Bay basin in the east of China which is the biggest depressed area. It has an oil-bearing area 0.91 km^2. The block reservoir is controlled by some fault systems trending NE and NS and

characterized by lithology mainly fine sand, siltstone and some shale content [6-9], (Figure 1).

Sedimentary facies

Zao 21 block store group a for shallow water delta deposits, work area is mainly delta front subfacies. Delta front mainly developed underwater distributary channel, distributary bay, mouth bar sedimentary microfacies [9-12].

The space is little, the study area well pattern density (57) 1.6 km^2 area of well spacing, thickness, single well facies classification larger workload, underwater distributary channel microfacies, reservoir area development so choose sand mudstone facies model for sand control constraint modeling instead of a phased modeling of sedimentary facies [12-18].

---

*Corresponding author: Agyare Peprah, China University of Geosciences, China
E-mail: knownpeps@gmail.com

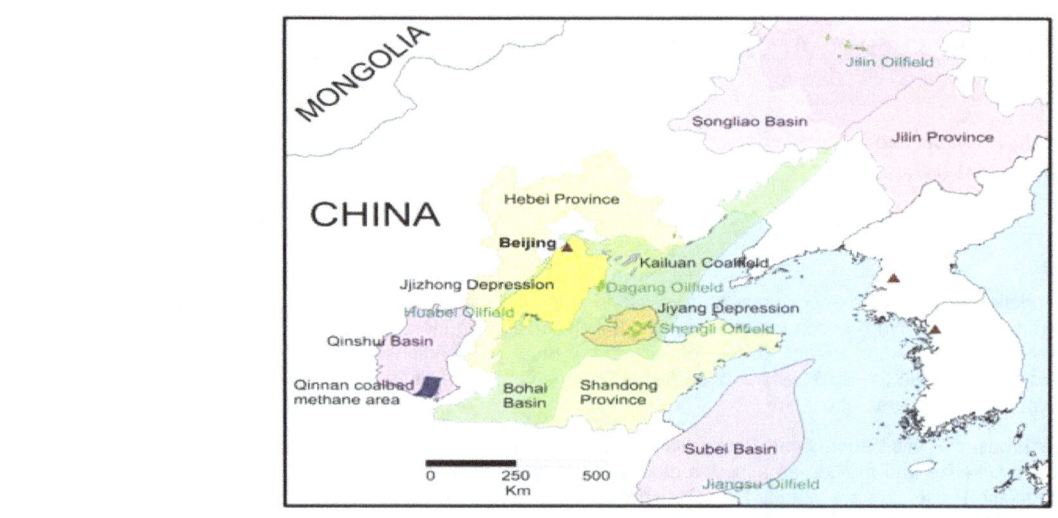

Figure 1: Location map of the study area.

The Modeling Approach

Reservoir model

In view of the necessity of dynamic simulation process and to arrive at a final well and production behavior, it was necessary to build a reservoir model that represented as closely as possible the sub-surface reality of Zao 21 block that have been encountered by most wells. The model of Zao 21 for the entire Zao 21 block in Dagang formation was built by integrating relevant sub-surface data and interpretation presented in the preceding sections. The seismic structural interpretation, lithological descriptions and facies interpretation, porosity, permeability and initial oil saturation from log analysis were used to build the reservoir model. The PETREL (Version 2009.1) suite was used in building the reservoir model. The structural and property model of the reservoir are briefly described as follows [19-22].

Structural Model

The structural model was based on seismic interpretation data. The input data consist of fault polygons and fault surfaces of interpreted faults. Fault modeling was the first step for building structural models with Petrel workflow tools (Figure 2) [23-28]. The process was used to create structurally and geometrically corrected fault interpretation within the horizon. The faults divided the model into 2 segments. Pillar gridding is a way of storing XYZ location to describe a surface which was used to generate a 3-D framework. A 3D-grid divided the space up into cells within which it assumed materials were essentially the same. The next step is Make Zones process in defining the vertical resolution of the 3D grid. The process creates zones between each horizon. The areal dimension of the grid cells was optimized at 50 × 50 m considering the reservoir description in Zao 21 prospect. The 3D reservoir model contained 22230 cells. Figures 3-19 shows the structural model of Zao 21 block [29-35].

Property modeling

Facies modeling: Facies modeling is an important aspect of the modeling process. The purpose is to simulate the sand bodies in the formation. The oil field is composed of mainly fine sandstone, silt stone and some amount of shale content [36-41]. The lithofacies were defined and calculated. The method of most of was used to average

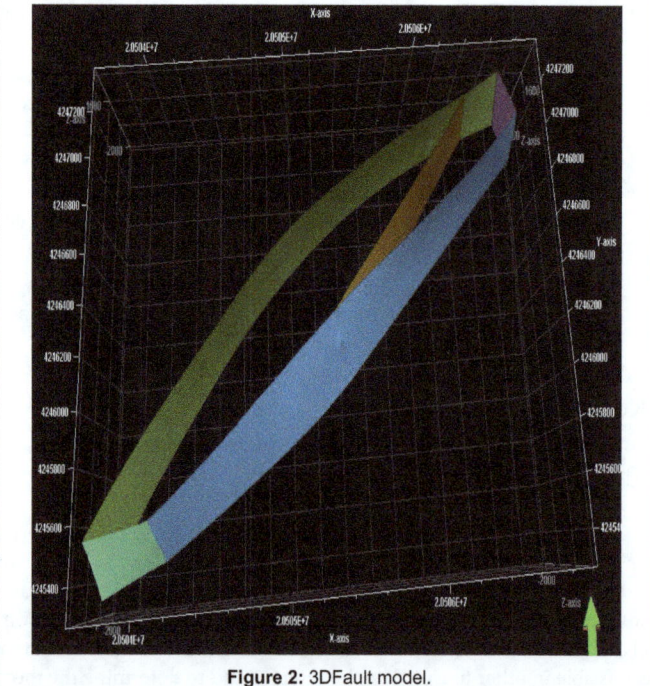

Figure 2: 3DFault model.

the facies. Sequential Indicator Simulation method was applied to simulate the sand bodies in the formation. The proportion of fine sand, silt and clay were 51.42%, 19.60% and 28.98%. This is stochastic method that combines variograms and target volume fractions. It is most appropriate with minimal well data, when either the shape of particular facies bodies is uncertain. It also allows easy modeling of facies environment where facies volume proportion vary vertically, laterally or both. Figure 6 shows the Lithofacies model.

Petrophysical modeling

Porosity model: Porosity is an essential property of an oil reservoir that determines the capacity of oil it can contain. The porosity model is based on porosity logs generated from the petrophysical interpretation of 41 wells. The well logs were scaled up using the method of arithmetic

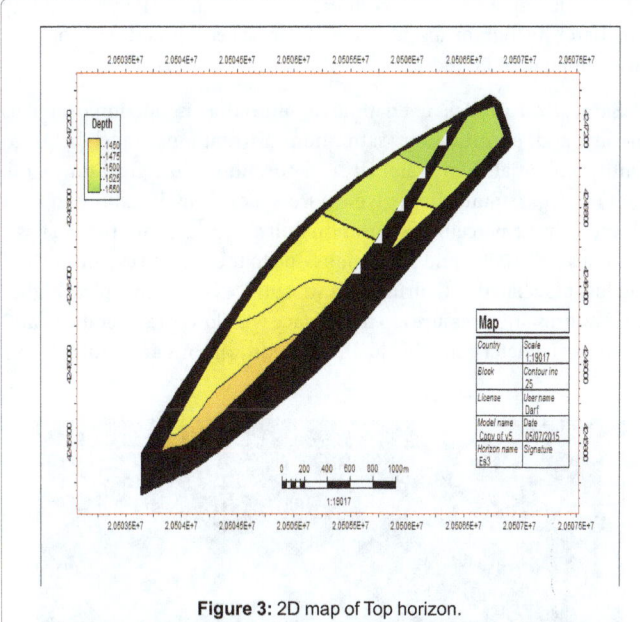

Figure 3: 2D map of Top horizon.

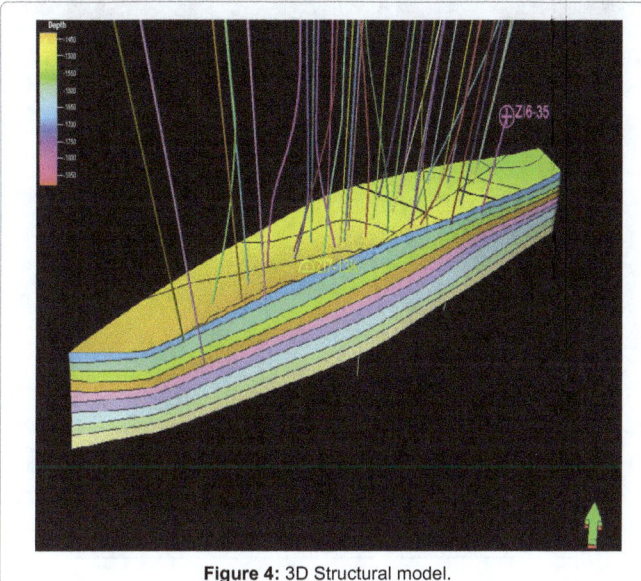

Figure 4: 3D Structural model.

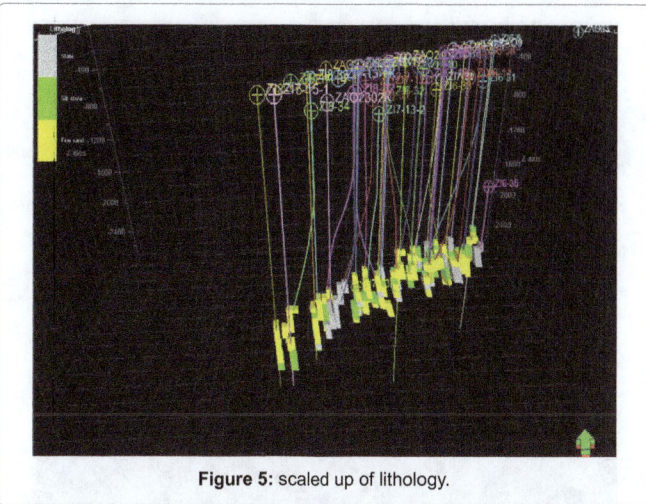

Figure 5: scaled up of lithology.

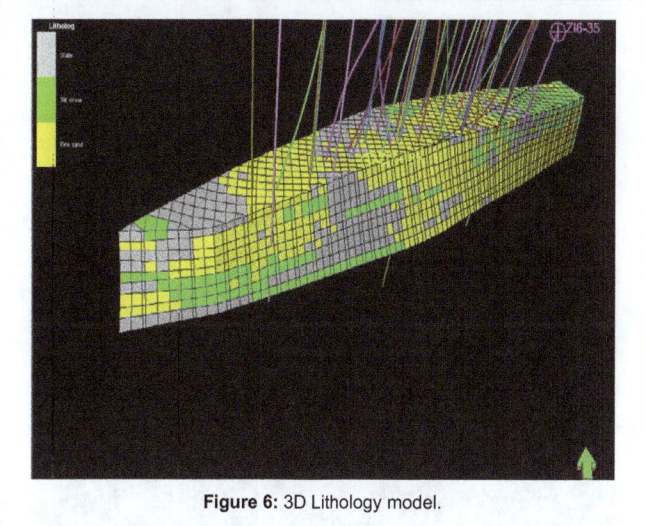

Figure 6: 3D Lithology model.

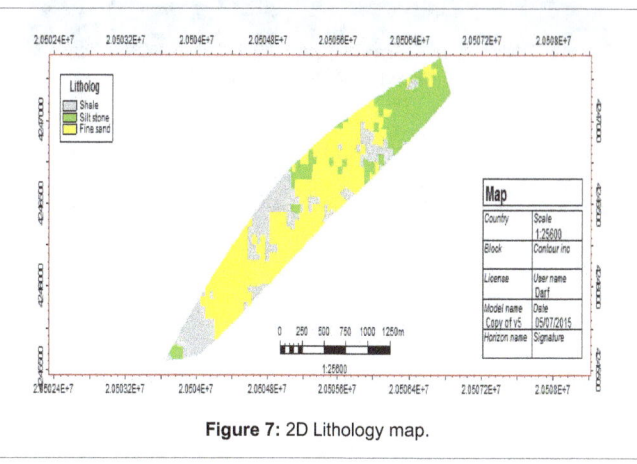

Figure 7: 2D Lithology map.

averaging. The porosity was distributed in the model using Sequential Gaussian simulation method. Data analysis tool provided variogram calculation and data transformation. This attribute is controlled by the distribution of Lithofacies in the reservoir. The porosity distribution is mainly concentrated between 12.5%-22.5% with an average porosity of 15.5%. Figures 8-10 show the scaled up porosity logs, porosity model and histogram showing the distribution.

Permeability model: Permeability is an essential characteristic of a Petroleum reservoir rock. It is a property of the porous medium that measures the capacity and ability of the formation to transmit fluids. The rock permeability is very important rock property because it controls the directional movement and the flow rate of the reservoir fluids in the formation [42]. The well logs were scaled up using harmonic averaging. Sequential Gaussian Simulation method was used. As a result of the relationship between permeability and porosity,

in the permeability modeling process, porosity was used as secondary variable. The method of collocated co-kriging was used which provides additional control parameter; the correlation coefficient between the primary and secondary variable. The permeability model shows that permeability of Zao 21 block is mainly concentrated between 40 mD~

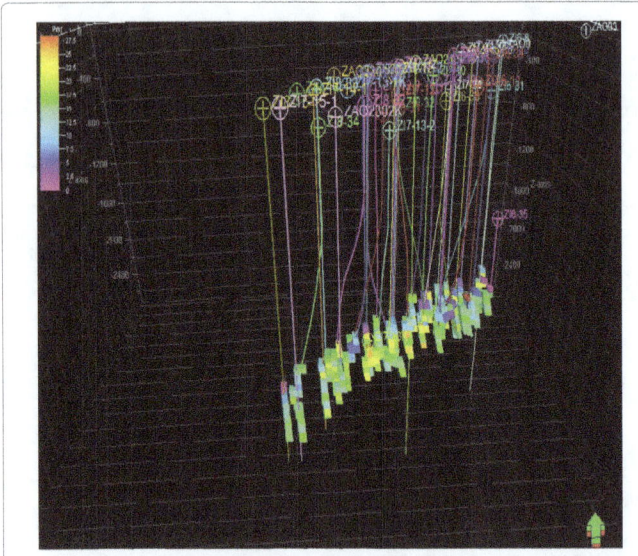

Figure 8: scaled up porosity log.

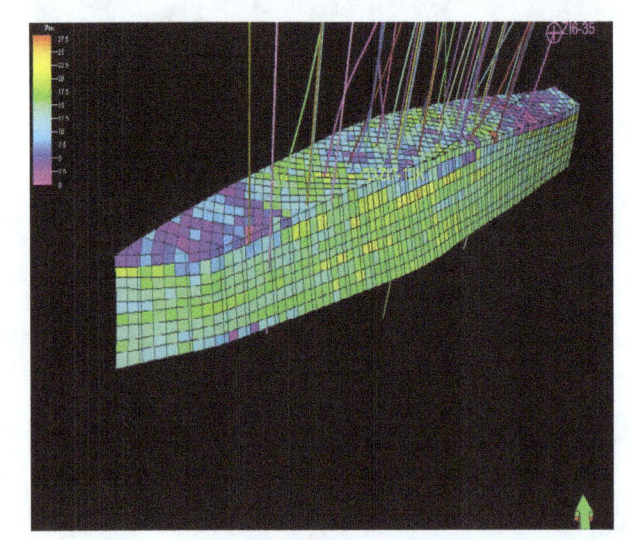

Figure 9: 3D Porosity model.

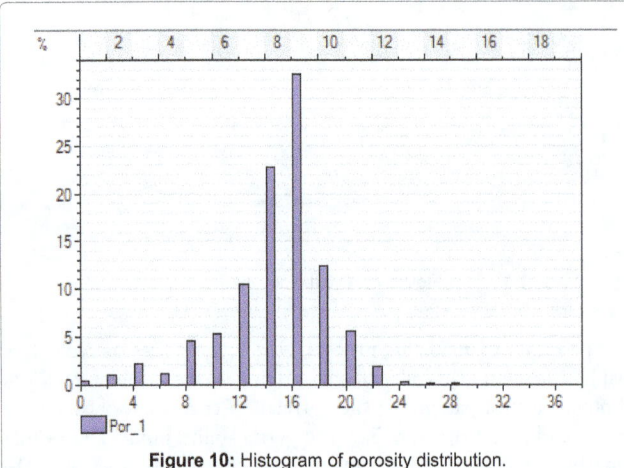

Figure 10: Histogram of porosity distribution.

110 mD, having an average permeability of 81 mD. The permeability model and histogram distribution of the scaled up and well logs are shown in Figures 11 and 12.

Saturation model: Even though saturation is not important as porosity and permeability, saturation distribution model helps to identify potential high water area. Saturation is the fraction of oil, water, and gas found in a given pore space. This is expressed as a volume/volume percent of saturation units. Typical saturation analysis does not show 100% fluid saturations due to the volume expansion and fluid loss associated with bringing a subsurface core with typical higher temperatures and pressures to the surface with lower temperatures and pressures. To determine the quality of hydrocarbons accumulated in a

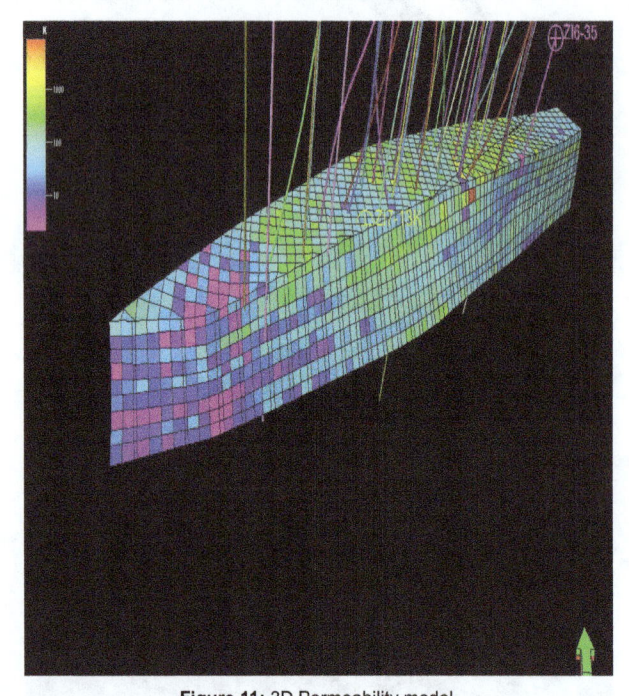

Figure 11: 3D Permeability model.

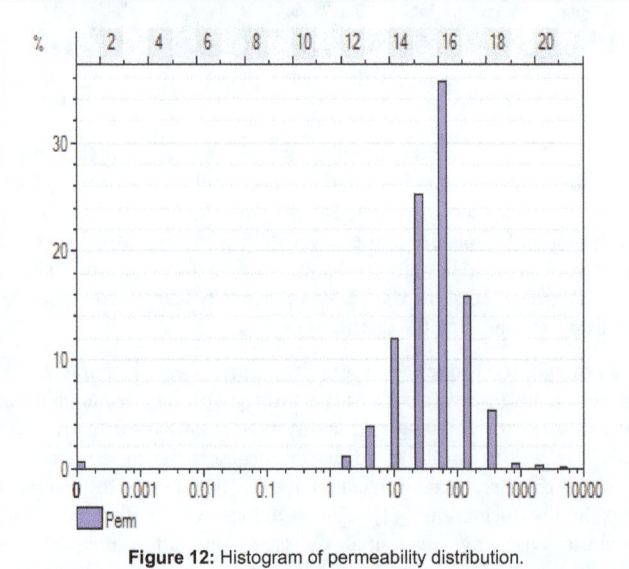

Figure 12: Histogram of permeability distribution.

porous rock formation, it's necessary to determine the fluid saturation (oil, water and gas) of the rock material. This study modeled the oil saturation of Zao 21 block which is shown in Figure 13. The method of arimethic averaging was used to scale up the well logs.

Porosity-permeability relationship of Zao 21 block: The amount of porosity is principally determined by shape and arrangement of sand grains and the amount of cementing material present, whereas permeability depends largely on size of the pore openings and the degree and type of cementation between sand grains. Although many

formations show a correlation between porosity and permeability, the several factors influencing these characteristics may differ widely in effect, producing rock having no correlation between porosity and permeability.

A cross plots of porosity-permeability of upscaled cells of Zao 21 block shows a strong correlation. The regression line is given by; $logK = 0.120\Phi - 0.989$.

Figure 15 shows a porosity-permeability cross plot for Zao 21 reservoir, where diagenetic effect is minimal at the reservoir depth resulting in some degree of heterogeneity. A correlation analysis between these two petrophysical parameters was done using a single cross-plot. Correlation coefficient for this method is 0.844.

Upscaling

High resolution Reservoir description models cannot be used directly to perform reservoir simulation study due to limitation of computer memory and speed. It is necessary to scale the high resolution reservoir description model to the coarser resolution of the production simulation. The result preserves representative simulation behavior. A grid dimension of 100×100 was used. Figures 16 and 17 shows the up scaled models.

Reservoir Volumetric

Reservoir volumetric is the process by which the quantity of hydrocarbon in a reservoir is estimated. This is very important because the exploration and development. After the reservoir model of Zao 21 was done (Table 1), the structural model and petro physical model built were used to calculate the reserves in terms of stock tank oil originally in place (STOIIP). Zao 21 block were estimated using the equation below.

$$STOIIP = 7758 \times A \times h \times \varnothing \times (1 - Sw) \times 1 / Bo$$

However, according to the 3D reservoir model, the following parameters were also calculated;

- Formation volume

- Hydrocarbon reservoir pore volume

- The volume of oil reserves

Results and Discussion

Interpretation and results

A. Structural model of Zao 21 block

Figure 4 indicates the system of different oriented growth faults with two major faults trending towards NE and NS. The other faults are categorized as minor faults. This model further buttresses the information gathered from the depth structural map.

B. Porosity model

A 3D perspective view of the porosity model is shown in Figure 9.

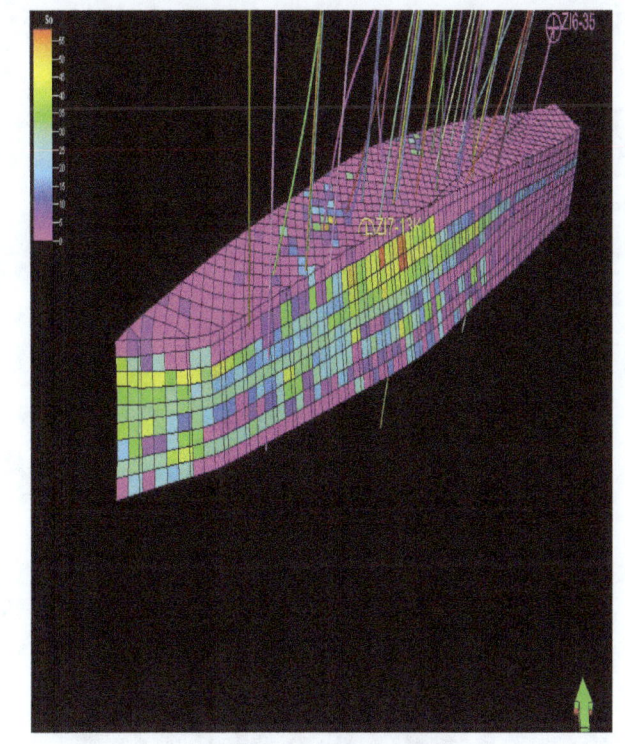

Figure 13: 3D Oil saturation model.

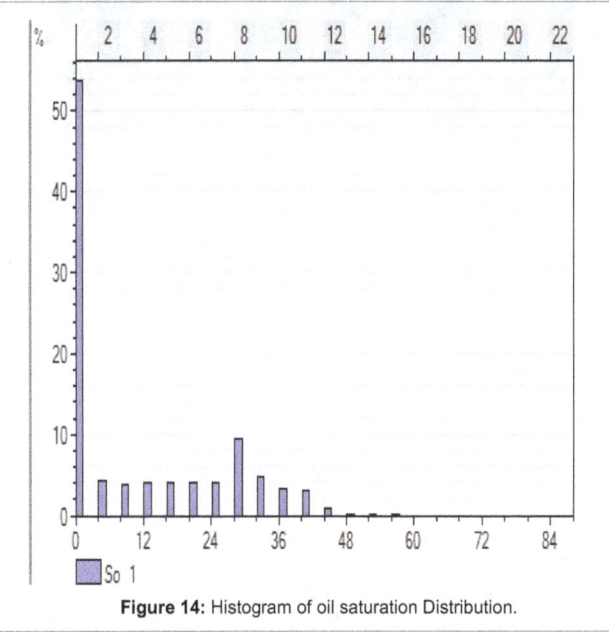

Figure 14: Histogram of oil saturation Distribution.

Fluid Properties	Parameters
Density of ground water	1.0 g/cm³
Oil density at reservoir condition	0.949 g/cm³
Crude oil formation volume factor at reservoir condition	1.114
Oil viscosity at reservoir condition	45.6 mPa.s

Table 1: Zao 21 Reservoir fluid properties parameters.

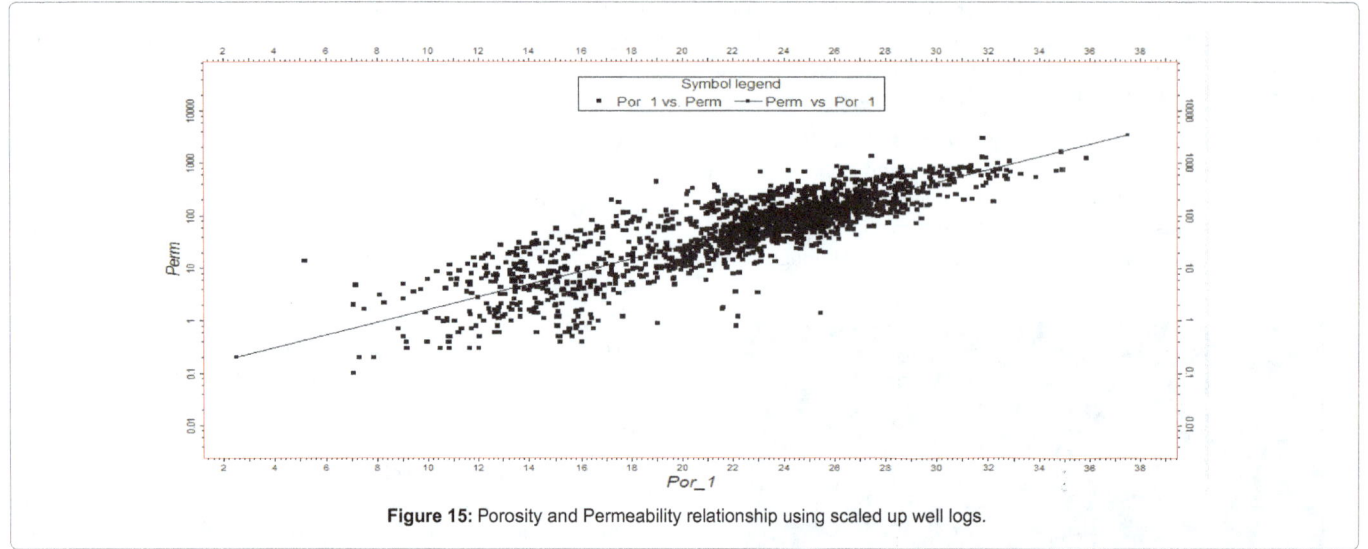

Figure 15: Porosity and Permeability relationship using scaled up well logs.

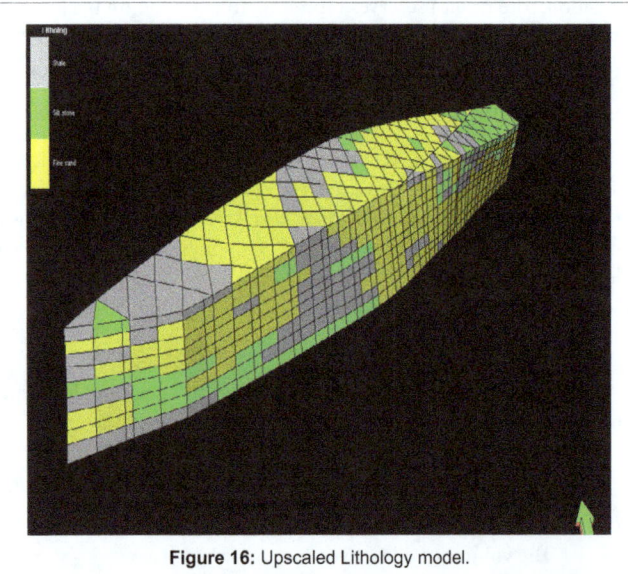

Figure 16: Upscaled Lithology model.

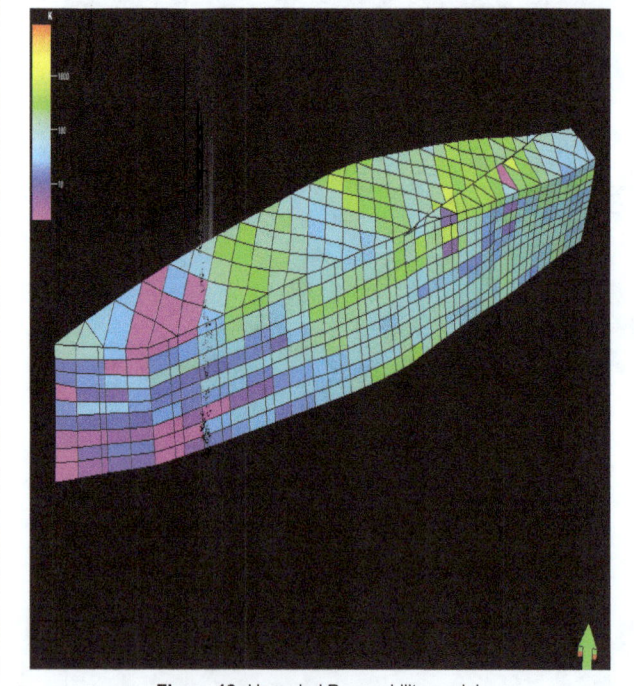

Figure 18: Upscaled Permeability model.

The formation porosity map shows the prominence of good porosity distribution which is mainly concentrated between 12.5%-22.5% of Zao 21 block. The central portion shows high porosity distribution. This indicates that the pore spaces have enough space to accommodate fluid. However, on a whole the average porosity value is 15.5% and according to Levorsen, it is a good reservoir rock as shown in Table 2.

A.   Permeability model

Figure 11 shows a 3D perspective view of the permeability model. The map underscores a permeability concentrated mainly between 40 mD to 110 mD shown in Figure 12 within the well areas of Zao 21 block with an average permeability of 81 mD within Zao 21 oil field. The value is a reflective of good connectivity of pore spaces of sand and their ability to transmit fluids. According to Levosen, it is a good

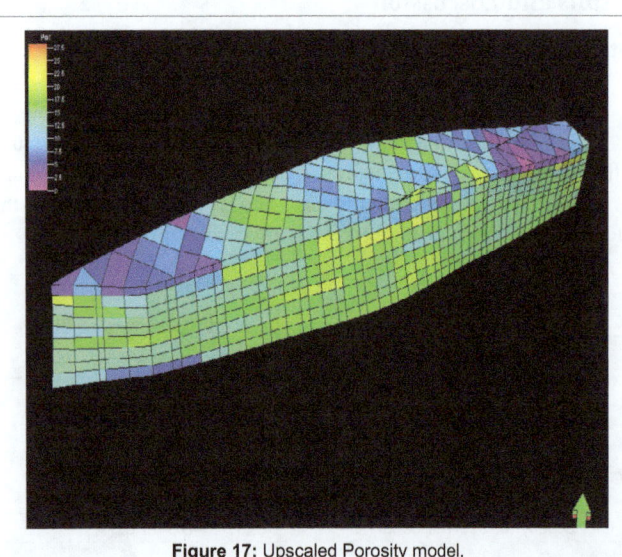
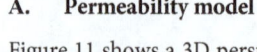

Figure 17: Upscaled Porosity model.

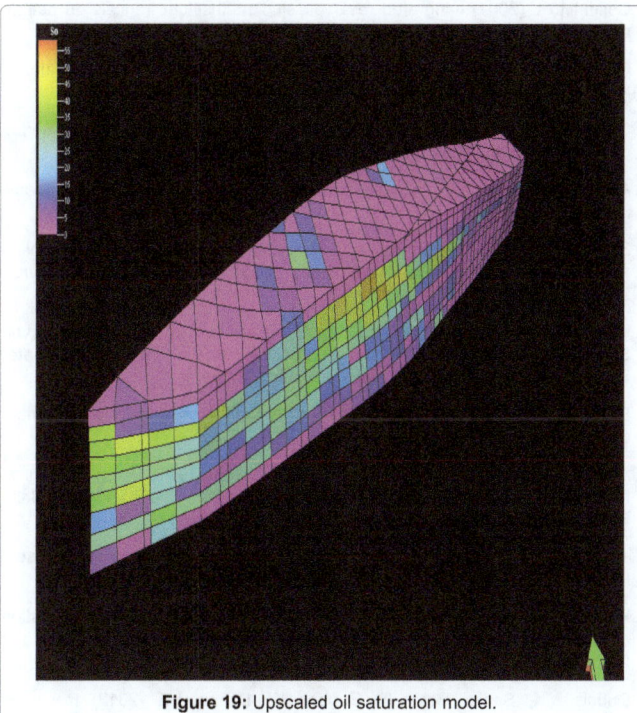

Figure 19: Upscaled oil saturation model.

Zones	Bulk Volume [x10⁶ m³]	STOIIP (in oil) [x10⁶ m³]
Zone 1	54.893112	0.646922
Zone 2	56.692663	0.668130
Zone 3	57.793266	0.681101
Zone 4	61.500136	0.724787
Zone 5	56.910949	0.670703
Zone 6	66.492047	0.783617
Zone 7	88.441250	1.042291
Zone 8	68.723252	0.809912
Zone 9	78.791877	0.928572

Table 2: Volumetric calculation results.

reservoir rock as shown in Table 3.

B. Oil Saturation model

Figures 3-8 shows a 3D perspective view of the oil saturation. The model reveals that the oil saturation distribution on Zao 21 block is highest (yellow, green to orange) at the central part (0.3-0.45) of the oil field than at the north eastern and south western. However, relatively, the north-eastern part shows better oil saturation than south-western part of the reservoir model.

C. Lithofacies model

A 3D perspective view of the facies model is shown in Figure 6. The model shows that the dominant face at the south-western part is dominated by fine sand. The central part shows a mixture of three facies, silt sand, fine sand and small proportion of shale while the north-eastern part shows good proportion of sand and clay with a small fraction of silt sand. The south-western part shows a good distribution of sand, however, the oil saturation is relatively low compared to the central part. This may be attributed to the diagentic factors in the formation.

Porosity (Φe, %)	Qualitative Evaluation
0-5	Negligible
5-10	Poor reservoir rock
10-15	Fair reservoir rock (general)
15-20	Good reservoir rock
20-25	Very good reservoir rock

Table 3: A Qualitative evaluation of Porosity (Levosen, 1967).

Permeability (mD)	Qualitative Description
<10.5	Poor to fair
15-50	Moderate
50-250	Good
250-1000	Very good
>1000	Excellent

Table 4: A Qualitative Evaluation of permeability (Levosen, 1967).

D. Reservoir volumetric

Table 4 reveals volumetric after modeling. The table shows the bulk volume, and STOIIP at each of the 9 Zones. Zone 7 shows the largest Vb of 88.441250 × 10⁶ m³ and STOIIP of 1.042291 × 10⁶ m³ whereas Zone 1 shows the least Vb and STOIIP, 54.893112 × 10⁶ m³, 0.646922 × 10⁶ m³. The Zao 21 block reservoir zones indicate that hydrocarbon of commercial value thus; the reservoir model could be as input for simulation and performance.

- Porosity and permeability are two distinct properties of the reservoir rock. The correlation analysis between porosity-permeability relationships (Figure 15) resulted in a correlation coefficient of 0.844. This shows good correlation, however, not a perfect one implying that this two quantities are closely related. The fluids are able to permeate through the reservoir rock by passing through the pores it contains, and greater the number and size of pores in the reservoir, easier it is for the fluids to pass through. Thus a higher porosity in the reservoir is likely to be accompanied by higher permeability. However, diagenetic process such as compaction, clay minerals such as montmorrillonite, smectite, illite, etc exist in the formation. This means the effect of these factors at the depth where the reservoir exist is minimal, hence, has small effects on the reservoir quality [43,44].

- Stochastic modeling methods was used due to incomplete information about dimensions, internal (geometric) architecture, and rock-property variability on all scales; the complex spatial distribution of reservoir building blocks or facies; difficult-to-capture rock-property variability and variability structure with spatial position and direction; unknown relationship between property value and the volume of rock used for averaging (scale problem).

Conclusion

1. This work shows the versatility of integrating seismic and well log data for reservoir modeling. The results of the comprehensive petro-physical analysis of 41 wells show one dominant reservoir across the wells in the field at different depth intervals.

2. This Zao 21 reservoir is very promising because of its good porosity and permeability values. The discrete properties gave the knowledge of the facies properties in the field while the continuous properties gave petro-physical properties of the field in terms of porosity, permeability and oil saturation. The volumetric calculation indicates that the reservoir has a reserve of 660 × 10⁴t. This analysis will

serve as a control of the reservoir during development.

3. The reservoir model of Zao 21 block has provided a better understanding of the spatial distribution of the discrete and continuous properties in the field. The study has developed a geological model for Zao 21 block that can be updated as new data are acquired for field development. The model can be exported for simulation to be run.

4. The study area is a heavy oil reservoir; the petro-physical properties (porosity, permeability and oil saturation) which control the oil storage and movement were modeled.

5. The reservoir properties are controlled by two main faults regimes formed due to tectonic activities in the region.

6. With reference to the model results, porosity, permeability and saturation models of the study area showed promising porosity and permeability properties at the southwestern and northeastern part, however, the oil saturation at the latter part is greater than the former.

7. This study shows a highest porosity value of 28% an average of 15.5%, peak permeability of 550 mD an average of 81 mD and highest oil saturation of 0.55 an average of 11.5%.

References

1. Cheng W, Wang K (2005) Oil bearing logging comprehensive evaluation of the reservoir in Zilaitun Oil field. China.

2. Tarek A (2010) Reservoir Engineering Handbook, Fourth Edition. Oxford: Gulf Professional Publishing, Elsevier.

3. Loveren AI (1967) Geology of Petroleum, Second Edition. AAPG Foundation.

4. Sattar A, Jim B, Rich J (2000) Computer-Assisted Reservoir Management. PennWell Corporation, Tulsa, Oklahoma.

5. Satter A, Ghulam MI, James LB (2008) Practical Enhanced Reservoir Engineering: Assisted with Simulation Software. PennWell Corporation, Tulsa, Oklahoma.

6. Bratvold RB, Terald S, Laris H, Kelly T (1995) STORM: Integrated 3D Stochastic Reservoir Modeling Tool for Geologist and Reservoir Engineers. Society of Petroleum Engineers.

7. Haldosen HH, Damsleth E (1990) Stochastic modeling. Bergen, Norsk Hydro, JPT.

8. Pettijohn FJ (1984) Sedimentary Rocks, 3rd edition. CBS Publisher & Distributors New Delhi.

9. Sonnel N (1988) Properties of Oil Reservoir Rocks of Boyabat/ Sinop Basin Units. Commun Fac Sci Univ Ank Series C 6: 289-299.

10. Clark NJ Elements of Petroleum Reservoirs. Dallas, Texas.

11. Wu XH, Stone MT, Stern D P, Lyons SL (2007) Reservoir Modeling With Global Scale-Up. Society of Petroleum Engineers, Manama, Bahrain.

12. Lukumon A, Onyekachi N, Olatinsu O, Fatoba J, Bello M (2014) Static Reservoir Modeling Using Well Log and 3-D Seismic Data in a KN Field, Offshore Niger Delta, Nigeria. International Journal of Geosciences 5: 93-106.

13. Buryakovsky L, Eremenko NA, Gorfunkel MV, Chilingarian GV (2005) Geology and Geochemistry of Oil and Gas, 1st Edition. Elsevier.

14. Nguyen HH, Chan CW (2005) Application of data analysis techniques for oil production prediction. Engineering Applications of Artificial Intelligence 18: 549-558.

15. Dave Garner (2014) The future of oil supply. Phil Trans R Soc A 372: 2006.

16. The University of Texas, A Dictionary for Petroleum Industry, second edition.

17. Khaled F (2006) Predicting production Performance using a simplified model. World Oil, Saudi Arabia.

18. Hossain MH, Hossain MAI (2011) A study of commonly used conventional methods for gas reserve estimation. Journal of Chemical Engineering 26: 54-69.

19. Schlumbeger (2007) Petrel Introduction Software book.

20. Cosentino L (2001) Integrated reservoir studies. Institute francais du petrole publications.

21. Benetatos C, Viberti D (2010) Fully Integrated Hydrocarbon Reservoir Studies: Myth or Reality. American Journal of Applied Sciences 7: 1477-1486.

22. Slider HC (1983) World-wide practical petroleum reservoir engineering methods. Pennwell publishing Co, Tulsa Oklahoma.

23. Ahmed T (2003) Reservoir Engineering Hand book. Gulf Publishing Company, Houston, Texas, USA.

24. Samson P, Jean-Michel G, Robbe O, Vivien de F, Rossi T, et al. 3D Modeling and Reservoir Uncertainties: A Case Study, Elf Exploration Production, Chevron Petroleum Technology Company.

25. Ma E, Ryzhov S, Gheorghiu S, Hegazy O, Banagale M, et al. (2014) Reservoir Simulation modeling of the World's Largest Clastic Oil Field - The Greater Burgan Field, Kuwait. International Petroleum Technology Conference.

26. Dan C (2014) Use of Reservoir Models and Dynamic Simulation in Development of Mississippian. AAPG, Oklahoma City, USA.

27. Ammer RJ, James CM, Thomas HM, George JK (1995) Using Geological Modeling and Reservoir Simulation to Increase Gas Storage Efficiency: A Case Study. Society of Petroleum Engineering, Morgantown, West Virginia.

28. Yan YS, Ma T, Wang TC, Xu ZK (2012) Difficulties and Strategies of Integrated Reservoir Studies. Copenhagen, SPE, EAGE.

29. Rainer T, Steven B, Robert R, Wierzbicki (2004) Deep Panuke: The integration of Geology, Geophysics and Reservoir Engineering for Field Appraisal. EnCana Corporation, CSEG National Convention, Canada.

30. Dubois MK, Senior Peter R, Eugene W, Dennis HE (2012) Reservoir Characterization and Modeling of a Chester Incised Valley Fill Reservoir. Pleasant Prairie South Field, Oklahoma City.

31. Fremming NP (2002) 3D Geological Model Construction Using a 3D Grid. Technoguide, 8th European Conference on the Mathematics of Oil Recovery, Oslo, Norway.

32. Peters EJ. Petrophysics. Austin, University of Texas, USA.

33. Schlumberger (2007) Petrel Introduction Course. Houston.

34. DeSorcy GJ (1979) PD12 (2) Estimation Methods for Proved Recoverable Reserves of Oil and Gas. World Petroleum Congress, Bucharest, Romania.

35. Trice ML, Dawe BA (1992) Reservoir Management Practices. JPT, Society of petroleum engineers 44: 1296-1305.

36. Abdus S, James EV, Hoang MUU T (1994) Integrated Reservoir Management. Texaco Inc, USA.

37. Christie MA (1996) Upscaling for reservoir simulation. JPT 48: 1004-1010.

38. Sacchi Q, Rocca V (2010) Gridding Guidelines for Improved Embedding of a Petrel Reservoir model into Visage. SIS 2010 Global forum, London, UK.

39. Mattax CC, Dalton (1990) Reservoir Simulation. In: Henry L, Doherty Memorial fund of AIME (Edn.). Society of Petroleum Engineers, Technology and Engineering: 1-173.

40. Carlson MR (2003) Practical Reservoir Simulation: Using, Assessing and Developing Results. Penn Well Corporation, Tulsa, Oklahoma.

41. Chen H, Chunquan Li, Hongwei P (2012) Reservoir Diagenesis and Quality Prediction. China University of Geosciences Press, Wuhan, China.

42. Haldersen HH, Dasleth E (1993) Challenges in Reservoir Characterization: GEOHORIZONS. AAPG Bulletin 77: 541-551.

43. Sheikhzadeh H, Haghparast GH. 3D Integrated Static Modeling Using Geostatistical Methods in Asmari Reservoir, Marun Oil Field Iran. 14th International Oil, Gas and Petrochemical Congress, Tehran 1389: 5-30.

44. Dean L (2007) Reservoir Engineering for Geologist. Part 3-Volumetric Estimation 11: 20-23.

9

Carbon Capture and Storage (CCS) and its Impacts on Climate Change and Global Warming

Aramesh Shahbazi and Behnam Rezaei Nasab*

Faculty of Law and Political Sciences, Allameh Tabatabaei University, Tehran, Iran

Abstract

From the beginning of the Industrial Revolution time period, the gas exterior from burning of fossil fuels and extensive clearing of forests has contributed to a increase in the atmospheric concentration of carbon dioxide and recently it has been estimated that, if greenhouse gas emissions continue at the present rate, Earth's surface temperature could exceed historical values as early as possible, with almost harmful effects on ecosystems, biodiversity and the living conditions of people all over the world. Therefore Global climate is maybe the most challenging environmental problem the world will be facing in future. To decrease the growth of greenhouse gases and its consequences, a set of CO2-limiting policies will be needed. Carbon capture and storage (CCS) technology is one of the most important technologies around the world that is considered as one of the options for reducing CO2 gas and decreasing the global warming although, some aspects of using this technology, especially those of regulatory issues on this aspects should be more considered by States all around the world. In this article we will consider the impacts of applying CCS on the reduction of air pollution and global warming and also survey the Side effects of this technology in the context of international environmental law.

Keywords: CCS; Climate change; Global warming; Environment; Fossil fuels; International Environment Law

Introduction

The Earth's atmosphere is being changed at an unprecedented rate, primarily by humanity's ever-expanding energy consumption, and these changes represent a major threat to global health and security. Sound policies must be quickly developed and implemented to provide for the protection of the planet's atmosphere [1].

As the scientists report, Global warming is defined as an increase in the average temperature of the Earth's atmosphere, especially an increase great enough to cause changes in the global climate conditions. The term global warming is synonymous with Enhanced greenhouse effect, implying an increase in the amount of greenhouse gases in the earth's atmosphere, leading to entrapment of more and more solar radiations, and thus increasing the overall temperature of the earth. The heating situation of the earth in itself causes the life of humanity to be in danger. Our world is characterized by fast moving geopolitical and natural changes and the scenarios drawn by climate change specialists are alarming. If we want to avoid dangerous climate change and its ample consequences for creatures all over the world, it is necessary to take actions right now.

There is now scientific consensus that human activities, and in particular the way we transform and use fossil fuel energy, are responsible for increased CO_2 concentrations in the atmosphere and climate change.[1] It's good to mention that the global temperatures are higher than they have ever been during the past years, and the amounts of CO_2 gas in the atmosphere have excessed all previous records of reports. The important discovery from examining different periods throughout Earth's history is that notable results amplify any warming at the first levels[2] [2]. This is why climate has changed so dramatically

in the past. Positive feedbacks take any temperature changes and amplify them. These feedbacks are why our climate is so sensitive to greenhouse gases, all of which CO_2 gas is the most important driver of climate change[3] [3]. On this way carbon capture and storage (CCS) is expected to play a key role in climate change mitigation strategies. CCS could be reduced from power stations using fossil fuel, such as coal, and it has been noted that no new power plants should be built without CCS facilities around. Carbon dioxide (CO_2) capture and storage (CCS) is a process including the segregation of CO_2 gas from industrial and energy sources, transport to a storage position and long-term isolation. It's good to point that increase in the amount of carbon dioxide will cause loss to the future generations because of its harmful effects accordingly. In this paper we would examine the impacts of CCS technology as one of the newest processes in order to minimize the global warming effect and the future of this technology and the way which it works. At the beginning of this paper it should be mentioned that thus paper includes some new and up to date information on the relation between technology of CCS and human common heritage as a worldwide concept that is new in itself.

Global Warming and Environmental Side Effects

Global warming

Increasing global temperature would result in the increasing the sea levels and will change the amount and pattern of environment,

---

[1] CO$_2$ capture and storage projects, Directorate-General for Research Directorate Energy, p. 6, 2007

[2] Denman KL, Brasseur G, Chidthaisong A, Ciais P, Cox PM, et al. (2007) Couplings Between Changes in the Climate System and Biogeochemistry. In: Climate Change 2007: The Physical Science Basis. Contribution of Working Group I to the Fourth Assessment Report of the Intergovernmental Panel on Climate Change, Cambridge University Press, Cambridge, United Kingdom and New York, USA.

[3] Durwood Zaelke and James Cameron, "Global Warming and Climate Change- an Overview of the International Legal Process", AM. U.J. Int'l L. & Pol'y 5: 249.

***Corresponding author:** Behnam Rezaei Nasab, Faculty of Law and Political Sciences, Allameh Tabatabaei University, Tehran, Iran
E-mail: Behnamrlaw1990@gmail.com

including an expanse of the desert regions. Some other effects include increases in the intensity of extreme weather conditions, changes in agricultural productions, glacier retreat, species extinctions and increases in the ranges of disease.

The greenhouse gases that cause climate change includes as the following: Carbon Dioxide, Methane and Nitrous Oxide are among the most noticeable gases. Carbon dioxide emissions therefore are the most important cause of global warming. CO_2 is created by burning fuels like oil, natural gas, diesel, petrol, organic-petrol, and ethanol.

Coal is the most carbon intensive fossil fuel. For every ton of coal burned, approximately 2.5 tons of CO_2 are produced.[4] Of all the different types of fossil fuels, coal produces the most carbon dioxide. Because of this and its high rate of use, coal is the largest fossil fuel source of carbon dioxide emissions. Coal represents one-third of fossil fuels' share of world total primary energy supply but is responsible for 43% of carbon dioxide emissions from fossil fuel use.[5]

Consequently increasing amounts of man-made greenhouse gases lead to an increase in the temperature on Earth. This temperature increase causes other effects, one of them being the increase of the amount of water vapor in the atmosphere. Although human activities don't directly add significant amounts of water vapor to the atmosphere, warmer air can contain more vapors. Because water vapor is a greenhouse gas, global warming will be again increased by the amounts of water vapor (Figure 1).

This is called a positive feedback or positive result. In other words:

- The current global warming is caused by man-made greenhouse gases (mainly CO_2, NOx and Methane).
- Global warming leads to a higher temperature on Earth.
- Because of the higher temperature, the air does contain more water vapor.

---

[4] Defra UK (2014) The 2014 Government greenhouse gas conversion factors for company reporting. London: U.K. Department for Environment, Food & Rural Affairs.

[5] http://whatsyourimpact.org/greenhouse-gases/carbon-dioxide-emissions (visited 2016/06/20)

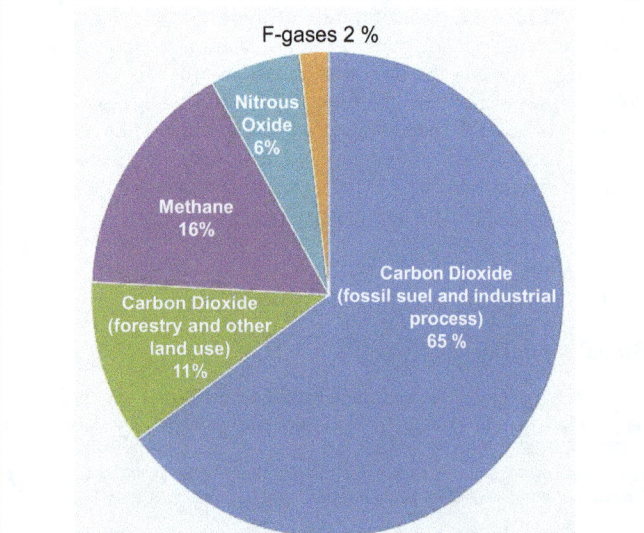

Figure 1a: Overview of greenhouse gases and sources of emissions. (Note: IPCC (2014) based on global emissions from 2010. Details about the sources included in these estimates can be found in the Contribution of Working Group III to the Fifth Assessment Report of the Intergovernmental Panel on Climate Change).

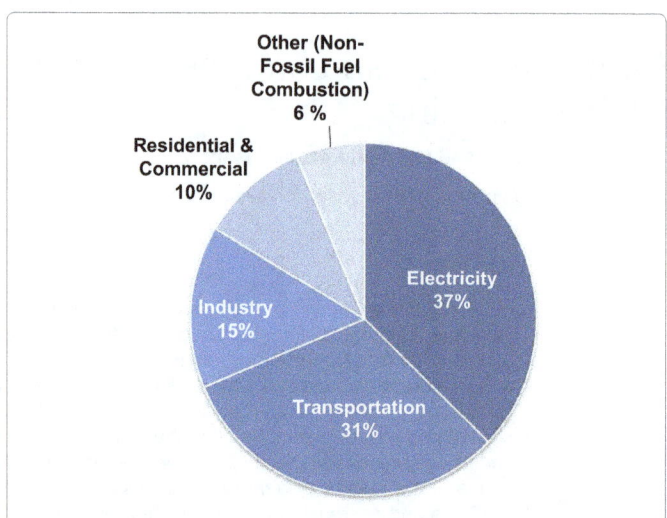

Figure 1b: U.S. Carbon dioxide emissions. (Note: All emission estimates from the Inventory of U.S. Green-house gas emissions and sinks: 1990-2014).

- This additional water vapor (a greenhouse gas) does again increase the effect of global warming (a positive feedback or secondary effect).

Global warming refers back to the recent increase in global average temperature near Earth's surface. It is caused almost by increasing concentrations of greenhouse gases in the atmosphere. Therefore increases in global temperatures have been accompanied by changes in weather conditions. Many places have been affected with changes in rainfall, resulting in more floods or droughts, as well as more severe heat waves. As such we could mention that Climate change refers back to any significant change in the measures of climate lasting for an increased times of year. In other words, climate change includes major changes in temperature or wind patterns that occur over several decades or longer and consequently climate change is expected to bring about major changes in freshwater availability, the productive capacity of soils, and patterns of human settlement. Accordingly climate change is intimately linked to human health either directly or indirectly. However, considerable uncertainties exist with regard to the extent and geographical distribution of these changes. Here undoubtedly it's good to note first that the agreed view on climate change and greenhouse gases is based on various lines of evidences. There includes basic physics, many different kinds of observations of both past and present climate conditions, and models that project future climate conditions.[6]

In the next part of this paper we'll discuss about the undesirable effects of climate changes and also global warming in the framework of environment international law.

The harmful effects of global warming for environment

The living conditions and desirable ways of creature lives depends upon the healthy and suitable environment and any manipulations and harmful attacks to the nature could lead to the undesirable conditions in living and accordingly would cause damages to the environment.

These damages to the environment are the direct causes of not suitable usages from energies available around us. This could lead

---

[6] IPCC, 2007: Climate Change 2007: The Physical Science Basis. Contribution of Working Group I
to the Fourth Assessment Report of the Intergovernmental Panel on Climate Change [Solomon, S., D. Qin, M. Manning, Z. Chen, M. Marquis, K.B. Averyt, M.Tignor and H.L. Miller (eds.)]. Cambridge University Press, Cambridge, United Kingdom and New York, NY, USA.

to the global warming and results in the climate changes. One of the most notable causes of this effect in the process of global warming is the excessive amounts of carbon dioxide in the atmosphere. Human activities are significantly increasing its concentration amounts in the atmosphere; consequently leading to Earth's global warming. Therefore there are both natural and human sources of carbon dioxide emissions in the atmosphere. Natural sources include decomposition, ocean release and respiration.

Human sources come from activities like cement production, deforestation as well as the burning of fossil fuels like coal, oil and natural gas. Due to human activities, the atmospheric concentration of carbon dioxide has been rising extensively since the Industrial Revolution and has now reached dangerous levels not seen in the last 3 million years. Human sources of carbon dioxide emissions are much smaller than natural emissions but they have upset the natural balance that existed for many thousands of years before the influence of humans. This is because natural sinks remove around the same quantity of carbon dioxide from the atmosphere than are produced by natural sources [4]. This had kept carbon dioxide levels balanced and in a safe range.

But human sources of emissions have upset the natural balance by adding extra carbon dioxide to the atmosphere without removing any. With reviewing the sources of CO_2 (as in the diagram) we would know that since the Industrial Revolution, human sources of carbon dioxide emissions have been growing. Human activities such as the burning of oil, coal and gas, as well as deforestation are the primary cause of the increased carbon dioxide concentrations in the atmosphere (Figure 2).

The important and also largest human source of carbon dioxide emissions is from the combustion of fossil fuels. This produces 87% of human carbon dioxide emissions. Burning these fuels releases energy which is most commonly turned into heat, electricity or power for transportation. Some examples of where they are used are in power plants, cars, planes and industrial facilities. Coal is the most carbon intensive fossil fuel. For every ton of coal burned, approximately 2.5 tons of CO_2 are produced.[7] Because of its high rate of use, coal is the largest fossil fuel source of carbon dioxide emissions. Anything involving fossil fuels has a carbon dioxide emission attached. Therefore burning these fuels releases energy but carbon dioxide also gets produced as a byproduct. This is because almost all the carbon that is stored in fossil fuels gets transformed to carbon dioxide during this process. It's important to note that the three main economic sectors that use fossil fuels are: electricity/heat, industry and transportation.[8]

It's better to understand that Carbon Dioxide should not be confused with carbon monoxide or CO, another odorless byproduct of combustion that is highly toxic to humans and animals. Carbon monoxide – not carbon dioxide or CO_2 – is the reason buildings must be properly ventilated, to prevent fireplace or furnace emissions from killing inhabitants. Volcanoes and deep sea vents also release notable amounts of CO_2. Indeed, they were the original sources of the CO_2 that helped launch life on earth. The periodic shifts in ocean current patterns in the southern tropical Pacific, known as El Niño and La Niña, also affect carbon dioxide levels. As scientific aspects, El Niño warms the sea waters and causes them to exhale huge amounts of CO_2 into the

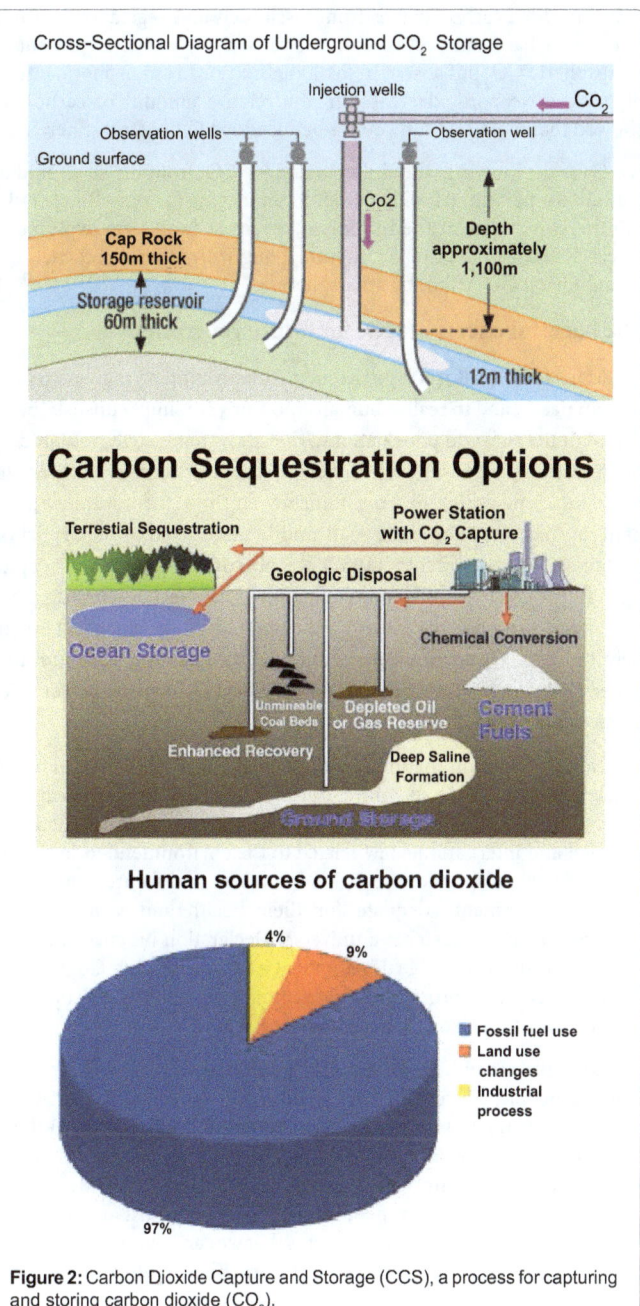

Figure 2: Carbon Dioxide Capture and Storage (CCS), a process for capturing and storing carbon dioxide (CO_2).

atmosphere; La Niña events cool waters and cause them to absorb more CO_2, which spurs the growth of oceanic algae.

As a result the increase in atmospheric CO_2 concentration is known to be caused by human activities because the character of CO_2 in the atmosphere, in particular the ratio of its heavy to light carbon atoms, has changed in a way that can be related to addition of fossil fuel carbon. Moreover, the ratio of oxygen to nitrogen in the atmosphere has declined as CO_2 has increased; this is as expected because oxygen is depleted when fossil fuels are burned. A heavy form of carbon, the carbon-13 isotope, is less abundant in vegetation and in fossil fuels that were formed from past vegetation, and is more abundant in carbon in the oceans and in volcanic emissions. The relative amount of the carbon-13 isotope in the atmosphere has been declining, showing

---

[7] Defra UK (2014) The 2014 Government Greenhouse Gas Conversion Factors for Company Reporting. London: U.K. Department for Environment, Food & Rural Affairs.

[8] International Energy Agency CO_2 Emissions from Fuel Combustion (2012) Paris: Organization for Economic Co-operation and Development.

that the added carbon comes from fossil fuels and vegetation. Carbon also has a bare radioactive isotope, carbon-14, which is present in atmospheric CO_2 but absent in fossil fuels. Prior to atmospheric testing of nuclear weapons, decreases in the relative amount of carbon-14 showed that fossil fuel carbon was being added to the atmosphere[9].

Over the long run, the ability to remove CO_2 from the air should be viewed as an essential tool in our kit for managing carbon-climate risks. We therefore need, at the minimum, a serious long term exploratory research effort to develop air capture along with other direct methods for removing CO_2 from the atmosphere.

The harmful effects of CO_2 for future generations

Many present efforts to guard and maintain human progress, to meet human needs, and to realize human ambitions are simply unsustainable - in both the rich and poor nations. They draw too heavily, too quickly, on already overdrawn environmental resource accounts (especially those un renewable) to be affordable far into the future without bankrupting those accounts. As Brundtland declared in her report on Sustainable development[10] (1987) we borrow environmental capital from future generations with no intention or prospect of repaying. They may damn us for our spendthrift ways, but they can never collect on our debt to them. We act as we do because we can get away with it: future generations do not vote; they have no political or financial power; they cannot challenge our decisions.

National and international law is being rapidly outdistanced by the accelerating pace and expanding scale of impacts on the ecological basis of development. Governments now need to fill major gaps in existing national and international law related to the environment, to find ways to recognize and protect the rights of present and future generations to an environment adequate for their health and well-being, to prepare under UN auspices a universal Declaration on environmental protection and sustainable development and a subsequent Convention, and to strengthen procedures for avoiding or resolving disputes on environment and resource management issues.

So while many believe that Our moral duties can extend only to existing people, Since future generations do not presently exist and therefore it is not possible to have any moral obligations toward them, there are many reasons to think this objection is mistaken. For one thing, it would also rule out any moral claims or responsibilities toward those in the past since past people do not presently exist anymore than future people do! Yet, it certainly seems that we can have moral concerns involving people in the past and in future. This is while today it has been to some extent accepted that we have moral and legal obligations toward future generations to provide a safe and secure world for them. [11] Global warming in result of burning of fossil fuel and reducing CO_2 gas, could harm both existing and future generations by some negative impacts on climate change and atmosphere stability.

Diagram below makes clear that with an increase in CO_2 concentration in the atmosphere, there are more CO_2 greenhouse molecules in the tropopause, they will radiate into space from a higher level (from Ze to Ze + ΔZe). Because of the adiabatic lapse rate, it will be colder there, so they will radiate (a lot!) less energy to space. This disturbs the equilibrium, so in order to restore that, the earth surface has to heat up (from Ts to Ts + ΔTs), so the adiabatic lapse rate (ALR) will move upwards and warm the tropopause, until the radiation into space is the same as before the CO_2 increase. In the drawings this theory is always illustrated with beautifully straight lines.[12] (Figure 3).

Some Possible Solutions

Now that the harmful effects of global warming on environment have been elucidated, we should also make clear and propose the possible solutions for facing with this notable challenge. In order to make choices that reduce greenhouse gas pollution amount, and preparing for the changes we can reduce risks from climate change.

There are several things we can do to deal with global warming. One answer is to stop making CO_2. We can do this by switching from oil, coal and gas to renewable energy. Next solution is to plant more trees. Trees absorb CO_2 and produce oxygen, which is not a greenhouse gas. Another idea is to use less energy and recycle more products. If we use less energy and be more environmentally friendly, the earth's temperature may not rise too much.[13] Obviously, the solutions are directed towards controlling the release of these harmful gases:

- One of the ways for this problem is reducing the usages of fossil fuels but unfortunately there is no way and the humanity should use one of the fuel sources available.
- Second way is using the alternative energy and clean energy sources and that's not cheaper though.
- And the middle way proposed is using CCS technology that simultaneously with CO_2 release, it will be captured and stored undergrounds and of course this is relatively middle way.

As I noticed for clarification in order to solve this danger, we have to reduce the consumption of energy and use the alternative energy

---

[12] http://www.climatetheory.net/ (visited 2016/06/26)

[13] T J, Rochelle (2016) Global Warming (Problem -Solution model essay).

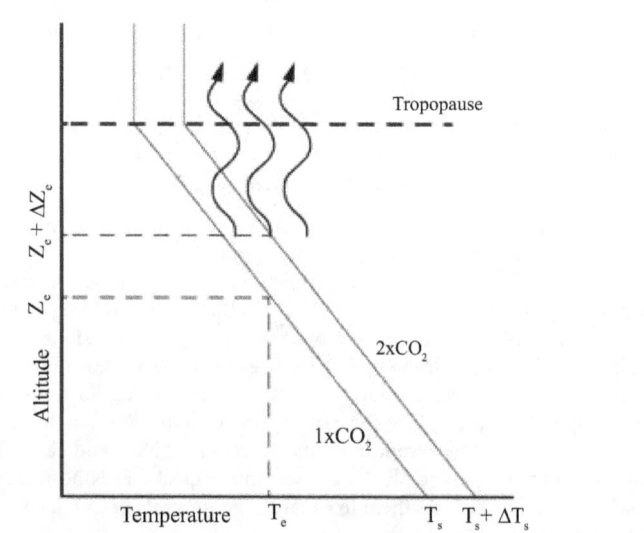

Figure 3: Schematic illustration of the change in emission level (Ze) associated with an increase in stn-face temperature (T_s) due to a doubling of CO_2 assuming a fixed atmospheric lapse rate. (Note that the effective emission temperature (T_e) remains unchanged).

---

[9] IPCC (2007) Climate Change 2007: The Physical Science Basis. Contribution of Working Group I to the Fourth Assessment Report of the Intergovernmental Panel on Climate Change [Solomon S, Qin D, Manning M, Chen Z, Marquis M, Averyt KB, Tignor M and Miller HL (eds.)] Cambridge University Press, Cambridge, United Kingdom and New York, USA.

[10] Report of the World Commission on Environment and Development: Our Common Future.

[11] Moral Obligations toward the Future

resources. If we calculate the present energy price, alternative energy must be more expensive than fossil fuels. However if we consider the negative price which is caused by global warming, this result might be different.[14]

It's better to mention that we can avoid the most severe results by implementing available clean energy and efficiency solutions at the local, state, and national levels. Many of these solutions provide immediate additional benefits including consumer savings on energy costs and cleaner air and water. We must begin enacting these available solutions today to impede the worst effects of global warming. We must get together to tackle the problem of global warming by managing the surroundings around our own houses.[15]

We need to find place to plant trees in the jungle. We should begin first with our own house. Gardens are necessary if we have to survive but we also need to device some other methods to cover the surface of buildings as is possible. Management of jungles has to be our top priority. Recycling of the trees in the jungles with cutting dry trees is very important because dry trees cause fire that not only destroy green trees but also cause release of harmful gases. Management of jungles is an issue that governments have to take care while each one of us can take care of surroundings around all of us. We also ought to reduce the consumption of electricity. As solar energy and wind energy is still too expensive to be used at mass scale.

Accordingly for this problem most of the states have proposed the underlined ways that some of them have been mentioned earlier and I'm taking into account them again in order to elucidate more:

- Renewable Electricity Standard: Great Lakes states in the world currently have standards and have significant economic and environmental benefits consequently.
- Increased Clean Energy Funding: The fund should be supported by a kilowatt hour charge on consumer bills to provide a guaranteed pool of public money for energy management programs.
- State Residential Energy Efficiency Building Code: The legislature should approve legislation to adopt such a code, which would yield consumer energy savings and help cut global warming emissions.
- Incentives for Cleaner Burning Fossil Fuel Generation: persuading the use of heat and power in a combination that produce both heat and electricity for a facility or surrounding community.
- Energy Efficient Lighting[16]

At last, making small changes now in the way we live means avoiding huge changes in the future. Scientists, governments and individuals must work together to overcome this serious threat.

Carbon Dioxide Capture and Storage (CCS)

The raison detre for carbon capture and storage (or sequestration) (CCS)[17] is to make the world use the limited conditions of having no or less carbon dioxide. Good points and benefits include economic competitiveness, energy security and a non-disruptive transition to low-carbon energy systems. We could consider CCS in the process

of mitigation actions for greenhouse gas concentrations. As we noted large point sources of CO_2 include large fossil fuel or biomass energy facilities, major CO_2-emitting industries, natural gas production, synthetic fuel plants and fossil fuel-based hydrogen production plants. Potential technical storage methods are: geological storage (in geological formations, such as oil and gas fields, unminable coal beds and deep saline formations[18]), ocean storage (direct release into the ocean water column or onto the deep seafloor) and industrial fixation of CO_2 into inorganic carbonates.

For CO_2 capture, the challenge is to separate CO or CO_2 from synthetic fuels derived from fossil fuels, or to separate CO_2 from the flue gas. This is respectively pre- and post-combustion capture. There are also alternative ways, like for instance having the combustion in almost pure oxygen (oxy-fuel combustion), which produces a stream of highly concentrated CO_2, but requires the separation of oxygen from the air.[19]

Large-scale CCS usages would need the creation of a regime to manage risks and supporting policies to facilitate technology investment. Within this frame, regulatory, legal, and public understanding considerations emerge as crucial factors that could either accelerate or inhibit CCS usages. Policy makers worldwide need to work towards a system of regulation and risk governance for CCS that is globally consistent nationally coordinated, and which adequately manages local risks. This policy shortly reviews regulatory issues post-capture, particularly the transport and geological storage of CO_2. It identifies key areas where relevant stakeholders should coordinate in international arena and proposes a model for development of national deployment and regulation, which includes jurisdictional specificities.

For practical implementation and working, CCS will need to be regulated as an industrial process, with regulations entered to each project stage: capture, transportation, site selection and permitting, site operations, site closure, and long-term stewardship. Despite the fact that all the elements of this industrial process exist, they are not yet developed to scale nor are they integrated. The structure of the future CCS industry deployment can take a number of possible forms in terms of the relationships between CO_2 producers, CO_2 pipeline operators, and geological storage site operators. This concise policy does not cover regulatory issues related to capture but it is worth noting this important point that, while the long-term potential for CCS is in capturing CO_2 at fossil-fired electric power plants, significant short-term potential lies in other industrial processes that already generate a concentrated CO_2 stream, such as natural gas or hydrogen production. Regulation of transport and geological storage must be designed to manage CO_2 from both electric utilities and from these other industries. A thorough CCS regulatory framework must balance competing needs and interests of local, national and international publics, CO_2 generators, CO_2 pipeline operators, geological storage site developers, financial institutions supporting the project, government agencies of safety and environmental requirements, and national and international agencies managing various climate regimes. The potential contribution of this technology will be influenced by factors such as the cost relative to other options, the time that CO_2 will remain stored, the means of transport to storage sites, environmental concerns, and the acceptability of this approach. The CCS process requires additional fuel and associated CO_2 emissions compared with a similar plant without capture.

[14] Solutions Of The Problem Of Global Warming

[15] http://www.ucsusa.org/sites/default/files/legacy/assets/documents/global_warming/ucssolutionilfinal.pdf

[16] http://www.ucsusa.org/sites/default/files/legacy/assets/documents/global_warming/ucssolutionilfinal.pdf
(accessed 2016/4/14)

[17] A note: CCS is a new technology, and issues of terminology are still in flux. Some practitioners use the phrase Carbon Capture and Sequestration while others prefer Carbon Capture and Storage. The EU, the IPCC, and the UNFCCC have adopted Carbon Capture and Storage.

[18] Saline formations are sedimentary rocks saturated with formation waters containing high concentrations of dissolved salts. They are widespread and contain enormous quantities of water that are unsuitable for agriculture or human consumption. Because the use of geothermal energy is likely to increase, potential geothermal areas may not be suitable for CO_2 storage.

[19] CO_2 capture and storage projects, Directorate-General for Research Directorate Energy, p. 6, 2007

Many factors will need to be taken into account in any comparison of mitigation options, not least that is making the comparison and for what purpose. In addition, there are broader issues, especially questions of comparison with other mitigation measures. Answering such questions will depend on many factors, including the potential of each option to deliver emission reductions, the national resources available, the accessibility of each technology for the country, national commitments to reduce emissions, the availability of finance, public acceptance, likely infrastructural changes, environmental side-effects and so on.

CO_2 from capture to storage

Capturing CO_2 typically involves separating it from a gas stream. Useful and suitable techniques were developed 60 years ago in connection with the production of town gas; these involved scrubbing the gas stream with a chemical solvent. Accordingly they were adapted for related purposes, such as capturing CO_2 from the flue gas streams of coal- or gas-burning plant for the carbonation of drinks and brine, and for enhancing oil recovery. It's important to mention that there are three main technology options for CO_2 capture in the generation of electricity and heat: post-combustion capture through chemical absorption, pre-combustion capture, and oxy- fuelling. The captured carbon dioxide must be compressed for transport and storage.

In the post-combustion process CO_2 is captured typically through the use of solvents and subsequent solvent regeneration, sometimes in combination with membrane separation. The basic technology has been used on an industrial scale for decades, but the challenge is to recover the CO_2 with a minimum energy penalty and at an acceptable cost. Pre-combustion capture processes can also be used in coal- or natural gas-based plant. The fuel is reacted first with oxygen and/or steam and then further processed in a shift reactor to produce a mixture of hydrogen and CO_2. The CO_2 is captured from a high-pressure gas mixture.

The oxy-combustion process includes the removal of nitrogen from the air in the oxidant stream using an air separation unit or, potentially in the future, membranes. At this time after the process of capturing the next stage takes place. This phase is transportation of carbon dioxide to the storage fields. Except when the emission source is located directly over the storage site, the CO_2 needs to be transported. Pipelines have been used for this purpose in the USA since the 1970s. CO_2 could also be transported in liquid form in ships similar to those transporting liquefied petroleum gas (LPG). For both pipeline and marine transportation systems of CO_2, costs depend on the distance and the quantity transported. For pipelines, costs are higher when crossing water bodies, heavily congested areas, or mountains. Compressed CO_2 can be injected into formations under the Earth's surface using many of the same methods. The three major types of geological storage are oil and gas reservoirs, deep saline formations, and un-minable coal beds. CO_2 could be physically trapped under a well-sealed rock layer or in the pore spaces within the rock. It can also be chemically trapped by dissolving in water and reacting with the surrounding rocks. The risk of leakage from these reservoirs is rather nothing in importance. Storage in geological formations is the cheapest and most environmentally acceptable storage option for CO_2. It's good to note that Oceans can store CO_2 because it is soluble in water. When the concentration of CO_2 increases in the atmosphere, more CO_2 is taken up by the oceans. Captured CO_2 can potentially be injected directly into deep oceans and most of it could remain there for centuries. CO_2 injection, however, can harm marine organisms near the injection point. It is furthermore expected that injecting large amounts would gradually affect the whole ocean.

Accordingly we could mention that storage of CO_2 in oceans isn't a good option for considering in this process. By chemical reactions with some naturally occurring minerals, CO_2 is converted into a solid form through a process called mineral carbonation and stored virtually permanently. This is a process which occurs naturally, although slowly. These chemical reactions can be accelerated and used industrially to store CO_2 in minerals artificially. However, the large amount of energy and mined minerals needed makes this option less cost effective. It is technically possible to use captured CO_2 in industries manufacturing products such as fertilizers. The overall effect on CO_2 emissions, however, would be very small, because most of these products rapidly release their CO_2 content back into the atmosphere. So we could describe three main mechanisms for CO_2 storage as followings:

- Physical trapping by immobilizing CO_2 in a gaseous or supercritical phase in geological formations. This can take two main forms: static trapping in structural traps and residual-gas trapping in a porous structure.
- Chemical trapping in formation fluids (water/hydrocarbon) either by dissolution or by ionic trapping. Once dissolved, the CO_2 can react chemically with minerals in the formation (mineral trapping) or adsorb on the mineral surface (adsorption trapping).
- Hydrodynamic trapping through the upward migration of CO_2 at extremely low velocities leading to its trapping in intermediate layers. Migration to the surface would take millions of years. Large quantities of CO_2 could be stored using this mechanism.[20]

Financial, legal and regulatory issues

A number of non-technical challenges need to be overcome if the full potential of CCS is to be achieved. These include:

- Financing near-term demonstration projects.
- Setting a long-term, sufficiently high and stable price for CO_2.
- Establishing legal and regulatory frameworks.
- Educating the public to foster awareness and acceptance.

These critical non-technical issues are discussed in this part of the paper.

Financing: In the newest fiscal and regulatory environment, commercial fossil-fuel power and industrial plants are unlikely to capture and store their CO_2 emissions, as CCS reduces efficiency, adds costs, and lowers energy output. Even in the European Union (EU), which has carbon constraints in, the benefits of reducing carbon emissions are not sufficient to outweigh the costs of CCS. These barriers can be partially overcome by government support in the form of tax credits and other related incentives. The wider penetration of CCS will require such support at all stages of project development.

Experience from early CCS projects will guide subsequent future commercial deployment and foster the learning needed to facilitate CCS for the power generation and industrial sectors. There are a variety of promising early opportunities for CCS, including expanding existing CO_2 capture in natural gas processing, or in ammonia or hydrogen manufacturing where the CO_2 is already separated, and developing EOR activities where there are financially attractive storage options.[21] CO_2-EOR offers a specially promising opportunity for early projects that are supported commercially by the values of additional recovered

---

[20] IPCC (Intergovernmental Panel on Climate Change) (2005) "Transport of CO_2", Chapter 4 of Special Report on Carbon Dioxide Capture and Storage, Cambridge University Press, Cambridge.

[21] Karstad O (2007) "CCS Business Models", IPIECA Workshop, Oslo.

oil. Large volumes of CO_2 are lately being captured and used for EOR in the United States, the Middle East and other regions and also it's vital to mention that with the right carbon pricing signals, the EOR market could provide more important early demand for CO_2. Consequently it's good to note that the most of CCS demonstration projects will need to be implemented and acted in the electricity generation sector. There is limited worldwide experience of carbon capture from coal-fired power plants, and also no experience of an integrated CCS project at a coal-fired power plant. There has been much debate and discussion about the minimum project size needed for meaningful demonstration of the relevant technologies.

Legal and regulatory issues: The expansion of CCS will present a number of legal and regulatory issues. The most important of these include: developing regulations for CO_2 transport; establishing jurisdiction among international, national, government actors; establishing ownership of storage-site resources and legal means for acquiring the rights to develop and use such resources, including access rights; developing clear guidelines for site selection, permitting, monitoring and verifying CO_2 retention; clarifying liabilities in a long run and financial responsibility for CO_2 storage operations; and, in the case of offshore CO_2 storage, complying with appropriate international marine environment protection instruments.

The safe and effective transportation of CO_2 requires the management of local environmental and safety risks and the mitigation of the potential impacts of CO_2 leakages on the global environment. There are different options for transporting CO_2 from capture sites to storage locations, including pipelines and pressurized road and sea tankers. Given the large volumes of CO_2 that are likely to need to be injected, pipelines offer the most cost-effective means of transport. As a result, most governments are focusing in the near-term on pipeline regulations.[22]. If other, non-pipeline transport mechanisms are used; they will require convenient regulatory frameworks to minimize safety and environmental risks levels. The most difficult issues in CO_2 pipeline regulations relate to funding, pipeline siting, and pipeline access.

Important point is that any regulatory and liability framework for CO_2 storage sites needs to define the roles and financial responsibilities of industry and government after site closure and permanent decommissioning. The level of risk associated CO_2 storage project will evolve as the project progresses along its life cycle.

Environmental Negative Effects of Using CCS

CCS is a means of separating out carbon dioxide when burning fossil fuels, and then dumping it - underground, or else at or under the sea bed.[23] CCS provides the greatest potential to reduce the greenhouse gases emitted by our stationary energy sector.[24] [5].

CCS technologies require approximately 15% to 25% more energy depending on the particular type of technology used, so plants with CCS need more fuel than conventional plants. This in turn can lead to increased 'direct emissions' occurring from facilities where CCS is installed, and increased 'indirect emissions' caused by the extraction and transport of the additional fuel.[25] The impact of carbon capture and storage (CCS) on environment is an important issue

in discussing whether this technology should be part of choices for facing with increasing greenhouse gas emissions, both nationally and internationally. On the other hand, CCS has the long term potential to make a substantial positive impact on the amount of CO_2 emitted into the atmosphere by the stationary energy sector.

The most substantial risk associated with CCS is the leakage of CO_2 from storage sites. While there is some experience with geological storage of CO_2 and natural gas for periods of approximately 10-20 years, long term storage over many hundreds or thousands of years has not been proven [6]. The IPCC Special Report on CCS suggests that the environmental risks associated with CO_2 capture and storage are low. As the IPCC stated well-selected geological formations are likely to retain over 99% of their storage over a period of 1,000 years.

Overall, the risks of CO_2 storage are comparable to the risks in similar existing industrial operations such as underground natural-gas storage and [EOR][26] [7].

Here it's good to mention that "Migration of CO_2 into neighbor geologic formations" is one of the related risks according to the projects of CCS that its probability of occurrence is very high and the direct and indirect consequences of that are as follows:

Lateral and/or descendent diffusion of CO_2 from the storage complex into neighbor formations (the caprock - top sealing rock layer is, by definition, impermeable to CO_2). CO_2 reactive processes with minerals of neighbor geologic formations (secondary trap mechanisms occurring at long-term storage).[27]

The environmental impact is highly dependent on the characteristics of underground geological formation for the purposes of lasting storage of the CO_2, partly because of overpressure issues of the reservoir and lithology adjacent to the storage reservoir. CO_2 stored in saline aquifers is absorbed and dissolved in the saline water, and also, eventually, part of the injected amount may have reacted with other dissolved minerals in the aquifer.

An eventual leakage from underground is a slow process that may last for decades or even centuries, depending on the diffusion capacity of the CO_2 through the geologic formations above the reservoir layer, until the CO_2 finally reach the surface. Although deep saline aquifers are considered to represent a huge potential for CO_2 storage and are geographically available all over the world, but in countries where hydrocarbons reservoirs are non-existent abandoned or unmineable coal seams represent a better potential location for lasting storage of CO_2. The safety of CO_2 sequestration depends on geological, both chemical and physical, trapping mechanisms for CO_2, which are different for saline aquifers and for coal. CO_2 naturally occurs in coal seams, associated with other gases Coal adsorbs CO_2 preferably to other gases, while in saline aquifers the injected CO_2 will not be adsorbed and will compete for underground space with brine, most probably causing its displacement accordingly. Because of this, overpressure of the storage reservoir is most likely to occur sooner in aquifers than in coal seams.

CCS and Its Future

Only geological storage of CO_2 is going to be considered to be environmentally acceptable in the world. Ocean storage above the

[22] MCMPR (Ministerial Council on Mineral and Petroleum Resources) (2005) Principles for engagement with communities and stakeholders, Melbourne

[23] http://www.greenpeace.org.uk/blog/climate/the-problem-with-carbon-capture-and-storage -ccs-20080103 (last visited 2016/06/20)

[24] Lacis AA, Schmidt GA, Rind D, Ruedy RA (2010) "Atmospheric CO_2: Principal Control Knob Governing Earth's Temperature. Science" 330: 356-359.

[25] http://www.eea.europa.eu/highlights/carbon-capture-and-storage-could

[26] United Nations Environmental Programme (UNEP) (2006) Can carbon dioxide storage help cut greenhouse emissions? A Simplified guide to the IPCC's 'Special report on carbon dioxide capture and storage'. p: 15.

[27] http://conferences.iaia.org/2012/pdf/uploadpapers/Final%20papers%20review%20 process/Oliveira,%20Gisela.%20%20Environmental%20Impact%20Assessment%20 of%20Carbon%20Capture%20and%20Sequestration.Pdf

seabed, for instance, is not considered acceptable. Several geological settings are envisaged as potential storage sites, oil and gas reservoirs, in exploitation or depleted, non-mineable coal seems, and deep saline aquifers. In all cases, the CO_2 should be under a hydrostatic pressure of more than 70 bars (that is deeper than 700 meters on-shore) to make sure that it is stored as a supercritical fluid, and not as a gas. From a storage potential point of view, estimates – also obtained in Framework Program research contracts – are that deep saline aquifers have the potential to hold more than all of the CO_2 which would be produced if we used all of the oil, gas and coal. These geological formations are – like coal reserves – quite evenly spread across the world. The challenge is therefore to make sure that the injection of CO_2 in these strata is a safe process, from the immediate health and safety issues associated with the injection process, to the CO_2 storage permanence required to effectively address climate change. From this point of view, CO_2 injection and storage, like any other engineering activity, will require a proper legislative and regulatory framework, norms and standards, good practice, and common sense. Mother Nature, which has stored oil, gas, water and CO_2 for million years, indicates that this should be feasible. Safety is also directly linked to the long term liability issue. The timescales required to effectively combat climate change are incompatible with the operations of a private company, so that liability transfer to the public authorities must take place sometime after the end of the injection. This can happen only if proper site certification, monitoring and verification methods are in place. Given appropriate emission reduction incentives, CCS offers a viable and competitive route to mitigate CO_2 emissions. CO_2 capture leads to an increase in capital and operating expenses, combined with a decrease in plant energy efficiency. With the recent development of a more robust methodology for storage capacity estimates, governments urgently need to conduct detailed evaluations of their national CO_2 storage capacity, working in partnership with bordering nations who share the same storage space.

The Role of Seabed Authority in Managing the Problem

UNCLOS (United Nations Convention on the Law of the Sea) is applicable to sub-seabed storage because its jurisdiction includes seabed and subsoil. UNCLOS specifies that the sovereignty of a coastal state over its territorial sea and contiguous zone extends to the "seabed" and "subsoil". According to article 56, Within its EEZ, a coastal state is provided sovereign rights for the purpose of exploring and exploiting, conserving and managing the natural resources of the "seabed and its subsoil". Within its continental shelf (the seabed and subsoil of the submarine areas extending beyond the territorial sea), and according to article 77 a state has sovereign rights for exploring and exploiting natural resources. Thus, the geologic formations that would be used for carbon dioxide storage fall within the jurisdiction of UNCLOS.

Accordingly "Dumping" as defined in article 1 by UNCLOS in this regard means the "deliberate disposal of wastes or other matter from vessels, aircraft, platforms or other man-made structures at sea". However UNCLOS does not define the term "wastes or other matter". Then it seems that any storage mechanism for carbon dioxide that was not a man-made structure at sea, such as a pipeline that transported the carbon dioxide from land directly to the sub-seabed point of injection, would not be "dumping" under UNCLOS.

Therefore Carbon dioxide storage using a vessel, platform, or man-made structure at sea would be defined as "dumping" under UNCLOS, but is not necessarily prohibited. Even if a carbon dioxide storage mechanism was used that fell under the UNCLOS definition of "dumping" (in the case the storage mechanism included a vessel,

platform, or manmade structure, and assuming arguendo that carbon dioxide was determined to fall under the UNCLOS definition of "waste"), the dumping is not necessarily prohibited by UNCLOS. Rather, in article 210, UNCLOS requires that states adopt laws and regulations to prevent, reduce and control pollution of the marine environment by dumping. These laws and regulations are expected to be based on rules, standards and recommended practices and procedures established by "competent international organizations;" and the London Convention, would be the appropriate source of international law in this case. Thus UNCLOS does not necessarily prohibit dumping of wastes or other matter, but rather would defer to the London Convention's interpretation of pollution by dumping.

Conclusion

Global warming is the result of increase in the earth's average surface temperature due to greenhouse gases like carbon dioxide and methane.[28] It is constantly resulting in extreme high temperature of the surface, reduction of snow cover, and rise in the water level and the increasing human activities are significantly contributing to the cause of global warming. The foremost activity among them is the burning of the fossil fuels by the industries, which forms the huge amount of the carbon dioxide and the Nitric acid into the air. In this way the environment is dealing with some of the very serious problems, which are having its considerable effect on the climate too. One of the main causes of global warming is increase in the amount of carbon dioxide in the atmosphere.

CO_2 is a naturally occurring atmospheric gas that is considered safe at levels below 0.5%. In addition to potential indoor exposure, high concentrations of CO_2 can collect outdoors. Outdoor exposure can occur where CO_2 is venting from below ground sources such as mining operations, natural gas production, and magmatic emissions.[29] It's good to note here that we should first recognize the sources of carbon dioxide in the environment that as the diagram shows: 87 percent of all human-produced carbon dioxide emissions come from the burning of fossil fuels like coal, natural gas and oil. The remainder results from the clearing of forests and other land use changes (9%), as well as some industrial processes such as cement manufacturing (4%)[30].

So it has become really important to reduce the process of warming as soon as possible by the way of reducing this hazardous gas in the atmosphere. Capturing carbon dioxide emissions (CCS) from power plants and storing it underground is seen as a crucial technology to reduce the global warming impact of fossil fuels such as coal and gas, on which the world will continue to rely for decades. According to International Energy Agency demand, CCS will need to lead somehow one fifth of emissions reductions, across both power and industrial sectors, so CCS plays an important role as part of an economically sustainable way to meet climate mitigation goals within the 2050 timeframe, that at the same time ensuring global and regional energy security. Finally it seems that CCS is currently one of the best available technologies to drastically reduce greenhouse gas emissions from certain industrial processes and it is a key technology option to decarbonize the power sector especially in countries with a high share of fossil fuels in electricity production. That's so necessary to install this equipment of CCS project because the main concern of the international society is to resolve and impede the harmful effects of greenhouse gases accordingly. However we should

---

28 http://www.conserve-energy-future.com/various-global-warming-facts.php

29 Health Risk Evaluation for Carbon Dioxide (CO_2) http://www.blm.gov/style/medialib/blm/wy/information/NEPA/cfodocs/howell.Par.2800.File.dat/25apxC.pdf

30 Le Quéré C, Jain AK, Raupach MR, Schwinger J, Sitch S, et al. (2012) "The global carbon budget 1959-2011" Earth System Science Data Discussions 5: 1107-1157.

not ignore the possible negative sides of this case. The future practice of States, specially developed states, in using CCS could help international community to provide a more proper Judgment about this technology and its advantages. Clearly we can mention to the role of sea bed authority in the way of reducing the amounts of carbon dioxide and decreasing global warming effects. We can dedicate the implied role of this authority related to this matter because there are no clear proofs for supporting this idea but with perusing the UNCLOS articles carefully we can make conclusions to the related roles of sea bed authority in decreasing the effects of global warming and its harmful consequences.

References

1. Matysek A, Ford M, Jakeman G, Gurney A, Fisher BS (2006) Technology: Its role in economic development and climate change. ABARE Research Report 06.6 Canberra: 100-101.

2. Denman KL, Brasseur G, Chidthaisong A, Ciais P, Cox PM, et al. (2007) Couplings between changes in the climate system and biogeochemistry. In: climate Change 2007: The physical science basis. Contribution of working group I to the fourth assessment report of the inter-governmental panel on climate change. Cambridge University Press, Cambridge, United Kingdom and New York.

3. Zaelke D, Cameron J (1990) Global warming and climate change- an overview of the international legal process. AM. UJ Intl L & Poly 5: 249.

4. Knutti R, Hegerl GC (2008) The equilibrium sensitivity of the earth's temperature to radiation changes. Nature geoscience 1:735-743.

5. Lacis AA, Schmidt GA, Rind D, Ruedy RA (2010) Atmospheric CO_2 principal control knob governing earth's temperature. Science 330:356-359.

6. TRU (2004) Energy Submission Country Women's Association of NSW: Friends of the Earth. Australia Submission 13: 7.

7. United Nations Environmental Programme (UNEP) (2006) Can carbon dioxide storage help cut greenhouse emissions? A Simplified guide to the IPCC's 'Special report on carbon dioxide capture and storage.

10

Effect of Combustion Chamber Shapes on the Performance of Mahua and Neem Biodiesel Operated Diesel Engines

Banapurmath NR[1]*, Chavan AS[2], Bansode SB[2], Sankalp Patil[1], Naveen G[1], Sanketh Tonannavar[1], Keerthi Kumar N[3] andTandale MS[2]

[1]*B.V.B College of Engineering and Technology, Hubli, Karnataka, India*
[2]*Dr. Babasaheb Ambedkar Technological University, Lonere, Raigad, MS, India*
[3]*B.M.S. Institute of Technology and Management, Bangalore, Karnataka, India*

Abstract

Shape of combustion chamber plays a major role in controlling combustion process and emission characteristics occurring inside internal combustion engines in general and diesel engines in particular. To optimize a combustion chamber for diesel engine applications, suitable design modifications are required that meet both emission norms as well as acceptable engine performance. In this context, experimental investigations were carried out on a single cylinder four stroke direct injection diesel engine operated in single fuel mode using Mahua oil methyl ester (MhOME) and neem oil methyl ester (NOME). Different combustion chamber shapes were designed and fabricated keeping the compression ratio same for the existing diesel engine. The existing engine was provided with hemispherical combustion chamber (HCC) shape. In order to study the effect of other combustion chamber shapes on the performance of diesel engine, cylindrical (CCC), trapezoidal (TrCC), and toroidal combustion chamber (TCC) shapes were designed and developed. Various engine parameters such as power, torque, fuel consumption, and exhaust temperature, combustion parameters such as heat release rate, ignition delay, combustion duration, and exhaust emissions such as smoke opacity, hydrocarbon, CO, and NOx, were measured. Results revealed that the TCC shape resulted in overall improved performance with reduced emission levels compared to other shapes tested. Total hydrocarbon emission (THC) and carbon monoxide (CO) were also decreased significantly compared to other combustion chambers.

Keywords: Biodiesel; Mahua oil; Neem oil; Emissions; Combustion chamber shapes

Introduction

Diesel engines are widely used for transport and power generation applications because of their high thermal efficiency, and their easy adoption for power generation applications as well. However, there is an increased impetus on improved engine performance, lower noise and vibration levels and lower emissions. Increasing energy demand, decrease in fossil fuel reserve in the earth crust and harmful exhaust gases have focused major attention on the use of renewable and alternative fuels. To overcome and meet these demands, use of renewable fuels such as biodiesels for diesel engines has gained greater momentum. To meet the challenge, it is essential in implementing new technologies and methods that improve the efficiency of diesel engine used for both transport and power generation applications. Renewable energy sources can supply the energy for longer periods of time than those of fossil fuels and have many advantages [1]. Liquid biodiesels are more suitable for diesel engine applications as their properties are closer to diesel.

A number of vegetable oils have been used for biodiesel production and their respective biodiesels are used as alternative fuels in diesel engines. Biodiesels derived from jatropha, honge (karanja), honne, palm, rubber seed, rape seed, mahua, and neem seed oils were used in diesel engine applications [2-14]. Slightly lowered performance with increased emissions and combustion studies was reported for Biodiesels engine operation by several researchers [2,3,8,9,15-17]. Effect of various engine parameters such as compression ratio (CR), injection timing (IT), injection pressure and engine loading on the performance and exhaust emissions of a single cylinder diesel engine operated on biodiesel and their blends with diesel were reported in the literature [18]. Changes in injection timings change the position of the piston and cylinder pressure and temperature at the injection. Retarded injection

timings showed significant reduction in diesel NOx and biodiesel NOx [19]. Cylinder pressures and temperatures gradually decreased when injection timings were retarded [20]. Other investigators also have performed experiments on CI engine with different vegetable oils and their esters at different injection pressures. Better performance, higher peak cylinder pressure and temperature were reported at increased injection pressures [20-23].

Mahua and Neem Biodiesel Operation

Biodiesels derived from Mahua and Neem oils have been used as potential alternative fuels to diesel by several investigator. Mahua biodiesel has been used as an alternative fuel for diesel engine application by several investigators [22,24,25]. Effect of injection timing, compression ratio on the performance of Mahua biodiesel has been reported [26]. They reported biodiesel could be blended with diesel fuel up to 20% at any compression ratio and injection timing for getting nearly same performance as compared with diesel. For neem biodiesel a slight drop in efficiency compared to diesel has been reported, while their blends B10, B20 showed performance closer to diesel operation [27,28]. CO emission increased with B100 and increased blends of B60, B80 due to the incomplete combustion. They suggested a change in

*Corresponding author: Dr. N.R. Banapurmath, B.V.B. College of Engineering and Technology, Hubli, Karnataka, India, E-mail: nr_banapurmath@rediffmail.com

injection pressure and combustion chamber design for better engine performance. Neem oil methyl ester resulted in lower emissions compared to diesel whereas neem oil gives lower CO_2 emission due to incomplete combustion while smoke opacity increased at part load and decreased at full load [28].

The combustion chamber of an engine plays a major role during the combustion of wide variety of fuels. In this context, many researchers performed both experimental and simulation studies on the use of various combustion chambers [29-31]. Improvement in air entrainment with increased swirl and injection pressure were reported [32,33]. Optimum combustion chamber geometry of the engine showed better performance and emission levels. Suitable combustion geometry of bowl shape helps to increase squish area and proper mixing of gaseous fuel with air [29,34]. Designing the combustion chamber with narrow and deep and with a shallow reentrance had a low protuberance on the cylinder axis and the spray oriented towards the bowl entrance reduced the NOx emission levels to the maximum extent [30,31,35]. The behavior of fuel once it is injected in the combustion chamber and its interaction with air is important. It is well known that nozzle geometry and cavitations strongly affect evaporation and atomization processes of fuel. Suitable changes in the in-cylinder flow field resulted in differing combustion.

From the literature survey it follows that very limited work has been done to investigate the effect of combustion chamber shapes on the performance, combustion and emission characteristics of diesel engine fuelled with Mahua and Neem oil methyl esters. In this context, experimental investigations were carried out on a single cylinder four stroke direct injection diesel engine operated on MhOME and NOME with different combustion chamber shapes adopted for this work.

Characterization of Mahua and Neem Oils

In the present study, Diesel, MhOME and NOME were used as injected fuels. Tables 1 and 2 shows the composition of MhOME and NOME, fatty acids contribution, chemical formula, structure and their molecular weight with their chemical structure. The properties of MhOME and NOME were determined experimentally and are summarized in Table 3.

Experimental Setup

Experiments were conducted on a Kirloskar TV1 type, four stroke, single cylinder, water-cooled diesel engine test rig. Figure 1 shows the line diagram of the test rig used. Eddy current dynamometer was used for loading the engine. The fuel flow rate was measured on the volumetric basis using a burette and stopwatch. The engine was operated at a rated constant speed of 1500 rev/min. The emission characteristics were measured by using HARTRIDGE smoke meter and five gas analyzer during the steady state operation. Experiments were conducted by using biodiesels selected for the study with four different combustion chamber shapes (cylindrical (CCC), trapezoidal (TrCC), and toroidal Combustion chamber (TCC) shapes). Figures 2 shows the different combustion chamber shapes. Finally the results obtained with biodiesel operation were compared with Diesel. The specification of the compression ignition (CI) engine is given in Table 4.

Results and Discussions

In the present work, diesel engine was operated on diesel, MhOME and NOME with different configurations of combustion chambers namely cylindrical (CCC), trapezoidal (TrCC), and toroidal combustion chamber (TCC) shapes. The results and discussions on the performance combustion and emission characteristics of diesel engine

SI No	Composition	Chemical name	Single/Double/ Triple bond	Structure	Saturated/ Unsaturated	Chemical formula	Weight (%)
1	Palmitic	Hexadecanoic	----	16:0	Saturated	$C_{16}H_{32}O_2$	16.0-28.2
2	Stearic	Octadecanoic	----	18:0	Saturated	$C_{18}H_{36}O_2$	20.0-25.1
3	Oleic	Cis-9 Octadecanoic	Single	18:1	Unsaturated	$C_{18}H_{34}O_2$	41.0-51.0
4	Linoleic	Cis-9,cis-12 Octadecanoic	Double	18:2	Unsaturated	$C_{18}H_{32}O_2$	8.9-13.7
5	Arachidic	Etcosanoic	-----	20:0	Saturated	$C_{20}H_{40}O_2$	0.0-3.3

Table 1: Fatty acid contribution of Mahua oils sample and its chemical structure.

SI No	Composition	Chemical name	Single/Double/ Triple bond	Structure	Saturated/ Unsaturated	Chemical formula	Weight(%) Or Molecular wt
1	Palmitic	Hexadecanoic	----	16:0	Saturated	$C_{16}H_{32}O_2$	16.0-28.2
2	Stearic	Octadecanoic	----	18:0	Saturated	$C_{18}H_{36}O_2$	20.0-25.1
3	Oleic	Cis-9 Octadecanoic	Single	18:1	Unsaturated	$C_{18}H_{34}O_2$	41.0-51.0
4	Linoleic	Cis-9,cis-12 Octadecanoic	Double	18:2	Unsaturated	$C_{18}H_{32}O_2$	8.9-13.7
5	Arachidic	Etcosanoic	-----	20:0	Saturated	$C_{20}H_{40}O_2$	0.0-3.3

Table 2: Fatty acid contribution of Neem oils sample and its chemical structure.

SI No	Properties	Diesel	Mahua oil	Neem oil	MhOME	NOME
1	Viscosity@40°C (cst)	4.59 (Low)	24.21	23.45	5.6	4.7
2	Flash point °C	56	212	210	129	118
3	Calorific value in kJ/kg	45000	36,100	38,100	36,900	40,000
4	Specific gravity	0.830	0.960	0.940	0.882	0.878
5	Density Kg/m³	830	960	940	882	878
8	Type of oil	----	Non edible	Non edible	Non edible	Non edible
9	Cetane number	42	----	----	52	52

Table 3: Properties of fuels tested.

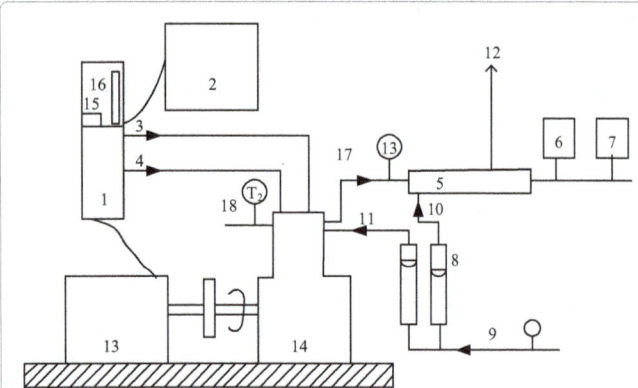

1-Control Panel, 2-Computer system, 3-Diesel flow line, 4-Air flow line, 5-Calorimeter, 6-Exhaust gas analyzer, 7-Smoke meter, 8-Rota meter, 9, 11-Inlet water temperature, 10-Calorimeter inlet water temperature,12- Calorimeter outlet water temperature, 13-Dynamometer, 14-CI Engine, 15-Speed measurement,16-Burette for fuel measurement, 17-Exhaust gas outlet, 18-Outlet water temperature, T1- Inlet water temperature, T2-Outlet water temperature, T3-Exhaust gas temperature.

Figure 1: Experimental set up.

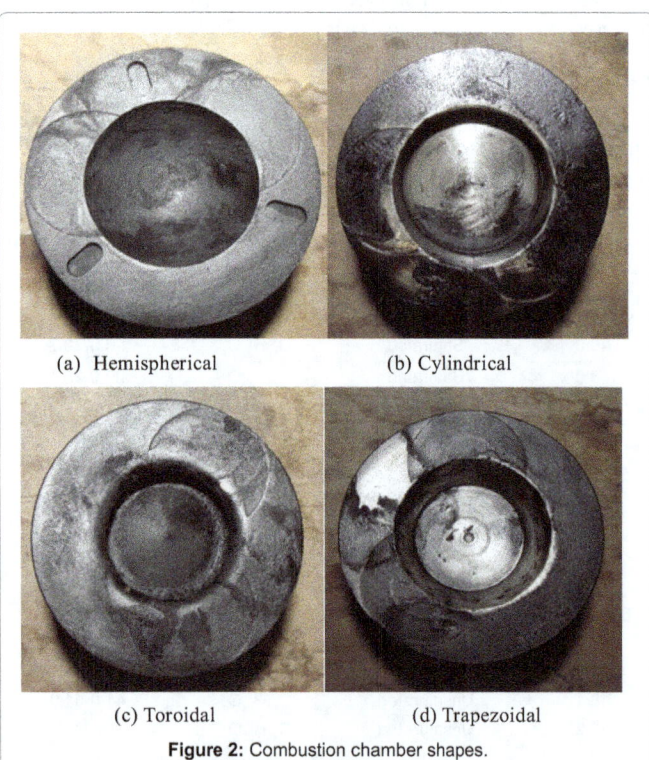

(a) Hemispherical   (b) Cylindrical

(c) Toroidal   (d) Trapezoidal

Figure 2: Combustion chamber shapes.

Sl No	Parameters	Specification
1	Type of engine	Kirloskar make Single cylinder four stroke direct injection diesel engine
2	Nozzle opening pressure	200 to 205 bar
3	Rated power	5.2 KW (7 HP)@1500 RPM
4	Cylinder diameter (Bore)	87.5 mm
5	Stroke length	110 mm
6	Compression ratio	17.5:1

Table 4: Specifications of the engine.

operating on two biodiesels is presented in the subsequent paragraphs.

Performance and emission characteristics

Figure 3 shows the variation of brake thermal efficiency (BTE) with brake power. It is observed that BTE for diesel fuel mode of operation was higher than both biodiesels of MhOME and NOME operation over the entire load range. This is mainly due to lower calorific value of the biodiesels and lower volatility as well. The study with different combustion chamber shapes showed that for biodiesel operation with TCC resulted in better performance compared to other combustion chambers. It may be due to the fact that, the TCC prevents the flame from spreading over to the squish region resulting in better mixture formation of biodiesel-air combinations, as a result of better air motion and lowers exhaust soot by increasing swirl and tumble. Based on the results, it is observed that the TCC has an ability to direct the flow field inside the sub volume at all engine loads and therefore substantial differences in the mixing process may not be present.

Figure 4 shows variation of smoke opacity with brake power. It is observed that the smoke opacity for diesel fuel operation was lower than MhOME and NOME biodiesels over the entire load range. This may be attributed to improper fuel-air mixing due to higher viscosity and higher free fatty acid content of biodiesels considered. However, TCC gives lower smoke emission levels compared to other combustion chambers. It may be due to the fact that, the air-fuel mixing prevailing inside combustion chamber and higher turbulence resulted in better combustion and oxidation of the soot particles which further reduction the smoke emission levels.

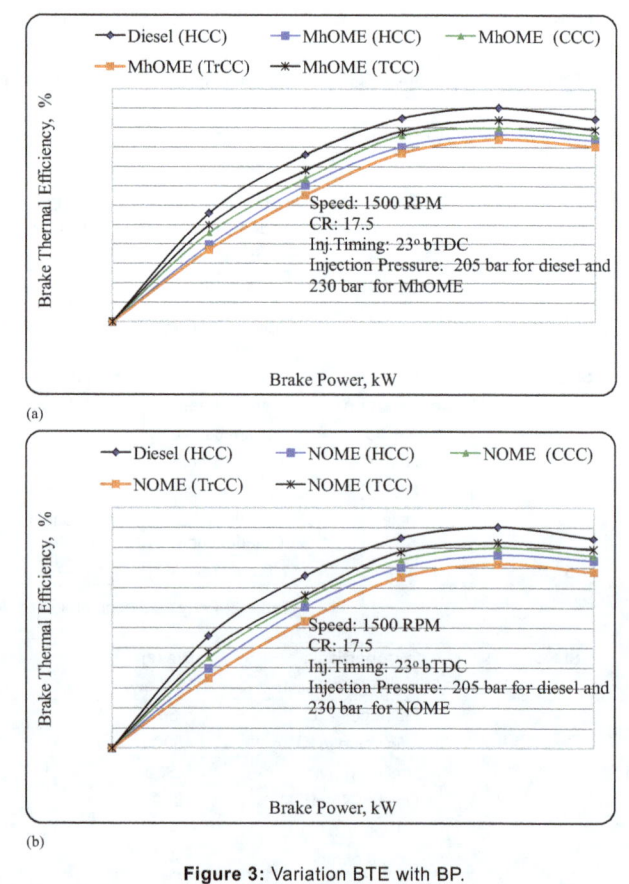

(a)

(b)

Figure 3: Variation BTE with BP.

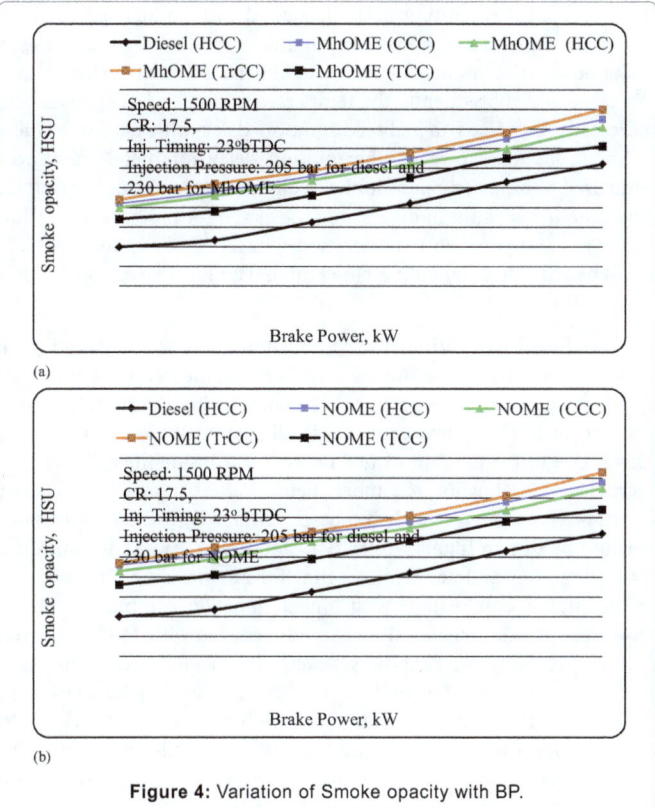

(a)

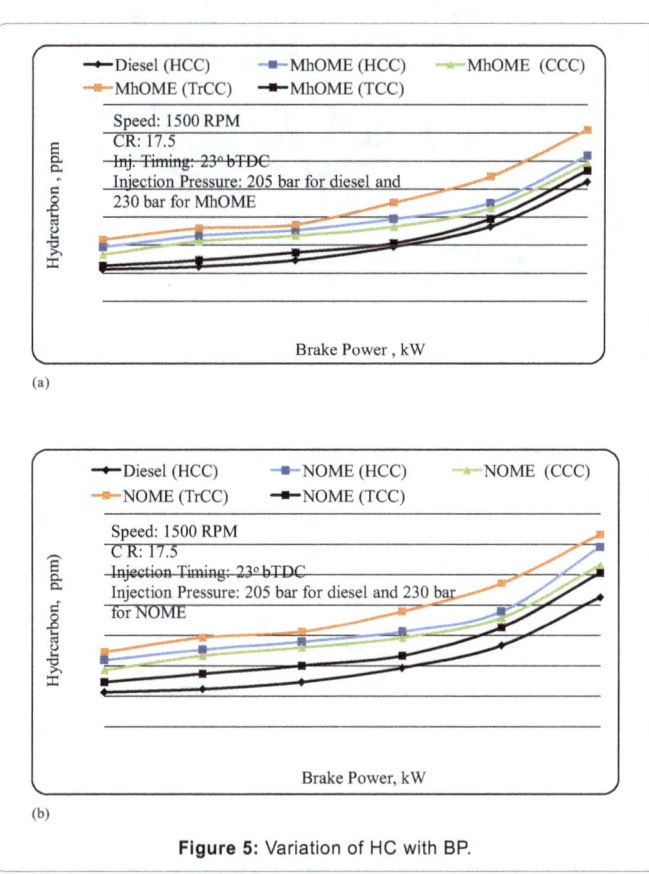

(b)

Figure 4: Variation of Smoke opacity with BP.

Figures 5 and 6 shows the variation of hydrocarbon (HC) and carbon monoxide (CO) emission levels for diesel, MhOME, and NOME at all loads. Both HC and CO emission levels were higher for MhOME, and NOME compared to diesel operation. Incomplete combustion of the MhOME, and NOME biodiesels is responsible for this observed trend. It could be due to their lower calorific value, lower adiabatic flame temperature and higher viscosity and lower mean effective pressures. However, TCC resulted in lower HC and CO emission levels compared to other combustion chamber shapes. It could be due to higher turbulence and comparatively higher temperature prevailing in the combustion chamber that resulted into minimum heat losses and better oxidation of HC and CO and hence reduced both emission levels. However, other combustion chambers may not contribute to the proper mixing fuel combinations.

The NOx emission levels were found to be higher for diesel fuel operation compared to biodiesel over the entire load range (Figure 7). This is because of higher heat release rate during premixed combustion phase observed with diesel compared to biodiesel operation. Slightly higher NOx resulted with TCC compared to other combustion chambers tested. This could be due to slightly better combustion occurring due to more homogeneous mixing and larger part of combustion occurs just before top dead center. Presence of oxygen in the biodiesels is also responsible for this trend. Therefore it is resulted in higher peak cycle temperature.

Combustion characteristics

In this section combustion characteristics of diesel engine fuelled with MhOME, NOME biodiesel has been presented.

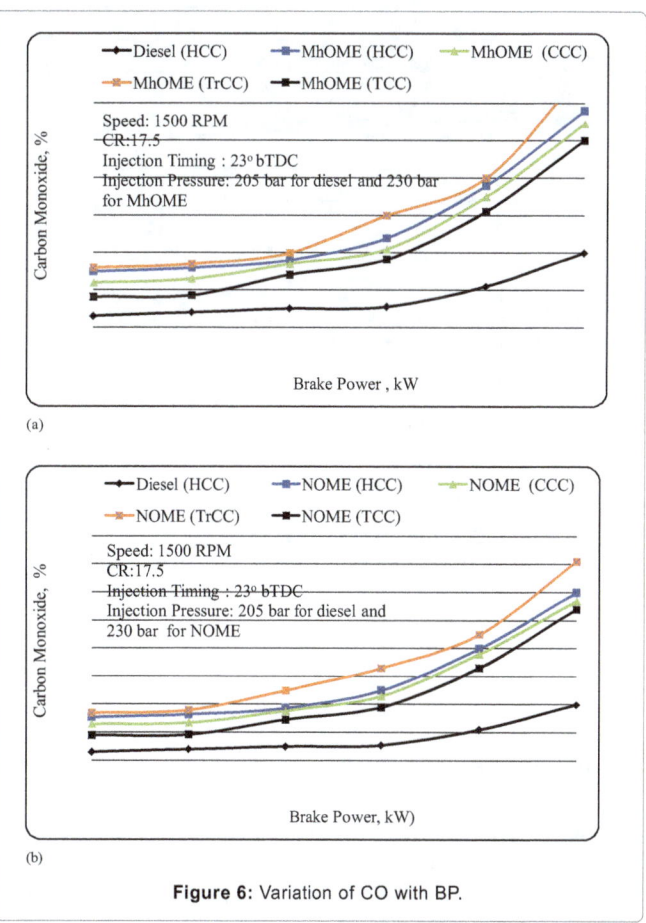

(a)

(b)

Figure 6: Variation of CO with BP.

(a)

(b)

Figure 5: Variation of HC with BP.

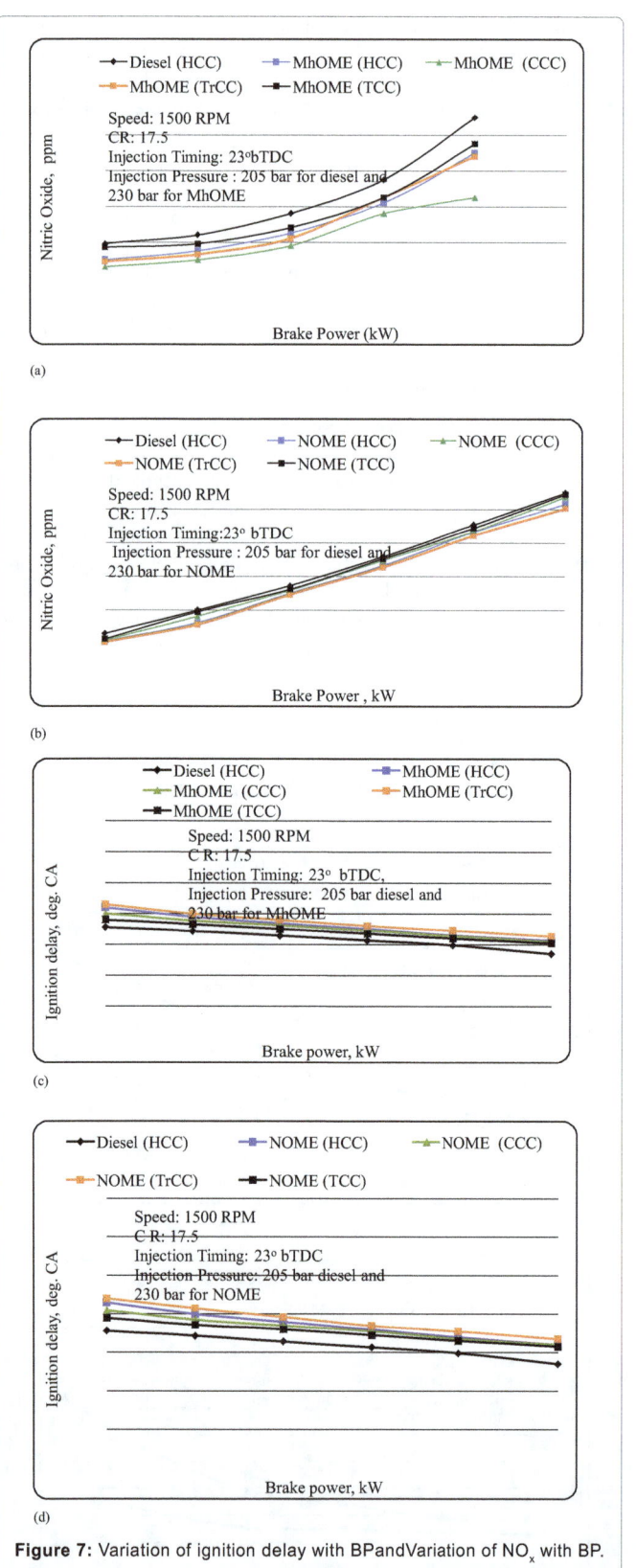

(a)

(b)

(c)

(d)

Figure 7: Variation of ignition delay with BP and Variation of NO$_x$ with BP.

Ignition delay: The variation of ignition delay with brake power for different combustion chamber shapes were shown in Figure 8. The ignition delay is calculated based on the static injection timing.

It is observed that ignition delay decreased with an increase in brake power for almost all combustion chamber shapes. With an increase in brake power, the amount of fuel being burnt inside the cylinder gets increased and subsequently the temperature of in-cylinder gases gets increased. This leads to reduced ignition delay with all combustion chamber shapes. However, the ignition delay for diesel was lower compared to biodiesel operation with all combustion chamber shapes. However, lower ignition delays were observed for biodiesel operation with TCC compared to the operation with HCC, CCC and TrCC. It could be attributed to better air-fuel mixing and increased combustion temperature.

Combustion duration: The combustion duration shown in Figure 8 was calculated based on the duration between the start of combustion and 90% cumulative heat release. The combustion duration increases with increase in the power output with all combustion chamber shapes. This is due to the amount of fuel being burnt inside the cylinder gets increased. Combustion chamber being same, higher combustion duration was observed with biodiesel compared to diesel operation. It could be due to higher viscosity of biodiesels leading to improper air–fuel mixing, and needs longer time for mixing and hence resulting in incomplete combustion with longer diffusion combustion phase. However with combustion duration was reduced with TCC compared to other combustion chambers tested. This could be attributed to improvement in mixing of fuel combination due to better squish. Significantly higher combustion rates with biodiesel operation leads to higher exhaust temperatures and lower thermal efficiency. However, biodiesel operation with TCC showed improvement in heat release rate compared to other combustion chamber shapes.

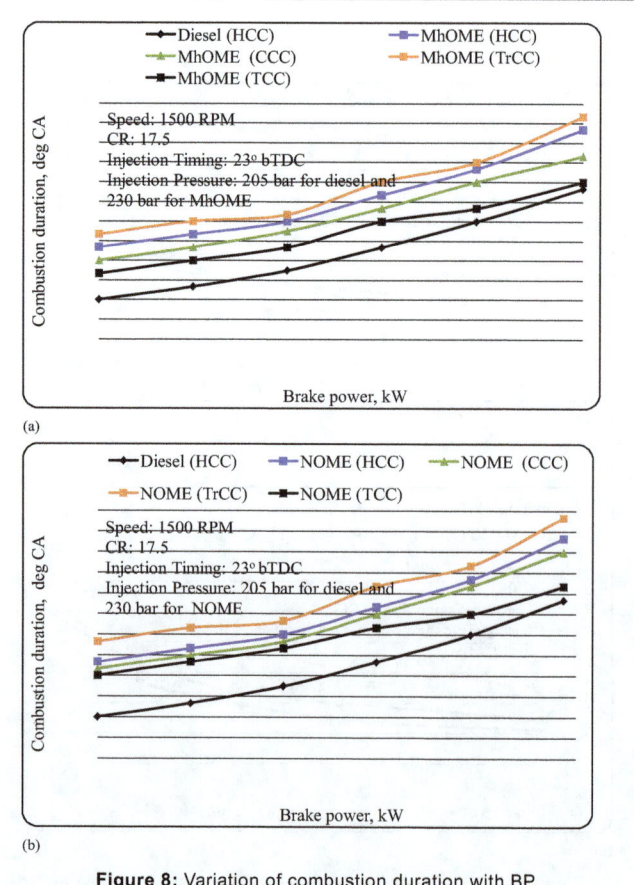

(a)

(b)

Figure 8: Variation of combustion duration with BP.

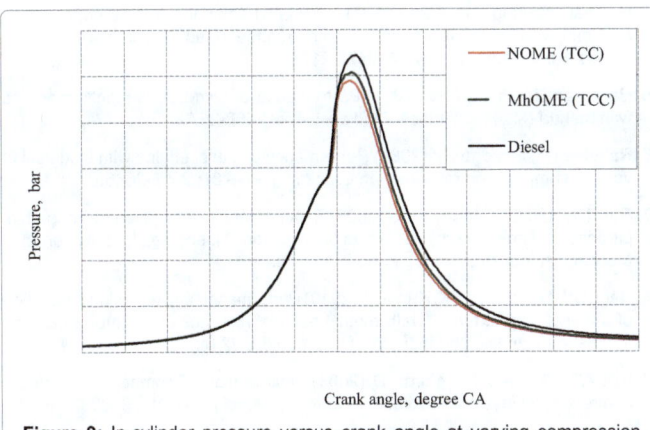

Figure 9: In-cylinder pressure versus crank angle at varying compression ratio for NOME and MhOME biodiesels at 80% load.

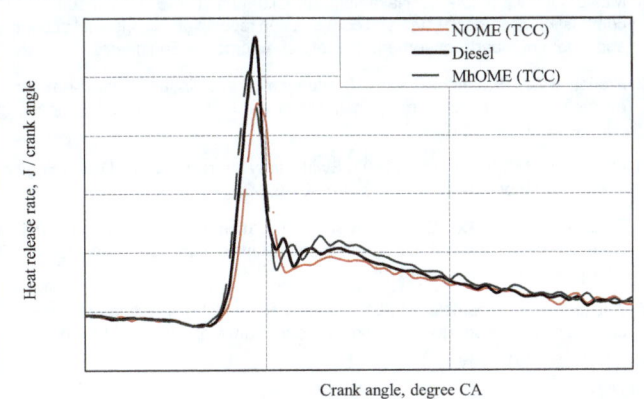

Figure 10: Rate of heat release versus crank angle at varying compression ratio for NOME and MhOME biodiesels at 80% load.

Cylinder pressure

Figure 9 shows the effect of combustion chamber shapes on the in-cylinder pressure operated on different fuel combinations. The peak pressure depends on the combustion rate and amount of fuel consumed during rapid combustion period. Mixture preparation and slow burning nature of biodiesel during the ignition delay period were responsible for peak pressure and maximum rate of pressure rise.

Results showed that biodiesel with TCC resulted in higher peak pressure as shown in Figure 9. The pressure for MhOME biodiesel operation with TCC was higher compared to NOME biodiesel tested. It could be due to the combined effect of longer ignition delay, lower adiabatic flame temperature and slow burning nature of the biodiesel operation. This could be attributed to incomplete combustion due to improper mixing of fuel combinations, reduction of air entrainment, and higher viscosity of biodiesel. The sharp increase in combustion acceleration showed increased cylinder pressure during the piston's descent and that the combustion energy as efficiently converted into work.

Heat release rate

Figure 10 shows rate of heat release versus crank angle for different two biodiesels with TCC combustion chamber shapes. Biodiesel operation for TCC resulted into higher heat release rate compared to the operation with other combustion chambers. Combustion chamber

being common NOME has higher second peak in the diffusion combustion phase compared to the MhOME operation with TCC operation.

Conclusions

From the exhaustive experimentation on the use of biodiesel in diesel engines with different combustion chamber shapes the following conclusions were made for the present study.

- Both biodiesels of Mahua and Neem resulted into inferior engine performance with increased emissions compared to diesel operation. The performance was improved with TCC combustion chamber provision.

- Improved air motion, better mixture formation with TCC compared to other combustion chambers was observed.

- Higher brake thermal efficiency with lower emission levels obtained with TCC.

- Combustion chamber optimization coupled with optimized injector position further improves the diesel engine performance.

References

1. Goldemberg J, Coelhobn ST (2004) Renewable energy- traditional biomass vs. modern biomass. Energy Policy 32: 711-714.

2. Banapurmath NR, Tewari PG (2008) Combustion and emission characteristics of a Direct Injection CI engine when operated on Honge oil, Honge oil methyl ester (HOME) and blends of Honge oil methyl ester (HOME) and diesel. International Journal of Sustainable Engineering 1: 80-93.

3. Banapurmath NR, Tewari PG (2009) Effect of biodiesel derived from Honge oil and its blends with diesel when directly injected at different injection pressures and injection timings in single cylinder water cooled compression ignition engine. Proceedings of Institute of Mechanical Engineers, Part A: Journal of Power and Energy. Professional Engineering Publications 223: 31-40.

4. Bari S, Lim TH, Yu CW (2002) Effect of Preheating of Crude Palm Oil (CPO) on Injection System, Performance and Emission of a Diesel Engine. Renewable Energy 27: 339-351.

5. Gajendra Babu MK, Chandan Kumar, Das LM (2006) Experimental Investigations on a Karanja Oil Methyl Ester Fuelled DI Diesel Engine. Society of Automotive Engineers, Paper No.: 2006-01-0238, USA.

6. Karnwal A, Naveen Kumar MM, Chaudhary HR, Siddiquee AN, Khan ZA (2010) Production of Biodiesel from Thumba Oil: Optimization of Process Parameters, Iranica. Journal of Energy & Environment 1: 352-358.

7. Onga HC, Mahlia TMI, Masjukia HH, Norhasyimab RS (2011) Comparison of palm oil, Jatropha curcas and Calophyllum inophyllum for Biodiesel. Renewable and Sustainable Energy Reviews 15: 3501-3515.

8. Ramadhas AS, Muraleedharan C, Jayaraj S (2005) Performance and emission evaluation of a diesel engine fueled with methyl esters of rubber seed oil. Renewable Energy 30: 1789-1800.

9. Ramadhas AS, Jayaraj S, Muraleedharan C (2005) Characterization and effect of using rubber seed oil as fuel in the compression ignition engines. Renewable Energy 30: 795-803.

10. Raheman H, Phadatare AG (2004) Diesel engine emissions and performance from blends of karanja methyl ester and diesel. Biomass and Bioenergy 27: 393-397.

11. Sahoo PK, Das LM, Babu MKJ, Naik SN (2007) Biodiesel development from high acid value polanga seed oil and performance evaluation in a CI engine. Fuel 86: 448-454.

12. Sahoo PK, Das LM, Babu MKJ, Arora P, Singh VP, et al. (2009) Comparative evaluation of performance and emission characteristics of Jatropha, karanja and polanga based biodiesel as fuel in a tractor engine. Fuel 88: 1698-1707.

13. Venkanna BK, Reddy CV (2011) Performance, emission and combustion characteristics of direct injection diesel engine running on *calophyllum*

inophyllum linn oil (honne oil). International Journal of Agricultural & Biological Engineering 4.

14. Sundraapandian S, Devaradjane G (2007) Experimental and Theoretical investigation of the performance of vegetable-oil-operated CI engine. Society of Automotive Engineers, Paper No: 2007-32-0067.

15. Nwafor OMI (2000) Effect of advanced injection timing on the performance of rapeseed oil in diesel engines. Renewable Energy 21: 433-44.

16. Nwafor OMI (2003) The effect of elevated fuel inlet temperature on performance of diesel engine running on neat vegetable oil at constant speed conditions. Renewable Energy 28: 171-81.

17. Scholl KW, Sorenson SC (1993) Combustion of soyabean oil methyl ester in a direct injection diesel engine. Society of Automotive Engineers, Paper No.: 930-934.

18. Gajendra Babu MK (2007) Studies on performance and exhaust emissions of a CI engine operating on diesel and diesel biodiesel blends at different injection pressures and injection timings. Society of Automotive Engineers, Paper No.: 2007-01-0613, USA.

19. Hountalas DT, Kouremenos DA, Binder KB, Raab A, Schnabel MH (2001) Using advanced Injection timing and EGR to Improve DI engine efficiency at Acceptable NO and Soot levels. Society of Automotive Engineers, Paper No.: 2001-01-0199.

20. Roy MM (2009) Effect of fuel injection timing and injection pressure on combustion and odorous emissions in DI diesel engine. Journal of Energy Resources Technology, ASME Transactions 131: 1-8.

21. Bari S, Yu CW, Lim TH (2004) Effect of fuel injection timing with waste cooking oil as a fuel in direct injection diesel engine. Proceedings of the Institution of Mechanical Engineers, Part D: Journal of Automobile Engineering 218: 93-104.

22. Puhan S, Vedaraman N, Ram BVB, Sankaranarayanan G, Jeychandran K (2005) Mahua oil (Madhuca Indica seed oil) methyl ester as biodiesel-preparation and emission characteristics. Biomass Bioenergy 87-93.

23. Rosli, Abu Bakar, Abdul Rahim Ismail Semin (2008) Fuel injection pressure effect on performance of direct injection diesel engines based on experiment. American Journal of Applied Sciences 5: 197-202.

24. Godiganur S, Murthy C (2009) 6BTA 5.9 G2-1 Cummins engine performance and emission tests using methyl ester mahua (Madhuca indica) oil/diesel blends. Renewable Energy 34: 2172-2177.

25. Raheman H, Ghadge SV (2007) Performance of compression ignition engine with mahua (Madhuca indica) biodiesel. Fuel 2568-2573.

26. Raheman H, Ghadge SV (2008) Performance of diesel engine with biodiesel at varying compression ratio and ignition timing. Fuel 87: 2659-2666.

27. Rao TV, Rao GP, Reddy KHC (2008) Experimental investigation of Pongamia, jatropha and neem methyl esters as biodiesel on C.I. engine. Jordan Journal of Mechanical Industrial Engineering 2: 117-122.

28. Ragit SS, Mohapatra SK, Kundu K (2010) Performance and emission evaluation of a diesel engine fuelled with methyl ester of neem oil and filtered neem oil. Journal of Scientific and Industrial Research 69: 62-66.

29. Risi AD, Donateo T, Laforga D (2003) Optimisation of combustion chamber of direct injection diesel engines. SAE International Paper No. 2003-01-1064.

30. Jaichander S, Annamalai K (2012) Performance and exhaust emission analysis on pongamia biodiesel with different open combustion chambers in a DI diesel engine. Journal of scientific and industrial research 71: 487-491.

31. Matsumoto K, Inoue T, Nakanishi K, Okumura T (1977) The effects of combustion chamber design and compression ratio on emissions, fuel economy and octain number requirement. Society of Automotive Engineers.

32. Pratiba BVV, Prasanthi G (2011) Influence of in cylinder air swirl on diesel engine performance and emission. International Journal of Applied Engineering and Technology 1: 113-118.

33. McCracken ME, Abraham J (2001) Swirl-Spray Interactions in a Diesel Engine. Society of Automotive Engineers 2001-01-0996.

34. Balawant ST (2012) Experimental investigation on effect of combustion chamber geometry and port fuel injection system for CNG engine. IOSR Journal of Engineering 2: 49-54.

35. Jaichander S, Annamalai K (2013) Combined impact of injection pressure and combustion chamber geometry on the performance of a biodiesel fueled diesel engine. Energy 55: 330-339.

11

Variation in the Carbon (C), Phosphorus (P) and Nitrogen (N) Utilization during the Biodegradation of Crude Oil in Soil

Oje Obinna A[1]*, Ubani Chibuike S[2] and Onwurah INE[2]

[1]*Department of Chemistry/Biochemistry/Molecular Biochemistry, Federal University Ndufu Alike Ikwo, Ebonyi State, Nigeria*
[2]*Department of Biochemistry, University of Nigeria, Nsukka, Enugu State, Nigeria*

Abstract

This study was aimed at determining the effect of varying concentration of crude oil pollution on the macro-nutrients in the soil. Various macro-nutrients (such as soil ammonium concentration, soil nitrate concentration, and available phosphorus), oxidizable organic carbon, and total petroleum hydrocarbon were determined. The result revealed that as the time increases, the oxidizable carbon, which is also a function of the organic matter decreases, which is as a result of the conversion of carbon to carbon (IV) oxide during cellular metabolism. This decrease showed that there was an increase in the activity of that leads to the breakdown of the carbon components in the soil. The soil phosphate concentration determination did not show any pattern in their increase or decrease, which shows that increase in crude oil concentration, did not significantly affect the phosphate concentration in the soil. The soil ammonium concentration increased from 24th hour to 168th hour but decreased before the end of the experiment. This increase could be attributed to ability of the Azotobacter *vinelandii* to fixed nitrogen as an innate responsibility, while *Pseudomonas* sp. which is known to contain nitrogen fixing genes. The result also showed that there is a constant increase in soil nitrate concentration which is affected as the concentration of the pollutant (crude oil) increases. This study has shown that the consortium of these organisms can be used as a bio-fertilizer as well as in bioremediation.

Keywords: Bioremediation; Ammonium; Nitrate; Phosphorus; *Pseudomonas* and *azotobacter*

Introduction

Crude oil is a complex mixture of various hydrocarbons, which are made up various aliphatic and aromatic hydrocarbons [1]. They also contain poly aromatic hydrocarbons (PAHs) which are recalcitrant. The dependency of crude oil as a major source of energy in Nigeria, has led to the pollution of different environment (land/soil, water and air). The causes of these pollutions include; exploration, exploitation, storage, transportation, vandalization, bunkering and gas flaring [2]. These have led to the introduction of various pollutants into the environments that have been implicated in the depiction of microbial flora in the soil, death of aquatic organisms, the depletion of the ozone layer and the formation of acid rain. Most of these pollutants such as the PAHs have been implicated in cancer, mutation etc [3]. In the soil and land environment, crude oil pollution has been known to affect agricultural yield as it hinders plant growth. Due to these effects of crude oil pollution, it then becomes necessary that oil spills are cleaned up as quickly as possible. One of the methods used in cleanup of the environment is the use of microorganism or products of microorganism known as bioremediation. The essence of bioremediation is not just to remove the pollutants, it also to restore the environment to its habitable form. Therefore physical methods of remediation are not employed. *Pseudomonas* species have been employed in the remediation of crude oil environment [4] due to its ability to biodegrade crude oil using it as a carbon source for generation of biomass and energy. But for the organisms to carry out their cellular activity, the supply of other macronutrient such as nitrogen and phosphorus are very necessary. Nitrogen been are major component of amino acid, purines and pyrimidines is required for the formation of proteins. RNAs and DNAs [5]. These control the functions of the organism to a large extent. Phosphorus have been involved in the formation of phospholipid which are involved in cell wall formation and also in the production of energy carriers, such as ATP, GTP, UTP, TTP, CTP etc. *Pseudomonas* are known to solubilize phosphorus from

the soil [6], while Azotobacter *vinelandii* fix atmospheric nitrogen into the soil [7], both organisms thereby act as a biofertilizer for the supply of these macronutrient. Therefore this study is aimed at determining how different concentration of crude oil in soil can affect carbon, nitrogen and phosphorus content in the soil.

Materials and Method

Crude oil

The Crude oil used was gotten from the Directorate of Petroleum Resources Port-Harcourt, Rivers State Nigeria,

Soil

The soil samples that were used in this study were obtained from the Agric Farm, Department of Agriculture, University of Nigeria, and Near Green House.

Microorganism

Two microorganisms were used in the course of the research. The *Pseudomonas* species was gotten from the culture collection Centre Department of Microbiology, University of Nigeria, Nsukka while the Azotobacter *vinelandii* was isolated from the soil around the

*Corresponding author: Oje Obinna A, Department of Chemistry/Biochemistry/ Molecular Biochemistry, Federal University Ndufu Alike Ikwo, Ebonyi State, Nigeria
E-mail: obinnaoje@yahoo.com

postgraduate laboratory, Department of Biochemistry, University of Nigeria, Nsukka using Azotobacter *vinelandii* specific media.

Determination of the remaining Total Petroleum Hydrocarbon (TPH) the modified methods of Ubani et al.

One gram (1 g) of soil was put in a test-tube and ten milliliter (10 ml) of Chloroform/Ethanol mixture (1:1) was added. The mixture was agitated for 5 minutes and then allowed to stand for 10 minutes. The sample was then filtered and the absorbance of the filtrate was taken at 520 nm using chloroform/Ethanol mixture (1:1) as a blank. The quantity of crude oil was estimated using a crude oil standard curve.

Determination of soil percentage oxidizible organic carbon, Total Organic Carbon and Organic Matter

Two grams (2 g) of dried soil was transfer to a 500 mL Erlenmeyer flask, and 10 mL of 0.167 M $K_2Cr_2O_7$ was added by means of a pipette. 20 ml of concentrated H_2SO_4 was and swirl gently to mix, (Excessive swirling was avoided to prevent the organic particles from adhering to the sides of the flask out of the solution). The mixture was allowed to stand on an insulation pad for about 30 minutes. Then 200 ml of distilled water was used to dilute the suspension so as to provide a clearer solution for viewing the endpoint. Then 10 ml of 85% H_3PO_4 was added and 0.2 g of NaF was also added. The H_3PO_4 and NaF are added to complex Fe^{3+} which would interfere with the titration endpoint. 10 drops of ferroin indicator was added and then titrated with 0.5 M Fe^{2+}. The color of the solution at the beginning is yellow-orange to dark green, depending on the amount of unreacted $Cr_2O_7^{2-}$ remaining, which shifts to a turbid gray before the endpoint and then changes sharply to a wine red at the endpoint.

The organic carbon and organic matter percentages where calculated thus:

a. Percentage easily oxidizable organic C

$$\%C = \frac{(B\text{-}S) \times M \text{ of } Fe^{2+} \times 12 \times 100}{\text{grams of soil} \times 4000}$$

B = ml of Fe^{2+} solution used to titrate blank

S = ml of Fe^{2+} solution used to titrate sample

12/4000 = milliequivalent weight of carbon in grams

To convert easily oxidizable organic C to total C, divide by 0.77 (or multiply by 1.30) or other experimentally determined correction factor.

b. Percentage organic matter (OM)

$$\% OM = \frac{\%C}{0.58} = \%C \times 1.72$$

Determination of phosphorus concentration (Ascorbic Acid Method Procedure)

Two grams (2 g) of soil was weighed in a boiling tube, 20 ml of the extracting solution was added and the soil mixture was agitated for 5-10 mins. It was allowed to stand for 30 mins with occasional agitation every 8 mins. At the end of the 30 mins, it was filtered and the filtrate was collected. 2ml of the filtrated was then. Add to 8 mL of working solution in a test – tube. It was thorough agitation and mixing occurs. The mixture was allowed to stand for 10 minutes for color development before taken the absorbance at 882 nm. Read percentage of transmittance or optical density on a colorimeter or spectrophotometer set at 882 nm. Color is stable for about 2 hours.

A standard curve was prepared by pipetting a 5 ml aliquot of each working standard, developing color and reading absorbance in the same manner as with the soil extracts. The absorbance was plot against concentration of working standards. The concentration in soil extracts was determined from absorbance and the standard curve.

Determination of soil ammonium

One milliliter 1 mL of solution soil extract was put into a test-tube; 5.5 ml of buffer solution was added and agitated for 5 minutes. 4 ml of salicylate/nitroprusside solution was added and also 2 ml of hypochlorite solution was added and mixed properly. The mixture was allowed to stand for 45 mins at 37°C. The absorbance was taken at 650 nm with 2 hours.

Determination of soil nitrate

Twenty grams (20 g) of soil was weighed into a 100 ml beaker and 50 ml of extracting solution was added. It was agitated for 5 minutes and the potential was read in millivolts (mV) using a millivolt meter, while the mixture is being stirred. The concentration of NO_3 - N, was determined using the standard curve.

Record the millivolt reading (if using a calibration curve technique) or read the NO_3–N concentration directly from a pH/ion meter.

Results and Discussion

The use of micro-organisms to decontaminate the environment (Bioremediation), is being increasingly seen as an effective, environment- friendly treatment for crude oil contaminated sites. Large quantities of organic and inorganic compounds are released into the environments every year as results of anthropogenic activities thereby causing serious environmental problems [8]. The result of this study reveals Azotobacter *vinelandii* possesses the ability to breakdown crude oil in the soil (Figure 1). Azotobacter *vinelandii* is an autotroph which has the natural ability to fix atmospheric nitrogen into the soil and improve soil fertility [9]. This it does with the aid of nitrogenase complex (EC1.18.6.1) [10]. Some Azotobacter sp. such Azotobacter *chroococcum* has been reported to breakdown crude oil, emulsifies waste motor oil and other fractions petroleum indicating its potentiality in utilization of various hydrocarbons [11]. Azotobacter sp are not known to possess the ability to breakdown crude oil but they have the ability to pick up plasmids from the environment [12], which confers on them the properties which they naturally do not possess. Most adapted strains of Azotobacter sp possess the ability to breakdown crude oil. Figure 1 also showed that the rate of breakdown

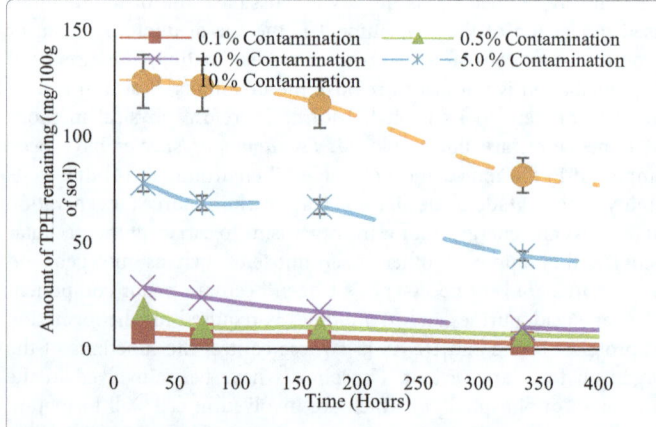

Figure 1: The concentration of TPH in the soil containing *Azotobacter vinelandii*.

of crude oil was low in the first 196 hrs. of the experiment; this is due to the time taken for the organism to produce the enzymes necessary for the breakdown of crude oil.

The TPH concentration in the soil seeded with *Azotobacter vinelandii*

Figure 1 shows the amount of total petroleum hydrocarbon (TPH) that is remaining in the soil. The figure shows that there is a gradual decrease in the TPH concentration with time. This decrease was found in all the groups from 0.1% contamination to 10.0% contamination.

The concentration of TPH in the soil containing *Pseudomonas* sp in Figure 2 showed that there is also a decrease in the amount of TPH in the soil with time. This is because of the ability of *Pseudomonas* sp to utilize crude oil as a carbon source [13,14].

During this process, biosurfactants are produced which helps to reduce the surface tension of crude oil thereby making the oil to be soluble in aqueous solution thereby making the crude oil available for microbial attack. *Pseudomonas* also produces lipase which helps in the degradation of lipids [15].

It have been reported that *Pseudomonas* sp possesses genes that code for enzymes such as catechol dioxygenase, alkane 1-monooxygenase and alkane sulfonate monooxygenase [16] that help in the breaking down of hydrocarbon chains.

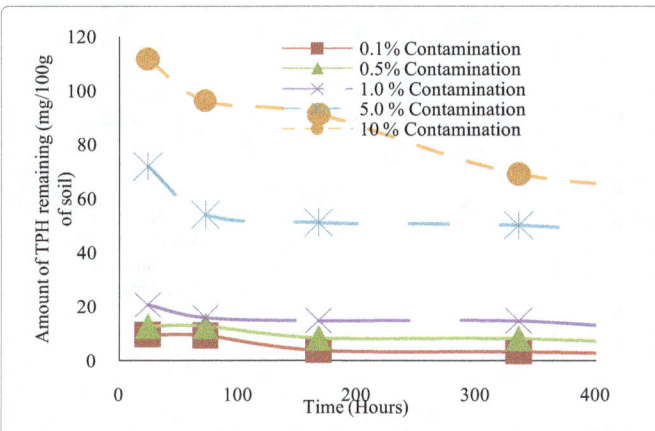

Figure 2: The concentration of TPH in the soil containing *Pseudomonas sp.*

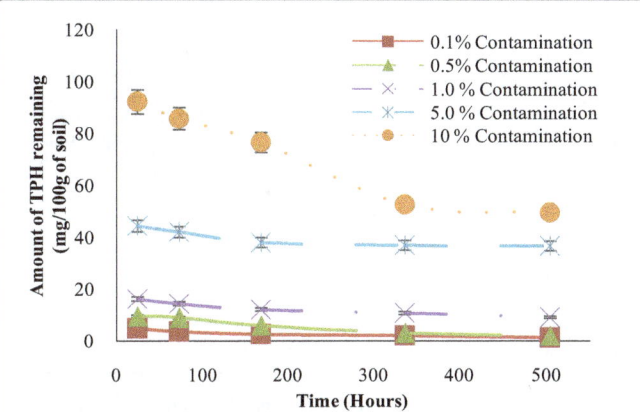

Figure 3: The concentration of TPH in the soil containing a consortium of *Azotobactervinelandii* and *pseudomonas sp.*

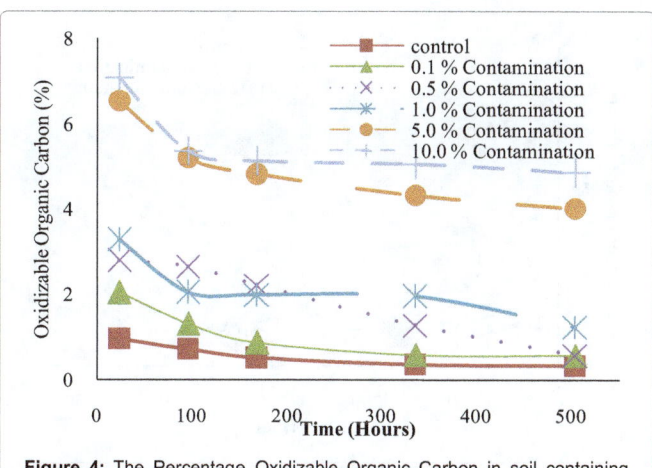

Figure 4: The Percentage Oxidizable Organic Carbon in soil containing *Azotobactervinelandii.*

The result also showed that *Pseudomonas* possess the ability to breakdown high concentration of crude oil thereby withstanding to some extent the toxicity of crude oil.

The concentration of TPH in the soil containing *Pseudomonas* in Figure 3 showed that there is also a decrease in the amount of TPH in the soil with time. The result shows that the consortium broke down more of the pollutants (crude oil) when compared with the individual organisms. This is as a result of the synergism that exist between the organisms [17].

The percentage oxidizable organic carbon in soil containing *Azotobacter vinelandii.*

The percentage of oxidizable organic carbon was observed to increase as the concentration of crude oil in the soil increases due to the fact that crude oil contains oxidizable carbons. As a result, the percentage of oxidizable carbon can be used to a parameter to determine the degradation ability of the microorganism [18,19].

Figures 4-6 show the gradual decrease in the percentage of oxidizable carbon as the time increases. It was also observed that the rate of decrease was high within the first 196 hrs of the experiment which is also corresponding to the period of rapid growth of the organisms. It could also be observed that at high concentration of the crude oil (10%, 5% and 1% contamination) there was a decrease in the rate of utilization of the oxidizable carbon (Figure 4). This might be as a result of the presence of some recalcitrant polyaromatic hydrocarbon PAH in the crude oil which might to be easily degraded be the organisms such as Azotobacter *vinelandii* or that the toxicity of crude oil increases with increase in percentage of crude oil contamination which might inhibit the metabolic activities in the organisms. Azotobacter *sp* are used as biofertilizers due to their ability to utilize oxidizable carbon for energy and fix nitrogen in the soil [20]. Probably, this may be the reason why Azotobacter sp is readily found in the soil environment.

The percentage oxidizable organic carbon in soil containing *Pseudomonas sp.*

The Figure 5 reveals a decrease in the percentage of oxidizable organic carbon in the soil containing *Pseudomonas* as was observed in Figure 4.

It was also observed that there was a continuous reduction in percentage of oxidizable carbon. Some species of *Pseudomonas* have

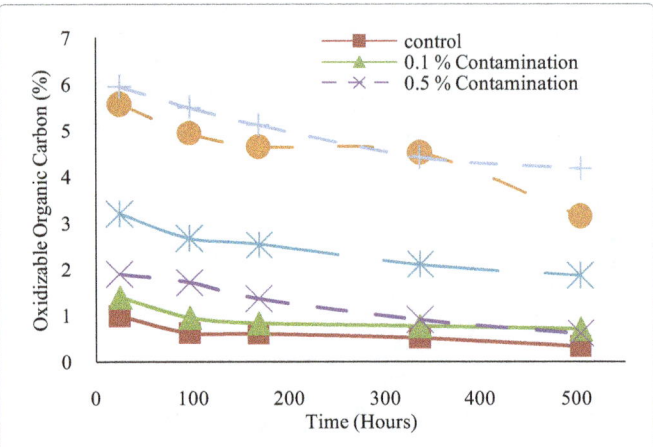

Figure 5: The Percentage Oxidizable Organic Carbon in soil containing *Pseudomonas sp.*

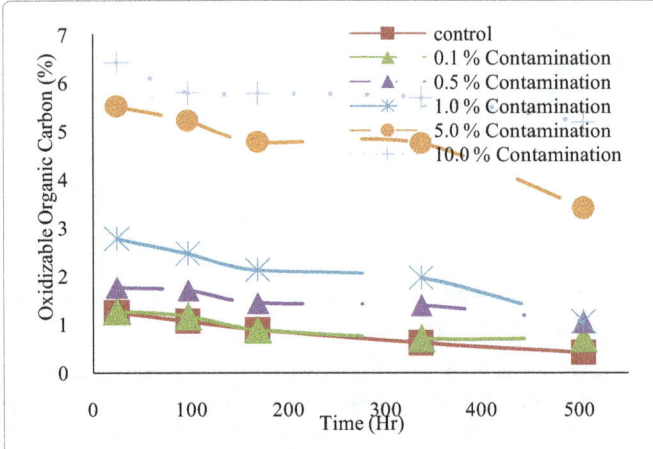

Figure 6: The Percentage Oxidizable Organic Carbon in soil containing a consortium of *Azotobactervinelandii* and *Pseudomonas sp.*

been found to remove organic carbon during denitrification reactions using them as electron donors during the process [21]. *Pseudomonas* sp have been known to breakdown long chain hydrocarbon utilizing them for growth and energy [22], therefore the continuous reduction in the percentage of oxidizable organic carbon could also be attributed to this ability the was naturally conferred on *Pseudomonas* sp.

The percentage oxidizable organic carbon in soil containing a consortium of *Azotobacter vinelandii* and *Pseudomonas sp.*

Figure 6 show the levels of oxidizable organic carbon in the soil containing a consortium of Azotobacter *vinelandii* and *Pseudomonas* sp. and the result reveals that there was a reduction in the organic carbon as the level of contamination increase although the result shows that at higher concentration of crude oil, that the change in the level of oxidizable carbon did not change significantly. Rathore [23] has shown that there is an existing synergism existing between *Azotobacter sp* and *Pseudomonas* sp in promoting the growth of plants. Therefore, one would have expected synergism between the organisms in reducing the oxidizable organic carbon but rate of reduction was not quite different when compared with the individual organisms. This probably show that to some extent that there are some level of inhibition existing between the organisms even though it did not affect the growth of the organism due to the fact that both organisms could be depending

on different substrates for energy and growth in which one organism might be depending on the product of the second organism.

The soil phosphate in soil containing *Azotobacter vinelandii*

The result in Figure 7 which shows the available soil phosphate concentration in the soil containing *Azotobacter vinelandii* revealed increasing concentration of phosphate in all the groups after 168 hours but decreased gradually in all the levels of contamination before 336 hours. The increase in phosphorus concentration also increases with in biomass (growth pattern) of the organism. Therefore as the biomass (*Azotobacter vinelandii*) increases, soil phosphate concentration increases because the organism possesses genes for phosphate solublization, even though they are expressed at a reduced rate. The initial rise in the soil available phosphorus was due to the ability of organism to solubilize phosphate in the soil. But as the concentration of phosphorus in the environment increases, there is a feedback inhibition of phosphatase activity by phosphate [24] which inactivates phosphorus solubilization.

The soil phosphate in soil containing *Pseudomonas sp*

The result in Figure 8 which shows the available soil phosphate concentration in the soil containing *Pseudomonas* sp. revealed higher

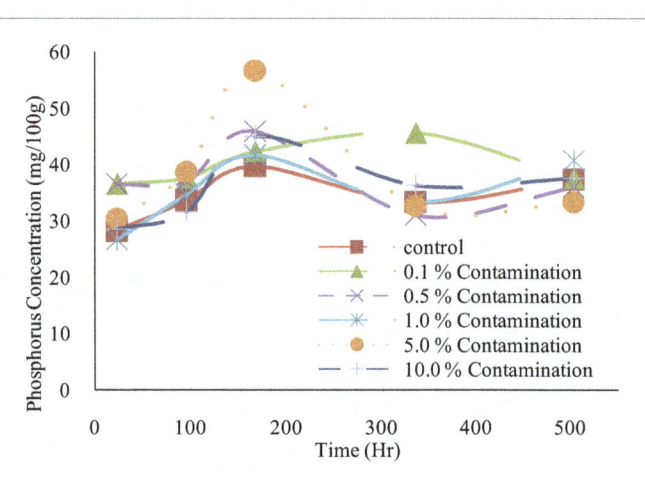

Figure 7: The concentration of Phosphate in soil containing *Azotobactervinelandii.*

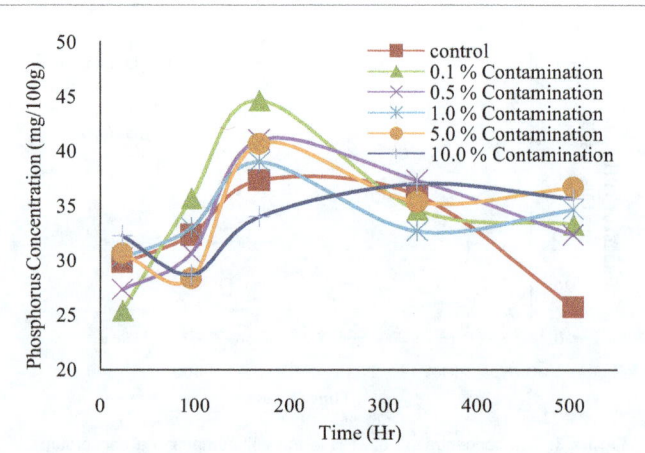

Figure 8: The concentration of Phosphate in soil containing *Pseudomonas sp.*

concentration of phosphate in all the groups after 168 hours but decreased gradually in all the levels of contamination. *Pseudomonas* sp are known to solubilize phosphate [6,25,26].

At the 168 hour, the concentration of available phosphate was lowest in soil sample with 10% contamination. This could be attributed to the toxicity of the pollutant. Also the control showed low available phosphate which could be attributed to the concentration of carbon source. According to Isolation and characterization of phosphate solubilizing bacteria (Klebsiella oxytoca) with enhanced tolerant to environmental stress [27].

The soil phosphate in soil containing *Azotobacter vinelandii* and *Pseudomonas sp*

The results (Figure 9) also show that there was an increase in the control group, 0.1% up to 168 hr and then the decrease sets in gradually. The variation at the 24th hour might be as a result of the variations in the localization of phosphate in the soil. But on the general note, it was observe that available concentration of phosphorus increase till the 168hr as was observed in the previous charts before a gradual decrease which could be attributed to microbial utilization of the free phosphorus.

The ammonium concentration in soil containing *Azotobacter vinelandii*

The concentration of the soil ammonium was found to increase in all the group up to about 96hours and then a gradual decrease was observed although the decrease did not follow a sequential other as seen in Figure 10.

The ammonium concentration in soil containing *Pseudomonas sp.*

The result in Figure 11 shows a sharp increase in the ammonium concentration in the soil from 24hours to 96 hours and after 168 hours, a gradual decrease was observed.

The ammonium concentration in soil containing a consortium of *Azotobacter vinelandii* and *Pseudomonas sp.*

Figure 12 also shows an increase in ammonium concentration in the soil from 24hour to 168 hour in all the groups but a sharp decrease was observed between 336 hour to 504 hours in all the levels of contamination.

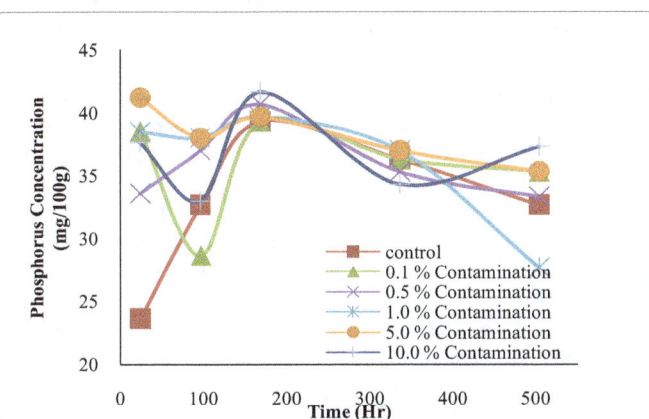

Figure 9: The concentration of Phosphate in soil containing a consortium of *Azotobactervinelandii*and *Pseudomonas sp.*

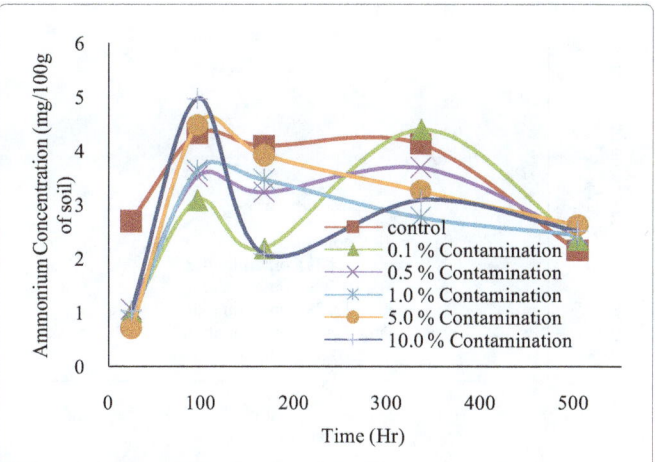

Figure 10: The concentration of Ammonium in soil containing a consortium of *Azotobactervinelandii.*

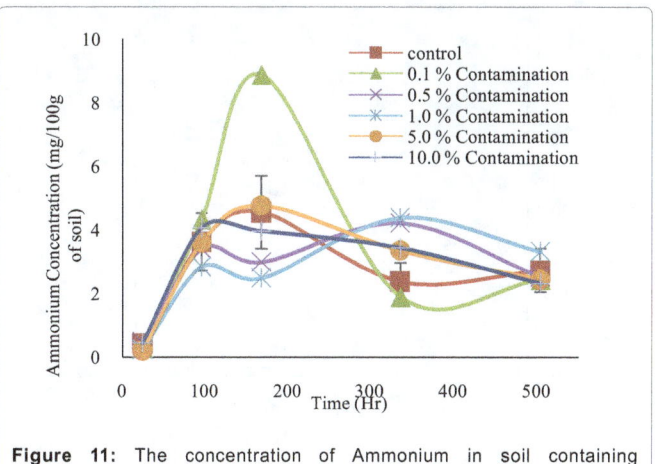

Figure 11: The concentration of Ammonium in soil containing *Pseudomonas sp.*

The nitrate concentration in soil containing *Azotobacter vinelandii*

The result of Figure 13 show that there was an increase in the nitrate concentration as the time increases but decreased with an increase in the percentage contamination.

The nitrate concentration in soil containing a consortium of *Pseudomonas sp.*

Figure 14 shows that there was a decrease in the nitrate concentration as the levels of contamination increase that is from 0.1% crude oil contamination to 10.0% crude oil contamination. It was also observed that the nitrate concentration increased with time. Although there is an unusual decrease at 5% crude oil contamination.

The nitrate concentration in soil containing a consortium of *Azotobacter vinelandii* and *Pseudomonas sp.*

Figure 15 above showed that there was a decrease in the soil nitrate concentration as the percentage of crude oil contamination increases. On the other hand, it was observed that was an increase in the nitrate concentration of the soil containing the consortium as the time increases. Although at the 336th hour in the consortium, there

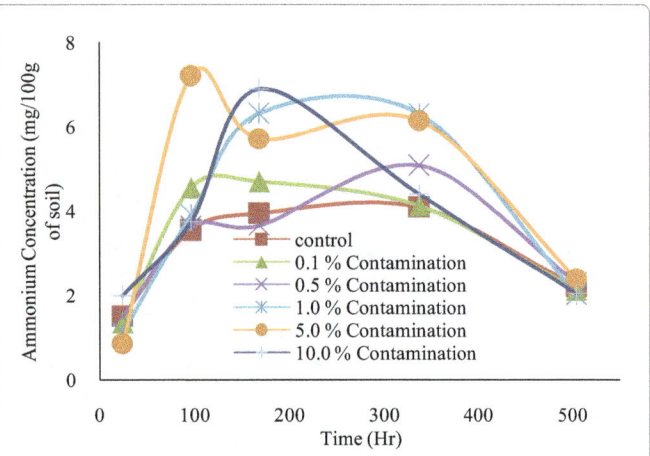

Figure 12: The concentration of Ammonium in soil containing a consortium of *Azotobactervinelandii* and *Pseudomonas sp.*

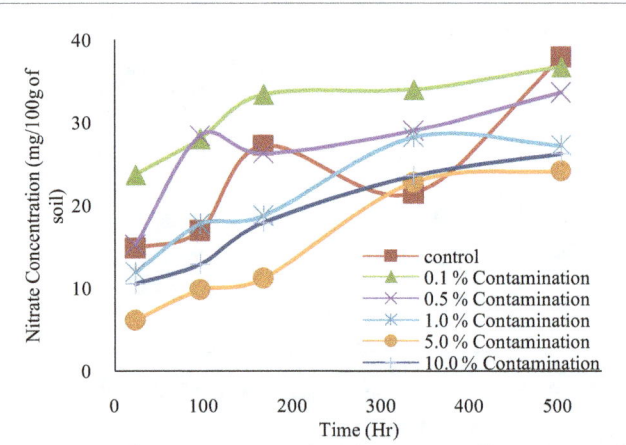

Figure 13: The concentration of nitrate in soil containing *Azotobactervinelandii.*

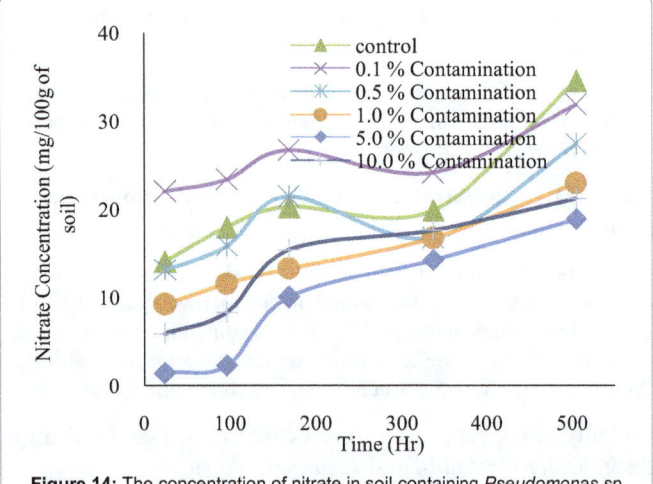

Figure 14: The concentration of nitrate in soil containing *Pseudomonas sp.*

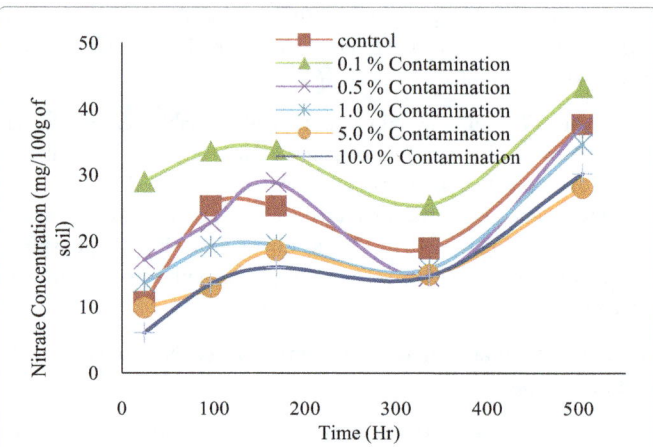

Figure 15: The concentration of nitrate in soil containing a consortium of *Azotobactervinelandii* and *Pseudomonas sp.*

utilization begins to reduce giving room to more nitrate in the soil.

Conclusion

Soil organic matter is made up of plant nutrients and carbon sources. Organic matter in soil are from natural materials such as plant materials, animal litters and microbial biomass. Soil organic matter influences the bioavailability of nutrients as well as soil enzymes produced by bacteria. Bioavailability is one of the most limiting factors in bioremediation of persistent organic pollutants in soil. Compared to the other soil characteristics, soil organic matter is the major factor which affects the distribution and bioavailability of petroleum hydrocarbons. The result revealed that as the percentage of contamination increases, there is also an increase in the above mentioned parameters. According to Liu et al., Soil organic matter increased significantly after an oil contamination. It was also observed that as the time (days) increases, the oxidizable carbon, total organic carbon and the organic matter decreases. This is as a result of the conversion of carbon to carbon (IV) oxide during cellular metabolism which is released into the atmosphere. It was observed in most cases, that the *A. vinelandii* showed lower concentration of the organic carbon when compared with *Pseudomonas* sp and the consortium. The decrease might be as a result of increase in the activity of that leads to the breakdown of the carbon components of the soil, even though there was low microbial growth. This might also be due to the large size of organism which might require more carbon for macromolecule synthesis.

The soil phosphate concentration was determined and the result did not show any pattern in their increase or decrease. This shows that increase in crude oil concentration did not significantly affect the phosphate concentration in the soil. *Pseudomonas* sp is known to solubilize phosphate in the soil and make them available. Azotobacter according to Rediers et al., should be classified under the genus as *Pseudomonas* since both of them have a lot of gene sequence in common. It is therefore possible that A. vinelandii also solubilizes phosphate in the soil and this should account for the slight increase in the level of phosphate found in the consortium.

Most *Pseudomonas* species are known for their ability to reduce of nitrates back into the largely inert nitrogen gas (N2), completing the nitrogen cycle. This process is performed by bacterial species such as *Pseudomonas* and Clostridium in anaerobic conditions. They use the

was an observed decrease in the nitrate concentration which might be as attributed to increase in the rate at which nitrate is utilized by the organism at the period which consider with the period of maximum growth. But as the amount of organisms begin to reduce, the rate of

nitrate as an electron acceptor in the place of oxygen during respiration. Some microorganisms possess some ability to undergo transformation especially in extreme conditions such extreme pH, temperature, and deficiency of nutrients. In aerobic conditions, some *Pseudomonas* species can help fix nitrogen by converting atmospheric nitrogen to ammonium and then to nitrate. The nif-Hphylogenies (which is responsible for fixing nitrogen is found in *Pseudomonas*) contains at least 42 genes to encode the denitrification apparatus' core structures. The genome also contains genes involved in nitrogen fixation, denitrification, chemotaxis and other functions that presumably give the *Pseudomonas* an advantage. The result of this research shows that the soil ammonium concentration increased from day 1 to day 7 but decreased before day 21. This increase could be attributed to ability of the Azotobacter *vinelandii* to fixed nitrogen as an innate responsibility, while *Pseudomonas* sp. which is known to contain nitrogen fixing activity. The increase observed tallies with the growth pattern of the organisms which shows that as the organism decrease in number, there is a decrease in the amount of ammonium in the soil. Although it should be noted that the organisms still use some quantity of nitrogen for cellular activity.

The result also showed that there is a constant increase in soil nitrate concentration from day 1 to 7 but this decrease with increasing concentration of the pollutant (crude oil). This shows that increase in crude oil also reduces nitrate concentration in the soil. At high concentration of crude oil, cellular activities are distorted due to toxicity introduced by the crude oil.

The result of this study showed a constant decrease in the percentage of organic carbon (OC) and the total petroleum hydrocarbon (TPH). This is evidence that organisms *Pseudomonas* and *Azotobacter vinelandii* have the ability to utilize the TPH as an alternative source of energy therefore able to remove this contamination and restoring the environment. It was observed that there was an increase in macro-nutrients (nitrogen and phosphorus) which are needed to improve the soil fertility. Therefore a consortium of the *Pseudomonas* spp and *Azotobacter vinelandii* can be used not only in the remediation but also as a bio-fertilizer.

References

1. Ghazali FM, Rahman RN, Salleh AB, Basri M (2004) Biodegradation of hydrocarbon in soil by microbial consortium. International Biodeterioration and Biodegradation 54: 61-67.

2. Ite AE, Ibok UJ, Ite MU and Petters SW (2013) Petroleum Exploration and Production: Past and Present Environmental Issues in the Nigeria's Niger Delta. American Journal of Environmental Protection 1: 78-90.

3. Le T, Qureshi A, Heisler J, Bryant L, Shah J, et al. (2014) Polycyclic aromatic hydrocarbons and small related molecules: Effects of Schizosacharomyces pombe morphlgy measured by imaging flow cytometry. Journal of Yeast and Fungal Research 5: 84-91.

4. Al-Wasify RS, Hamed SR (2014) Bacterial Biodegradation of Crude Oil Using Local Isolates. International Journal of Bacteriology 2014: 1-8.

5. Katoch R (2011) Qualitative and Quantitative Estimations of Amino Acids and Proteins. Analytical Techniques in Biochemistry and Molecular Biology 93-147.

6. Delgado M, Mendez J, Rodríguez-Herrera R, Aguilar CN, Cruz-Hernández M, et al. (2014) Characterization of phosphate-solubilizing bacteria isolated from the arid soils of a semi-desert region of north-east Mexico. Biological Agriculture and Horticulture: An International Journal for Sustainable Production Systems 30: 211-217.

7. Swain H, Abhijita S (2013) Nitrogen Fixation and Its Improvement through Genetic Engineering. Journal of Global Biosciences 2: 98-112.

8. Battikhi MN (2014) Editorial: Bioremediation of Petroleum Sludge. J Microbiol Exp 1: 00011.

9. Onwurah INE, Ogugua VN, Onyike NB, Ochonogor AE, Otitoju OF (2007) Crude Oil Spills in the Environment, Effects and Some Innovative Clean-up Biotechnologies. International Journal of Environmental Research 1: 307-320.

10. Seyhan E, Kirwan DJ (1979) Nitrogenase activity of immobilized Azotobacter vinelandii. Biotechnol. Bioeng 21: 271-281.

11. Thavasi R, Jayalakshmi S, Balasubramanian T, Banat IM (2006) Biodegradation of Crude Oil by Nitrogen Fixing Marine Bacteria Azotobacter chroococcum. Research Journal of Microbiology 1: 401-408.

12. Guo C, Sun L, Kong D, Sun M, Zhao K (2014) *Klebsiella variicola*, a nitrogen fixing activity endophytic bacterium isolated from the gut of *Odontotermes formosanus*. African Journal of Microbiology Research 8: 1322-1330.

13. Chikere CB, Ekwuabu CB (2014) Culture-dependent characterization of hydrocarbon utilizing bacteria in selected crude oil-impacted sites in Bodo, Ogoni land, Nigeria. African Journal of Environmental Science and Technology 8: 401-406.

14. Almansoory AF, Idris M, Abdullah SRS, Anuar N (2014) Screening For Potential Biosurfactant Producing Bacteria From Hydrocarbon Degrading Isolates. Advances in Environmental Biology 8: 639-647.

15. Azhdarpoor A, Mortazavi B, Moussavi G (2014) Oily wastewaters treatment using Pseudomonas sp. isolated from the compost fertilizer. Journal of Environmental Health Science & Engineering 12: PP.

16. Patel PA, Kothari VV, Kothari CR, Faldu PR, Domadia KK, et al. (2014) Draft genome sequence of petroleum hydrocarbon-degrading Pseudomonas aeruginosa strain PK6, isolated from the Saurashtra region of Gujarat, India. Genome Announc 2: e00002-14.

17. Anitha G, Kumudini BS (2014) Isolation and characterization of fluorescent pseudomonads and their effect on plant growth promotion. Journal of Environmental Biology 35: 627-634.

18. Mrayyan B, Battikhi MN (2005) Biodegradation of total organic carbons (TOC) in Jordanian petroleum sludge. J Hazard Mater 120: 127-134.

19. Okoro SE, Adoki A (2014) Bioremediation of crude oil impacted soil utilizing surfactant, nutrient and enzyme amendments. J Bio Env Sci 4: 41-50.

20. El-Lattief EA (2013) Impact of integrated use of bio and mineral nitrogen fertilizer on productivity and profitability of wheat (*Triticum aestivum* L.) under Upper Egypt conditions. International Journal of Agronomy and Agricultural Research 3: 67-73.

21. Guo H, Chen C, Lee DJ, Wang A, Ren N (2013) Sulfur-nitrogen-carbon removal of Pseudomonas sp. C27 under sulfide stress. Enzyme Microb Technol 53: 6-12.

22. Alsulami AA, Altaee AMR, Al-Kanany FNA (2014) Improving oil biodegradability of aliphatic crude oil fraction by bacteria from oil polluted water. African Journal of Biotechnology 13: 1243-1249.

23. Rathore P (2014) A Review on Approaches to Develop Plant Growth Promoting Rhizobacteria. International Journal of Recent Scientific Research 5: 403-407.

24. Traoré L, Nakatsu CH, DeLeon A, Stott DE (2013) Characterization of six phosphate-dissolving bacteria isolated from rhizospheric soils in Mali. African Journal of Microbiology Research 7: 3641-3650.

25. Rajasankar R, Manju-Gayathry G, Sathiavelu A, Ramalingam C, Saravanan VS (2013) Pesticide tolerant and phosphorus solubilizing Pseudomonas sp. strain SGRAJ09 isolated from pesticides treated Achillea clavennae rhizosphere soil. Ecotoxicology 22: 707-717.

26. Goteti PK, Desai S, Emmanuel LDA, Taduri M, Sultana U (2014) Phosphate Solubilization Potential of Fluorescent Pseudomonas spp. Isolated from Diverse Agro-Ecosystems of India. International Journal of Soil Science 9: 101-110.

27. Walpola BC, Arunakumara KKIU, Yoon M (2014) Isolation and characterization of phosphate solubilizing bacteria (Klebsiella oxytoca) with enhanced tolerant to environmental stress. African Journal of Microbiology Research 8: 2970-2978.

12

Safe Practices in Drilling and Completion of Sour Gas Wells

Mahmood Amani* and Mohamed Almodaris

Texas A&M University, Qatar

Abstract

This paper examines the current status of the global market of sour fields. In particular, it clarifies on the existing methods, standards, and technologies applied for drilling, cementing, and completing a well with high H_2S concentration, with view of recommending the minimum health, environmental, and safety (HSE) protocols or standards required when dealing with sour environments.

The overall approach for this study was literature review. This will involve exploration of various exploitation case studies from the region and the globe meant for improving the economics of developing sour reservoirs in this cost sensitive time for the industry. Literature on current status of methods and technologies applied for drilling, cementing and completing high H_2S concentration wells will be reviewed. Standards and recommendations set by the American Petroleum Institute (API) and National Association of Corrosion Engineers (NACE) on the minimum Health, Environmental, and Safety (HSE) for use in sour environments will also be reviewed.

The findings of this study informed of the fact that there is very small even any room for error when it comes with exploiting high H_2S concentration well reservoirs. The results of the study established that dealing with sour fields is a challenge the industry will face more often, and in increasing magnitudes over time. In particular, it was established that the best approach for meeting this challenge is for companies in the industry to acquire knowledgeable personnel, an element that dictates for consulted investing in the proper training of employees. In order to ensure proper training of personnel, core areas of the training should encompass best practices on the selection of materials, planning operations, and the swift planning and executing of operation plans and incidents.

This study will have implication for unlocking the full potential of the sour environments frontier for the region. The findings of this study will provide drilling and completions engineers with insightful knowledge on best practices in dealing with sour fields.

Keywords: Drilling; Gas-wells; Petroleum; Environmental; Oil; Bacteria

Introduction

Sour service refers to a well environment containing significant amounts of Hydrogen Sulfide (H_2S). H_2S is hazardous to human health, living organisms, and the environment in general. Failures in sour environments are a major concern to Oil & Gas companies due to the higher risk, cost, and lost time associated with these failures [1]. There is no set cut off point to classify whether a field is sour or sweet. The threshold differs from one area to another. In majority of areas, gas is qualified as sour gas if H_2S makes for more than 2.5% of gas contents. The Middle-East region isn't strange to highly sour fields with H_2S up to 30% in some fields. Canada is one of the first locations to discover sour fields in and is notorious for high H_2S content with one well containing up to 90% H_2S. H_2S isn't restricted to a certain type of field. It can be found in oil and/or gas fields, onshore and offshore, HPHT and conventional, etc. Even if a field starts with no H_2S content, the H_2S content keeps on increasing as the field is aging.

The International Energy Agency (IEA) published in their 2014 medium term gas market report a 1.2% growth in global natural gas demand over the span of 2013. And BP forecasts in their energy outlook an increase in global natural gas demand by an average of 1.9% per year to 2035. With increasing demand of gas worldwide, some highly sour oil and gas reservoirs are being explored, mainly in Russia, the Middle East, China, North America, and are now more and more associated with complex well profiles – such as deep reservoirs or extended reach wells. As time passes, more of the previously uneconomical sour fields will become viable development projects. The Oil and Gas industry will need to step up its game and address the challenges associated with developing such fields.

Reservoir souring

The process of introducing or increasing H_2S in a reservoir is called "Reservoir Souring." Reservoir souring can be attributed to a number of factors. It can happen due to poor microbial treatment in pipeline or topsides facilities, the addition of bacteria to the reservoir through poorly treated injection water, and use of sour gas for gas lift [2]. When injection water isn't treated properly, it can introduce Sulfate-Reducing Bacteria (SRB). SRB are present almost everywhere including injection water (seawater). In seawater, SRB is inactive due to the oxidizing environment but under reservoir conditions, SRB becomes active. SRB are bacteria that breaks down organic material while reducing sulfate and producing H_2S instead. SRB are tolerant to high pressures but are less tolerant of high temperatures. SRB can survive under pressures up to 7500 psia but thrives in pressures below 4000 psi. Most common SRB can't grow in temperatures above 113° F. However, there have been reports from oil fields about SRB that can grow in temperatures above 158° F [3]. Even if reservoir temperature is above SRB limit, SRB would grow near the injectors and continue to generate H_2S (Figure 1) [3].

Once SRB is introduced to the reservoir it is impossible to fully treat it. If the biocides even leave a small fraction of SRB in the reservoir, it's

*Corresponding author: Amani M, Texas A&M University, Qatar
E-mail: mahmood.amani@qatar.tamu.edu

still enough to sour it. Under favorable conditions, a bacteria colony can double size in 20 minutes. Still, it's useful to maintain biocide injection especially in injection fluid and packer fluid to decrease SRB effect. One of the recent methods to control reservoir souring is injecting nitrates with seawater (injection fluid) [3]. This method encourages Nitrate-reducing bacteria (NRB), a competitive bacterium, that produces nitrogen instead of H_2S. NRB will consume the organic compounds, denying SRB from them. This method is widely used yet its potential corrosion side effect is still under study.

Pitting corrosion and sulphide stress cracking

H_2S in produced fluids reacts with steel to form a semi-protective film of iron sulphide (FeS). Unfortunately iron sulphide is rarely uniform and can be removed by flow exposing fresh metal to H_2S. The exposed metal rapidly corrodes causing pitting similar to Figure 2. Pitting corrosion is a localized form of corrosion by which cavities or "holes" are produced in the material. Pitting is considered to be more dangerous than uniform corrosion damage because it is more difficult to detect, predict and design against. Corrosion products often cover the pits. A small, narrow pit with minimal overall metal loss can lead to the failure of an entire engineering system.

H_2S can also cause sulphide stress cracking (SSC). SSC is a form of hydrogen stress cracking. The role of H_2S is to provide hydrogen at the metal surface by corrosion and to prevent hydrogen from escaping into the production fluid. The hydrogen then finds an alternative route by

migrating through the metal structure; this is possible due to small size of the hydrogen atom. In the absence of H_2S, normally, hydrogen atoms would combine to form the larger hydrogen molecule and just bubble off instead of migrating through metal. This migration is temperature dependent, the higher the temperature the easier the migration. Under low stress blistering will occur while under high stress cracking occurs [3]. Brittle metals are more prone to cracking. SSC is more critical than pitting corrosion caused by H_2S because it happens suddenly while corrosion is a relatively slower process. While coating and continuous corrosion inhibition through chemical injection can lower H_2S pitting and cracking, The National Association of Corrosion Engineers (NACE) doesn't qualify them as mitigation measures due their low reliability. Coating can be removed and inhibitor injection can stop due to low supply or downhole injection valve failure. NACE provides a guideline to labeling sour service. Sour service severity is split into four regions based on pH versus H_2S partial pressure as in Figure 3.

a. Region 0: H_2S partial pressure below 0.05 psi is considered to be non-sour.

b. Region 1: characterized with low partial pressure and relatively high pH is considered mildly sour.

c. Region 2: is considered moderately sour and requires API N80 and C95.

d. Region 3: the highly sour level which requires API L80 and C90.

Materials

The metallurgy of the drilling equipment has evolved over many years and keeps on evolving to meet new challenges. Proper material selection is the main line of defense against H_2S pitting and cracking. A prime example of the importance of correct material selection would be the recent pipe leak in Kashagan field in Kazakhstan. After only a few weeks from inauguration of the field, production was shut after discovering H_2S leakage from the pipelines. The solution is to replace 200 km of leaking pipe which is estimated to cost more than \$3.6 billion [4]. The most important drilling equipment selections in a sour environment are drill pipe, tool joints, drill collars, blowout preventers, and wellheads. These equipment have the highest chance of suffering from corrosion or cracking during drilling. The emphasis for these equipment is on strength (pressure containment) and fracture resistance (SSC and corrosion resistance). Choosing the correct pipe strength and composition is a very case explicit choice. There is no one specific grade to cover all possible applications and being over conservative can make a project uneconomical. For example of grade selection diversity, SSC reduces at high temperature and low-alloy steels have different temperature constraints. L80 pipe is suitable for sour service (Region 3) under all temperatures whilst P110 only above 175° F and Q125 only above 225° F. These temperature constraints render P110 and Q125 generally unsuitable for sour service tubing but useful for liners and lower sections of production casing strings [3]. Another example, while L-80 is suitable for sour service, it isn't suitable in the presence of high CO2 concentrations. The combination of H_2S and CO_2 creates a harsher environment than L80 can deal with. That's why a (2Mo-5Ni) 13 Cr pipe was designed to cope with the combination of H_2S and CO_2. On the other hand, the more costly alloys a pipe contains, such as chrome and nickel, the more expensive it becomes. Table 1 presents the pros and cons of different types of drill pipe material available in the market.

When designing tubing for sour environment, one must consider the increasing levels of H_2S; especially in water flooding reservoirs. NACE

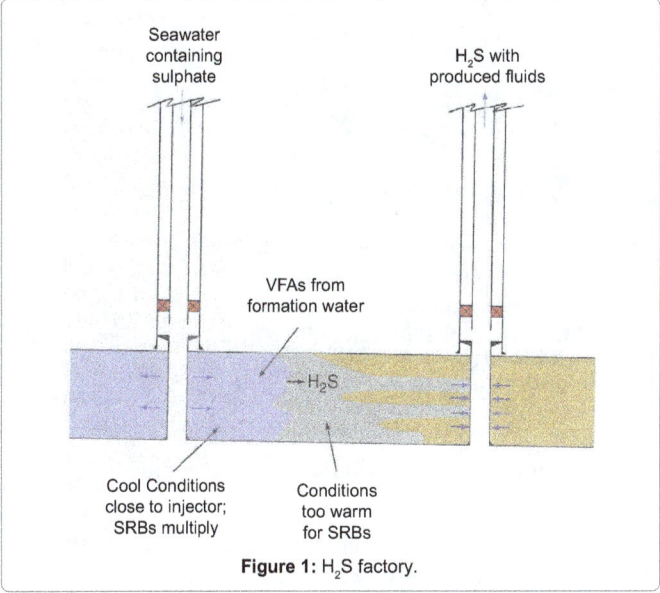

Figure 1: H_2S factory.

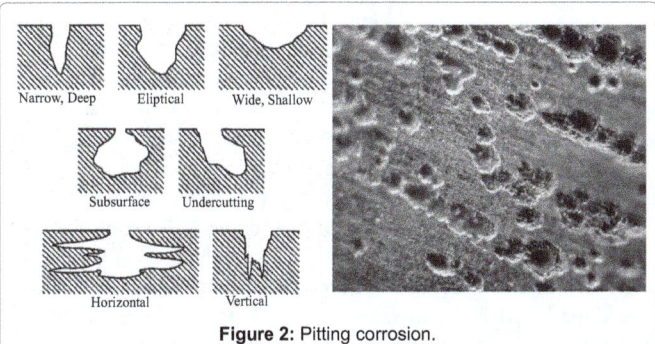

Figure 2: Pitting corrosion.

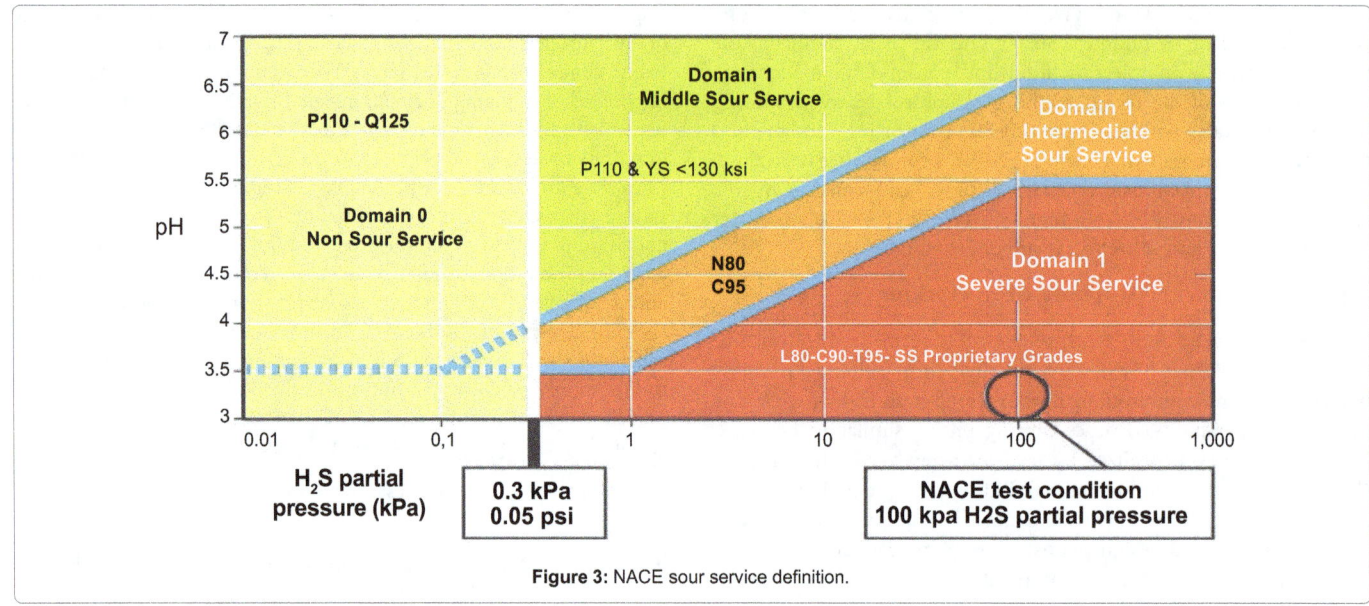

Figure 3: NACE sour service definition.

	Steel Drill pipe	Aluminum Drill pipe	Titanium Drill pipe
Advantages	• Strength • Low Cost • Multiple Grades	• Lighter weight • Deeper drilling with mixed strings • Better fatigue life than steel • More resistant to H_2S and CO_2	• High strength • Light weight • Corrosion resistance to most drilling muds • Resistance to H_2S and CO_2 • Excellent fatigue life • Good high temperature strength
Disadvantages	• Susceptible to corrosion • Susceptible to SSC • Brittle fracture • Wear	• Low resistance to mud with pH > 10.5 • Susceptible to pitting in chloride • Susceptible to chloride stress cracking in oxygenated brine muds • Limited to 250 F • Galvanic corrosion • Lower abrasion resistance	• High cost • Wear

Table 1: Pros and cons of different drill pipe materials.

has developed a standard governing the material recommendations for sour environment in their NACE MR0175/ISO 15156 document. The standard is divided into three parts. Part one is general, part two covers carbon and low-alloy steels and part three covers corrosion-resistant alloys (CRAs). Adhering to the recommendations in this document is crucial for safe and long life equipment. In the United States, this standard is legally enforced.

Completions

Tubing material and design

In tubing, a design philosophy that ensures that the safety of the immediate areas and that of the well bore should be considered. Next, one should come up with the largest possible practical tubing that can achieve the highest possible flow rates [5]. The most suitable solution for the tubing material and design is the use of corrosion resistant alloy tubing string with corrosion resistant alloy components. This is the optimal solution because the tubing will exceed the field's productive life and it will lead to a low maintenance and low risk production. Although this selection is expensive when compared to other tubing options, the net result is comparatively cheap because it will not require frequent maintenance [6].

Among other corrosion resistant alloys, the Alloy C-276 can be selected. However, this should not be restrictive. Other alloys can be selected as long as they are corrosion resistant. The material should

have a superior strength to weight ratio. It should also be resistant to the corrosion and stress cracking. The potential failures in the wells are from corrosion and chloride stress cracking. As well, in wells with temperatures less than 180 degrees Fahrenheit, embrittlement of the alloys can occur. As such it is vital to pay attention to the selection of materials for the tubular goods. The materials should be cracking and corrosion resistant. This will prevent mechanical failures [5].

Tubing connections and field handling procedures

Handling of the joints of the tubing, the torque requirements, the rig supervision and the cleaning of pipe threads should be given close attention. These should be maintained at the highest level. This will ensure proper makeup and protection of the tubing when run into the hole [5]. A coupled and threaded tubing connection should be used. It should make use of the metal to metal nose flank seal and the interference fit threads. A secondary seal should be provided by a Teflon ring in threads. Most of the tubing used in this case will be gall sensitive. There is a likelihood of galling during the connection makeup. It should thus be noted that the engineers should use proper makeup procedures and connection preparations to eliminate the galling problem. An anchor pattern should be applied to the box and the pin surfaces through sand blasting. This will help smooth the rough edges. It will also provide a textured surface that will help retain the pipe dope [6]. It is worth remembering that the proper handling of tubing is essential to ensure the integrity of the sour well. If the pipes are handled poorly, this could lead to the failure of the tubing or the connection. There should be no metal contact between the tubes. The connections should be protected using nonmetal thread caps.

Tree, tubing hanger and production choke

The upper sections of trees and the tubing spools should be manufactured from materials that are suitable to resist the sulfide stress cracking. This is outlined in the NACE Standard number MR-01-75. The valve bodies should be fabricated using stainless steel with gates, seals and packing materials that are compatible with sour environment and with the amines used for inhibition. In case of wells with high pressure and situated in sensitive environmental areas, a surface circulating system should be constructed at wellheads. This will provide a way of controlling the wellhead from remote location if the gas release occurs. Such a system should have three separate lines. The lines can be used to pump fluids in the well for purposes of reduction of pressure and to kill the wells. Behind the production choke, a pump in should be made. It should also be made into the casing annulus or into the injection string. Surface system should be monitored for corrosion information, safety status and wellhead data. This should be carried out using a scheduled manual inspection and electronically. Vital information that should be monitored includes flowing and injection pressure, pneumatic power supply, injection pump status and temperatures [5].

As well, it should be noted that the tree should be fitted with an Alloy C-276 autoclave needle piping and valves. The tree valves should be manufactured with metal single slab seats and gates. A heated Alloy 718 can be used as the material for stems, gates and other internal components of the valves. This is because of their corrosion resistance and high strength. The hangers should be constructed using Alloy 718 material. They should be sealed below and above with a plastically deformed metal to metal seal [6].

Design of the down-hole production seals and PBRs

The design philosophy to be adopted here is that there should be minimized cases of leaks in this system over the well's life. This helps maintain the seals of the tubing in a static position in all production conditions and routine operation. For safe and reliable operations, the tubing casing annulus should be sealed downhole via the use of polished bore receptacle system. The system's elements should have a continuous polished bore run. It should also have a set of elastomeric seals and a locator on the completion tubing. The polished bore receptacle should be constructed using Alloy C-276 material. This should help in corrosion resistance and durability. This system is one of the most reliable in the completion components [6].

The packer design should be considered. The packers should be used to isolate the casing annulus from the corrosive gas produced from reservoirs. It should also be used to isolate the annulus from pressure. Here, the use of long seal elements on the tubing string is desired. With high surface temperatures and high bottomhole experienced during the production, there should be compensations for the large tubing movement. Still, allowances should be made in the seal length to compensate fracturing, potential acidizing and reverse flowing [5].

Corrosion control and surface control systems

There are problems associated with sour wells and they include corrosion caused by carbon dioxide, hydrogen sulfide, precipitation of the scales and the SSC. Problems caused by SSC should be solved using proper selection of the materials used for wellheads, tubulars and surface facilities. The presence of carbon dioxide and free water in the production wells causes the formation of low pH surroundings. These surroundings produce conditions that are extremely corrosive. The presence of carbon dioxide also causes erosion phenomenon that result from the high pressures production. The high pressure production causes high velocities that cause the erosion. This problem can be combated using inhibitors and internal coatings. As well, the use of internal coatings can help provide extra protection in the hot sour environments. The coatings should however be used alongside good inhibitor programs. The inhibitors that should be used in such wells include water dispersible and oil soluble systems. Oil soluble system is much desired because it provides a film that is more tenacious that the other systems. An oil soluble amine can be used in conjunction with hydrocarbon carriers [5].

As stated by McDermott & Martin III (1992), the nearness of the wells to areas that are environmentally sensitive, the corrosive nature of the gas and the need for the safety of the workers calls for enhancements to the surface control systems. The wellhead needs to be fitted with electronic sensors for measuring the tubing and the casing pressures and the temperature of the well. The sensors should be remote. The use of this system is vital because it helps come up with early detection of problems in the wellbore. It also comes up with the necessary information about the next course of action should there be a problem. There should be a complete kills system that can help facilitate an immediate reaction to a wellbore problem. Such a system should have permanent piping, high pressure pumps and weight fluid.

Displacement technique

Oil or fresh water that is treated with a corrosion inhibitor, oxygen scavenger or bactericide should be used as the annulus fluid. These fluids should be used to offer protection to the production casing made up of the carbon steel. They also prevent external corrosion of the corrosion resistant alloy tubing string for the entire life of the well. After the installation of the tree and the landing of the tubing, the pressure should be applied down the tubing. This will help expel the plug and thus allow the well to flow [5].

H_2S

H_2S is an extremely toxic and flammable gas. Inhalation at certain concentrations can lead to injury or death. The human body can cope with small amounts of H_2S but an air concentration of 100 ppm is currently considered as Immediately Dangerous to Life and Health (IDLH) by the American Conference of Governmental Industrial Hygienists. H_2S is colorless, has a rotten-eggs odour and is heavier than air. At levels as low as 10 ppm, eye irritation can occur and nausea and dizziness can occur at higher levels. Even though the distinctive smell of H_2S can be an indicator of its presence, it mustn't be depended on as it can lead to rapid paralysis at high concentrations. That's why most H_2S detectors are set to alarm at concentration of 10 ppm. One of the mitigations to H_2S risk is orientating of the rig heading according to the prevailing wind to locate living quarters upwind of the well. Another mitigation is through adequate training of onsite personnel. API recommends a minimum training that discusses the hazards, proper use of H_2S detectors, emergency response, breathing equipment utilization, wind direction awareness, and importance of pre-job meeting. This training need to be given to visitors just as it's given to regularly assigned personnel, no exceptions. This training needs to be taken on yearly basis to keep all personnel's awareness at highest level. In addition, API emphasizes the importance of regular H_2S drills. The aim of the drills is to familiarize the crew with the necessary steps during an emergency [7].

RIG

A rig used for drilling sweet fields can't be used to drill high H_2S wells without upgrading and adjusting it to mitigate the H_2S risk. This part of the paper considers offshore rigs because of the additional complication and risk. However the points in this section still apply to onshore rigs.

The living quarter is the most crucial facility to upgrade due to it being the nerve system of the rig with majority of on-board personnel residing in it. The fact that its occupants might be sleeping makes it even more vulnerable. Every HVAC intake inside the living quarter needs to be fitted with an H_2S detector. The living quarter also needs to be tight in order to fight gas ingress. This can be done by ensuring a pressure greater than the outdoor atmosphere by minimum of 50 pa throughout the living quarters. The living quarter doors the main and weakest points against gas ingress. Thus, optimizing the number of doors, renewing the seals on the doors and adding a second sealed door behind each door can greatly increase the protection against gas ingress [7].

While most rigs are equipped with H_2S detectors, BA sets, and gas & H_2S detection system, their amount needs to increase to suite sour service operations. Number of H_2S detectors need to enough to cover every personnel exposed to mud flow or reservoir fluids. A complete breathing air cascade system composed of compressors, airlines, and breathing apparatus (BA) sets needs to be implemented to cover all critical areas. One of these critical areas is inside the lifeboat. The breathing air manifold will greatly increase the time of breathing air in case of lifeboat boarding during rig abandonment. All these detectors and breathing equipment require regular tests to ensure proper function and calibration. The same requirements regarding living quarters, detection system and breathing system must be implemented on supply vessels as well [7].

Recommended practices

API recommends a few additional measures than the ordinary while performing the following operations in an H_2S environment [1].

Venting operations: Any venting operation that has a likely release of H_2S should use correct piping to vent to a remote location.

Wireline operations: The minimum lubricator equipments are wire-line valve (blowout preventer), lubricator riser, pressure bleed valve, and stuffing box. If pressure bleeding was necessary and can't be vented to a remote location, then all personnel must wear breathing protection equipment. If permissible, wire-lines and slick-lines should be treated with corrosion inhibitor before running in the hole. After operation completion, wireline should be inspected on site for pitting, surface damage, and embrittlement. Regarding positioning the swabbing unit, wind direction should always be taken into consideration first. The swabbing unit needs to be placed upwind from well, swabbing tanks, and mud pits.

Snubbing operations: Snubbing operations should be limited to daylight hours only unless an emergency snubbing intervention is required. The number of personnel should be kept to essential personnel only. Every worker on the snubbing basket must have a BA set and an escape device.

Coiled tubing operations: The placement of coiled tubing unit should be upwind of the well. The unit should be effectively secured in order to avoid harmful movement. It is recommended to consider a pump cross and a second ram preventer below the pump cross when operating underbalanced. Wellbore fluids shouldn't be circulated back to the coiled tubing unit.

Valve drilling and hot tapping operations: All equipment used in these operations should have a working pressure rating higher than the anticipated pressure inside tapped equipment. The lubricator assembly should be equipped with two valves in series on each bleed-off port. The inner valve is for emergency use and the outer valve is mainly operated instead to preserve the inner valve. These valves should be suitable for sour service and pressure rated to the same or above working pressure of the lubricator.

Coring operations: All members involved in coring operations should don protective breathing equipment at least 10 stands before the core barrel reaches surface. H_2S detectors need to be used on the core sample. Sample containers must be made of H_2S resistant materials and labeled accordingly.

Well evaluations and testing operations: A pre-job meeting focusing required PPE, no-smoking rule, and emergency procedure is a must. Personnel performing the operations should be kept to a minimum and carry H_2S detectors at all time. All produced gas should be safely vented and fluids samples are handled and stored using H_2S resistant materials.

Wellbore fluids: The recommended practices to mitigate sulfide stress cracking using the mud system are:

a. Minimizing formation fluid influx.

b. Using H_2S scavenger and constantly monitoring its level.

c. Maintaining the mud system at pH of 9 or higher.

d. Adding diesel oil or other protective fluid to the mud.

Coiled tubing in sour environment

Introduction: Coil tubing has been used in sour well environments for over two decades. The advances made in the coil tubing technology make it an economical alternative when compared to other alternatives in the oil field operations. This is the result of research whose aim is to understand the response of coiled tubing (CT) when exposed to sour down-hole conditions [8,9]. This paper analyzes the specifications, limitations, new improvements, tests, and applications of CT.

Specifications: The purchase specification is the heart of the management system and it is also the chronological starting point, At this point, the strings ordered are assumed to be fit for use in the sour wells. The specifications are also important because strings that are obtained specifically for sour service are thought to be less susceptible to catastrophic sour degradation. Generally, there are two types of strings purchased for work-over operations [10]. The first is the non-taper 400 m standard string which has a wide range of uses such as acid stimulation, milling operations, wellbore cleanouts, gas lifting, cementing work and fishing. The second is the tapered string. For both of them, different purchasing specifications are recommended. After a scientific testing, the following purchasing specifications of pipes were introduced into the market. For standard strings, the maximum allowable fatigue is 70% while the taper strings have fatigue allowable being up to 60%. For Sour fatigue, current industry standards allow up to 20% for standard strings but for the tapered strings the specification is 48%. As for sour jobs, the standard requirement is 19% when using standard string while taper string requirement is 87%. Corrosion fatigue requirements are a standard of 25% for normal string while taper requirement is 40%. The standard requirement for acid pumped is 117 m^3 while the taper requirement is 1450 m^3 [8]. Normally, the standard number of jobs a CT should carry out is 37. However, when using tapered string, the requirements restrict it to 15 jobs. The standard time of service is 5 months for standard non-tapered strings while for taper strings requirements allow up to 20months. Finally, standard field requirements for inspection stand at 2 while for taper string they stand at 10.

Limitations: Initial JIP considered only CT strength grades ranging between 70 and 80 ksi and the resulting conclusion was that sour service CT strings should be limited to a maximum strength of 80ksi yield. Though the 80 grade sour CT limit served well, it imposed undesirable limitations on sour well interventions which resulted in requirements for higher strength CT. In earlier publications, the use of butt welded joints had been advised on the basis of engineering judgments in sour service strings [8,9]. Based on the current JIP testing results, it is clear that 100% H$_2$S sour environments reduce the range of bend fatigue performance for butt welds in ranges of 35% to 75% of the achievable performance for these welds in sweet conditions.

New improvements: As a result of the growth in technology, there have been new improvements in the CT field. The first improvement is the use of electronic inspection. Electronic inspections have proven to be effective in identifying both the actual and potential flaws that can result in pipe failure. The nature of H$_2$S results in a magnification of some pipe flaws that could not have escalated in non-sour conditions. The pipe inspections currently form an integral part of the management system [11]. Current efforts in the electronic inspection area are geared towards identifying the criticality of different flaw types, orientation with reference to tube axis orientation, and the rate of occurrence. The second improvement is the use of inhibition. Different inhibitors are applied to the tubing to protect it from sour attack. Currently, there are a number of application methodologies that have been identified. In most cases, the inhibitor is applied externally to the tubing as it enters into the well. Before any treatment, a slug of inhibitor is also pumped through the string [12]. The decisions on how the slug is applied are dependent on the nature of work and exposure. Another development is in the area of fatigue monitoring. Today, it is common practice to apply low-cycle fatigue monitoring of CT. There are a number of guidelines that have been applied to limit the acceptable fatigue life of CT [13]. The suggestions are that the maximum fatigue life allowable should be a fraction of the sweet service. Another development is in the area of stewardship. The introduction of stewardship in the area of management was prompted by a high number of CT failures which pointed fingers more to human errors than industrial standards. As a result, the introduction of the management system required that only one or two individuals should be involved in the process of ordering strings, maintaining strings, and selecting strings for a given operation. These individuals, stewards, acquire an intrinsic understanding of strings and the type of service that they have seen. This has resulted in the stewards gaining inherent knowledge of strings to an extent that their intuitive decisions correspond to analytical decisions made in laboratories. For example, laboratory tests can recommend the removal of a string only to find that the stewards have already retired it.

Tests: The main test protocol is one that includes a custom test procedure that involves either a single-sided or double-sided sour exposure of around 1.25 m length of full size CT. Single-sided exposures were found to be more representative of field exposure. One common test is constant load test without inhibitors. This is performed on the higher strength grades for high concentrations of 100% H$_2$S in a NACE "A" solution. The maximum permissible working load used is equal to von Mises strength of 80% of SMYS [14]. Using new CT90 grades, tensile failure was realized after 48 hours for combined constant hoop and 44 hours for axial tensile hoop. These were half of the desired survival period. The second test performed was incubation times from AE Corrosion Cell Tests. In the test, external services of cycled CT70 showed spots of hydrogen blistering which was likely to have been caused by plane AE signals. For cycled CT80, an exterior crack on the radial direction of the outer CT surface was observed.

Exploitation of high H$_2$S fields

Developing high H$_2$S fields isn't a new thing. It is resourceful to learn from peer experiences around the world. This section aims to provide a couple of cases and present the challenges faced and actions taken.

Example 1: As argued by Malik et al., the supper Giant Kashagan oil field is one of the most important sources of crude oil in the world. Discovered in 2000 in Caspian Sea, it is known to be the most significant discovery in the world for the last 30 years. The contents of this field include light oil made up of 5% carbon dioxide, 15% hydrogen sulfide as well as huge amounts of associated gas. The most significant challenge faced by the consortium of companies developing this field is how to manage the enormous volumes of associated gas, which are highly sour. To overcome this challenge, the consortium had two basic options. One of the options was to evacuate the sour gas to the field's shore. This gas is treated before being sold out to interested parties. Although this was found to be technically unchallenging, it is an expensive approach. The second option, re-injecting the sour gas into the reservoir, was found to be cheaper despite being technically challenging.

When embracing the gas injection option, the consortium found it important to understand the cap rock integrity in order to avoid its breach. Geo-mechanical studies are necessary in ascertaining how the *in-situ* stress is distributed. The stress distribution was better understood through collection of well data by conducting a series of injection, leak-off and mini-frac tests. Additionally, geo-mechanical models were used to determine the best injection pressure at the bottom of the hole. Laboratory tests were also important in determining the possible effects of gas injection on cap rock integrity as well as detection of any present micro-fractures that can lead to gas slippage. These studies included mechanical failure, threshold pressure, gas bubble migration, geochemistry and petrography. The initial field data about the oil well was used together with laboratory measurements to develop geo-technical models important in evaluation of gas injection, natural depletion and end members at very high pressures.

Since there was the risk of asphaltene precipitation and deposition, the consortium conducted an asphaltene stability study. This gave them an opportunity to evaluate the well's fluid stability as far as asphaltene precipitation was concerned. They were also able to assess how the asphaltene behavior is affected by the sour gas flood. It is through the same stability study that the consortium was able to do validation of thermodynamic modeling that revealed how asphaltene stability is affected by the gas injection. The asphaltene stability study was made possible through dead oil screening and live oil depressurization test. Reservoir management is another important activity at the Super Giant Kashagan. Two aspects are used in this management to make sure that miscible gas injection is possible. Major aspects of this management include injection gas front monitoring and gas shut-offs comprising mechanical and chemical options.

Example 2: As presented by Hrncevic et al., Croatia had developed large natural gas fields by the beginning of 1980's, including Kalinovac, Stari Gradac and Molve. For all these fields, the engineers were faced by the challenge of great initial reservoir pressures which exceeded 450 bars. They were also faced by the problem of high temperature and high levels of carbon dioxide and hydrogen sulfide. Additionally, presence of additional non-carbon compounds such as mercury and mercaptans in high levels was a great challenge. Table 2 describes the fluid composition of each reservoir. Currently, the Croatian petroleum industry has great experience on how to process sour natural gas with minimal effects to the environment.

The mentioned challenges were experienced by the individuals working in the Croatian petroleum industry as well as international technologists and experts. Among the challenges, the most significant one was how to determine and apply appropriate mechanism to prevent corrosion when dealing with chloride, mercury, hydrogen sulfide and carbon dioxide. This necessitated the need to device ways to avoid chloride stress cracking, sulfide stress cracking hydrogen sulfide corrosion and carbon dioxide corrosion. This was made possible through the use of injection valves at specific depths to continually inhibit this corrosion. However, this strategy produced poor results since there was corrosion of the valve after a short period of time. The corrosion was caused by some aromates present in the inhibitor. Additionally, the effect of the inhibitor was reduced by the high temperature through disintegration. Since the anticorrosion protection was associated with great damage, the experts were forced to replace the production equipment. This was followed by the use of high alloy steel grades to make it possible to resist various types of corrosion under diverse conditions of pressure, stress and temperature. The materials used for the production casing were the following:

a. Production casing above permanent packer was carbon steel.

b. Production casing below the packer was made of duplex steel with 25% Cr in gas wells and 13% Cr in gas condensate wells.

c. Production tubing was made of high-alloy duplex steel with 25% Cr in gas wells (Molve gas field) and high-alloy steel with 13% Cr; Hardness HRC < 23 in gas condensate wells (Kalinovac and Stari Gradac gas field).

d. Subsurface equipment (packer, SCSSV) is made of Incoloy 925 material.

e. The thread joint of the production string and production casing are gas tight premium thread with three to four sealing elements, one of them being metal to metal.

Just like in the Giant Kashagan oil field, the natural gas extracted from the Podravina's gas fields could not be used directly due to the presence of noxious substances. This led to the construction of the CGS Molve (Molve Central Gas Station) consisting of three plants for gas processing. The three plants have different gas processing capacities and rely on a number of technological procedures to achieve natural gas purification. The technologies are also relied on when preparing the gas for distribution. The specific process of natural gas treatment on CGS Molve involves separation of natural gas, hydration of natural gas, carbon dioxide and hydrogen sulfide extraction. Furthermore, the treatment involves mercury removal and extraction of hydrogen sulfide through its oxidation into elemental sulfur.

One of the plants (GPP Molve III) uses an amine solution to get rid of acid gases (carbon dioxide and sulfur dioxide). The other two plants use potassium carbonate instead of the amine solution. After removal,

Composition	Molve	Kalinovac	StariGradac	Goladuboka
Methane	69.22%	69.97%	66.5%	41.04%
Ethan	3.26%	6.76%	7.19%	1.76%
pro ane	1.02%	2.35%	2.83%	0.66%
Isobutene	0.20%	0.63%	0.92%	0.17%
n-butane	0.23%	0.75%	1.21%	0.18%
iso-pentane	0.09%	0.39%	0.67%	0.05%
n-pentane	0.06%	0.34%	0.63%	0.08%
Hexane and C_6+	0.53%	5.26%	9.09%	0.02%
Nitrogen	1.64%	1.3%	0.94%	2.38%
Carbon dioxide	23.75%	12.17%	9.02%	53.64%
Hydrogen sulfide	80 ppm	100 ppm	400 ppm	900 ppm
Mercury	1000-1500 ug/m³			
Mercaptans	20-30 mg/m³			

Table 2: Fluid composition of Molve, Kalinovac, Stari Gradac and Gola Duboka reservoirs.

these gases are directed into a unit meant for elementary sulfur recovery. This is where hydrogen sulfide is converted to elemental sulfur and water. The purified natural gas is then directed into a dehydration processing system in which water vapor is eliminated by solid desiccant-molecular sieves. It is important to note that natural gas extracted from the well has several natural liquids that need to be removed. These liquids are known to have a higher value when used as separate products. This explains why they are isolated from the well's gas stream. Cryogen is an important substance in the removal of these liquids at Molve III. The natural liquids are then converted into the basic components through fractionation. Fractionation is used due to the fact that the natural liquids have hydrocarbons with different boiling points.

Conclusion

Dealing with sour fields is a challenge the industry will face more often. A challenge that will keep on increasing in difficulty. The key to facing this challenge is knowledgeable personnel who have received the proper training. The proper training to select material and plan operations. The proper training to execute a plan safely and react swiftly. There is very little room for error when it comes to dealing with H_2S.

References

1. American Petroleum Institute (API) (2001) Recommended practice for drilling and well servicing operations involving hydrogen sulfide. (3rdedn): Recommended Practices 49.

2. Marsh Z, Marsh J (2012) Sour service assessment for aging assets and pipelines. SPE International Conference and Exhibition on Oilfield Corrosion. Aberdeen, UK.

3. Bellarby J (2009) Well completion design. pp: 419-450.

4. Farchy J (2014) Leaking pipelines to add up to $4bn in costs to Kashagan oil project.

5. Huntoon GG (1984) Completion practices in deep sour tuscaloosa wells. J Petr Tech 36: 79-88.

6. McDermott Jr, Martin-III BL (1992) Completion design for deep, sour norphlet gas wells offshore mobile, Alabama. SPE 24772, Washington, DC pp: 4-7.

7. Frattini L, Madera A, Ferrante A (2011) H₂S risk management plan for a semi-submersible drilling RIG. OMC Ravenna.

8. Stadlweiser J, Howes M, Cawston R (1994) Coil tubing plug retrieval/fishing operations on deep, high pressure sour wells: J Canad Petro Tech 3: 27-30.

9. Stadlweiser J, Best J, Willson D, Bloor B (1994) Coil tubing fishing operation on deep, high pressure, sour gas wells. J Canad Petro Tech 33: 15-19.

10. Crabtree A, Gavin W (2005) Coiled tubing in sour environments: Theory and practice. SPE.

11. Luft H, Padron T, Kee E (2007) Sour-well serviceability of higher-strength coiled tubing. SPE/ICoTA Coiled Tubing and Well Intervention Conference and Exhibition, 20-21 March, The Woodlands, Texas, USA.

12. Szklarz k, Luft B, Padron T (2009) Limits of use for coiled tubing in sour service. Corrosion 2009, 22-26 March, Atlanta, Georgia.

13. Malik Z, Charfeddine M, Moore S, Francia L, Denby P (2005) The Super-giant Kashagan field: Making a sweet development out of sour crude. International Petroleum Technology Conference, 21-23 November, Doha, Qatar.

14. Hrncevic L, Simon K, Kristafor Z, Malnar M (2010) Long-lasting experience in environmental protection during sour gas reservoir exploitation. SPE Deep Gas Conference and Exhibition, 24-26 January, Manama, Bahrain.

13

Impact of Multi-ion Interactions on Oil Mobilization by Smart Waterflooding in Carbonate Reservoirs

Awolayo AN* and Hemanta K Sharma

Department of Chemical and Petroleum Engineering, University of Calgary, 2500 University Drive, N.W. Calgary, Alberta, Canada

Abstract

The injected brine composition has been observed to have intense effect on efficiency of waterflooding in carbonate reservoirs. This process is known as smart waterflooding and has proved to be an effective process in improving oil recovery. Different approaches have been tested in carbonate reservoirs due to the complexity of the process. Based on these approaches, different mechanisms have been proposed with some level of uncertainties. This has led to several arguments on the chemical mechanisms responsible for such feat achieved. One of the approaches is discussed in this paper, however with much interpretation considering all factors influencing the oil-brine-rock interactions. Therefore, this paper presents the influence of multi-ion interactions during smart water flood on carbonates. Sequential flooding of formation brine and smart brines and the effluent ion analysis were conducted to confirm the multi-ion interactions leading to improved recovery. In addition, zeta potential measurement was conducted to examine the alteration process and correlated with the core flood results. The results from zeta potential measurement showed that multi-ion interaction alters the rock surface charge, which led to more water-wetness. Significant improvement in oil displacement efficiency was observed beyond the secondary waterflood and effluent ionic analysis demonstrated that these multi-ionic interactions led to the observed alteration.

Keywords: Smart water; Zeta potential; Displacement; Alteration; Electrostatic forces; Ionic analysis

Nomenclature

BV: Bulk Volume; D: Diameter; IRF: Incremental Recovery Factor; IS: Ionic Strength; K_L: Liquid Permeability; K_o: Permeability To Oil; Kppm: Thousands Part Per Million; L: Length; OOIC: Original Oil In Core; PV: Pore Volume; RF: Recovery Factor; Swirr: Irreducible Water Saturation; TDS: Total Dissolved Solids; XRD: X-Ray Diffraction; Ø: Porosity; Θ: Contact Angle

Introduction

Waterflooding has been an effective and relatively inexpensive process of improving recovery beyond the primary natural drive. In carbonate reservoirs, which are always mixed/strongly oil-wet, waterflooding seems ineffective as water cut escalates leaving behind high residual oil saturation (ROS). In such case, rock wettability do not favour the optimal condition for oil displacement by the injection water.

Through the centuries in a reservoir, a thermodynamic equilibrium has existed between the brine, oil and the rock and there is no doubt that injecting different brine would considerably change this equilibrium. This is a major reason why the brine injected during convectional waterflooding is usually selected based its compatibility with existing formation brine. This change is linked with the interaction between the injected brine and the rock surface. However, decades of research has confirmed that the change is more positive towards improving oil recovery [1-4]. The brine-rock interaction has been observed to be prompted by the reactivity of the ions in the injected water, which is crucial in creating a surface charge alteration [5].

Furthermore, during smart waterflooding, it is predicted that as the injected water displaces the oil ahead, rock wettability is favourably altered along the way. Therefore, waterflooding is termed smart when the injected water's ionic composition and salinity are manipulated – selective addition or removal of certain ions.

It is evident that a mechanistic study of smart water flood on lime stones (carbonates) is much more complicated as compared to its application on sandstones and chalk [2, 6-13] probably as a result of divergent mechanisms working together with each individual influence. Sandstones possess less interaction between the polar components in the oil and its minerals and chalk having a higher surface area compared to lime stones [14-17].

Recently, there has been a rapid growth in the publications of smart waterflood laboratory experimental results on carbonates with very limited field trials and different mechanisms have been suggested to be accountable for the improved displacement efficiency [18]. Awkwardly, most of these suggested mechanisms are debatable, probably because many factors are involved which are associated with the reservoir fluids, rocks and the injection fluid. Although, wettability alteration is the widely accepted mechanism by researchers, other physical mechanisms suggested are fines migration, multi-ion exchange, surface charge alteration, double layer expansion, etc. The different mechanisms are associated with the different approaches that have been implemented in smart water flood on carbonates [19], which include;

I. Brine salinity reduction

a. Brine dilution.

b. Reduction of water hardness – Ca^{2+}, Mg^{2+}.

c. Reduction/Removal of non-active ions – Na^+, Cl^-.

*Corresponding author: Awolayo AN, Department of Chemical and Petroleum Engineering, University of Calgary, 2500 University Drive, N.W. Calgary, Alberta, Canada. T2N 1N4, E-mail: adedapo.awolayo@ucalgary.ca

II. Ionic modification

a. Surface interaction ion concentration increment – SO_4^{2-}, PO_4^{3-}, BO_4^{3-}.

b. Potential determining ions, PDI- (SO_4^{2-}, Mg^{2+}, Ca^{2+}) concentrations.

In previous works by Fathi et al. [20], the potential of smart waterflood has been explored through the second approach to enhance oil mobilization. Although this technique has a number of merits, its potential was thoroughly investigated in carbonates to achieve optimal concentrations and particularly in the high-temperature (110°C), high-pressure (20 MPa) and high-salinity (200-250 kppm) environment. This work is therefore focused on the first approach to better understand the underlying chemical mechanisms through experimental studies and then breed a database on the impact multi-ion interaction in a smart waterflood conducted under reservoirs conditions.

Experimental Section

Materials

Core plugs: Carbonate cores with consistent properties were selected based on routine core analysis, which was first conducted to measure the dimensions, air permeability, porosity, and pore volume of the core plugs. Table 1 shows their detailed properties. Core plates cut from plugs into rectangular shape dimension (1.27 cm × 1.905 cm × 0.762 cm) were used for the contact angle measurements. Part of the rock was crushed to conduct X-ray diffraction (XRD) analysis; the result showed in Figure 1 reveals that the main rock composition is calcite with little trace of minerals like dolomite and quartz.

Fluid properties: All synthetic brines were prepared according to the geochemical formulations listed in Table 2 by dissolving specified amounts of reagent-grade salts in deionized water. The brines consisted of formation water named as "FW" which is classified as the base brine, used in establishing irreducible water saturation and secondary flooding; and various smart brines, which are injected after base brine and possibly to trigger incremental recovery. Likewise, smart brines were prepared based on seawater compositions by brine dilution and

Brines/Ions [kppm]	Na⁺	Ca²⁺	Mg²⁺	K⁺	SO₄²⁻	HCO₃⁻	Cl⁻	TDS	IS
FW	76.68	19.12	3.35	0.08	0.11	0.06	161.81	261.21	5.19
SW	13.64	0.39	1.73	0.01	2.85	0.01	24.65	43.28	0.87
DSW	1.00	0.06	0.18	0.00	0.32	0.01	1.95	3.52	0.07
SW0NaCl	1.51	0.54	1.73	0.01	3.08	0.01	6.04	12.91	0.35
DSW0NaCl	0.22	0.06	0.18	0.00	0.34	0.01	0.71	1.52	0.04

Table 2: Brine geochemical compositions.

removal of NaCl; their nomenclature imitates the relative alteration in their composition. In this work, seawater is termed as "SW", Ten times diluted seawater as "DSW", seawater with selective removal of NaCl as "SW0NaCl" and Ten times diluted seawater with selective removal of NaCl as "DSW0NaCl.

Core preparation: The core plugs were first subjected to solvent cleaning using Dean-Stark extraction and dried at 100°C for about 24 hours prior to petro-physical measurements. Then the plugs were saturated with FW under vacuum for around 72 hours in order to establish equilibrium. The pore volume (PV) was calculated from weight difference and compared to the one estimated from the helium porosity measurement (comparison gave an error of < 2%). Field dead oil (with about 5PV) was then injected in each direction into the cores to establish irreducible water saturation, and the cores were aged using steel aging cell under experimental temperature (110°C) and around 13.8 MPa for 6weeks. Before starting the flooding test, more than 1PV dead oil was injected into the core to displace the oil used during aging and the effective permeability to oil at irreducible water saturation was recorded.

Core flooding and chemical analysis: The experimental set-up is similar to the one discussed in our previous work [19]. Each waterflood test commenced with the base brine at a constant injection rate of 0.25cc/min. Once the oil production ceased, it was followed by sequential injection of smart brines. All experiments were conducted at 100°C, 20 MPa pore pressure, and 10.3 MPa net confining pressure with horizontal orientation. The ionic concentrations of Ca^{2+}, Mg^{2+}, SO_4^{2-}, Cl^- and Na^+ were analysed by an ion-exchange chromatograph pre and post coreflooding tests, in order to better interpret the observed results.

Contact angle monitoring: Advancing contact angles were measured on core plates to verify the wettability alteration process by smart water. First, the polished core plates were placed under vacuum for at least two hours and aged in the base brine for another 24 hours. Then the plates were aged in the field dead oil at reservoir temperature of 110°C for at least 6 weeks to restore the wettability. Because of the limitations to water boiling at 100°C, the test was carried out at 95°C, so the brines were preheated to 95°C before commencement. Then, the plates were exposed to different brines. The monitoring were performed immediately after aging as a reference, and then at different periods of time that the plates had been exposed to the different brines as described by Awolayo [20].

Zeta potential experiment: The zeta potential of brine/rock interface of the aqueous rock suspension was measured by using a Zeta Electroacoustic Spectrometer. A representative solution was prepared using the base brine and pulverized core sample according to procedures described by Awolayo [19]. The initial charge at the interface was then measured before being re-measured as different smart waters were mixed with the representative solution. Solution pH was also reported with the zeta potential measurement. In order to better understand the relationship between surface charge and oil recovery, this study is significant as several studies reported little could be done to alter oil/brine interface charge from strongly negative (Table 3).

Sample ID	L [cm]	D [cm]	Ø [%]	K_L [mD]	PV [cm³]	OOIC [cm³]	Swirr %
P#1	6.37	3.84	30.78	9.91	22.32	17.80	21.10
P#6	6.22	3.81	29.16	6.11	20.55	17.00	17.27
P#8	7.11	3.86	25.34	18.35	20.48	15.00	26.8

Table 1: Core plugs petro-physical properties.

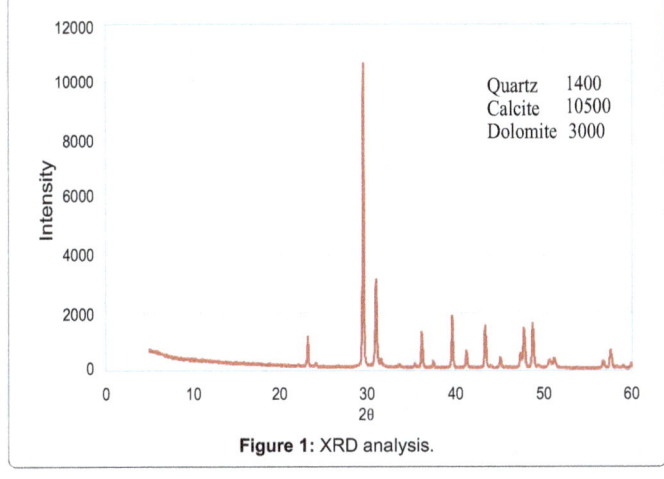

Figure 1: XRD analysis.

Quartz 1400
Calcite 10500
Dolomite 3000

Results and Discussion

This section summarizes the results complemented by critical cognitive discussions on the experimental studies where FW was set as a baseline to compare the performance of other smart brines.

Surface charge of carbonates

The charges at oil/brine and brine/rock interfaces primarily control the stability of the water film between the oil and rock, hence the rock wetting state. If the two interfaces possess similar surface charges, then a strong repulsive force is created and if opposing charges, a strong attractive force is generated. A strong electrostatic repulsion between the interfaces will create a stable and thick water film which would results in water-wet rock [19]. The rock wettability could change towards oil-wet if weak electrostatic repulsion exists between the interfaces leading to unstable and thin water film. Therefore, rock wettability depends on the sign and magnitude of the electrical charges at the two interfaces due to the electrostatic attractive or repulsive forces generated between them. The results showed positive surface charges at the rock/brine interface at all range of salinity and they are discussed below;

Surface charge with salinity dilution: The baseline condition (SO_4^{2-} : NaCl = 1: 2129.4), gave a positive zeta potential value of 4.49 mV at a solution pH of 7.33; SW (SO_4^{2-} : NaCl = 1: 11.6), gave a positive value of 2.2 mV at a solution pH of 8.08. While introduction of DSW (SO_4^{2-} : NaCl = 1: 11.6) gave a zeta potential value of 2.26 mV at a solution pH of 7.79. Figure 2 presents effect of brine dilution on zeta potential as a function of pH.

The result indicates that brine salinity/ionic strength reduction due to brine dilution results in increasing the rock surface charges which is influenced by decreased in solution pH. This result trend means so far the ratio between the active (SO_4^{2-}) and the non-active ions concentration remain the same, no further alteration of charges could be obtained. This is due to the fact that the active ions responsible to decrease the brine/rock surface charge were in low concentrations.

Surface charge with non-active ions depletion: Also, analysis of the case where brine salinity/ionic strength was reduced by non-active ions depletion, the baseline (SO_4^{2-} : NaCl = 1: 2129.4) gave a zeta potential value of 4.49 mV at a solution pH of 7.33. Then with

SW (SO_4^{2-} : NaCl = 1: 11.6), gave a zeta potential value of 2.2 mV at a solution pH of 8.08. Introduction of SW0NaCl (SO_4^{2-} : NaCl = 1: 2.5) gave a zeta potential value of 1.54 mV at a solution pH of 7.86 as shown in Figure 3a. This same trend was also observed with DSW (SO_4^{2-} : NaCl =1: 11.6) which gave a zeta potential value of 2.26 mV at a solution pH of 7.79 while DSW0NaCl (SO_4^{2-} : NaCl =1: 2.7) gave a zeta potential measurement of 2.1 mV at a solution pH of 7.92 as shown in Figure 3b.

The result shows that brine/rock interface charge is significantly altered with the removal of non-active ions from the smart brine. Also the removal reduces the brine salinity and the ratio between the active and non-active ions, which prompted the electrical double layer (EDL) expansion as there is increased repulsion between the two interfaces. Likewise as sulphate is more electronegative compared to chloride, depleting smart brines in chloride as a non-active ion would tend to create more reactivity for sulphate ion towards the rock surface. Just as

Fluids	FW	SW	DSW
Density [g/cm³] @ 20°C	1.1728	1.0306	1.001
Viscosity [cP] @ 70°C	0.815	0.463	0.435
Fluids	**SW0NaCl**	**DSW0NaCl**	**Oil**
Density [g/cm³] @ 20°C	1.0089	0.9992	0.8376
Viscosity [cP] @ 70°C	0.471	0.429	1.927

Table 3: Brine and dead oil properties.

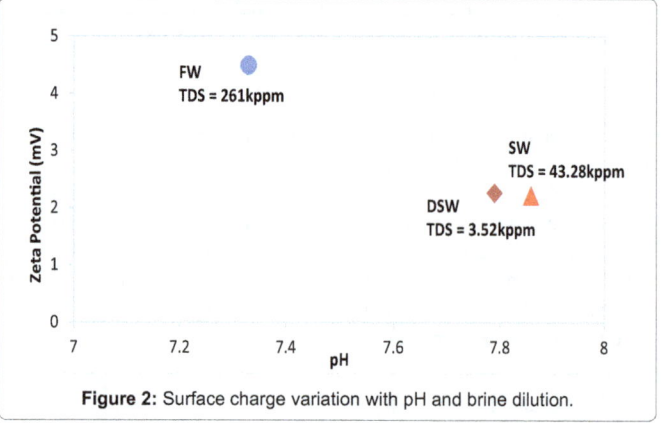

Figure 2: Surface charge variation with pH and brine dilution.

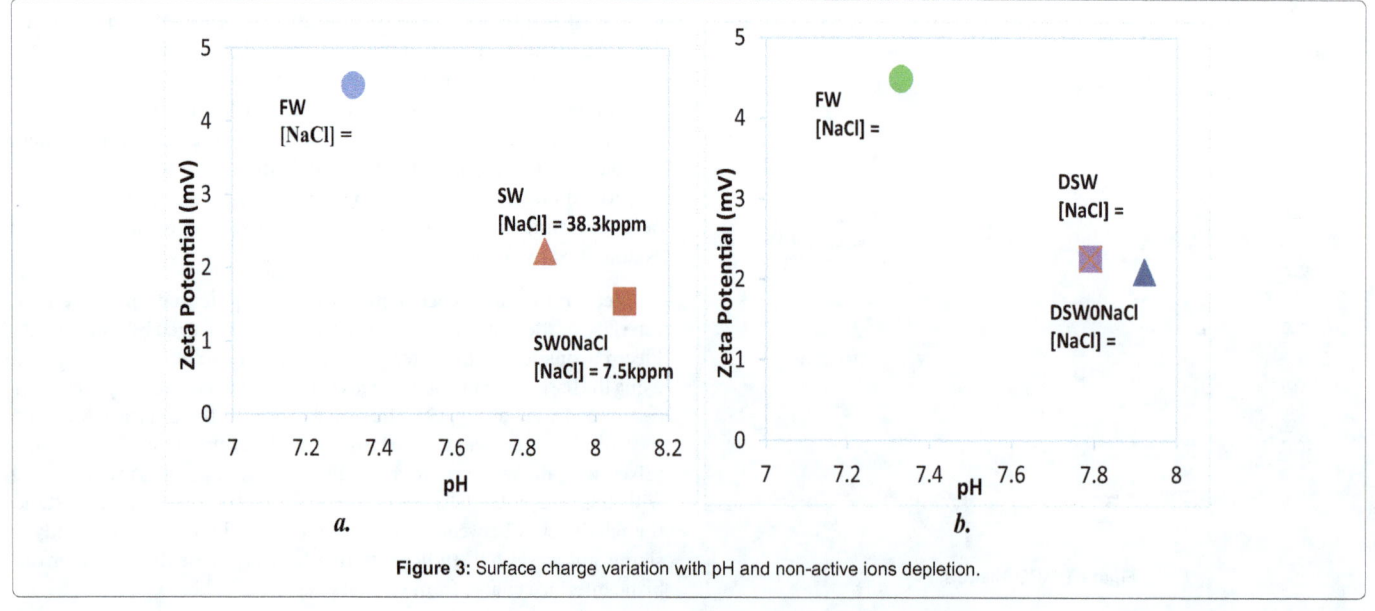

Figure 3: Surface charge variation with pH and non-active ions depletion.

discussed earlier, it can be noted that reducing brine salinity through the removal of non-active ions could favourably increase recovery.

Core flood and ionic analysis

Corefloods were conducted to study the impact of the ionic interaction in smart brines on oil recovery. The displacement efficiency and differential pressure were recorded. The effluents were collected for ionic analysis and the results were normalized against the injected composition and correlated with the observed oil displacement efficiency to further reaffirm the postulated mechanisms. For the ionic analysis, much emphasis is placed on the tertiary injections.

Core flood 1: After aging P # 1, it was flooded sequentially with FW, SW and SW0NaCl. A longer period of flushing was to ensure any remaining mobile oil is displaced before commencing the injection of subsequent brines. This crucial step was to examine the effect of smart water on residual oil saturation and ultimately on displacement efficiency. A displacement plateau of 77.8% OOIC was established by about 8PV FW injection and no mobile oil was further displaced. Before breakthrough, 67% OOIC was recovered after about 5.7PV injection. This established that residual oil is what was left in the core, which cannot just be recovered by FW. The initial oil saturation was observed to decrease to 17.5% by about 11PV injected.

Then, the injection fluid was switched to SW and an additional 2.4% OOIC was recorded. Thereafter, SW0NaCl was injected and oil displacement efficiency increased to 87% OOIC. The displacement by SW and SW0NaCl gave about 1.95 and 5.4% reduction in ROS respectively as shown in Figure 4. As smart water of lower salinity was introduced, the differential pressure decreased and later stabilized. This same observation was reported by Yousef [15]. The observed pressure drop reduction, in line with their argument, is due to the reduced capillary forces encountered as water saturation increases to displace the residual oil left in the sample. Similarly according to Gupta et al. [21], the pressure reduction could also be as a result of additional oil recovery owing to the injection switch to brines of lower viscosity which increases the mobility of the brine. The pressure stability was then as a result of one-phase flow of brine after displacing all mobile oils from the core.

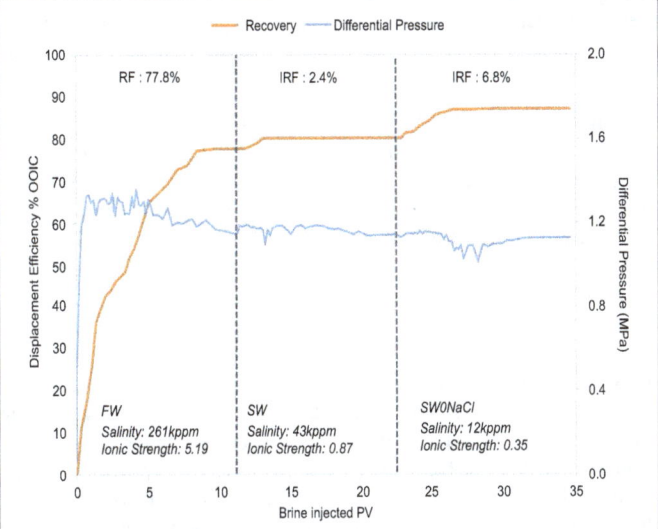

Figure 4: Displacement efficiency, injection rate and pressure drop as function of brine injected PV for P#1.

Physical observation of the effluent shows less turbidity. During the SW injection (Figure 5), the $[Ca^{2+}]$ increased immediately and started declining due to the dilution between FW and SW, but didn't not return to 1 as all normalized concentrations are expected to drop to 1 if no reaction, ionic exchange or dissociation was occurring. Also the $[Mg^{2+}]$ reduced in the effluent and gradually increase to the injection level, perhaps because Mg^{2+} displaced Ca^{2+} from the rock surface at high temperature (110°C). This could mean that $[Ca^{2+}]$ not returning back to normalized concentration of 1 was as a result of ionic exchange process and calcite dissolution as $[Ca^{2+}]$ didn't return to the injection level. Then $[SO_4^{2-}]$ reduced denoting easy penetration of SO_4^{2-} through the diffuse layer into the stern layer as the non-active ions (Na^+ and Cl^-) concentrations in the effluent increased.

The non-active ions were high initially and steeply declined until almost constant at an equilibrium value which is due to the dilution of FW with SW (containing less non-active ions) as observed in Figure 6. At the stage where the non-active ion concentrations in the effluent were high, SO_4^{2-} adsorbed strongly to the rock surface with increased co-adsorption of Mg^{2+}, thereby releasing the oil. This change in ion potential and adsorbed components alters the rock surface wettability and stimulates further imbibition of water, leading to improved oil mobilization. This trend align with the proposed mechanism by Zhang et al. as they identified Mg^{2+} and SO_4^{2-} as responsible for the alteration at high temperature [12].

Prior to the injection of SW0NaCl, the concentration of SO_4^{2-} in the effluent increased due to the fact no further wettability alteration occurred and this was when the recovery plateau had been established. During the SW0NaCl injection (most of its non-active ions depleted), similar to what has been explained above, but intensely higher was the non-active ions concentration in the effluent. The dilution of SW with SW0NaCl generated the increased concentration which monotonically decreased but not to the injection level as the injected PV of SW0NaCl increased.

Since non-active ions have been displaced from the diffuse layer, SO_4^{2-} gained better access to the stern layer and onto the rock surface. This caused its reduction in the effluent while it remained below the injection level. Reduced $[Ca^{2+}]$ in the effluent at the onset of SW0NaCl injection signifies that more Ca^{2+} attached to the rock surface due to simultaneous increased SO_4^{2-} adsorption; while $[Mg^{2+}]$ went to settle at 1.15 times the SW slug which could create a small surface dolomite dissolution as XRD analysis proved presence of dolomite in the core. In this case, reduction of the salinity/ionic strength of injected fluid were achieved by reducing the concentrations of the non-active ions (Na^+ and Cl^-). The increased oil recovery by SW and SW0NaCl are related to their abilities to reduce the surface charge and dissolve minerals at the core surface.

Therefore, it can be recalled that FW contained very high concentration of non-active ions, Na^+ and Cl^-. These ions are not considered active in wettability alteration process but majorly dominate the diffuse layer, which can create a barrier for the active ions to easily access the rock surface. As SW was introduced, with reduced concentration of NaCl, the active ions get more access to the surface. Then at the reservoir temperature, sulphate adsorption created more electrostatic repulsion and facilitated the substitution of Ca^{2+} by Mg^{2+} at the rock surface, thereby leading to a substantial incremental recovery [12]. The recovery plateau was established once the diffuse layer was re-occupied by the non-active ions and barred the entrance of the active ions.

Then there is composition change in the diffuse layer by the introduction of smart water depleted in NaCl, which increases the

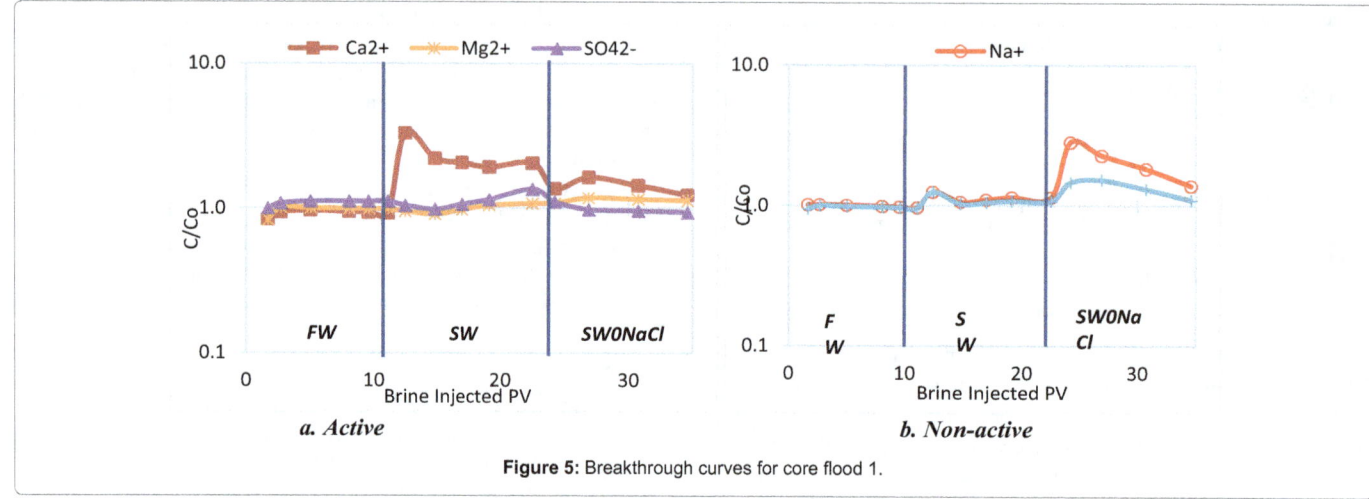

Figure 5: Breakthrough curves for core flood 1.

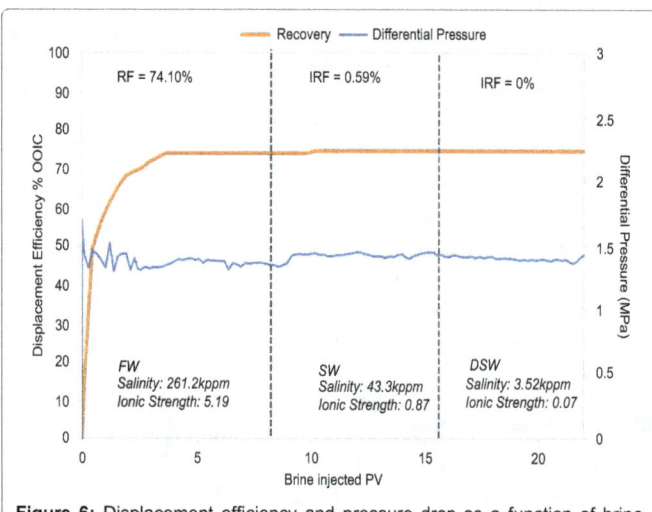

Figure 6: Displacement efficiency and pressure drop as a function of brine injected PV for P#6.

activity of the active ions. This led to the active ions possessing better access to the rock surface. Increased adsorption of sulphate resulted, which lowered the positive charge on the rock surface and changed the established thermodynamic equilibrium. Then the availability of excess Ca^{2+} and Mg^{2+} at the rock surface, allowed formation of a complex compound with the organic component in crude oil and led to the desorption of more organic materials. The response from this double layer effect is wettability alteration towards more water-wetness and increased oil recovery as observed by Fathi et al. on chalk core materials [22]. Although, chalk (2 m^2/g) has been reported to possess significant larger surface area compared with limestone (0.3 m^2/g) but both have similar chemical composition, $CaCO_3$ [5,14].

Core flood 2: Same procedures as mentioned above was followed only that P#6 was flooded sequentially with FW, SW and DSW. The baseline injection reached a recovery plateau of 74.1% OOIC in about 2.89PV before oil production ceased. The initial oil saturation was 82.7% and decreased to 21% by waterflooding. Then the injection fluid was changed to SW and an incremental recovery of 0.59% OOIC was recorded. The reduced incremental recovery observed here as compared to first case was probably due to the differences in rock characteristics and heterogeneities. Then SW was replaced by DSW and in about 5.62PV, no substantial oil recovery was observed (Figure 6).

It can be inferred that reducing only the salinity/ionic strength without altering the concentrations of the active/non-active ions (SW – 43 kppm to DSW – 3.52 kppm), could not mobilize additional oil. The pressure drop profile showed a decrease in pressure and later stabilization as the brine salinity reduces just as discussed in the first case.

The relative ion profile is shown in Figure 7a shows that during SW injection, $[Ca^{2+}]$ increased in the effluent and later decreased but couldn't return back to the injection level, probably because of rock dissolution which produces more Ca^{2+} into the effluent. $[SO_4^{2-}]$ reduced in the effluent coinciding with high concentration of non-active ions in the effluent while $[Mg^{2+}]$ was constant. SO_4^{2-} could penetrate easily, then adsorbed to the rock surface and release the oil but little displacement was observed due to inability of the divalent ions to co-adsorb. This agrees with Zhang and Strand. where they concluded that SO_4^{2-} must act together with Ca^{2+} and Mg^{2+} in order to obtain substantial recovery [12,23]. While the rock dissolution might have occurred in the pores already swept by the FW and denied the divalent ions the ability to co-adsorb.

Nil production observed during the DSW injection is as a result of the low concentration of active ions found in the injected brine. Then, the non-active ions concentration in the effluent settled around the normalized concentration value of 1.2 (Figure 7b). This indicates that sufficient amount of non-active ions were observed in the diffuse layer; thereby the active ions (Ca^{2+}, Mg^{2+} and SO_4^{2-}) were prohibited access to the rock surface. This led to increase concentration of the active ions in the effluent.

For the period of DSW injection, the non-active ion concentrations increased and then declined until they reached an equilibrium value at the injection level. Correspondingly, $[SO_4^{2-}]$ was at 1.2 times the DSW slug and didn't return to the injection level, which indicates no interaction with the rock surface. Then $[Ca^{2+}]$ and $[Mg^{2+}]$ settled above and around the normalized concentration respectively. This case differs from the previous case due to the fact that here salinity was reduced by diluting SW. As for the baseline injection, the displacement efficiency was similar to that observed in the previous case. Then the efficiency of SW flooding in this case is comparatively low as likened to previous case probably due to independent adsorption of SO_4^{2-} observed during the effluent ionic analysis. Sulphate multi-ion exchange process was thought to be responsible in this case, which might have caused marginal recovery due to inability of divalent ions to help sulphate compete with the carboxylic group [17]. In a study by Gupta and Mohanty [24], they also concluded that SO_4^{2-} can alter rock wettability to a larger degree in the presence of Mg^{2+} and Ca^{2+} than in their absence.

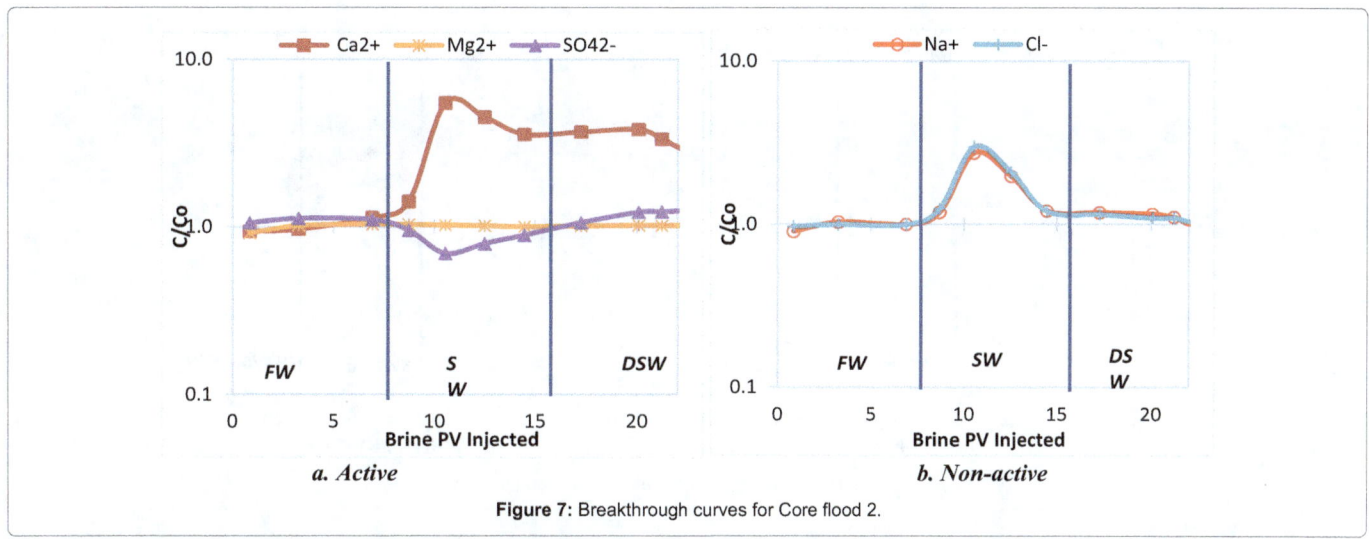

Figure 7: Breakthrough curves for Core flood 2.

Core flood 3: This core flood was carried out to further confirm the observation made during coreflood 2 and here, the core, P#8, was flooded sequentially with FW, SW0NaCl and DSW0NaCl. Injection of regular FW recovered 64.63% OOIC and oil production ceased in about 6.6PV. No further production was observed for continuous injection of FW (about 5PV more), which was to ensure all the mobile oil was thoroughly flushed out. The oil saturation was reduced from 74.2% to 25.91% which slightly differs from the previous cases.

Then the injection fluid was changed to SW0NaCl (lower ionic strength and salinity) and an incremental recovery of 1.33% OOIC was documented, reaching its recovery plateau after 2PV injection. Then the injection fluid was replaced with DSW0NaCl and after 5PV injection, no additional oil recovery was observed as in Figure 8. Before the experiment, fractures were observed through the core sample and this limited the displacement efficiency as observed in other coreflooding results. This resulted in the little displacement efficiency observed during SW0NaCl flooding as compared to the Coreflood 1. This is due to the fractures diverting most of the flow to the swept zone thereby resulting in mineral dissolution of the calcite surface.

Similar observation in the breakthrough curves as shown above was made compared to coreflood 2, except that the effects are more pronounced. This case placed more emphasis on the fact reducing brine salinity by just brine dilution couldn't result in improved oil recovery (Figure 9). The inability to recover additional oil could be attributed to the following reasons:

a) Presence of sufficient amount of non-active ions in the diffuse layer which could deny the relatively small concentration of active ions access to the rock surface [22].

b) A film was formed by the residual oil after SW/SW0NaCl injection that prohibited DSW/DSW0NaCl from interacting with the rock surface [25].

This result once again proves that diluting injection water could not recovery additional oil, which agrees to conclusions from Gupta, Fathi, RezaeiDoust and Austad et al. [21,22,26,27]. But seems to conflict with those from Alotaibi et al. Yousef et al., Zahid et al., Yi and Sarma and Chandrasekhar and Mohanty [15-17,28,29], where they maintained that diluting injection water have a significant impact on oil recovery accompanied with high pressure differential. Moreover, the dissimilarity could be as a result of different rock typing. Then Austad

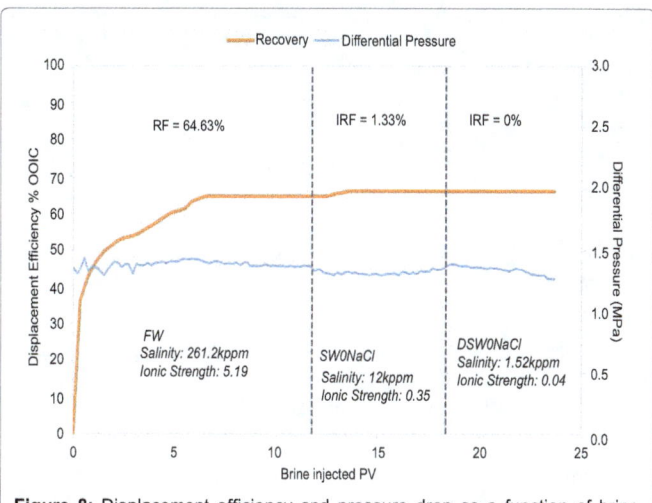

Figure 8: Displacement efficiency and pressure drop as a function of brine injected PV for P#8.

described that brine salinity reduction by diluting injected water could only give incremental recovery in case there is anhydrite present in the carbonate core. However in this case XRD analysis showed the absence of anhydrite from all core samples used, which would be the reason for not observing incremental recovery during DSW injection. Gupta reported that the effectiveness of smart water with lower salinity was mainly due to the reduction of the divalent cation concentration rather than dilution of the injected water [21].

Wettability monitoring

Principally, this study was to investigate the wettability alteration by smart water as the principal mechanism responsible for improved oil recovery and further support evidence presented by the surface charge alteration. The influence of various smart brines on wettability was tested by evaluating the contact angle of oil droplet on the core plates and plotted against exposure time as shown in Figure 10.

Exposure to high saline brine (FW) could only maintain the rock wettability in the range of preferentially oil-wetting state. Then low saline brine (SW) drastically decreased the contact angle from 170.5° to 148.2° during the first 40hrs and with time further decreased to an equilibrium value of 119.8° which seems to adjust the rock's wettability

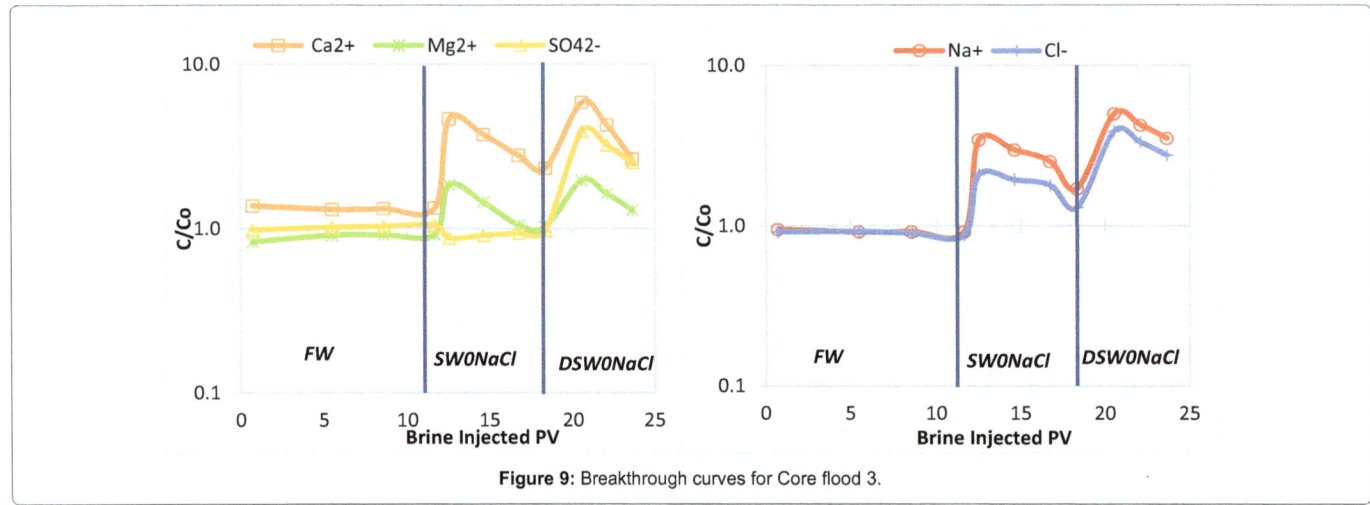

Figure 9: Breakthrough curves for Core flood 3.

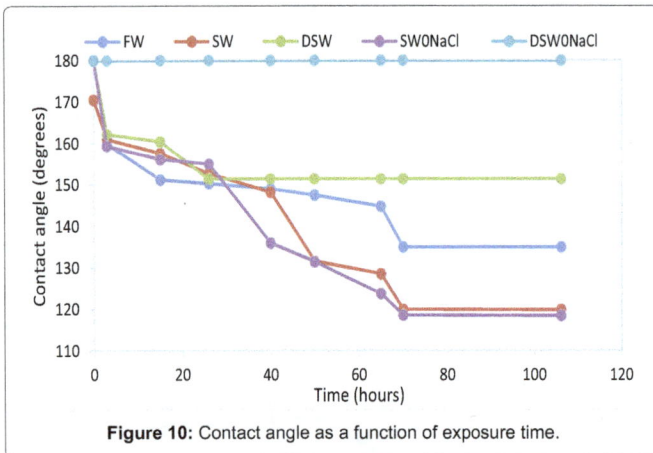

Figure 10: Contact angle as a function of exposure time.

Observation here shows that FW couldn't change the wettability beyond the strongly oil-wet zone and SW decreased the wettability from strongly oil-wet to weakly oil-wet. Then considering seawater with its non-active ions depleted (SW0NaCl), contact angle decreased from 180° to 118.4°, which indicates that the rock plate becomes relatively more water-wet as compared to SW. This shows that removal of NaCl salt from the invading brine could actually improve the water-wetness of the rock and improve oil recovery. This can be observed in coreflood 1 and coreflood 3 where injection of SW0NaCl explicitly increased recovery beyond SW and FW respectively.

Conclusions

Finally, the multi-ion interaction leading to improved recovery by smart waterflooding was thoroughly investigated through series of experiment on candidate carbonate reservoirs. Based on the experimental findings, the following conclusions are drawn:

a) Direct relationships exist between NaCl depletion and zeta potential. For surface charge alteration to take place, there must a reduction in the ratio of SO_4^{2-} to NaCl in the injected smart brine.

b) Ultimate oil displacement increased relative to formation brine when NaCl was selectively removed from smart brine as compared to brine dilution.

c) Effluent concentration analysis presents evidence to support the multi-ion interaction occurring to influence wettability alteration through non-equilibrium state caused by surface charge and mineral alteration.

d) For effective Brine salinity reduction, a certain ion-ratio between the active and non-active ions should be maintained.

e) The degree of wettability alteration seems to be a function of the magnitude of surface charge alteration at the two interfaces: removal of NaCl gave a better alteration compared to brine dilution.

Acknowledgment

The authors appreciate the financial support provided by The Petroleum Institute (PI) in conducting this research work and the permission to present this paper. We thank ZADCO UZFDS Laboratory for helping with effluent sample analysis.

References

1. Jadhunandan PP (1990) Effects of brine composition, crude oil, and aging conditions on wettability and oil recovery. Department of Petroleum Engineering, New Mexico Institute of Mining & Technology, Mexico.

from preferentially oil-wet state to weakly oil-wet state. Then with DSW, the contact angle measured with time was found within the preferentially oil-wet state which indicates that DSW was unable to alter wettability of the rock plate. Besides, SW0NaCl gave a more water-wet nature with time as compared to DSW0NaCl which couldn't alter the rock wettability.

It is quite interesting to note that during the experiment core flood 2, DSW invading after SW couldn't increase the recovery since SW gave a more water-wetness compared to DSW. The ratio of ions in the diluted brines (DSW) was found to be same in SW while the ionic strength decreased by the dilution factor. Under this condition it was impossible to improve oil recovery by DSW due to the fact that there is reduced concentration of active ions as the brine was diluted. This same findings was reported by Fathi when no incremental recovery was observed when chalk core was flooded with diluted seawater because it gave low water-wet fraction compared to the original seawater [22].

But this findings seems to conflict with results from Yousef and Yi and Sarma [15,16], they confirmed wettability alteration with dilution of invading brine salinity. They regarded this result to be due to rock microscopic dissolution, also in our case, rock dissolution seems to occur with the rock plates but the oil droplets vanished which made it hard to quantify. The discrepancies between our study and the reported results could be ascribed mainly to the particularity in rock characteristics, initial fluids composition, experimental conditions and procedures [30].

2. Tang GQ, Morrow NR (1997) Salinity, temperature, oil composition, and oil recovery by waterflooding. SPE Reservoir Engineering 12: 269-276.

3. Zhou X, Morrow NR, Ma S (2000) Inter-relationship of wettability, initial water saturation, aging time, and oil recovery by spontaneous imbibition and water flooding. SPE Journal 5: 199-207.

4. Zhang P, Austad T (2006) Wettability and oil recovery from carbonates: Effects of temperature and potential determining ions. Colloids and Surfaces A: Physicochemical and Engineering Aspects 279: 179-187.

5. Strand S, Austad T, Puntervold T, Høgnesen EJ, Olsen M, et al. (2008a) "Smart water" for oil recovery from fractured limestone: A preliminary study. Energy & Fuels 22: 3126-3133.

6. Jadhunandan PP, Morrow NR (1995) Effect of wettability on water flood recovery for crude-oil/brine/rock systems. SPE Reservoir Engineering 10: 40-46.

7. Lager A, Webb KJ, Black C, Singleton M, Sorbie K (2008) Low salinity oil recovery-an Experimental investigation. Petrophysics. 49:1

8. Agbalaka CC, Dandekar AY, Patil SL, Khataniar S, Hemsath JR (2009) Core flooding studies to evaluate the impact of salinity and wettability on oil recovery efficiency. Transport in Porous Media 76: 77-94.

9. Austad T, Strand S, Høgnesen E, Zhang P (2005) Seawater as IOR fluid in fractured chalk. SPE-93000-MS. Proceedings of the SPE International Symposium on Oilfield Chemistry, The Woodlands, Texas.

10. Høgnesen E, Strand S, Austad T (2005) Water flooding of preferential oil-wet carbonates: Oil recovery related to reservoir temperature and brine composition. SPE-94166-MS. Proceedings of the SPE Europe/EAGE Annual Conference, Madrid, Spain.

11. Zhang P, Tweheyo MT, Austad T (2006) Wettability alteration and improved oil recovery in chalk: The effect of calcium in the presence of sulfate. Energy & Fuels 20: 2056-2062.

12. Zhang P, Tweheyo MT, Austad T (2007) Wettability alteration and improved oil recovery by spontaneous imbibition of seawater into chalk: Impact of the potential determining ions Ca^{2+}, Mg^{2+}, and SO_4^{2-}. Colloids and Surfaces A: Physicochemical and Engineering Aspects 301: 199-208.

13. Strand S, Puntervold T, Austad T (2008b) Effect of temperature on enhanced oil recovery from mixed-wet chalk cores by spontaneous imbibition and forced displacement using seawater. Energy & Fuels 22: 3222-3225.

14. Austad T (2013) Understanding of the EOR Potential Using "Smart water" enhanced oil recovery field. Case Studies pp. 301.

15. Yousef A, Al-Saleh S, Al-Kaabi A, Al-Jawfi M (2010) Laboratory investigation of novel oil recovery method for carbonate reservoirs. SPE 137634-MS. Proceedings of the Canadian Unconventional Resources and International Petroleum Conference, Calgary, Alberta, Canada.

16. Yi Z, Sarma H (2012) Improving water flood recovery efficiency in carbonate reservoirs through salinity variations and ionic exchanges: A promising low-cost smart-water flood" approach. SPE 161631-MS. Proceedings of the Abu Dhabi International Petroleum Conference and Exhibition, Abu Dhabi, UAE.

17. Chandrasekhar S, Mohanty K (2013) Wettability alteration with brine composition in high temperature carbonate reservoirs. SPE 166280. Proceedings of the SPE Annual Technical Conference and Exhibition, New Orleans, Louisiana, USA.

18. Yousef A, Liu J, Blanchard G, Al-Saleh S, Al-Zahrani T, et al. (2012) Smart water flooding: Industry's first field test in carbonate reservoirs. SPE-159526. Proceedings of the SPE Annual Technical Conference and Exhibition, San Antonio, Texas, USA.

19. Awolayo AN, Sarma HK, AlSumaiti AM (2016) An experimental investigation into the impact of sulfate ions in smart water to improve oil recovery in carbonate reservoirs. Transport in porous media 111: 649-668.

20. Awolayo AN, Sarma HK, Al-sumaiti AM (2014) A laboratory study of ionic effect of smart water for enhancing oil recovery in carbonate reservoirs. SPE 169662. Proceedings of the SPE EOR Conference at Oil and Gas West Asia, Muscat, Oman.

21. Gupta R, Smith Jr P, Willingham T, Lo Cascia M, Shyeh J, et al. (2011) Enhanced water flood for middle east carbonate cores–impact of injection water composition. SPE 142668. Proceedings of the SPE Middle East Oil and Gas Show and Conference, Manama, Bahrain.

22. Fathi SJ, Austad T, Strand S (2010) "Smart water" as a wettability modifier in chalk: The effect of salinity and ionic composition. Energy & Fuels 24: 2514-2519.

23. Strand S, Høgnesen EJ, Austad T (2006) Wettability alteration of carbonates - Effects of potential determining ions (Ca^{2+} and SO_4^{2-}) and temperature. Colloids and Surfaces A: Physicochemical and Engineering Aspects 275: 1-10.

24. Gupta R, Mohanty K (2008) Wettability alteration of fractured carbonate reservoirs. SPE 113407-MS. Proceedings of the SPE/DOE Symposium on Improved Oil Recovery, Tulsa, Oklahoma, USA.

25. Nasralla RA, Alotaibi MB, Nasr-El-Din HA (2011) Efficiency of oil recovery by low salinity water flooding in sandstone reservoirs. SPE 144602. Proceedings of the SPE Western North American Regional Meeting, Anchorage, Alaska, USA.

26. RezaeiDoust A, Puntervold T, Strand S, Austad T (2009) Smart water as wettability modifier in carbonate and sandstone: A discussion of similarities/differences in the chemical mechanisms. Energy & Fuels 23: 4479-4485.

27. Austad T, Shariatpanahi S, Strand S, Black C, Webb K (2011) Conditions for a low-salinity enhanced oil recovery (EOR) Effects in carbonate oil reservoirs. Energy & Fuels 26: 569-575.

28. Alotaibi MB, Azmy R, Nasr-El-Din HA (2010) Wettability challenges in carbonate reservoirs. SPE-129972-MS. Proceedings of the SPE Improved Oil Recovery Symposium, Tulsa, Oklahoma, USA.

29. Zahid A, Shapiro A, Skauge A (2012) Experimental studies of low salinity water flooding in carbonate reservoirs: A new promising approach. Paper SPE-155625l presented at the SPE EOR Conference at Oil and Gas West Asia, Muscat, Oman.

30. Tang GQ, Morrow NR (1999) Influence of brine composition and fines migration on crude oil/brine/rock interactions and oil recovery. Journal of Petroleum Science and Engineering 24: 99-111.

14

Review on the Fundamental Aspects of Petroleum Oil Emulsions and Techniques of Demulsification

Souleyman A Issaka*, Abdurahman H Nour and Rosli Mohd Yunus

Faculty of Chemical and Natural Resources Engineering, University Malaysia Pahang, Malaysia

Abstract

This review is aimed to introduce a comprehensive survey on the most prominent and sustainable techniques and methods that could abate the environmental worries as well as financial insecurities in treating petroleum emulsions, since the existence of water is not desired because of the paramount troubles it may cause on the processing streamlines, as well as financial cost associated with transporting water mixed with petroleum. Currently, the most commonly used method for treating petroleum emulsion is the application of chemical additives, known as demulsifiers. Althogh, there are many other methods that are claimed to be more favorable from economic and environmental perspectives, yet, have not being fully put into real life practice, because of the drawbacks and disadvantages. In this review, several techniques have been surveyed including, chemical, electrical, membrane, centrifuge, bacteria, air floatation, ultrasonic, and microwave. Based on this Theoretical survey, silcone based demulsifiers were reported to be very effective and environmental friendly, but expensive. Also microwave and ultrasonics were reported to be very effective in treating petroleum emulsion and could be recommended as future ulternatives for treating petroleum emulsions.

Keywords: Petroleum emulsions; Demulusification techniques; W/O emulsion separation; Stability

Introduction

This review paper would investigate the related literature and previous studies on the petroleum emulsions, and techniques of separating water from crude petroleum oils. since water separation from petroleum is vital from both tchnical as well as environmental view points.

Emulsion formation and stability

Basically all types of emulsions consist of two immiscible liquids, one of which is dispersed as droplets called internal phase or droplet phase) into another, that is normally called continuous phase. These two phases are also normally stabilized by a third phase, called emulsifying agent (also known as surfactants, surface active agent, or emulsifiers). In water-in-crude (w/o) emulsions, the dispersed water droplets (internal phase) are encapsulated by an oil matrix (external or continuous phase). Without the emulsifying agent, emulsions made of the individual pure components of only oil and water phases are not stable, given enough time, they would quickly split back into their original phases (water and oil). that is due to the elevated interfacial tension between the oil and water phases. High interfacial tension would induce the attractive forces to attact the neighboring water droplets, wich will soon merge together and eventually separate to two distict phases (water and oil). Therefore a surface active agent or emulsifier is needed to stabilize the interfacial films, and eventually the whole system. Naturally, petroleum crude oil emulsions are stabilized by the indigenous surfactants, namely asphaltenes and resins which are partially soluble in both systems and have a strong tendency to migrate and sediment at the interface between the water droplet and the oil phase forming what so called interfacial barrier [1]. However, most researchers use experimental emulsions, which is produced in laboratory via commercial serfactants to mimic the natural emulsion. Stable Water-in-crude Emulsions are generated during crude oil mining, because water and oil coexists in the resevoir, the come to contact and mix while pumped out on the wellbore, Indeed, the existence of the indigenous surfactants, as such asphaltenes and resins,

heavy metals, fatty acids and clays would strongly lower the interfacial tension between the oil and water, and lead to the formation of stable barriers between the water-oil interface that resist the coalescence of water droplets and eventually produce stable water-in-crude oil emulsions [2,3]. Crude oil constitudes of various components, mainly, carbon, hydrogen. Minor quantities of others gases such as nitrogen, oxygen, together with heavy metals, such as sulfur, and trace amount of other common metals (iron, vanadium, and nickel). These elements would facilitate the formation of hydrogen bonding as well as polar interactions among the asphaltene components, and that would give the interfacial film its elastic behavior and strength [4]. Plenty of researches was conducted on the emulsion preparation and stabilization process in the field of crude oil emulsion. Various compositions, combinations, as well as preparation conditions, have been reported. since emulsions are multiphase systems of mainly Oil, water and emulsifying agent, that moulded together by the means of mechanical mixing, therefore stability study is aimed to identify the right variables, including, type of chemicals, mixing power and speed, mixing duration (emulsification time), mixing or emulsification temperature (EMT). Knowlege of these parameters is vital, in producing any type of stable experimental emulsion. Tron eric Havre and Johan sjoblom, have investigated the effectiveness of combined surfactants Phases and Asphaltene particles in producing stable emulsions. Solution of heptane/Toluene mixture, that contains Paraffinic and aromatic components was prepared to represent the crude oil phase. They extracted the asphaltene from an original crude, this asphaltene is then

*Corresponding author: Issaka SA, Faculty of Chemical and Natural Resources Engineering, University Malaysia Pahang, Malaysia, E-mail: sahame@gmail.com

combined with aforementioned commercial surfactants phase to represent the emulsifying agent. Ultra Turrax T25 mixture was used at speed of 9500 rpm for 30 seconds (emulsification time). Emulsification was brought about in water bath at 70°C. Their prepared emulsion was investigated based on stability as well as droplet size analysis. Their study revealed that water-in-crude oil emulsions can be stabilized by means of combined surfactants as such naphthenic acid, naphthenate and asphaltene particles [5]. Lixin Xia also investigated the effect of asphaltene and resin in jet kerosene continuous phase w/o emulsion, but at different emulsification conditions, he used hand shaking as mechanical mixing source. He characterized his emulsions using optical microscopic visualization, and water resolution techniques. His result recomended the water resolution technique as the best way to study the stability of emulsions [6]. In another work Brujic, have produced a transparent Oil-in-water emulsion, the intended to measur the inter-droplet forces that govern these micron sized emulsion droplets. Silicone Oil and Sodium dodecyl Sulphate (SDS) were selected as oil phase and emulsifying agents respectively, Mixing speed was 7000 s⁻¹, They produced emulsion with mean droplet size of 3.4 µm, ranging from (1-10 µm), which was reported to be stable over a period of at least one year [7]. Guo et al., have investigated the significance of alkaline-surfactant-polymers as emulsifiers to stabilize emulsions made from model oil. Their model oil, consisted of jet fuel as the dispersing medium, mixed with fractions from Gudong crude oil (from Shengeli oil field in China) to mimic the natural crude oil composition. Emulsions were prepared via simple hand mixing at 60°C. Emulsion stability was determined by visually monitoring the rate of water resolution. The interfacial active components in the Asphaltene were found to be ionized when they are mixed with the alkali solution, thus, the ionized and non ionized groups can make hydrogen bonds, which lead to the formation of the acidic soaps. These soaps act as emulsifiers and promote stable emulsion on petroleum wells during enhanced oil recovery where the alkaline-surfactant-copolymers are injected in the reservoir to increase the productivity [8]. Stability effect of the alkaline-surfactant-copolymer also investigated by Minguan et al. on stabilizing Daqing crude oil from China. They comfirmed the ability of alkali surfactants to produce stable emulsions. Model oil was prepared from jet fuel mixed with some polar fractions from daqing crude oil. The water to oil ratio was 50-50% (v/v), emulsion was prepared by handshaking at 45°C [9]. Venezuelian crude oil having API of 24.5° (sour crude oil), was found to produce be very stable emulsions, even without emulsifiers. Thus, in the subsequent studies, researchers have diluted the original Venezuelan crude oil with various amount of heptanes (Poor solvent for asphaltene), several emulsions were made from the diluted oil by vigorous mixing at 40°C. results proved that the Alifatic/Aromatic ratio of the crude oil can tremendously affect the stability of the emulsion. This affirm the theory that attributing the asphaltene and resin activities to be strong function of the Alifatic/aromatic fractions of the petroleum. Hence, when the heptane (alifatic) is added to the venezuelan crude oil (dilution), it was found to induce the asphaltene precipitation out of the crude oil. This precipitation increases the size and poly-dispersity of submicrone asphaltene particles that lead eventually to reduce the stability of the emulsion, because of the attractive forces [10]. Nave Aske et al., have developed an electric cell that can supply an electric field to the emulsion, it was used to predict emulsion stability, since the electric field can break the emulsion and separate the water. They have successfully tested that device on emulsions made from 21 different crude oils with different water cuts (30% and 20%), and from their reported results, they have categorised the stability into several groups, mainly, stable and unstable, and that was imparted that to the asphaltene contain, its aggregation

state and the interfacial elasticity [11]. In another work, bitumen was dispersed in water phase, and produced a stable bitumen-in-water emulsion at elevated pH values [12]. Minguan Li et al. have investigated the effects of alkaline surfactant-polymers on stabilizing crude oil emulsion that is originated from Daqing oil field (China), however, their results had proved the ability of theses Alkaline surfactant to produce stable w/o emulsions inside the reservoir when they are injected during enhanced oil recovery. Their sample was composed of 50-50% (w/o), and prepared by simple hand shaking [9].

In another work, Andreas et al. have studied the effects of silica nano-particles on stabilizing model oil emulsions. Asphaltene and resins were extracted from the Brazilian sour crude oil and used as coating on the surface of the silica particles to render them amphiphilic solid emulsifiers. Heptanol-toluene mixture was choosen as model oil phase. ultraturax mixture was used at speed of 22000 rpm for three minutes. The stability study was carried out within a period of 30 minutes. The percentage of water resolved from the emulsion within this 30 minutes was used to assess the emulsion stability. Their results showed that, coatings has greatly enhanced the model emulsion stability [13]. Elsewhere, the interfacial activity of asphaletene were measured, a model oil emulsions were prepared by mixing Heptane, Toluene, Asphaltene, Resin, and Native solids. The resulting emulsions were characterised by calculating the emulsion surface area from droplet sizes distribution. The Asphaltene and Resin used in their study were extracted from two different crude oils, although, their emulsion contained 40% water, mixed at 17000 rpm, for 7 minutes, yet it was observed to separate within few minutes form it is first preparation [12]. Far from crude oil emulsions, other researchers reported the polymer blind mixture. However, in comparision to emulsions, surfactants-like additives are used as stabilizers in polymer blends. theses additives are called Compatibilizers, and used to stabilize the non-compatible polymers. Moran et al., have studied the role of Naphthenic acid in crude oil emulsions, since naphthenic acid is one of the natural indigenous components of the crude oil. Thus, they have produced model oil (Heptane/toluene) emulsions that was stabilized by Naphthenic acid based surfactant, namely, Sodium Naphthenate, at different concentration ranges (0-10%). sonicator was used to mix the emulsion sample [14]. Others, have investigated the effects of physicals parameters, including internal phase volume fractions, Mixing speed (rpm), mixing (emulsification temperature) on the rheological properties of water-in-crude oil emulsions [15]. Rosli Daike et al., studied the effect of types and concentrations of various commercial emulsifiers in stabilizing Liquid natural rubber (LNR) based w/o emulsions. Various concentration of emulsifiers were used (ranged from 1-6%). Emulsions were prepared at room temperature with continuous stirring for 30 minutes. Their results proved that, Liquid Natural Rubber with lower molecular weight had produced emulsions with smaller droplet sizes, high viscosity and more stable emulsions. Whereas, the opposite pattern was observed with the lower molecular weight (LNR). In the same fashion, Adam Macierzanka et al. have investigated the microstructure of the internal phase of emulsions stabilized by Acylpropyleneglycols containing C_{16}/C_{18} fatty acids.The oil phase was mixture of Paraffin oil and paraffin wax. Emulsions were prepared in a glass thermostate emulsor, mechanical two blade agitator was used at mixing speed of 500 rpm, the phase composition of the emulsions was varied from 20-80% (water-oil% respectively). Lei Zhang and others, have studied the correlation between the Hydrophilic-Lipophilic Balance (HLB) of the surfactants and the emulsion properties. They used surfactants with various HLB numbers to prepare experimental water-in-oil emulsions, two blade stirere was used at mixing speed of 500 rpm for 15 minutes. Span 80 and Tween 80 were chosen to be the major stabilizers for their study. various other fractions with different HLB ranging from 4.5-15 were produced by blending [16]. Oppositely Svetlan et al. have reported a comprehensive investigation on emulsion Rheology, but focused their attentions mainly on the effect of the individual components on emulsion rheology of dilute emulsions of oil base and polymer blends [17]. Also Minguan Li et al. studied the stability of model emulsions that was stabilized by the indigenous incompatible components of the crude oil, thus the stability was investigated by means of various apparatus such as IR, UV, and GC-MS, hence the interfacial active components (Surfactants) were extracted from two different crude oils that originated from Saudi Arabia and Shengli (China). The interfacial active components (Asphaltene and resins) are normally extracted from the crude oils by SARA fractionation techniques. SARA is stand for Saturate, Aromatic, Resin, and Asphaltene, the main components in crude oil. In SARA extraction process, crude oil is first mixed with n-hexane in the ratio of 1:30 (v/v) respectively. The mixture is left to settle for 24 hours. The precipitated asphaltene could be filtered out and dried. The percentage of asphaltene in the crude oil is calculated based on the wheight of this precipitated asphaltene. the n-hexane soluble part, is then poured in a silica column after evaporating the solvent to percolate the remaining SARA fractions by their respective dissolving solvents [18]. Beside asphaltenes and resins, stability of petroleum emulsions also reportedly, depends on combination of several factors such as molecular size of the interfacial active components, aromaticity or aromatic condensation of the crude oil itself, and carbonyl group concentrations [18]. Marco A. Farah, odserved the dependence of water-in-crude oil emulsions on the other physico-chemical properties of the system such as dispersed phase (water) volume fractions, temperature, average droplets size, shear rate, droplets sizes distributions, viscosity and density of the of the oil. They varied the water content of the emulsion as 10, 20, 30, 40, and 60%. Mixing speed was fixed at 10,000 rpm for three minutes. Stability pattern was verified by measuring the amount of water resolved from the stable emulsion within settling duration of four hours [19]. In emulsion field, every new emulsion sample of what ever application, has to undergo some stability test before proceeding to its ultimate application. It was observed that, water-in-crude oil emulsions of high API (≥ 38) crude oils, was very difficult to produce stable emulsions especially at high water volume fractions. Also increasing the water volume fraction in emulsion would increase the temperature at which emulsion shows Newtonian behavior. Below wax appearance temperature (WAT), emulsions present rheological behavior of a bingham plastic in certain shear range, indeed variation of kinematic viscosity of the emulsion with temperature was observed to correspond with the ASTM equation [19]. Sanfeld and others, investigated the repulsive interaction within droplets of dense w/o emulsions, and found that, the thickness of the repulsive electrical double layer is inversely proportional to the internal phase (water) Volume fractions [20]. Indeed Kristofer Paso et al., have observed the stability as well as flow properties of two Brazilian crude oils. Their emulsions were composed of 30, 50, and 70% water volume fractions. Ultra-turax homogenizer was used for mixing at 24000 rpm for 2 minutes. Emulsification temperature also were varied as 4, 25 and 60°C. They concluded that at low water volume fractions, emulsion can resists coalescence in a better manner than its high water volume fraction counter part. Another observation was that, the waxy crude oil was able to form stable emulsion with water cut as high as 70%, while the heavy oil forms stable emulsions with water cut as high as 50%. This superior stability of the waxy crude oil might be attributed to the greater

abundance of Asphaltene and lesser abundance of resins [21]. In another report, Christophe Dicharry et al., have investigated the viscoelastic character of the interfacial film at the water-oil interface by means of the oscillating drop tensiometer method. They concluded that, interfacial parameters such as interfacial tension and elasticity modulus, can help to predict emulsion stability. Stability was a found to increase proportionally with increasing the interfacial gel strength and glass transition temperature of the gel [22].

Einar Eng Johnsen et al. have had proposed a method for measuring the viscosity of emulsions under pressurized conditions, their apparatus that consists of rotating wheel that would make the fluid inside to move to the opposite direction, hence, the torque acting on the wheel shaft is measured and transformed via a calibration model to a viscosity of the fluid, they also found that the variation of the viscosity increases with increasing the water cut [23].

Demulsification and demulsification techniques

Demulsification is the process of breaking the emulsion into its individual incompatible phases, mainly water and oil. Demulsification process is very important process in petroleum industries, wherein, emulsions almost always occur either naturally or deliberately (man made emulsions). In petrochemical industries and petroleum refineries, separation of water from the crude oil is required before oil refining. currently chemical additives called emulsion breakers are massively used to break the water-in-oil emulsions. Technically speaking, the resistance of a w/o emulsions to coalescence and their response to the demulsification techniques such as thermal, mechanical, electrical or chemicals depends mainly on the Physico-chemical structure of the oil from which they are formed, emulsification conditions, and Aging. This means, the effort and Strategies for optimizing the w/o emulsion demulsification may vary from one oil field to another [1]. Thus far, various methods for breaking emulsions were introduced. These including, Chemical, Thermal, centrifugal, freeze/thaw, and electric demulsifications. chemical demulsification is the most widely used in oil fields [3,24]. In chemical demulsification, massive amount of chemicals are used to separate the water from the crude oils. However, most of these chemicals are expensive, toxic and un-ecofriendly. Therefore, other alternatives and sustainable demulsification methods are still in demends. In the following sections, various literature servey on the existing and potential future demulsification techniques is discussed.

Chemical demulsification

Chemical demulsification is one of the very crucial techniques of resolving water-in-oil emulsions, and it is massively applied in the petroleume industries. Basically, massive amount of surfactants can be prepared by just manipulating the existing surfactants, one way is by changing the acceptor, composition, quantity, and sequence of hydrophobic and hydrophilic groups in the commercial long chain polymeric surfactants [25]. The very basic fundamental of chemical demulsification mechanism of what so ever type of emulsion is that, the demulsifiers gradually replace the emulsifiers within the water-oil interfacial film, and that would eventually cause tremendous changes on the interfacial viscosity and elasticity [26].

Hafiz et al. had synthesized some novel demulsifiers for treating water-in-crude oil emulsions, they prepared cationic polymer of of diethanol amine easters by condensation and polymerization of the diethanolamine to diethanol amine polymers esterification reaction was carried out at temperature range of 140-160°C using different types of catalysts. Their demulsifiers were found to be effective in treating emulsions of refinery wastewater. The purity of the treated water was determined by turbidity measurement [27]. Alejandro et al. have altered structure of some commercially available demulsifiers (Alkylphenol polyalkoxylated resin and polyurethanes). The efficiency of the newly formulated products were assessed by several techniques such as: bottle test, rheometry, equilibrium interfacial tension, and transient changes in drop sizes distribution that was measured using Nuclear Magnetic Resonance. Emulsifiers were tested on experimental emulsion that was prepared by simply dispersing the brine in the oil without emulsifiers. that means, the asphaltene and resins originally existing in the crude oil did the job of the emulsifiers. Their results showed that, the best and the highest separations were observed when emulsions were injected with resin having intermediate polyoxyethylene and polyoxypropylene moieties [28]. Jiangying Wu and others have investigated the property and performance of 20 blocked copolymer from four different surfactant families on breaking model emulsions. Model emulsions were prepared by first diluting bitumen (5%) in toluene, and centrifuging for five minutes to remove suspended indigenous small particles. This mixture was mixed with water and homogenised. polytron homogenizer was applied at 20 000 rpm for of three minutes. Their emulsion was observed to be stable without water resolution for three consecutive weeks. This emulsion was then broken via the aforementioned demulsifiers. Their resulted revealed that, the sequential block copolymer with more than 40% ethyleneoxide percentage was more effective than the others [25]. Wanli Kang et al. have studied the effect of several demulsifiers, namely, Phenol-formaldehyde resin polyoxyethylene, polyoxypropylene and polyoxyethylene polyoxypropylene polymers. Their model emulsion was prepared by mixing purified kerosene and dewatered crude oil in a ratio of 7:3, then water was added drop wise. The demulsification process was carried out at 45°C. Demulsification process was monitored in two distict samples, one without demulsifiers and other with demulsifiers. Results showed that, the strength and life time of the interfacial film was decreased when demulsifiers were added. Hence, the curve of the interfacial elasticity was found to decrease with increasing the demulsifiers' concentration, but came to a plateau of constant values beyond the critical concentration of demulsifiers. Masatto et al. have used Shirasu-porous-glass membrane to break water-in-crude oil emulsion, shirasu is a type of ash that is very rich in metal oxide mainly Silicone dioxide, however their results showed that Shirasu-porous-glass membrane could efficiently break water-in-oil emulsions with droplet sizes bigger than the average membrane pore diameter [24]. Delphine et al. have studied the effectiveness of non-toxic silicone based demulsifiers of polysiloxane copolymers on splitting water-in-crude oil emulsions. The test was carried out on experimental w/o emulsions, their emulsions which composed of 30% water were prepared at room temperature. polytrone homogenizer was used for mixing. To breack the emulsions, 2% (w/w) of demulsifiers (polyoxyethylene-silicone triblock copolymers) were first dissolved in Xylene/methanol solution (75/25 (w/w) before injecting them into the emulsions [3]. Similar experimental work also was reported by dalmazone et al. who investigated the efficiency of various formulations of these aforementioned non-toxic silicone based surfactants. The efficiency of the different formultions was determined by visual observation and measurement of water separation from emulsion, and dynamic interfacial tension measurement. They have tested silicone based formulation on two types of crude oils originated from North Sea and French oil fields. Their emulsification and demulsification methods were same as mentioned earlier [3]. Their results revealed that, poly siloxane demulsifirs were very efficient in breaking water-in-crude oil emulsions in comparison to some commercial demulsifiers.

Furthermore, some blend of silicone-silicone and silicone-organic was selected for their sustainability and versatility reason. Althogh silicone demulsifiers are claimed non-toxic and good potential alternatives, yet they are not lucrative to be commercialized for economic and cost reasons. However, the current increase in ecological constrains may favor them to be used in the near future or to be integrated in the formulation of the demulsifiers [29]. Svetlana et al. have investigated the effects of modified Chitosan derivatives in resolving o/w type emulsions. The study focused on the effectivness of hydrophobically modified (HM) Chitosan derivative demulsifiers copared to the commercial cationic polyacrylamide flocculant, and Unmodified (UM) Chitosan. The molecular weight of the demulsifiers was also modified. To test the demusifiers, experimental emulsions were made in laboratory. Sodium Dodecyl sulphate (SDDS) was selected as emulsifiers, emulsion was mixed for 10 minutes using ultrasonic mixture. was used to provide the mixing energy for an emulsification duration of 10 minutes. Results showed that, hydrophobically modified (HM) Chitosan derivatives with various molecular weights had shown to cause great separation compared to the others [30]. Abolfazl et al. have used the microfiltration membrane technology in separating the emulsion. They have tested the efficiency of the hydrophobic Polytetrafluoroethylene (PTFE) membrane having pore sizes of 0.45 micron meter in separating w/o emulsions. The experimental water-in-oil emulsion was prepared in laboratory using span 80 (oil soluble surfactant) as emulsifying agent. Emulsions were prepared according to the agent in oil method, in which Span 80 was first dissolved in the crude oil then water was added gradually as drops, while agitation is going on. The water addition rate was 25 ml/min. The mixing was carried out using pitched curve blade blender having 6 fins. Their results revealed that, the permeate flux from the membrane was found to be rich in emulsifiers and poor in water. Furthermore, the permeate flux was found to decrease with increasing pressure. The effect of the temperature was indeed obvious [31]. Ing Harald Auflem and others have studied the effect of pressure on separating w/o emulsions from North Sea crude oils. They prepared their emulsion by first mixing the emulsifiers (Silverson LURT emulsifiers) with the continuous phase (North Sea crude oil) then the aqueous phase was added gradually (Agent in oil method). A mixing speed of 2000 rpm was applied for one minute (emulsification time). The demulsification process was carried out by introducing the aforementioned emulsion into a high pressure demulsification rig. Their results showed that the emulsion stability was to some extent related to the demulsification pressure. This stability-pressure relation is imparted to the fact that, when the rig pressure was reduced beyond the bubble point pressure, the low molecular weight portion of the crude oil could evaporate and form gas bubbles that evaporate from within the sample to the top part of the system. This massive movement of the bubbles and density differrences would put the heavy components in a floatation state. This floatation effects would induce the emulsifiers residing at the water-oil interface to dissipate in the gas phase (the bubbles). This dissipation would eventually render the droplets free from their encapsulating emulsifiers matrixs, hence, the neighboring droplets would merge together (coalesce) and eventually separate into two phases (water at the bottom and oil at the top). another important finding of their investigation was that, by diluting the emulsion with toluene, the stability was also decreased and that is because dilution also decreases the viscosity as well as the density of the emulsion and also cause asphaltene to aggregate [32]. Nael Zaki et al. have used the carbone dioxide to break the w/o emulsions of various actual, as well as experimental emulsions. The dense CO_2 was assumed to induce the Asphaltene fluctuation, flocculation and precipitation. Thus, the absence of asphaltene the strength of the interfacial film by causing the film thinning and film rupture and eventually droplets collision and phase separation. The experimental emulsion samples used in their study was prepared by mixing the two phases at 1500 rpm using via Ultra-high speed Virtishear cyclone IQ Homogenizer for 2 minutes. emulsion stability was tested by visually opserving the water separation from the emulsion within duration of 24 hour. Then only the stable emulsions were chosen for demulsification via the aforementioned CO_2 injection. Their results proved the possibility of using CO_2 to treat water-in-crude emulsions of different characteristics and compositions [4]. Zhiaing et al. have investigated the demulsification effects of Poly (ethylene oxide (PEO), Poly (propylene oxide (PPO). They varied the molecular weight by altering the composition. Thus, PPO/PEO ratios were varied through the anion polymerization process. the effectiveness of these modified copolymers on demulsification of the water-in-crude oil emulsions were tested on experimental emulsion coposed of 50-50% (w/o). The experimental emulsion was prepared in laboratory by mixing the aqueous phase with petroleum oil taken from Shengli oil field (China). Emulsion was mixed at 1200 rpm for 5 minutes by HT-2 Homogenizer. For the demulsification process, Temperature and concentration of demulsifiers was fixed at 50°C, and100 mg/L respectively. The separation efficiency of the demulsifiers was determined by recording the water resolution for a settling period of 3 hrs. Result showed that, demulsification process can be enhanced with the reduction of the PEO fraction in the surfactant formulation [33].

Abdurahman et al. had made a comprehensive study on water-in-crude oil stability as well as demulsification. The investigated various stability parameters including, crude oil types, demulsifiers concentration, mixing speed, water volume fraction, and temperature. crude oils used in their study were originated from Iran and Malaysia. Several experimental emulsions were prepared according to the agent in oil technique, in which the crude oil is normally mixed with the emulsifiers first then water would be added gradually in drop wise fashion while mixing process goes on. The mixing energy was provided by three blade propeller, mixing time was fixed at 10 minutes. It was found that, the Iranian crude oil can produce more stable emulsion than the Malaysian Miri light crude oil [34]. Same authors, have studied the effect of the different types of the commercially available demulsifiers in splitting the aforementioned stable w/o emulsions. they have tested four different groups of demulsifiers, namely, Amine group, Polyhydric alcohol group, Sulphonate group, and polymeric group. beside that, the effects of the alcohol in the performance of demulsifiers also investigated [34]. Their results shwoed that, amine group demulsifiers were more effective in breaking the emulsion, followed by Acid then polymeric demulsifiers. From within the amine group demulsifiers, Hexylamine and diacylamine were the most effective commercial demulsifiers based on their their study [34].

Electrical demulsification

Electrical demulsification is a process of breaking of either water-in-oil, or oil-in-water emulsions by means of electrodes. Electrical demulsification devise is normally consists of two electrodes placed in opposed direction to each other. Emulsion is placed between the two electrodes. When, DC voltage is introduced to the system, electric field voltage is generated within the emulsion residing between the electrodes, that because the water in the emulsion is conductive. This will break the energy barrier between droplets , and water molecules shall immediately be hydrolyzed into hydrogen and oxygen gases molecules. The hydrolysis reaction is normally indicated through by hydrogen bubbles that evolve from the Positive electrode (cathode),

and oxygen bubbles that evolve from the negatively charged electrode (anode). It was observed that, when the anode materials are made of metals with less oxidation energy than the system, these materials would dissolve in the system to produce metal ions instead of the evolving oxygen bubbles. Then these metal ions would react with hydroxyl ions that produced during hydrogen generation, to give metal hydroxides.

An electrochemical reactor consists of tank with two sets of metal electrodes (anodes and cathodes). These two electrodes are normally placed at certain specific distance apart from one another and both of which should be immersed in the system under treatment for oil contamination. The cathodes and anodes are mounted on the negative and positive outlets of a dc power supply of the reactor, respectively. Hence During the operation (emulsion treatment), the decaying metal ions are hydrolysed, such as Fe (II) (ferrous) and Fe(III) ions in the case of Iron metal cathodes [35]. For safety reasons, the electrodes are normally insulated from the emulsion to prevent short circuit [36]. A lot of parameters were reported to be affecting the performance and efficiency of the process. Harpur et al. have studied the ffect of 50 Hz sonosoidal electric field to treat and remove water from w/o emulsion. they used a horizontal rectangular container (electrostatic coalescer). Seven electrode modules were used, each model was 50 cm in length, and generating an electric field of 25-60 KV/cm. Their results showed that the application of the electric can cause a significant growth in drop sizes. However, increasing temperature was found to accelerate the aggregation of the droplets [36]. Woo-in Jang et al. have also studied various parameter that affect the electrical demulsification techniques, Including, types of the electrode materials, polarization of the reactor, electrode size, temperature, oil concentration in the emulsion, mixing rpm, pH, and electric power consumption.

The oil sample used in their study was waste of metal cutting oil, from Korea Houghton Corporation. The oil consisted of 80% mineral oil and 20% surfactants plus some other chemicals such as anti foaming, bactericide, and anti corrosion. Experimental emulsion of the oil was prepared. The emulsion sample was introduced to the electrostatic coalescer and treated. Sample of the treated emulsion was taken from the reactor and analyzed after 30 minutes. Results showed that, aluminum reactor was performed better than the iron reactor, because, it gave high separation efficiency, and consumed less energy. Furthermore, water removal removal rate and energy consumption increased with increasing the applied potential. However, maximum separation of 94.05% was acheived at 60 volte, whereas, 85.50% separation was acheived at 30 Volts. Indeed, a wider electrode gave higher water removal and less energy consumption compared to narrower electrode, beside that, the best operating temperature was found to be 50°C, and variation of the oil volume fractions were found to have no effects on the demulsification process. Furthermore, agitation found to reduce the separation rate, while low pH was observed to be more effective than high pH. However, when the treated sample was analyzed, it was observed that there is a gradual increase in the particle sizes with time, hence their general conclusion had conformed the ability of the electrocoalescence process to treat the industrial oil-in-water emulsion [37]. Gary Sams and Zaouk have investigated all the upstream production parameters of petroleum wells, and their effects on emulsion separation. results revealed that, more vigorous and powerful electrostatic techniques are needed to treat and resolve the tight emulsions [38]. Simone less and others have used a device named Aibel vessel electrostatic coalescer, to break petroleum emulsions. The separation efficiency of the devise was evaluated in terms of percentage water resolution, and droplets sizes distribution, before and after the treatment. However, their results showed that, this system can effectively reduce the water content in the emulsion. But, chemical addition was found to greatly accelerate the separation process. Tsunki Ichikawa and others have applied electric field to break oil-in-water emulsion. Their experimental emulsion of various volume fraction was prepared by simple hand shaking, around 100 times. Their electrostatic device was consisted of two stainless steel plates fixed in opposite position inside glass container. Emulsion sample was placed in the device and the external electric field is applied. Their results revealed that, higher concentration of the electric had given high separation rate, but increasing the amount of ionic surfactants does not increase the separation rate. Although, Their devise was very good in breaking oil-in-water emulsions, yet it could not break the w/o emulsion even at elevated electric field, and frequency as high as 10 KHz [39]. Harld Fordedal and co-worker have studied the percolation behavior of water-in-oil emulsions under the influence of the electrical field. Percolation refers to the abrupt increase in dielectric constant of the emulsion, when its volume fraction attains a maximum values at constant temperature. They have prepared two experimental emulsions for this reason, one of which is stabilized by the naturally occurring surfactant (Asphaltene and resin), and the other is stabilized by the commercial surfactants. Their results proved that the dielectric constant of the emulsion was found to increase with the increase of strength of the electric field; this was observed with the water-in-oil emulsion stabilized with the indigenous interfacial active surfactants, wherea, the reverse effect was observed with the emulsion stabilized by the commercial surfactant. Thus, they concluded that, the dielectric properties of the commercial surfactants are by far different than that of the indigenous surfactant (Asphaltene and resins) [40]. Junji et al. have applied the electrostatic techniques to break water-in-oil emulsion hence they used some kind of metal oxide chemicals that are assumed to induce the increase in the coalescence rate of the crude oil droplet under the influence of the electric field, thus the performance of the device was assessed via video microscop from which a digital picture was captured and transferred to a computer for further analysis, their study revealed that, the shapes of the droplets were observed to change from spherical shapes to hemispheres when adsorbed at the electrode surface also the metal ions used were observed to be oxidized indeed the ion concentrations of the metal ions was observed to increase at the oil-water interface, also the image analysis of the coalescence of the neighboring droplets was observed to depend on the electrical potential applied [41]. Bailes et al. have combined the air bubbling process with electrostatic demulsification. However, their results concluded that, bubbling had enhanced the separation efficiency. Iindeed, separation rate also was observed to increase with increasing the air flow rate [42]. Tsuneki Ichikawa et al. have used a low electrical field to destabilize oil-in-water emulsion, hence the low electric field was assumed the reduce the thickness, and weaken the strength of the repulsive double layer, due to movement of the ions [43]. Seiji Kanazawa et al. have developed an apparatus for electrostatic coalescence using the electrostatic atomization techniques. The system was tested have tested on two types of emulsions. One was composed of water droplets dispersed in silicone oil, and the other was a stable emulsion that was prepared without using emulsifiers. The test was excuted by monitoring the movement of the positively charged oil droplets as a response to the applied electric field. thus they have first observed the marked drops in the transparent oil, then after the DC electric field was applied, the individual droplets were observed to aggregate and their traveling distance from the electrode, and the traveling energy were monitored [44]. Others have investigated the electrodynamic mechanism of the electrostatic demulsification process, via simulating the effect of the field strength on an individual water droplet [45]. Yakhkeshi et al. have used a uniform electric field to treat oil-in-water emulsion. They aimed to observe how the electric current could induce the scattered find droplets within the emulsion to aggregate and come close to one another, and produce bigger sized droplets. Their emulsion was consisted of benzene as the oil phase (model oil) and water. Emulsion was mixed by a magnetic stirrer for 15 minutes. (emulsification time). SDDS was used as emulsifying agent in a dosage range of 2-4 ppm. The dispersed Phase volume fraction was 4% of total emulsion volume. Their result again consolidated the usefulness of the electrostatic demulsification process on breaking the oil-in-water emulsion. The influencial parameters were, type of the insulator at the surface of the electrode, types of the oils, oil volume fraction in the emulsion, temperature and the electrical voltage [46]. Mir et al. have used non-uniform electrostatic techniques to treat simulated waste water emulsions. They formulated the experimental waste emulsion by mixing Xelene, water and metals, and surfactants. The volume fractions of the oil ans surfactants were 2% and 1% respectively. Ultrasonic mixture was applied for 6 minutes. For demulsification process, The electrostatic apparatus consisted of two electrodes that mounted oppositely in glass chamber. Three types of the electrode metals, namely Iron, Tungsten and copper were tested. However, their results proved that, Copper electrode was the best and performed better than the others. Indeed, beside the the type of electrode, voltage, temperature, and time were also crucial parameters. The highest separation was observed at electric power of 5400V and temperature of 45°C. Akuma Oji et al. have applied the High Voltage Dielectric Current (HVDC) having voltage range of 4- 16 KV to break real oil field emulsions from Obagi oil field. They investigated the effects of variating the distance between electrodes made up of Zinc. The distance was altered gradually from 60 to 244 mm. The operation time also ranged from 10-40 minutes. The best operation condition was observed to be 8 KV voltage and 122spacing distance [47].

Byoung et al. have used a high AC field dehydrator to treat water-in-oil emulsions. They used 2-20 KV Voltage, and frequency range of 60-2000 Hz was applied [48]. They studied several parameters, including chemical demulsifiers, operating temperature, operating time and residence or contact time in terms of the feed flow rate. Their experimental model w/o emulsion of 20% (v/v) water contain was used via electromagnetic stirrer (Heidolph RZR 205, Germany), at 1000 rpm and 10 minutes emulsification time . The electrostatic hehydrator device was made up of glass chamber, and the electrodes that were insulated with glass material to prevent possible short circuiting. Prior to the demulsification, certain amount of chemical demulsifiers were added to the emulsion and mixed for 5 minutes at 1000 rpm, then transferred into water bath to attain the desired temperature before transferring to the electrostatic coalescer. the flow rate was monitored to give a residence time of three minutes. Their results showed that, separation efficiency increased with increasing the electric field. also adding the demulsifiers was found the increase the separation efficiency furthers to 80%, while the temperature and the residence time were not that effective [48].

Thermal demulsification of emulsions

Thermal treatment is referred to the use of temperature to break petroleum emulsions. Conventional hot plate is used in lab scales to provide the optimal temperature. Beside that some researchers had treated the emulsion by reducing their temperature up to beyond freezing point then rising the temperature gradually, this is known as the freeze/thaw method. However, in most cases, thermal treatment is applied jointly with chemicals to improve the efficiency, since temperature can tremendousely reduce viscosity of emulsions.A group of researchers have reported the usefulness of the thermal demulsification techniques in breaking crude oil emulsions, fore instance Abduraman et al. have investigated the use of hot-plate in breaking w/o emulsions in comparison to the microwave demulsification techniques. The test was carried out in an experimentally prepared emulsion sample of two types of pure crude oils that were donated by PETRONAS (A Malayisan based oil and gas company). The experimental emulsion was prepared in 500 ml beaker with different volume fraction of oil and water phases. Mixing energy was supplied by standard three blade propeller at 1600 rpm, for 5 minutes at ambient temperature of 28°C. the volume fraction of the water in the emulsion was varied as 30, 40 and 50% (V/V). Befor demulsification, emulsions were tested for stability and quality, and only stable emulsion were considered for treatement. Their results revealed that, microwave have shown great separation efficiency over the conventional hotplate heating, and that is because of the heating patterns. since microwave heat materials volumetrically according to their dielectric properties. Thus, its energy is generated within the molecules of the materials unlike the conventional hotplate, in which heat is transferred from the surface of the hotplate to the bottom surface of the sample holder, then to the surface of the sample, then to the bulk of the sample. This makes the conventional heating slow and non-uniform. Another thing is that, microwaves are electromagnetic waves, so they have high potential to neutralize the electromagnetic repulsive barrier (Zeeta potential) between the droplets, this together with reduced viscosity as temperature increase, would ease the droplets aggregation and coalescence [34]. Raman Morales Cherbrand et al. have tested the combination of three distinct techniques, namely, Heating, Enzymatic and centrifugation to break emulsion that is originated from soybean oil. Soybean oil was extracted by dispersing the soybean powder in 2L of aqueous water phase (water comes in contact with oil) at 200 rpm. the pH was increased to 8 by adding certain amount of sodium hydroxide. the stirring process had continued to 15 minutes, after which three distinct phase were separated by centrifugation. Befor centrifugation, soybean cream was initially mixed with water to adjust its pH, then 1% w/w of the enzyme was added then shaked at 15 rpm and 50°C for three hours. After that Enzyme were deactivated by heating at 95°C. After that the sample was centrifuged to separate the oil. In another step, the enzymetic treated sample was heated at 95°C for 30 min, then cooled in a chilled water bath and centrifuged. In a third step, a fresh set of enzyme-treated cream held in a freezer at -18°C for 24 h, then thawed at 30°C for 3 h prior to centrifugation (Ramon Morales). Tov et al. have studied the effect of heat and chemical demulsifiers on w/o emulsion resolution. they investigated the effect conbing the two techniques instead of any of them alone. Experimented oil spill emulsion was formulated for testing by rotating flask methods. Their study had led to the following conclusions, w/o emulsions of paraffinic petroleum could break faster than that of the high asphaltene oil. Heat treatment could break the emulsion slowly but after adding chemicals the separation rate had increased. indeed for viscous emulsion it had better to introduce the chemical demulsifiers at 10°C, whereas, for highly viscous emulsion, it had better to add the demulsifiers after heating the emulsion. Moreover, emulsion formed from the distillation residue could be broken with moderate heating, while that with diesel oil were not broken even at high temperature. also as emulsion breaks down, its viscosity reduces [49]. Chantal et al. have used the insitu emulsion burning techniques to treat the oil spill emulsions. Their intension was to minimize the oil spill pollution. They investigated certain specific parameters that include the mechanism of the burning process, the techniques used for ignition, the environmental

disturbance that burning could cause, the burning process was excuted on iced medium.Their results had drawn so many ideas and findings. Fore instance, they identified the characteristics of the specific igniter, that should be used to ignite the crude emulsion burning. Thus, Helitourch was identified as one of the most effective ignition techniques for crude oil emulsion burning. Beside that the basic requirement for any igniter for this purpose must be composed of lightly fueled front part, and heavily fueled back part, and must be provided with quick sparking to start the ignition. Indeed, anti-foaming additives are required to improve the flame spreading capability. The burning efficiency acheived by this method, had reached up to 75%. The burning residue consisted of heavily viscous and tar-like component with some minor raw emulsion [50]. Bernard et al. have studied effect of temperature on separating w/o emulsion that stabilized by wax particles. Experimental emulsion was prepared by first dissolving the wax particles in squalane. Emulsion types was checked by drop test, and conductivity measurement. Stability test was performed by testing the amount of water or oil resolved from the emulsion within a fixed period of time. Temperature increased the extend of coalescence, that is because at elevated temperature, the wax particles at the interface would fuse, melt vanishes from frfrom the interface, and this would induce the droplets to easily come together and merge (means separation). The barrier collapsing temperature is normally the same as the melting temperature of the pure wax particles alone. However, if emulsions were prepared at emulsification temperature that is close to wax melting temperature, then it was observed to be stable even at elevated temperature [51]. Taylor et al. have studied the effect of temperature on breaking bitumen emulsions that is originated from Wolf-lake (Canada). the experimental emulsion was stabilized by ethoxylated nonylphenol surfactant.Their results showed that, as the temperature of the system getting closer to the cloud point, emulsion could break gradually. Indeed, the bitumen-water interfacial tension was found to decrease with increasing temperature, and found to be very small at cloud point temperature [52]. Basically, the mechanism behind the effect of temperature on emulsion separation is that, at low temperature below the cloud point, the system was observed to poses micelles plus free surfactants that would stabilize the bitumen-water interface, but at elevated temperature the ethoxylated groups could be dehydrated, and consequently the hydrophilicity of the surfactants could be reduced, which means their tendency to dissolve in the continuous water phase would reduce, and eventually would causes the phase separation. although at high temperature, ethoxylate group have the tendency to dissolve in the oil phase, but in their study, they mentioned that the complex structure of the bitumen will hinder the ethoxylate group to dissolve in the bitumen (oil phae), so they would aggregate as separate phase letting the two phases to separate into pure bitumen and water [52]. Guohua and Gaohong have studied the freeze/thaw method, which involves the cooling of the sample to a very low temperature (below the freezing temperature -40 for exemple), then increase the temperature backward gradually. However, their emulsion was real emulsion taken from the plan, Small amount of sample was placed in a centrifuge bottle, and cooled to freezing temperature, then thawed back to certain temperature, and then centrifuged for 5 min at 400 rpm. The volume of the water separated from the emulsion was read off through scales on the centrifuge tube (sample holder). They found that, demulsification of the water-in-crude oil emulsion depends strongly on certain parameters including, the original water contents, freezing temperature, Freezing period as well as thawing speed and temperature.The optimal freezing temperature in their results was found to be around -40°C, for the oil sludge taken from the used oil. indeed high initial water volume fraction and milder thawing would lead to increase in the volume of water separation. Beside that, the optimal thawing conditions were found to be either Air at its ambient temperature, or water bath at temperature below 20°C. and the separated water was found to contain some organic contaminent, during the freeze/thawing process, the formation of the surfactant micelles also was observed. theses micelles were greatly affected the quality of the separation [53]. Also Chang Lin et al. have used the freeze/thaw method to break water-in-oil emulsion that produced from emulsion liquid membrane. However, their experimental emulsion was produced using the agent in oil method, wherein, the surfactant (Span 80) was dissolved in the oil phase, then the de-ionized water was dispersed while the mixing was going on at a mixing speed of 1000 rpm to coursen the system first, then the mixing speed was increased to 2000 and eventually to 6000 rpm for 20 minutes, to get the submicron sized droplets emulsions. and from their results they have revealed that the freeze/thaw method was very effective and easy to handle process to separate w/o emulsions. and the mechanism behind the freeze/thaw method follow the collision theory, in which drops are moved close to each other gradually when the mechanical barrier capturing them is gradually removed. The fundamental mechanism behind the freeze/thaw demulsification method is that, during the transformation of the water phase from liquid to ice the volume expansion of the droplet could take place, and that would cause partial or full coalescence. Hence, these coalesced small droplets could diffuse into the big droplets, because of the interfacial area shrinking that occurs during the heating (thawing) process. Again, the most effective parameters in freez/thawing process, were the amount of the water content in the emulsion. Also, freezing with dry ice was found to be more effective than that with the refrigerator. but preferentially the optimal freezing temperature should be below the solidification temperature of the droplets, to achieve a suitable and uniform crystallization [54].

Xiaogang yang et al. have studied the stability pattern of asphaltene stabilized emulsions, and the effect of the Freeze/thawing on the demulsification process. the parameters of concern were the surface tension, Zeeta potential, the water droplets and the water separation from emulsion. Their experimental emulsion was prepared by choosing the dodecane as a model oil and deionized water as an aqueous phase. the indigenous surfactants (Asphaltene and resin) used in their study were extracted from crude oil via the n-alkane precipitation method, then the experimental emulsion which contain 30% water was prepared by mixing the dodecane with the deionised water and emulsifiers (Asphaltene and resins) [55]. The concentration of the Asphaltene was varied in the range of 3 to 9 g/l, and that of Resin was varied in a range of 0.47 to 3.3 g/l. The system was mixed 300 time by simple hand shaking method. Their results revealed that, emulsion stability is very strong function of Asphaltene and Resin concentration, because their adsoption at the oil-water interface would produce a film that prevent coalescence. However, their surface activities were observed to decay with increasing asphaltene concentration. Also Zeeta potential of the asphaltene and resin stabilized emulsions was observed to be negative, and increases with asphaltene and resin concentrations. for demulsification by thawing process, the separation was more easy in resin stabilized emulsions than their asphaltene stabilized counterparts. Further more, the air thawing was found to remove more water than the water thawing, indeed, the thawing with microwave heating was found to be very effective than thawing with air and water [55].

Other Demulsification Methods

This section would consider mostly the mechanical methods that

are in common use for treaiting crude oil emulsions. That including, centrifugal, membrane and so on. Habn et al. have investigated the effects of ultracentrifugal demulsification on breaking w/o emulsions. the emulsions were made in laboratory by the use of several emulsifiers such as Gantrezan 119, Tween 80, and sponto 221, each with various concentration ranged between 0.05-0.01%. The oil volume fraction in the emulsion was fixed at 30%, the mixing energy was supplied by Waring Blender for a mixing period of 30s. For the demulsification process, the centrifuge efficiency was investigated by direct measurement of the droplet sizes before and after the processing [56]. Kashmiri et al. have investigated the effect of ultracentrifugal method on demulsifying o/w emulsions. They produced several experimental emulsions with different concentrations of the emulsifying agent, namely, Tween 20, Triton x-100. The oil to water volume fractions were fixed at 50-50%. The demulsificatiom process was carried out by high speed ultracentrifuge, which was operated at the speed of 36,460 rpm, at ambient temperature of 25°C. However, result showed that the rate of separation was found to be independent of emulsifiers' concentrations [57]. Beside that Abdurraman et al. have investigated the feasibility of using the high speed centrifuge in separating the coconut milk emulsion into pure water and coconut oil. the parameters of interest were the speed of the centrifuge (rpm), which was in the range of 6000 to 12000, and centrifugation time that was ranged between 30 to 105 min. Their results revealed that, among the four centrifugational speeds (6000, 8000, 10000, and 12000) used, only 12000 rpm had given the optimal separation efficiency. Centrifuge separate emulsions in accordance with the general centrifugation theory, which tells us that, the suspended particles always move in the opposite direction of the centrifugal force, and separation occurs because of the density difference. The processing time was also segmented into 6 different parts, 30, 45, 60, 75, 90, and 105 minutes. Temperature was fixed at 30°C, and there again the best separation efficiency was observed with the centrifugation time of 105 min [34]. Indeed Kocherginsky et al. have used the microfiltration membrane to break w/o emulsion. The fundamental theory behind that was the oil droplets would fluctuate at the membrane pore. Factors of concern were, types of membrane, membrane pore size, the transmembrane pressure, membrane thickness and the initial water concentration. Their liquid membrane emulsion was prepared by mixing kerosin and alkali aqueous solution and di-2-ethylhexyl phosphoric acid then the mixture was agitated vigorously at 1000 rpm via mechanical mixture for an emulsification period of 1 h, at ambient temperature. Their results revealed some interesting fact including the possibility of using the polymer membrane as a good choice to break w/o emulsion. though some parameters such as membrane material, Pore sizes, and transmembrane pressure, had very major effects on the separation process. also another important observation was that, not all the membrane types are suitable for treating w/o emulsions. Hence, according to their results, hydrophilic membrane is the only one that could break the w/o emulsions with pore sizes smaller than the droplet sizes indeed increasing the transmembrane pressure also was found to increase the separation efficiency, but membrane thickness was found to be not effective in the separation efficiency, hence the separation process is based on the interaction between the droplets and membrane not based on the sieving effects [58]. Nalin Nadarajah et al. have used some sorts of bacterial culture to break water-in-petroleum oil emulsions. They have collected the bacteria from crude oil contaminated soil. After the isolation, the bacteria is kept in cyclone fermenter on mineral salts, then various bacteria colonies were extracted and purified. A model (kerosene) emulsion was prepared by first dissolving the emulsifiers in the aqueous phase. For demulsification experiment, 1 ml of the bacterial colonies was added to 9 g of model oil emulsion and mixed for 20 s, and incubated at 50°C under static condition.The demulsification efficiency of the bacteria was determined by simply measuring the water separated from the bacteria treated emulsion, and compared with that of the control sample. From their results, they have observed that, it was difficult to prepare a model emulsion of the crude oil that mimic the actual crude oil emulsions because of the complexity found in the crude oil, the best separation efficiency for the biological demulsification was observed at 50°C, and that was expected since the elevated temperature reduces the viscosity of the oil and emulsion, and increases the density difference between the two phases, and weaken the interfacial barrier which leads automatically to droplets coalescence. The culture age was found to be not important in the demulsification. However, culture cultivated from crude oil medium had given high demulsification efficiency than culture cultivated from motor oil and diesel [59].

Tereza Neuma et al. have applied the microemulsion system to break w/o emulsion of Brazilian crude oil. A moded microemulsion system was prepared in laboratory, the aqueous phase contained 5.2% HCL, the oil phase was toluene, and the emulsifying agent was isopropyl alcohol. The demulsification experiment was carried out by directly contacting the microemulsion and w/o emulsion in centrifuge tube for a duration of 30 min, with continuous mixing and fixed temperature of 70°C. The separation efficiency was calculated as the percentage of water separated with respect to the original amount of water used. From the results, microemulsion was proven to be a suitable method to break w/o emulsions formed during the mining process of the Brazilian oil. The original water content of the system had a great effect on the process [60]. Del Colle et al. have used the alumina tubes as a filter material to split water-in-sunflower oil emulsions using tangential filtration process. They used a ceramic tube that was made of alumina. The tube was first sintered at 1450°C, then impregnated by zirconic material then calcinated and heat treated at temperature range of 600-900°C to discard the light organic materials, and convert the zirconium to zirconium oxide that was impregnated in the aluminia in the forms of nanoparticle aggregates. This new microporous membrane was tested in a microfiltration of aqueous system to assess its performance. The parameters investigated were, the transmembrane pressure, and input emulsion flow. Results showed that, the ziconia impregnated model was found to be very effective and could operate at very high temperature of 900°C [61].

Mohtada Sandrazadeh et al. have used the PTFE membrane to treat o/w emulsion that was originated from wastewater. A simulated waste water model emulsion was prepared experimentally by mixing gas oil and distilled water. Emulsion was mixed using a pitched curved blade blender with 6 fins, for 10 min. Their results showed that, increasing temperature, pressure, as well as flow rate would affect the permeate flux. No fouling was observed and the water content of the permeate was to increase with increasing the temperature and pressure and decrease with increasing the flow rate [62]. Aileen Lozan et al. have used salts to induce the flocculation of oil-in-water emulsion. that was stabilized by anionic emulsifiers Sodium dodecyl sulfate (SDDS). o/w nano emulsion was prepared experimentally by two distinct steps: in step one, the low energy emulsification method used, wherein no heat or vigorous mixing is required, only mixing of oil, water, surfactants, alcohol and appropriate amount of salt and aqueous phase to make a bicontinuous emulsion system. In the second step, that bicontinuous emulsion system was injected in large amount of water with slow mixing to produce nanoemulsion system of fixed drop sizes which was further diluted in sodium dodecyl sulfate solution to get the desired oil-in-water emulsion [63].

Microwave demulsification

Electromagnetic spectrum laying within 300 MHz to 300 GHz is known as microwave. Microwave have electric and magnetic properties. thus, when they are projected to materials they obey the optical rules of transmission, absorption and reflection depending upon the medium characteristics, hence, the applied field will induces a polarization effect to the medium.The wave length vary from 1mm to 1m, in accordance with aforementioned frequency which is between (300 MHz to 300 GHz). Beside heating and scientific research purposes, some frequencies are reserved for specific purposes, including these of cellular phone, radar, and television satellite communication, by Federal Communication commission (FCC). However, the most frequently used frequency for Industrial, scientific and medical (ISM) purposes are 915 and 2450 MHz, with 915 is used in this research [64]. Microwave is commonly preferred in material processing over its conventional counter part, due to its volumetric heating. The mechanism of heating in conventional heating occurs through diffusion from the surface of the material to the bulk. whereas, in microwave heating, the temperature gradient in different location within the sample is almost invariant, another important phenomenon is that different material possess different heating pattern, and that is due to the variation in material's microwave absorption capacity which in turn depends on dielectric properties [64].

Equation of heat transfer in microwave

The general form of heat transfer equation consists of three principal forms which are conductive convective and irradiative form given bellow.

$$g_{MW} = \frac{hA}{V}\left(T_m - T_a\right) + \frac{\varepsilon A\sigma}{V}\left[\begin{array}{c}\left(T_m + 273.15\right)^4 \\ -\left(T_a + 273.15\right)^4\end{array}\right] + \rho C_p\left(\frac{dT}{dt}\right) \quad (1)$$

Unless this is not the case in microwave heating, wherin heat is generated within the material and it is transfer property is directly linked to the thermophysical properties of material as reported by curet and others. the detailed heat transfer mechanisms in microwaves will be explained in the following equations.

$$\rho c_p \frac{\partial T}{\partial t} = \mathrm{div}\left(k\nabla T\right) + Q \quad (2)$$

Where: ρ and C_p are density and heat capacity of sample being heated, $\frac{\partial T}{\partial t}$ rate of temperature increase and Q is the internal energy acquired by the sample from the dissipated microwave power and could be calculated for Z directed electromagnetic wave by an equation relating power drop from surface to bottom which is:

$$Q_z = \frac{F_0}{d_p(z)}\exp\left(-\int_0^z \frac{dz}{d_p(z)}\right) \quad (3)$$

F_0: Microwave surface flux and d_p is the penetration depth which is strongly related to dielectric properties of the sample designated as the distance at which the microwave represents 1/e of the surface power flux and its mathematical expression is:

$$d_p = \frac{c_0}{2\pi f}\left[2\varepsilon'\left(\left(1 + \left(\frac{\varepsilon''}{\varepsilon'}\right)^2\right)^{\frac{1}{2}} - 1\right)\right]^{-\frac{1}{2}} \quad (4)$$

While surface flux at the same conditions as Maxwell equations is calculated by the following relation.

$$F_0\left(x,0\right)=\frac{2P_{surf}}{ab}\sin^2\left(\frac{\pi\,x}{a}\right)$$ (5)

Where (a, b) are the dimensions of the wave guide

However, based on Maxwell equation perspective, the magnitude of microwave power absorbed by the dielectric samples can be estimated by the local electric field strength on the sample being treated as

$$Q_{abs}=\omega.E^2.\varepsilon.\varepsilon^{"}$$ (6)

The electric field for plane wave propagation mode is solved by the following equation

$$\nabla\times\left(\mu_r^{-1}\,\nabla E\right)-\left(\varepsilon_r-j\frac{\sigma}{\omega}\,\varepsilon_0\right)k_0^2\,E=0$$ (7)

ω: Pulsation of microwave radiation (Rad/s)

E: electric field intensity (v/m)

ε : Complex permittivity

σ : electrical conductivity (s/m)

μr : relative electromagnetic permittivity of material (H/m)

k : thermal conductivity (w/m.K)

According to some previous researchers, in the above equation the radiative and convective terms are very minor and that is because sample container has very low dielectric constant therefore it does not generate much of heat and hence most of the waves are transmitted through the glass sample container to the emulsion. The density and heat capacity of emulsion are calculated from the following mixing rules:

$$\rho_m=\rho_w\,\varphi+\rho_o\left(1-\varphi\right)$$ (8)

$$C_{p,m}=C_{p,w}\,\varphi+C_{p,o}\left(1-\varphi\right)$$ (9)

Where:

ρ_m,ρ_w,ρ_o : are the densities of emulsion, water, and oil respectively

$C_{p,m},\,C_{p,w},\,C_{p,o}$: are the heat capacity of emulsion, water, and oil respectively.

Some literatures on microwave demulsification

Different materials have different heating capabilities when exposed to the incident microwave energy. Hence, in opaque materials, the waves are totally reflected, while in transparent materials, they are transverse, but in dielectric materials, waves are absorbed and transformed to heat. However, introducing some specific metal powder (Susceptive) to non conducting solvent would render them microwave active and they can be heated [65]. The treatment of crude emulsion is of major importance in the petroleum industries, since most of the world's crude oils are produced in the forms of emulsion. Water is normally a good microwave absorber, because of its high dielectric properties, however, in some repports, at large samples, the heat transfer in oil was faster than that in the water and that is imparted to the thickness of the oil, but this convention does not work with small sample hence for small samples, the heating rate of the water was faster [66].

Emulsion was first treated via microwave by Wolf who was first started the concept of microwave demulsification. Since then, more and more researches have been devoted to this field. The efficiency of microwave on emulsion was found to induce some effect on the treated emulsion such as, the increase in the temperature shall lead to reduction in viscosity, and this would increase the mobility of the water droplet, that may in turn neutralizes the zeta potential of the dispersed droplets, and also break the hydrogen bonding between the water and the surfactant molecules. Beside that, the electromagnetic wave is believed to increase the internal pressure of the water droplets, leading to reduction of the interfacial film thickness and charges surrounding water droplets, and that would set the water droplets free to move toward each other, and downward by gravitational force [67]. the advantage of applying microwave energy over its conventional counter part is the ability to heat sample more uniformly than the conventional heating, although in some cases local overheating may occurs causing Hotspot or thermal runaway within the sample. Crude oil consists of vast numbers of components that differ in their conductivity and polarity, and absorptive patterns with asphaltene being the dominant and the major charge carrying component [68,69].

In another report the effects of container types and water to oil ratios on emulsion stability was studied, the types of containers used in that experiment were non reflective and partially reflective ceramic-metal complex plates with emulsion being placed in the open exposed part. The maximum power distribution (resonance R) is determined by plotting the power calculated against various parameters, including dispersed phase volume fraction, emulsion sample thickness and plate types. The averages power absorption for o/w emulsion was greatest in sample heated with reflective surface, and decrease with increasing oil volume fraction. The thermal runaway was estimated by plotting the difference between the maximum and minimum temperature of the samples (o/w emulsion), plotted versus irradiation time. It was found to increase with increasing dispersed phase volume fraction [69]. Another observation was that unlike the o/w emulsion, in w/o emulsion there was a decrease in the amplitude at metal plate-emulsion interface, this situation is claimed to arise because of small dielectric loss of the later case, beside that also it does not respond much to water volume fraction variation. Numerous investigations have been done in the field of microwave heating and material processing. Christian et al. have studied the absorption mechanism for emulsion containing aqueous and nano droplets, and found that absorption is attributed to the types of ions, their concentrations, as well as their polarization [70]. Others have studied the effect of microwave in treating industrial waste water emulsions, and admitted that microwave can be used as an alternative to the existing demulsification methods, however, addition of small amount of acid was reported to increase the efficiency, moreover, the effect of aqueous phase composition on microwave absorption was investigated by Montserrat et al., [71].

Bjerdalen et al. have applied the microwave and ultrasonic techniques to treat the solid particles deposited at the petroleum well. These particles consist mainly of asphaltene, resins , and precipitated wax, together with other indigenous solid particles, such as silica, bentonite and gypsium. The parameters of concern were, temperature, Viscosity, and concentration of crude oil. The microwave device used was sumsung Mini-Chef MW 101°C that had a frequency of 60 Hz, and output power of 120 V. The temperature measuring probe was of type Greenlee THH-500. While the ultrasonic device was Medson mysono 201 A19001213 that had power output of 15V and frequency of 3.5 MHz. The crude sample was originated from SANTA BARBARA. The processing time for microwave, was 80 and 180 seconds. Whereas, for ultrasinc treatment it was varied as 30, 60, 180 seconds. Results proved that, microwave irradiation could elevate the temperature of petroleum crude oil and asphaltene and decrease the viscosity, while with ultrasonic treatment the viscosity was observed to increase with increasing the irradiation time [72]. Abdurahman et al. have investigated the separation of water from crude oil by sing conventional heating and microwave heating comparatively. the study was carried out on two different types of crude oils, and two different water to oil ratios of 30-70% and 50-50% w/o emulsions. Experimental emulsions were prepared by mixing the two phase vigorously via the standard three blade propeller, at mixing speed of 1800 rpm, in ambient temperature for 5 min. Three types of emulsifiers, namely Low sulfur wax residue (LSWR), Triton-x100 and Span 83 were used to produce the most stable w/o emulsion. Elba domestic microwave oven was used for demulsification. The temperature profile of the sample was monitored using three thermocouples that were inserted in the sample at different location (Top, Middle and Bottom). Microwave processing time was varied as 30, 60, 90, 120,150, 180, 210 Seconds. Their results revealed effectiveness of the microwave heating over the conventional heating in breaking w/o emulsions. Vladan Rajakovic et al. have used the combined Freeze/Thaw and microwave irradiation method to spilt oil from oil-in-water emulsion that was originated from metal working oil. The splitting process was carried out via the ultrafreezer device at three distinct temperatures -20, -40 and -60°C. The sample was first freezed 10h, then thawed back in Air at 20°C, then water bath at 40°C or microwave at 95°C, 800w, 2450 MHz. Their results showed that, the effectiveness of the freeze/thawing and microwave demulsification techniques depend mainly on the oil content of the emulsion sample, freezing time, thawing steps, and thawing time. Indeed, microwave also believed to induce molecules to acquire more energy because of super heating, and volumetric heating. Therefore, microwave demulsification is proved to be the most effective method in breaking emulsions [73]. Nour et al. have investigated the batch microwave oven with 2450 MHz frequency in treating and breaking w/o emulsion that was stabilised by some commercially available emulsifiers. The commercial emulsifiers used in this study were Triton-X100 and Low Sulfur Wax Residu (LSWR). Their experimental emulsion was consisted of 20-80 and 50-50% water to oil ratios. The crude used was given by PETRONAS (Malaka refinery). The mixing speed was fixed as 1800 rpm for 8 min. However, their comprehensive study have proved that, microwave can provide very fast and uniform heating and have a very strong potential to be used as an alternative demulsification technique for petroleum emulsion treatment [74]. Abdulbari et al. have studied the stability as well as microwave demulsification of petroleum emulsion. Various commercial emulsifiers used, including, Span 83, Triton-x100, Low sulfur wax residue (LSWR), and sodium dodecyl sulfate (SDDS). the targeted parameters were, the surfactant types, surfactants concentration, water volume fractions in the emulsion (10-90%), emulsification temperature and agitation speed. The raw crude petroleum oil were originated from Iran and Malaysian oil fields. Experimental emulsions were prepared through the agent in oil method. The emuksification speed was 1600 rpm. The demulsification experiment was carried out using the elba microwave oven whose power output was 900w, and frequency of 2450 MHz. their results showed that Microwave gave rapid separation over the conventional gravitational settling [75]. Montserate Fortuny et al. have investigated the important parameters for microwave demulsification, namely, salinity, temperature, water content and pH. Their experimental emulsion was prepared by mixing the crude oil with saline water, and homogenized using two steps emulsification techniques. In the first step, the mixture was mixed by simple hand shaking and certain amount of water was added in a drop-wise fashion while the mixing is going on. In the second step, the system was vigorously mixed using the ultra-turax T-25 homogenizer fitted with a S25-25G dispersing tool. The water volume fraction was varied as (25, 35, 45%). Emulsion were mixed at 17500 rpm for 10 min respectively. The demulsification experiment were performed using the commercial microwave, at processing time of 15 min. Results showed that, increasing microwave power could increase the demulsification process, beside that, high water contained emulsions, were found to give high separation efficiency and that is because of the dielectric heating. Also high salt content or high pH had reduced the demulsification rate. Indeed, the best separation rate was observed with emulsion having water volume fraction of 45% and pH of 7 [76].

Yunus et al. have conducted a laboratory experiment to examine the performance of microwave heating technique in breaking water-in-crude oil emulsion in comparison with the conventional demulsification techniques, and their results had concluded that, microwave gave better separation than the conventional heating. They claimed that, microwave does not need any chemicals addition, and could be used as an alternative green technique to treat petroleum emulsions [77].

Ilia Anisa et al. have also used the microwave demulsification techniques to treat w/o emulsions, and optimize the operating conditions to fix the best operating conditions for microwave demulsification by using the RSM software [75,78]. Lixin Xia et al. have studied the stability of crude oil emulsion and its relation to the presence of Asphaltene and resins (the natural surfactants), followed by microwave demulsification. Their experimental emulsion was produced by using jet kerosene as model oil phase mixed with certain amounts of asphaltene and resin. The asphaltene and resin were extracted from Daqing crude oil field (China). The concentration of the asphaltene in the bulk emulsion was varied in the range of 0.3 to 0.9% (w/w), and that of the resin was in the range of 1-4% (w/w). Emulsion was prepared by simple hand shaking. The demulsification process was carried out by using either oil bath at 90°C, or microwave oven with power output frequency of 850 w and 2450 MHz respectively. Their results revealed that, stability was strongly depended on the amount of the asphaltene and resins in the daqing crude oil, and that is because the adsorption of these interfacial active components at the interfaces is believed to form a mechanical film between the droplets, also microwave demulsification had given better separation result compared to the conventional demulsification methods. Abdurahman et al. have proposed a continuous microwave demulsification method for water-in-crude oil emulsion, and found that the temperature rise within the sample at given location was linear, and the rate of temperature increase of emulsion decreases at elevated temperature due to the decrease of dielectric loss of water [79].

Ultrsonic demulsification

Ultrasonic is the branch of physics that is dedicated to the studies of sounds, including, generation, transmission, control, reception as well as the effects of mechanical waves in solids, liquids and gases [80]. Ultrasonic are used in various fields, some owhich are environmental, architectural, musical and engineering acoustics. Ultrasonic or Accostic is the study of sound with a frequency higher than audible sound. Environmental acoustics deals with noise control. Architectural acoustics is the study of how sound waves and buildings interact. Musical acoustics deals with the design and use of musical instruments and how they affect the listener. Engineering acoustics concerns the recording and reproduction of sound. Thus, ultrasound are well established in most engineering application and material processing such as sonochemistry, metal working, cleaning and many more. Each application have certain range of frequency. The basic requirement of fabricating an ultrasonic equipment are transducer (a device that can convert electric charges to sound wave) and medium within which sound could propagate (Mostly liquids), for sonochemical application the medium is mostly water, while, transducers vary with application. In the following paragraphs some criteria and types of the existing transducers would be highlighted [81]. Gautam et al. have applied the low intensity ultrasonic field to separate oil phase that is suspended in an aqueous solution. However, the system consisted of rectangular chamber that was filled with matrix of porous medium within which the oil drops was assumed to be filled, three types of mesh were used that were glass beads, aluminum mesh and polyester. Rectangular piezo-electric (PZT) transducer and stainless steel reflector were mounted in the chamber in paralleled position to each other. and the distance between the transducer and reflector was adjusted carefully to generate standing wave on the emulsion sample that was filled in chamber (between the Transducer and reflector). The acoustic field energy was produced by energizing the transducer at 680 KHz frequency, using a continuous sonisoidal signals generated and amplified by KROHN- HITE 2100 A signal generator, and ENI 240 L power amplifiers respectively.Their experimental oil-in-water emulsion was produced by using the soybean oil and deionized water, with 50-50% oil to water ratios. Emulsion was first agitated for one minute then sonicated for 8 minutes to produce a stable emulsion having around 1 to 10 μm sized droples, that was further diluted prior to feeding in the acoustic chamber for demulsification to ease the visual observation of the droplets' response to the ultrasonic field. Results showed that the ultrasonic separation efficiency depends strongly on the feed flow rate, path length as well as the electrical power [82].

Riera-Franco de Sarabia et al. have reported the application of ultrasound to separate the suspended particles from fluid. The basic fundamentals behind the ultrasonic particles separation, depends normally on the material to be treated. for-instance, in the case of gas systems the very fine suspended solids particles have to be removed via the agglomeration process, whereby, particles collect together to increase their sizes, then separate out of the suspending medium, while in the liquid systems the agglomeration rate is less efficient than that of the gas phase, but generally its very useful to split and separate the fine particles, from their suspending medium [83]. Riera et al. have studied the possibility to apply the ultrasonic agglomeration process to treat suspension and separate the suspended particles from industrial fumes. They observed that, the agglomeration process was not very good in liquid systems [84]. Ye Guoxiang et al. have investigated the ultrasonic electric desalting and dewatering techniques in treating petroleum emulsions. their system was consisted of ultrasonic device and electric desalting and dewatering device that are connected in such a way that make the feed emulsion had to pass through the ultrasonic tube in order to enter the electric desalting and dewatering units. The ultrasonic equipment was designed with very specific geometry to produce an ultrasonic standing waves field. The frequency used was either 10 or 20 KHz. The initial salt content of the petroleum oil was as high as 40-70 mg/L. The process was carried out in three distinct steps that are, crude oil pretreatment, ultrasonic irradiation, and electric desalting and dewatering. During pretreatment, crude oil was treated to render it less viscous and flowable, then certain amount of water was added in the oil containing demulsifiers in the pipeline, then the mixture was agitated by static agitator. In the second stage, emulsion was treated with ultrasonic standing wave field. In the third stage, emulsion was treated with electric desalting and dewatering process, and this caused the water droplets to coalesce and settle down

as separate phases. Results showed that, ultrasonic electric united process increased the dewatering and desalting process of the crude oil emulsion. The optimum treatment conditions were found to be 5% by volume of injected water, 30 μg/g of demulsifiers, 0.45 MPa of Mixing pressure drop, 10 KHz and 150 w of ultrasound (standing waves field), 3.4 min ultrasound irradiation time, 1.2 Kv/cm electric intensity, 20 L/h of emulsion flow rate, and processing temperature of 80°C. Their system was reported to be very effective, hence beside water separation, the salt concentration was also found decrease from 67.5 mg/l to 3.97 mg/l at 80°C [85].

Sanjay et al. had studied the mechanism of aggregation of fine particles in porous medium under the influence of the standing waves field; the parameters of interest were, the bulk fluid flow rate, ultrasonic power, and the feed concentration. Ultrasonic device was a rectangular chamber that consisted of two rectangular leads on parallel position to each other and piezoelectric transducer. A porous mesh was placed between the transducer and reflector, their results showed that the separation efficiency had reached up to 70-80% by using the porous media. Separation efficiency was found to be directly proportional to the flow rate of the suspension, while the particle concentration was observed to have no effect on the separation efficiency [85].

Nii et al. have used a 2 MHz ultrasonic frequency generating devices to treat emulsions that were originally prepared from canola oil and water. The parameters investigated were the ultrasonic power and processing time. However, their experimental emulsion was produced by mixing deionized water and canolla oil, mixing power was supplied by an ultrasonic horn sonifier (BRANSON DIGITAL). The demousification experiment was carried out by doting a drop from the stable emulsion on microscopic lens, then the lens was soaked in an ultrasonic water bath that was consisted of glass cylinder with stainless steel plate at the bottom and transducer was mounted on the steel plate. Results proved that, the action of the ultrasonic power pushed the oil droplets and lead them to aggregation and eventually to massive coalescence and phase separation. beside that a remarkable increase in the velocity of the oil droplets was observed after sample was irradiated via the ultrasonic [86]. Garcia et al. have applied the ultrasonic standing wave resonanting chamber to treat o/w emulsions. They have used an ultrasonic frequency range of between 1-2 MHz, and the apparatus used was standard research system ENI 310L RF power amplifier Tektronix oscilloscop, the experimental emulsion used in their study was prepared using an Omni-Ruptor 250 Homogenizer to provide the mixing power. The oil volume fraction in the emulsion was fixed at 10%. For demulsification experiment, several transducer types were examined to generate the standing wave in the emulsion. However, once the standing wave was generated, The oil droplets were observed to aggregate as response to the effect of the standing waves, and thus the standing wave treated o/w emulsion sample had shown a great separation efficiency over the control or the gravitationally separating sample. It was also observed that, at elevated ultrasonic power, acoustic streaming and heating could be produced. Furthermore, directly contacting the emulsion sample with the transducer was found to be very effective [87].

Summary

Crude petroleum is almost always exists in the forms of emulsions of various types including, w/o, o/w, w/o/w, o/w/o. With w/o being the most prominent type encountered in petroleum institutions, that is imparted to the fact that most of the indigenous surface active agents such as asphaltene, resin, long chain fatty acids are oil soluble amphiphilics. Water comes into contact with petroleum oil either inside the well, or during pumping from the reservoir, as well as during desalting process, this latter is very important in petroleum industries wherein water is used as a cost effective solvent to remove the water soluble salts from the petroleum prior to refining process. Indeed water also can be used to reduce the viscosity in the case of heavy oil transportation in pipelines.However, the presence of water in the crude petroleum causes other problems such as corrosion, quality reduction of the refined products. Therefore, water must be removed or reduced to minimum amount before refining would start. Various demulsification techniques are currently used to remove water from crude oil, including chemicals, conventional heating, electrical and centrifugal, with chemical being the most prominent and widely used, but as reported in literature, it has a lot of disadvantage in many aspects, including, safety, cost and environmental effect, therefore, more and more versatile, lucrative and sustainable techniques are needed, and that is where the objective of the current work comes in. Moreover, emulsion formation and stability varies from one location to another, even between two different wells in the same field, Currently chemical demulsification is the most dominating technique in action. In chemical demulsification, Amphiphilic copolymers called demulsifiers that categorized into anionic, cationic, nonionic and ziwitrionic are injected to the emulsion with respect to their major groups. Microwave demulsification have gained a considerable attention in resent years. it was first introduced to this field of petroleum emulsion by Wolf in his various lab scale experiments. Recently the application of ultrasonic in crude oil teatment was also reported. Most researches in crude emulsion were performed in experimental emulsion that mimic the real emulsions in their characteristics. However, there are some chalenges in prepaering the suitable experimental emulsion, since every researcher uses his own formultion and characters. Table 1 below reveals some of the formulation and parameters used to produce stable experimental petroleum based emulsions.

Conclusions and Recomendations

This essay had surveyed the ultimate techniques and methods of preparing and treating crude oil emulsions and water separation (demulsification). Based on chemical demulsification, polysiloxane demulsifiers claimed to be very efficient and environmental friendly, compared to the commercial demulsifiers. However, they have not being used because of economical reason. Microwave and ultrasonic techniques were found to be very effective in separating water from petroleum. Microwave is commonly preferred in material processing over its conventional counter part, due to its volumetric as well as dielectric heating. The mechanism of heating in conventional heating occurs through diffusion of heat from the surface of the material to the bulk, which the gradient in heating rate. whereas, in microwave heating, the temperature gradient in different location within the sample is almost invariant, that means microwave heat materials uniforly and volumetricaly. another important phenomenon is that different material possess different heating pattern, and that is due to the variation in material's microwave absorption capacity, which in turn depends on dielectric properties. Both microwave and ultrasonic are more environmental friendly and costeffective, but have not beeing commercialized yet. Other minor techniques such as, centrifuge, membrane and bacterial treatment also reported in the literature.

As mentioned earlier, chemical demulsification is not really recomended from environmental view point, hence some suggstion are given next:

- Developing more green chemical that is cost effective, and environmental friendly

- Commercialization of microwave demulsification process to operate in mass treatment of petroleum

- Developing the ultrasonic techniques for petroleum treatment

No	Oil type	Surfactants	Mixture type	RPM (S^{-1})	Mixing time (minutes)	Reference
01	Heptane/toluene	Asphaltene/resin	Ultra-turax T25	9500	0.5	Havre and Sjöblom [5]
02	Jet Kerosen	Asphaltene/resin	-	-	-	-
03	Silicon oil	SDS	-	7000	-	Burjic [7]
04	Heptane/toluene	Asphaltene/resin	Ultra-turax T25	22000	30	Andreas et al. [13]
05	Heptane/toluene	Asphaltene/resin	-	17000	7	Rodiguez et al. [12]
06	Paraffin oil	Span 80,Tween20	-	500	15	Zhang et al. [16]
07	-	-		10000	30	Fara et al. [19]
08	Petroleum oil	-	Ultra-turax	24000	2	Paso et al. [21]
09	Bitumen/toluene	-	Polytron	20000	3	Wu et al. [25]
10	-	SDDS	Ultrasonic	-	10	Bratskaya et al. [30]
11	Petroleum oil	Silverson lurt	-	20000	1	Auflem [32]
12	-	-	Vertishear cyclone	1500	2	Zaki et al. [4]
13	Petroleum oil	-	ht-2 homogenizer	1200	5	Zhiqing [33]
14	Petroleum oil	-	3 blade propeller	2000	10	Nour [34]
15	Benzene	SDDS	Magnatis stirrer	-	15	Yakhkeshi and Hosseina [46]
16	Xelene	-	Ultrasonic	-	6	Oji and Opara [47]
17	-	-	Electromagnetic	1000	10	Yun Kim [48]
18	Petoleum oil	-	3 blade propeller	1600	5	Nour [34]
19	-	Span 80	-	6000	20	Lin et al. [54]
20	Dodecane	Asphaltene	Hand shaking	-		Yang et al. [55]
21	-	Twee 20	blender			Mittal and Vold [57]
22	Kerosen	-	-	1000	60	Kocherginsky et al. [58]
23	Gas oil	-	6 blade blender	-	10	Sadrzadeh et al. [62]
24	-	-	Ultra-turax T25	17500	10	Fortuny et al. [71]

Table 1: Some importants parameters, for emulsion formulation.

- Integration of the above mentioned techniques

- Combining chemical with microwave or ultrasound to reduce the dosage of chemicals.

References

1. Kristiansen TS, Lewis A, Daling PS, Nordvik AB (1995) Heat and chemical treatment of mechanically recovered w/o emulsions. Spill Science and technology 2: 133-141.

2. Auflem IH, Kallevik H, Westvik A, Sjoblom J (2001) Influence of pressure and solvency on the separation of water-in-crude-oil emulsions from the North Sea. Petroleum Science and Engineering 31: 1-12.

3. David DD, Pezron I, Dalmazzone C, Noık C, Clausse D, et al. (2005) Elastic properties of crude oil/water interface in presence of polymeric emulsion breakers. Colloids and Surfaces A: Physicochem Eng Aspects 270: 257-262.

4. Zaki NN, Carbonell RG, Kilpatrick PK (2003) A Novel Process for Demulsification of Water-in-Crude Oil Emulsions by Dense Carbon Dioxide. Ind Eng Chem Res 42: 6661-6672.

5. Havre TK, Sjöblom J (2003) Emulsion stabilization by means of combined surfactant multilayer (D-phase) and asphaltene particles. Colloids and Surfaces A: Physicochem Eng Aspects 228: 131-142.

6. Xia L, Lu S, Cao G (2004) Stability and demulsification of emulsions stabilized by asphaltenes or resins. Colloid and Interface Science 271: 504-506.

7. Burjic J (2003) Measuring the distribution of interdroplet forces in a compressed emulsion system. Physica A 327: 201-212.

8. Guo J, Liu Q, Li M, Wu Z, Christy AA (2006) The effect of alkali on crude oil/water interfacial properties and the stability of crude oil emulsions. Colloids and Surfaces A: Physicochem Eng Aspects 273: 213-218.

9. Li M, Lin M, Wu Z, Christy AA (2004) The influence of NaOH on the stability of paraffinic crude oil emulsion. Fuel 84: 183-187.

10. Kumar K, Nikolov AD, Wasan DT (2001) Mechanisms of Stabilization of Water-in-Crude Oil Emulsions. Ind Eng Chem Res 40: 3009-3014.

11. Aske N, Kallevik H, Blom JS (2002) Water-in-crude oil emulsion stability studied by critical electric field measurements. Correlation to physico-chemical parametersand near-infrared spectroscopy. Petroleum Science and Engineering 36: 1-17.

12. Valverde MARG, Vı´lchez MAC, Duenas AP, Ivarez RHA (2003) Stability of highly charged particles: bitumen-in-water dispersions. Colloids and surfacees: physicochem Eng Aspects 222: 233-251.

13. Hannisdal A, Ese MH, Hemmingsen PV, Oblom JS (2006) Particle-stabilized emulsions: Effect of heavy crude oil components pre-adsorbed onto stabilizing solids. Colloids and Surfaces A: Physicochem. Eng. Aspects 276 :45-58.

14. Moran K, Czarnecki J (2007) Competitive adsorption of sodium naphthenates and naturally occurring species at water-in-crude oil emulsion droplet surfaces. Colloids and Surfaces A: Physicochem Eng Aspects 292: 87-98.

15. Quintero CG, Noïk C, Dalmazzone C, Grossiord JL (2008) Modelling and characterisation of diluted and concentrated water-in- crude oil emulsions: comparison with classical behavior. Rheol Acta 47: 417-424.

16. Zhang L, Que G (2008) Influence of the HLB parameter of surfactants on the dispersion properties of brine in residue. Colloids and surfaces A: physiochemical and engineering aspects 320: 111-114.

17. Bratskaya S, Avramenko V, Schwarz BS, Philippova I (2006) Enhanced flocculation of oil-in-water emulsions by hydrophobically modified chitosan derivatives. Colloids and Surfaces A: Physicochem Eng Aspects 275: 168-176.

18. Li M, Xu M, Ma Y, Wu Z, Christy AA (2001) The effect of molecular parameters on the stability of water-in-crude oil emulsions studied by IR and UV spectroscopy. Colloids and Surfaces A: Physicochemical and Engineering Aspects 197: 193-201.

19. Farah MA, Oliveira RC, Caldas JN, Rajagopal K (2005) Viscosity of water-in-oil emulsions: Variation with temperature and water volume fraction. Journal of Petroleum Science and Engineering 48: 169-184.

20. Sanfeld A, Steinchen A, Mishchuk (2005) Energy barrier in dense W/O emulsions. Colloids and Surfaces A: Physicochem Eng Aspects 261: 101-107.

21. Paso K, Silset A, Sorland G (2009) Characterization of the Formation, Flowability, and Resolution of Brazilian Crude Oil Emulsions. Energy and Fuels 23: 471-480

22. Dicharry C, Arla D, Sinquin A, Graciaa A, Bouriat P (2006) Stability of water/crude oil emulsions based on interfacial dilatational rheology. Colloid and Interface Science 297: 785-791.

23. Johnsen EE, Rønningsen HP (2003) Viscosity of 'live' water-in-crude-oil emulsions: experimental work and validation of correlations. Petroleum Science and Engineering 48: 169-184.

24. Kukizaki M, Goto M (2008) Demulsification of water-in-oil emulsions by permeation through Shirasu-porous-glass (SPG) membranes. Journal of Membrane Science 32: 196-203.

25. Wu J, Xu Y, Dabros T, Hamza H (2004) Effect of EO and PO positions in nonionic surfactants on surfactant properties and demulsification performance. Colloids and Surfaces A: Physicochem. Eng. Aspects 252: 79-85.

26. Kang W, Jing G, Li HZM, Wu Z (2006) Influence of demulsifier on interfacial film between oil and water. Colloids and Surfaces A: Physicochem. Eng. Aspects 272: 27-31.

27. Hafiz AA, El-Din HM, Badawi AM (2005) Chemical destabilization of oil-in-water emulsion by novel polymerized diethanolamines. J Colloid Interface Sci 284: 167-175.

28. Alejandro AP, George J, Hirasaki, Miller CA (2005) Chemically Induced Destabilization of Water-in-Crude Oil Emulsions. Ind Eng Chem Res 44: 1139-1149.

29. Dalmazzone C (2005) Mechanism of crude oil-interface destabilization by silicone demulsifiers. SPE International symposium on oilfield chemistry SPE 80241

30. Bratskaya S, Avramenko V, Schwarz S, Philippova I (2006) Enhanced flocculation of oil-in-water emulsions by hydrophobically modified chitosan derivatives. Colloids and Surfaces A: Physicochem. Eng. Aspects 275: 168-176.

31. Ezzati A, Gorouhi E, Mohammadi T (2005) Separation of water in oil emulsions using microfiltration. Desalination 185: 371-382.

32. Auflem IH (2002) Influence of asphaltene aggregation and pressure on crude oil emulsion stability, PhD thesis, Department of chemical engineering, Norwegian University of science and technology, Trondheim

33. Zhiqing Z (2004) Characterisation and demulsification of poly (ethylene oxide)-block-poly (propylene oxide)-block-poly (ethylene oxide) copolymers. Colloidal and Interface Science 777: 464-470.

34. Nour AH (2009) Demulsification of Virgin coconut oil by centrifugation method: A feasibility study. International Chemical Technology 1: 59-64.

35. Yang CL (2007) Electrochemical coagulation for oily water demulsification. Separation and Purification Technology 54: 388-395.

36. Harpur G (1997) Destabilization of water-in-crude oil emulsions under the influence of an AC electric field: experimental assessment of the performance. Electrostat 40: 135-140.

37. Jang W, Lee Y (2000) Removing oil from oil-in-water emulsion using electrical demulsification method. Industrial and Engineering Chemistry 6: 85-92.

38. Sams GW, Zaouk M (2000) Emulsion Resolution in Electrostatic process. Energy and fuel 14: 31-37.

39. Ichikawa T, Itoh K, Yamamoto S, Sumita M (2004) Rapid demulsification of dense oil-in-water emulsion by low external electric field :I. Experimental evidence. Colloids and Surfaces A: Physicochem Eng Aspects 242: 21–26.

40. Fordedal H, Sjoblom J (1996) Percolation Behavior in W/O Emulsions Stabilized by Interfacially Active Fractions from Crude Oils in High External Electric Fields. Colloid and interface science 181: 589-594.

41. Yoshida J, Chen J, Aoki K (2003) Electrochemical coalescence of nitrobenzene j water emulsions. Electroanalytical Chemistry 553: 117-124.

42. Bailesa PJ, Kuipa PK (2001) The efect of air sparging on the electrical resolution of water-in-oil emulsions. Chemical Engineering Science 56: 6279-6284.

43. Ichikawa T (2007) Electrical demulsification of oil-in-water emulsion. Colloids and Surfaces A: Physicochem Eng Aspects 302: 581-586

44. Kanazawa S, Takahashi Y, Nomoto Y (2008) Emulsification and Demulsification Processes in Liquid–Liquid System by Electrostatic Atomization Technique. Industry applications, IEEE transactions 44: 1084-1089.

45. Haifeng G, Ye P (2008) Polarization Characteristic of Droplet of Water-in-Oil Emulsion in a High Uniform Electric Field. International Workshop on Modelling, Simulation and Optimization 139-142.

46. Yakhkeshi A, Hosseina M (2010) Demulsification of Benzene-in-water emulsion by electric field. Middle east Journal of scientific Research 5: 57-60.

47. Oji A, Opara CC (2012) Electrocoalescence of Field Crude Oil using High voltage Direct Current. International Journal of Engineering Science and Technology.

48. Yun Kim B (2007) Demulsification of water-in-crude oil emulsions by a continuous electrostatic dehydrator. Separation Science and Technology 37: 1307-1320.

49. Dalling S (2003) Norwegian testing of emulsion properties at sea-the importance of oil type and release conditions. Spill science and technology Bulletin 8: 123-323.

50. Gunet CC, Sveum (1995) In situ burning of emulsions R&D in Norway. Spill science and technology Bulletin 2: 75-77.

51. Binks BP, Rocher A (2009) Effects of temperature on water-in-oil emulsions stabilised solely by wax Microparticles. Colloid and Interface Science 335: 94-104.

52. Taylor SE (2011) Thermal destabilisation of bitumen-in-water emulsions–A spinning drop tensiometry study. Fuel 90: 3028-3039.

53. Chen G, He G (2003) Separation of water and oil from water-in-oil emulsion by freeze/thaw method. Separation and Purification Technology 31: 83-89.

54. Lin C, He G, Li X, Peng L, Dong C, et al. (2007) Freeze/thaw induced demulsification of water-in-oil emulsions with loosely packed droplets. Separation and Purification Technology 56: 175-183.

55. Yang X, Tan W, Yu Bu (2009) Demulsification of Asphaltenes and Resins emulsions via freeze/thaw method. Energy and fuels 23: 481-486.

56. Habn AU, Mittal KL (1979) Mechanism of demulsification of oil-in-water emulsion in the centrifuge. Colloid and polymer sci 257: 959-967.

57. Mittal KL, Vold RD (1972) Effects of initial concentration of emulsifying agent on the centrifugal stability of oil-in-water emulsions. American oil chemistry society 491: 527-532.

58. Kocherginsky NM, Tan CL, Lub WF (2003) Demulsification of water-in-oil emulsions via filtration through a hydrophilic polymer membrane. Membrane Science 220: 117-128

59. Nadarajah N, Singh A, Ward OP (2002) Evaluation of a mixed bacterial culture for de-emulsification of water-in-petroleum oil emulsions. World Journal of Microbiology and Biotechnology 18: 435-440.

60. Dantas TNC, Neto AAD, Moura EF (2001) Microemulsion systems applied to breakdown petroleum emulsions. Petroleum Science and Engineering 32: 145-149.

61. Colle RD, Longo E, Fontes SR (2007) Demulsification of water/sunflower oil emulsions by a tangential filtration process using chemically impregnated ceramic tubes. Membrane Science 289: 58-66.

62. Sadrzadeh M, Gorouhi E, Mohammadi T (2008) Oily wastewater treatment using polytetrafluoroethylene (PTFE) hydrophobic membranes. Twelfth International Water Technology Conference.

63. Lozsan A (2012) Salt-induced fast aggregation of nano emulsions: structural structural and kinetic scaling. Colloid and Polymer Science 290: 1561-1566.

64. Thostenson, Chou (1999) Microwave processing: fundamentals and applications. Composite A 30: 1055-1071.

65. Camelia (1998) Dielectric parameters relevant to microwave dielectric heating. Chemical Society Reviews 27: 213-224.

66. Barringer (1994) Effect of sample size on the microwave heating rate: oil vs. Water. AIChE journal 40: 1433-1439.

67. Montserrate (2007) Effect of Salinity, Temperature, Water Content, and pH on the Microwave Demulsification of Crude Oil Emulsions. Energy and Fuels 21: 1358-1364.

68. Helene (2009) Dielectric properties of crude oil components. Energy Fuels 23: 5596-5602.

69. Samanta SK, Basak T, Sengupta B (2008) Theoretical analysis on microwave heating of oil–water emulsions supported on ceramic, metallic or composite plates. International Journal of Heat and Mass Transfer 51: 6136-6156.

70. Dicharry C, Arla D, Sinquin A, Graciaa A, Bouriat P (2006) Stability of water/crude oil emulsions based on interfacial dilatational rheology. Journal of Colloid and Interface Science 297: 785-791.

71. Fortuny M, Oliveira CBZ, MeloRLFV, Nele M (2007) Effect of Salinity, Temperature, Water Content, and pH on the Microwave Demulsification of Crude Oil Emulsions. Energy Fuels 21: 1358-1364.

72. Bjorendalen N (2004) The effect of microwave and ultrasonic irradiation on crude oil during production with a horizontal well. Petroleum science and engineering 43: 139-150.

73. Rajakovic V, Skala D (2006) Separation of water-in-oil emulsions by freeze/thaw method and microwave radiation. Separation and Purification Technology 49: 192-196.

74. Nour AH (2010) Demulsification of water-in-crude oil (w/o) emulsions by using Microwave radiations. Journal of Applied sciences 10: 2935-2939.

75. Anisa ANL, Nour AH (2011) Destabilization of heavy and light crude oil emulsions via Microwave heating Technology: An optimization study. Journal of Applied Science 11: 2898-2906.

76. Nour AH, Yunus RM (2006) A continuous microwave heating of water-in-crude oil emulsions: An experimental stud. Journal of Applied science 6: 1868-1872.

77. Yunus RM, Nour AH (2005) Water-in-oil demulsification via microwave irradiation. Proceedings of the International Conference on Recent Advances in Mechanical & Materials Engineering 30-31.

78. Nour AH, Anisa ANI (2012) Demulsification of water-in-oil (W/O) emulsion via microwave irradiation: An optimization. Journal of Scientific Research and Essays 7: 231-243.

79. Abdulbari HA, Abdurahman NH, Rosli YM, Mahmood WK, Azhari HN (2011) Demulsification of petroleum emulsions using microwave separation method . International Journal of the Physical Sciences 6: 5376-5382.

80. Nguyen NT (2005) Wu ZG Micro mixers : A review. Micromech Microeng 151-16.

81. Mason TJ, Lorimer JP (2002) Use of power ultrasound in chemistry and processing. Journal of Applied sonochemistry.

82. Pangu GD, Feke DL (2004) Acoustically aided separation of oil droplets from aqueous emulsions. Chemical Engineering Science 59: 3183- 3193.

83. Sarabia ER, Gallego-Jua´rez JA, Corral GR, Segura LE (2000) Application of high-power ultrasound to enhance fluid/solid particle separation processes. Journal of Ultrasonics 38: 642-646.

84. Guoxiang YE, Xiaoping LU, Peng F, Pingfang H, Xuan S (2008) Pretreatment of Crude Oil by Ultrasonic-electric United Desalting and Dewatering. Chinese Journal of Chemical Engineering 16: 564-569.

85. Gupta S, Feke DL (1997) Acoustically driven collection of suspended particles within porous media. Journal of Ultrasonics 35: 131-139.

86. Nii S, Kikumoto S, Tokuyama H (2008) Quantitative approach to ultrasonic emulsion separation. Ultrasonics Sonochemistry 1: 145-149.

87. Lopez A, Gand Sinha DN (2008) Enhanced Acoustic Separation of Oil-Water Emulsion in Resonant Cavities, The Open Acoustics Journal 1: 66-71.

15

Remove Organic and Inorganic Contaminants in the Waste Oil Using MBBR, Nano Reactor and Physical Methods

Maryam Shirinkar* and Masoud Zolanvar

North Branch of Islamic Azad University of Tehran, Iranian Offshore Oil Company, Iran

Abstract

The aim of this study was to evaluate the performance of pilots physical, biochemical and chemical, in series and together for waste water treatment plants, processing oil in the Persian Gulf's oil and gas. For this purpose a Skimmer tank, MBBR reactor, Nano reactor, UV tank was used. Incoming wastewater system effluent desalination unit. Hydraulic retention time for various different system is intended. If you reduce the retention time of 18 to 12 hours, removed COD from 91% to 76% reduced. It was also found that the total removal of carbon materials in part MBBR 78.2% and 21.8% of the biofilm floating microorganisms. The results indicate that the whole system, which consists of 4 pilot that each part of its eliminates COD in waste water, the ability to remove 91% of COD. Solution input load 1.766 Kg COD/m^2 was signed. Finally, after the passage of treated wastewater and filters in 4 stages, has a relatively uniform quality and desirable.

Keywords: MBBR; Skimmer tank; Organic; Nano reactor; COD; Wastewater

Introduction

Waste water is a problem in oil industry. Over the last decades there has been a growing interest in biofilm processes for waste water treatment. Moving bed biofilm reactor (MBBR) is an efficient alternative for organic carbon and nitrogen removal which combine the advantages of both the active sludge process and a biofilm reactor by incorporating free-floating carriers that provide a large surface area for colonization with no need for biomass recycling [1].

There are several reasons for the fact that biofilm processes often are being increasingly favored instead of activated sludge processes, such as the following:

a. The treatment plant requires less apace.

b. The final treatment result is less dependent on biomass separation since the biomass concentration to be separated is at least 10 times lower.

c. The attached biomass becomes more specialized (higher concentration of relevant organisms) at a given point in the processes train, because there is no sludge return.

There are already many different biofilm systems in use, such as trickling filters, rotating biological contactors (RBC), fixed media submerged biofilters, granular media biofilters, fluidized bed reactors etc.

They have all their advantages and disadvantages. The trickling filter is not volume-effective.

Mechanical failures are often experienced with the RBC`S.It is difficult to get distribution of the load on the whole carrier surface in fixed media submerged biofilters. The granular media biofilters have to be operated discontinuously because of the need for backwashing and the fluidized bed reactors show hydraulic instability [2].

Among various semiconductors, TiO_2 is one of the most efficient photocatalyst for the degradation of pollutants in aqueous suspension through the photogenerated strong oxidizing agents like hydroxyl radical and superoxide radical anions under UV light irradiation. It is pertinent to mention here that TiO_2 can only be excited by UV light due to its large band gap energy (3.2 ev for anatase) which is not ideal to absorb visible light.

To overcome this problem research groups working in the area of photocatalysis are trying to extend the optical response of photocatalyst from UV to visible region by doping it with metal or non-metal into the TiO_2 lattice.

The various methods used for doping TiO_2 involve ion implantation, sol-gel reaction, hydrothermal reaction, solid-state reaction, etc.

Among these, one of the sol-gel processed is undoubtedly the simplest and the cheapest one which provides control on the size and shape of nano particles [3].

Few studies relating to the synthesis of molybdenum (Mo), Manganese (Mn) and Zinc (Zn) doped TiO_2 using different methods and its photocatalytic activity for the degradation of pollutants have been reported earlier. For example, in two separate studies Mo-doped TiO_2 has been synthesized using thermal hydrolysis and sol-gel method, In another study by Devi etal. Mn-doped TiO_2 was synthesized using sol-gel process and its photocatalytic activity was investigated by studying the degradation of a dye derivative amaranth. On the other hand, synthesis of Zn-doped TiO_2 using different methods and their activity for the degradation of dyes and organic pollutant has been reported in the literature [3].

Waste water with high levels of organic matter (COD) Phosphorous (P) and Nitrogen (N) cause several problems, such as eutrophication, oxygen consumption and toxicity, when discharged to the environment.

Biological processes are a cost-effective and environmentally sound alternative to the chemical treatment of waste water.

*Corresponding author: Maryam Shirinkar, North Branch of Islamic Azad University of Tehran, Iranian Offshore Oil Company, Iran, E-mail: Shirinkar.Maryam@yahoo.com

Biological processes based upon suspended biomass (i.e., activated sludge processes) are effective for organic carbon and nutrient removal in municipal waste water plants. But there are some problems of sludge settleability and the need for large reactors and setting tanks and biomass recycling. Biofilm reactors are especially useful when slow growing organisms like nitrifies have to be kept in a waste water treatment process.

During the past decade, it has been successfully used for the treatment of many industrial effluents including pulp and paper industry waste, poultry processing waste water, cheese waste, phenolic waste water, dairy waste water and municipal waste water [4].

This paper proposes a new methodology for treatment waste water. This study, there for, set out to assess the effect of Moving Bed Bio Reactor (MBBR), and the effect of Nano particle of TiO$_2$ doped Mn, Mo, Zn and UV lamp to removing oil in waste water.The present research explores, for the first time, the effect of Skimmer tank and MBBR and Nano particle in reactor and UV lamp (4 system) on waste water of oil industry. Due to practical constraints, this paper cannot provide a comprehensive review of removing oil from waste water with 4 systems. It is my experience of working with skimmer tank, MBBR, Nano particle, UV that has driven this research.

This paper begins by introducing Skimmer tank, it will then go on to MBBR system and Nano reactor, then UV, finally it will deal with removal of free oil and decreasing COD in waste water.

Material and Methods

A Schematic of pilot process concept is shown in Figure 1, where the treatment train consists of 2 tanks and 2 reactors.

The skimmer tank

In the Skimmer tank, remove oil free from the surface of waste water. The input current is entered at rate of 1 lit/hr.

The biofilm reactor (MBBR)

Carriers: In the MBBR, the biomass grows on carriers that move freely in the water volume by aeration [5]. The biofilm carriers are made of high-density polyethylene (density 0.95 g/cm^3). The size of the carrier varies from lengths of 7-15 mm and diameters of 10-15 mm. The carrier filling fraction (percentage of reactor volume occupied with carriers in empty tank) is normally 60-70% [5]. The Kaldnes carriers have a specific gravity of 0.96 with a specific biofilm protected surface area of 500 m^2 per m^3 bulk volume of carriers. The Kaldnes biofilm carrier element is illustrated in Figure 2 [6].

Lab-scale reactor and waste water: A Laboratory scale Plexiglas with total liquid volume of 2 lit in the study (Figure 3). Diffusers were used for oxygen supply and mixing. The dissolved oxygen (Do) concentrations in the MBBR ranged from 0.2 to 5.00 mg o$_2$/lit depending on the influent organic loading rates. The temperature and PH in the reactor varied from 28 to 31 c and 6.72 to 7.88, respectively [6] and the pressures varying between 0.1 and 0.5 bar [5].

Waste water: Effluent entering the fourth stage was the output unit 2 Ft in height desalter of Kharg Island in Iranian offshore Oil Company. In Figure 4 and Table 1, GC taken combined effluent shows. Synthetic oil was mixed with waste water in laboratory. The effluent from the first stage of the pilot project, from the skimmer tank to sign the final phase, removal and cleaning after several phases and the water from final phase was uniform and satisfactory quality and standard of environment. The calculated COD/N/P ratio of the syntheticwaste water was 100/5/1 [6]. The waste water was enriching with the macro-nutrients by adding

Figure 1: Skimmer tank.

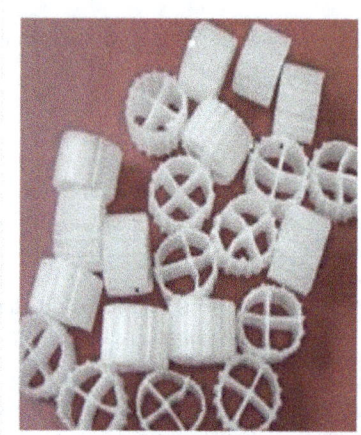

Figure 2: Media K2.

Figure 3: MBBR reactor.

urea as nitrogen source and K$_2$HPO$_4$ as phosphorus source [4]. NH-N ranged from 25-125 mg/lit and po4-p ranged from 5-25 mg/lit were prepared and used as feed to the systemBy the way, vermin-compost leachate was used for injection and enrichment of waste water.

Nano reactor

Synthesis TiO$_2$ doped Mn, Mo, Znnano particle: TiO$_2$ nano particles were synthesized by sol-gel method using Titanium tetra

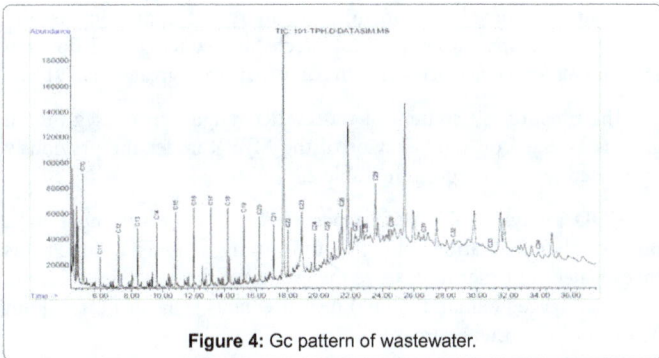

Figure 4: Gc pattern of wastewater.

No	Compound Name	RT (min)	Units	Amount
1	C10	4.863	ppm	1.304
2	C11	5.978	ppm	0.420
3	C12	7.167	ppm	0.617
4	C13	8.398	ppm	0.736
5	C14	9.622	ppm	0.732
6	C15	10.83	ppm	0.825
7	C16	11.996	ppm	0.884
8	C17	13.116	ppm	0.921
9	C18	14.19	ppm	0.883
10	C19	15.217	ppm	0.876
11	C20	16.198	ppm	0.798
12	C21	17.139	ppm	0.732
13	C22	18.032	ppm	0.603
14	C23	18.897	ppm	1.448
15	C24	19.727	ppm	0.459
16	C25	20.522	ppm	0.544
17	C26	21.439	ppm	1.759
18	C27	22.263	ppm	0.686
19	C28	22.931	ppm	0.122
20	C29	23.558	ppm	2.662
21	C30	24.556	ppm	0.623
22	C31	26.611	ppm	0.034
23	C32	28.485	ppm	0.049
24	C33	30.853	ppm	ND
25	C34	33.895	ppm	0.021
26	C35	0	ppm	ND

Table 1: The output unit 2 Ft in height desalter of Kharg Island in Iranian offshore Oil Company.

isopropoxid (TTIP) as titania precursor and doped with different concentrations of Molybdenum (Mo), Manganese (Mn) and Zinc (Zn) (0.5-1%) and characterized by standard analytical techniques such as x-ray diffraction (XRD) and scaning Electron Microscopy (SEM). The XRD analysis shows the partial crystalline nature and Rutile phase (Figure 5). The SEM images of doped TiO_2 at different magnifications also show the partial crystalline nature with rough surfaces (Figure 6).

Coating TiO_2 doped Mn, Mo, Zn in nano filter: TiO_2 doped nano particles were coated on filter plates Whatsman N.34 and placed in an oven for 24 hours at a temperature of 34-42 c, pages was ready to install in reactor.

Nano rector: A schematic of pilot process concept is shown in Figure 7. Doped nanoparticles on pages tray tower, when in fact these pages are filters that were built for the job, coats and waste water passes on these plates, such as fluid flow in the distillation tower. On plates waste water dealing with these nano particles, mass transfer and removal of organic contaminants and treatment is better.

UV tank

A schematic of pilot process concept is shown in Figure 8. Treated wastewater from step 1 to 3, enter UV-Tank, and in the presence of UV light rays with a wavelength of 280 – 100 nm, refined and wastewater output has reached that limit environmental standards and can be used for watering plants and entering the environmental and sea. Data management and analysis were performed using SPSS, Xpert and Excle.

Results and Discussion

Experimental work in the laboratory was carried out in order to evaluate the efficiency of the system for the removal of organic matter and relationship between organic removal and abserved yield using a high loaded MBBR and Nano reactor.

Effluent total COD and COD removal rates versus time in Skimmer tank, MBBR, Nano reactor, UV-tank are show in Figure 9.

The XRD, SEM patterns of doped TiO_2 with Mo, Mn, Zn calcinated at 550 c for 4 h were analyzed [3].

All deped TiO_2 particles showed that particle crystalline nature and Rutile phase [3].

The average rutile crystallite size of doped and undoped TiO_2 nano particles was determined by Debye Scherer formula:

$$D = K^* \gamma / B^* Cos Q$$

Where D is the crystallite size, K the shape factore, γ the wavelength, Q the diffraction angle and B is the full width at half maximum [3].

The XRD, SEM is shown in Figures 5 and 6.

COD analysis

COD analysis in skimmer tank: An oil skimmer continues to remove oils as long as they are present. Depending on oil influx rate and the oil skimmer's removal rate, residual oil in the water maybe as low as a few parts per million. When residual oil reaches this level and further reduction is required, it may be more pratical to use a secondary removal method following skimming, such as MBBR.

COD analysis in MBBR: In order to observe the quality of the aqueous solution, chemical oxygen demand (COD) measurements were also carried out before and after the MBBR. As demonstrated in Figure 10 increasing inlet concentration up to 800 mg/l did not significantly affect the performance of the MBBR in COD removal and the efficiencies were over 91% for this parameter, although further increasing inlet concentration showed an adverse effect on the removal

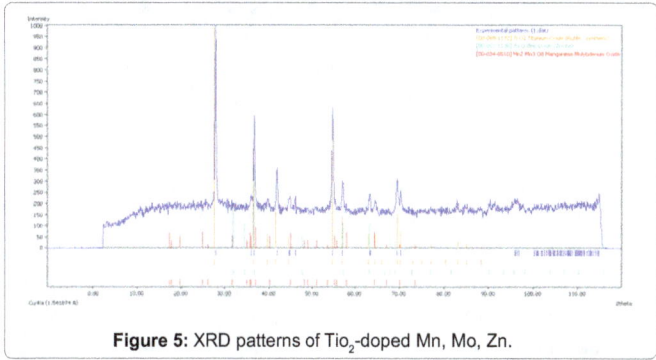

Figure 5: XRD patterns of TiO_2-doped Mn, Mo, Zn.

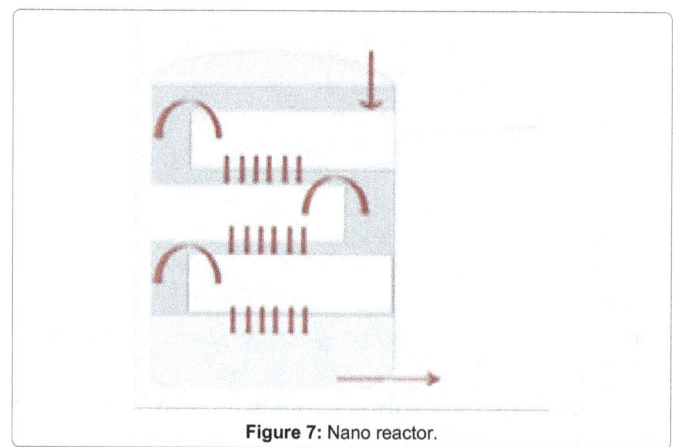

Figure 6: SEM images of Tio$_2$-doped Mn, Mo, Zn.

efficiency particularly, increasing inlet concentration to 1000 and 1400 mg/l resulted in decreasing COD removal 91% respectively.

Therefore the optimum surface loading rate based on inlet concentration (at HRT of 18 h) on the MBBR was found to be 6.62 g COD/m^2.day. Accordingly, the MBBR could effectively remove organic and inorganic contaminants in waste water, (Figures 10 and 11).

Figure 10 shows that organic and inorganic contaminants and COD removal efficiencies were not affected by reducing HRT down to 18 h and removal efficiencies of both parameters were greater than 91%.

The effect of salt content of wastewater ranging from 10 g/l to 70 g/l was assessed on the behavior of the MBBR under the previously optimized conditions given in Table 2.

COD analysis in NANO reactor: A COD measurement by a photocatalytic oxidation sol-gel method using nano TiO$_2$ doped was investigated. In order to observe the quality of the aqueoues solution, chemical oxygen demand (COD) measurements were also carried out before and after the treatment.

A significant decrease in the COD values was observed, which clearly indicates that the photocalytic method offers good potential for the removal of organic and inorganic contaminants wastewater.

COD analysis in UV-tank: UV-VIS spectroscopy was suggested as fast and versatile monitoring tools for BOD and COD in water samples. This spectroscopy has also advantage of limiting measurements time for BOD from 5 days to few minutes and also limiting the usage of a large amount of expensive reagents. Absorbance of UV in wastewater to removal organic and inorganic contaminants shown in Figure 11.

Conclusion

4-steps process studied in this research, including pilot Skimmer tank, MBBR, Nano reactor and UV-tank that is, the ability to remove the COD has been quite successful.The amount MLSS rise, the resistance reactors against fluctuations will increase. The COD and organic and inorganic contaminants in wastewater much reduced in this 4 steps and process applicable on an industrial scale operation.

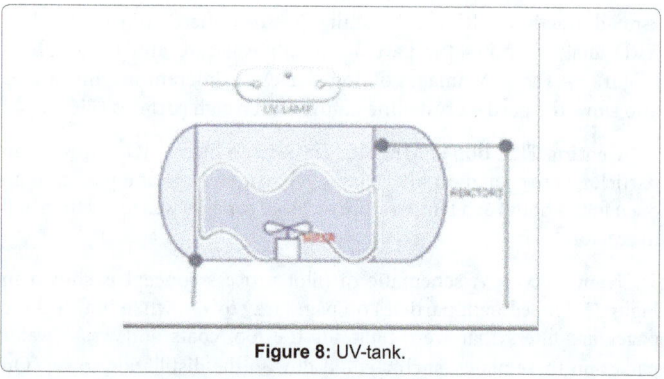

Figure 7: Nano reactor.

Figure 8: UV-tank.

Experimental phases and MBBR operation timing schedule					
Phase	Day	Operation	C_{in} (mg/L) COD	Salt content (g/L)	HRT(b)
1	0-90	Biomass acclimation	107.5-1075	0-30	-
2	91-150	Effect of C_{in}	430-2580	30	24
3	151-190	Effect of HRT	1720	30	8-24
4	191-245	Effect of salt content	1720	10-70	18
5	246-247	Response to organic shock loading	-	30	18
6	248-249	Response to hydraulic shock loading	1720	30	-
7	250-251	Response to salt shock loading	1720	-	18

Table 2: COD analysis in MBBR.

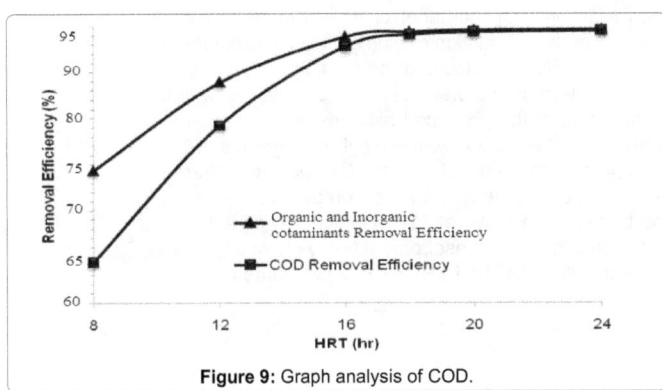

Figure 9: Graph analysis of COD.

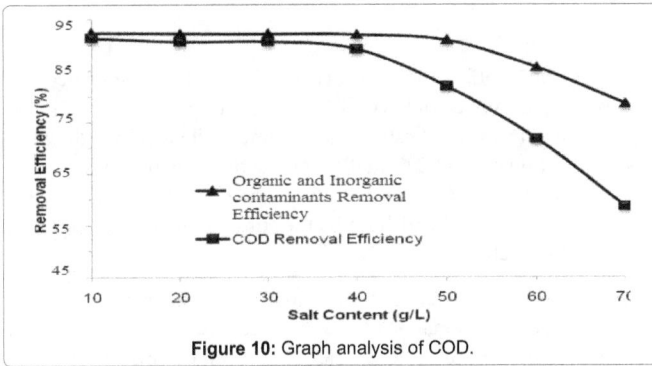

Figure 10: Graph analysis of COD.

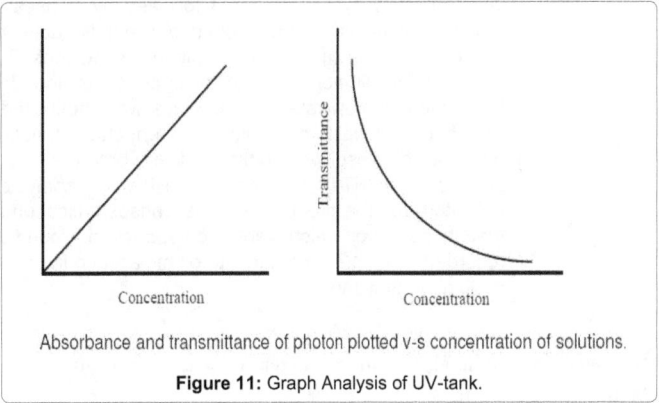

Absorbance and transmittance of photon plotted v-s concentration of solutions.

Figure 11: Graph Analysis of UV-tank.

Acknowledgements

Authors would like to acknowledge Iranian Offshore Oil Company; Nano Company for the financial support provided this work.

References

1. Zinatizade AAL (2012) Separation of nitrophenols using cellulose acetate nanofiltration membrane: Influence of surfactant additives. Separat Purif Technol 85: 147-156.

2. Qdegaard H (1999) The Moving Bed Biofilm Reactor. Hokkaido Press. pp: 250-305.

3. Umar K (2013) Mo, Mn and La doped Tio_2: Synthesis, characterization and photocatalytic activity for the decolourization of three different chromophoric dyes. Journal of Alloys and Compounds 578: 431-438.

4. Kermani M, Bina B, Movahedian H, Amin M, Nikaein M (2008) Application of Moving Bed Biofilm Process for Biological Organics and Nutrient Removal from Municipal wastewater. American Journal of Environmental Sciences 4: 675-682.

5. Leiknes T (2005) The development of a biofilm membrane bioreacto. Desalination 202: 135-143.

6. Aygun A (2008) Influence of High Organic Loading Rates on COD Removal and Sludge Production in Moving Bed Biofilm Reactor. Environmental Engineering Science 25: 1311-1316.

16

Production of Biodiesel from Waste Vegetable Oil via KM Micro-mixer

Elkady MF[1,2]*, Ahmed Zaatout[3] and Ola Balbaa[3]

[1]*Chemical and Petrochemical Engineering Department, Egypt-Japan University of Science and Technology, New Borg El-Arab City, Alexandria, Egypt*
[2]*Advanced Technology and New Materials and Research Institute (ATNMRI), City of Scientific Research and Technological Applications, Alexandria, Egypt*
[3]*Chemical Engineering Department, Faculty of Engineering, Alexandria University, Alexandria, Egypt*

Abstract

The production of biodiesel to substitute fossil fuels has been challenging to apply in many countries especially developing countries. Given the importance of this fact, this study includes the production of biodiesel from waste vegetable oils by pre-treatment followed by transesterification reaction with methanol using a KM micro-mixer reactor. KM micro-mixer happened to give noticeable enhancement for the production of biodiesel quality compared to the normal batch reactor at optimum conditions. The parameters affecting biodiesel production process such as alcohol to oil molar ratio, catalyst concentration, the presence of tetra-hydrofuran (THF) as a co-solvents and the volumetric flow rates of inlet fluids were optimized. The properties of the produced biodiesel were compared with its parent waste oil through different characterization techniques. The presence of methyl ester groups at the produced biodiesel was confirmed using both the Gas chromatography–mass spectrometry (GC-MS) and infrared spectroscopy (FT-IR). Moreover, the thermal analysis of the produced biodiesel and the comparable waste oil indicated that the product after the transesterification process began to vaporize at 120°C which makes it lighter than its parent oil which started to vaporize at around 300°C. The maximum biodiesel production yield of 97% was recorded using 12:1 methanol to oil molar ratio in presence of both 1% NaOH and THF/methanol volume ratio 0.3 at 60 mL/h flow rate.

Keywords: Biodiesel; KM micro-mixer; Transesterification; Methyl esters; Waste oil

Introduction

The idea of using alternative fuels has been widely spreading for many years now as a replacement for Fossil fuels. The importance of this idea came from the large scale of utilization of Fossil fuels in mechanical power generation various sectors, like agriculture, commercial, domestic and transport, also the fact of the continuous rise in Fuels cost and their eventual vanish [1].

The use of vegetable oils and their derivatives was found one of the reasonable solutions. However, the direct use of Vegetable oils in diesel engines was found impractical due to several factors, such as the high viscosity, acid composition and free fatty acid content. Accordingly, they require further modifications for effective use [2]. Undergoing transesterification reaction is the most favorable for decreasing oil's viscosity and producing what so called "Biodiesel fuel" [3]. Biodiesels are mono-alkyl esters of long chain fatty acid derived from renewable lipid feedstock. The interest of this alternative energy resource is that the fatty acid methyl esters, known as biodiesel, have similar characteristics of petro-diesel oil which allows its use in compression motors without any engine modification [4]. However, using vegetable oil to replace fuel caused the food versus fuel issue all over the world [5]. So the idea of using waste vegetable oil (WVO) has been introduced as an economical solution which also gives a waste management solution [6].

Transesterification is a process of transforming triglycerides in vegetable oils into a mixture of fatty acid esters using alcohol and catalyst to speed up this reaction to the right side and to obtain high biodiesel yields. Methyl or ethyl esters are obtained, with much more similar properties to those of conventional diesel fuels. The main by-product obtained is glycerol. Figure 1, shows the general equation of transesterification reaction. The most common alcohol used for biodiesel production is methanol because of its price and conversion rates. Other alcohols can be used too, such as plant based ethanol, propanol, isopropanol and butanol [3]. In presence of excess alcohol, the foreword reaction extends beyond the reverse reaction. Many catalysts could be utilized in the process, however, it was confirmed that transesterification is completed faster using an alkali catalyst [7]. The mechanism of transesterification shows some challenges regarding this process, starting from the limitation of reaction rate by mass transfer between the immiscible oil and alcohol besides the reversibility of the transesterification itself which limits the conversion and consequently increases the reaction time and cost [4].

These challenges of transesterification reaction happened to appear clearly using conventional batch reaction processes. Many alternatives have been proposed to undergo the reaction in a more effective way through improving mixing rate, enhancing heat and mass transfer of the reaction and decreasing cost and time consumed [8]. For instance, changing the process performance using super critical conditions through applying high temperature and pressure enhance the process mass transfer [2]. Moreover, proposing different catalysis approaches such as heterogeneous or enzyme catalysis improve the process reaction rate. Also, changing the process design and mixing concepts such as using ultrasonic homogenizers increase both the process mass and heat transfer [9]. The methods mentioned were found quiet effective for solving the problems facing transesterification like time consumption, soap formation, etc. However, energy consumption rate increases significantly and therefore the total cost of the process increases. Another proposed change in the process design was the

*****Corresponding author:** Elkady MF, Advanced Technology and New Materials and Research Institute (ATNMRI), City of Scientific Research and Technological Applications, Alexandria, Egypt, E-mail: marwa.f.elkady@gmail.com

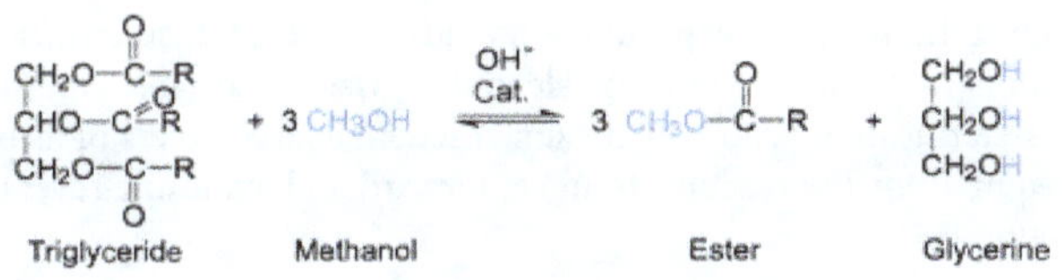

Figure 1: General equation of transesterification.

use of microreactors for achieving transesterification reaction within short time. Generally, micro reactors are micro structured reactors with micro-channels, they have various shapes and different structures designed for better mixing and completing the reactions. Simple micro-scale capillaries were the first reported microreactors used in biodiesel synthesis [10]. Other advanced microreactors were later fabricated using wide variety of materials and different manufacturing techniques [10]. As previously discussed, the mass transfer of the reacting triglycerides from the oil phase towards the methanol/oil interface limits the rate of methanolysis reaction and controls the kinetics at the beginning of the reaction [11]. Also the droplet size highly affects the methyl ester yield in this reaction. Accordingly, microreactors were utilized at the transesterification reaction holding the advantage of high volume/surface ratio, short diffusion distance, fast and efficient heat dissipation and mass transfer [12]. By this role microreactors promote the overall volumetric mass transfer coefficient of methyl esters due to the increase of the specific interfacial area by decreasing the droplet size. This eventually results in the increase of reaction rate for triglycerides [13]. The KM micro- mixer has been tested for mixing two immiscible fluids and was found superior over other mixers designs. Also it provides high throughput and stable operation in a wide range of flow rate ratios for the two reactant fluids [14]. In this investigation, a KM micro-mixer has been used as a microreactor for transesterification of waste vegetable oil with methanol in presence of NaOH as catalyst. The influences of transesterification process variables such as alcohol to oil molar ratio, catalyst concentration, volumetric flow rate and effect of an organic co-solvent presence were optimized. GC-MS analysis was utilized for characterization and identification of the produced biodiesel.

Materials and Method

Materials

Waste Vegetable oil was purchased from a local restaurant as a source of triglycerides for transesterification reaction. The alcohol selected was methanol (99.8%, Sigma-Aldrich.). Other utilized chemicals for transesterification process are from analytical grades such as sodium hydroxide (99%, Sigma-Aldrich), acetic acid (98%, Sigma-Aldrich) and tetra-hydrofuran (anhydrous 99%, Sigma-Aldrich).

Waste vegetable oil pretreatment

The waste vegetable oil (WVO) was first filtered to remove bits of food residues using a glass Büchner funnel filtration system then it was subjected to an acid catalyzed esterification process in order to maintain free fatty acid content lower than 1% [15].

Experimental setup

The KM micro-mixer proposed for this investigation consists of 3 stainless steel plates, inlet, mixing and outlet plates holding fourteen micro-channels fabricated for fluid streams. Dimensions of the micro-mixer are shown in Table 1. The mixer has 2 inlets for two different

reactant fluids. The fluids are transferred to the mixing plate through annular channels where fourteen micro-channels are present.

Micro-channels are fabricated by Micro Electric Discharge Machining (μ-EDM). The stream of each fluid was divided into half of the total number of micro channels. The two divided fluids meet at the center of the mixing plate and are immediately mixed. The diameter of the mixing zone was found to be 220 μm. Finally, the outlet plate has a hole for the exit of the mixed fluid at the center of the plate, the exit fluid hole (200 μm) is smaller than the diameter of the mixing zone to accelerate the mixing process [14].

The experiment is set as shown in Figure 2. The KM mixer is immersed in a water bath to provide the required reaction temperature. Two Syringe pumps (KD Scientific, KDS100, USA) were used for feeding the inlet reactant fluids.

Biodiesel production process using KM mixer

Two reactant fluids were fed via syringe pumps into the designed experimental setup, the first is the preheated oil at specific temperature, and the second is the mixture of methanol, sodium hydroxide and THF. The proper amount of sodium hydroxide was dissolved completely in methanol to avoid clogging the micro channels at the KM mixer with solid particles. The amount varied from 0.5% to 2% (wt/wt of oil) to elucidate the most suitable amount that attains the highest biodiesel production yields. The reactants feeding rates were changed over a wide range from 20 mL/h to 200 mL/h to investigate the influence of residence time on the biodiesel production process. The reactants molar ratio was optimized to determine the most proper mixing ratio. The KM mixer that includes the process reactants was maintained at specific water bath temperature 70°C. The experiments were usually repeated two times to determine the experimental error. At the KM mixer outlet the product is collected after reaching steady state in a beaker containing appropriate amount of acetic acid to neutralize the excess alkaline catalyst and stop the reaction. The product of the reaction is placed in a separating funnel to be separated into two clear phases. The biodiesel which is the main product separated as upper light-colored phase while the lower dark phase is mainly glycerol. The upper phase was washed after separation with distilled water for excess catalyst and glycerol removal then it was heated up to 70°C to vaporize the excess solvent. The remaining main product was then characterized using GC-MS analysis to confirm oil conversion and identify biodiesel production yield.

Characterization of produced biodiesel

In order to characterize the quantity and the quality of the produced biodiesel several techniques were utilized. The volume of biodiesel product was first measured and the volume yield percentage was calculated according to the following equation (1):

Volume Yield %= (volume of product/volume of oil fed) X 100   (1)

The fatty acid methyl esters (FAMEs) in the produced biodiesel were then characterized and identified using Gas chromatography mass spectrometry (GCMS-QP2010Ultra, Shimadzu, Japan) fitted with 5MS column (30m, 0.25 mmID, 0.25 μm). GC-MS analysis mainly identifies the quality and quantity of the produced biodiesel resembled in the methyl esters present in the product sample. This analysis technique also gives the distribution area for each component in the produced sample. Table 2 shows the GC-MS configuration used for biodiesel analysis. The total yield from the biodiesel was finally calculated according to equation (2):

Internal sketch	Number of channel	Channel width	Diameter of mixing zone	Diameter of Outlet
	14	50 μm	220 μm	200 μm

Table 1: Dimensions of KM micro-mixer.

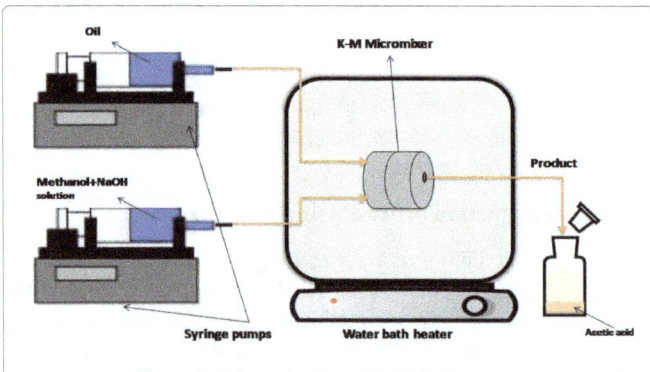

Figure 2: Schematic view of the KM mixer system.

	Injector
Inlet temperature	200°C
Sample size	2 μl
Split ratio	50
	Column temperature program
Initial temperature	50°C
Rate 1	15°C /min to 180°C
Rate 2	7°C /min to 230°C
Rate 3	10°C /min to 280°C
	Detector
Type	Mass Spectrometer
Interface temperature	200°C
	Column
Type	HP-5MS (30 m, 0.25 mmID, 0.25 μm)
Flow rate	3 mL/min

Table 2: GC-MS configuration.

Biodiesel yield% = FAMEs percentage from GC analysis X Volume yield   (2)

Thermal analysis of the prepared biodiesel and its parent waste oil were carried out using thermal gravimetric analysis (TGA-50, Shimadzu, Japan) with a heating rate of 20°C/min under the flow of nitrogen gas, starting from ambient condition up to 800°C.

Fourier transform infrared spectroscopy (Vertex 70, Germany) was utilized to compare the main function group presence in the produced biodiesel and its parent waste oil. The I.R. spectrum was scanned through a wave length range of 4000-400 cm⁻¹. Both the TGA and FT-IR techniques were used for quantitative and qualitative analysis respectively of the product and to confirm GC-MS results.

Results and Discussion

Biodiesel production using KM mixer

In order to optimize the processing parameters affecting biodiesel production according to the main equation of the transesterification process using KM mixer. The influence of molar ratio of methanol to oil, catalyst concentration, volumetric flow rate and presence of THF were investigated and the product was analyzed using GC-MS analysis.

Effect of methanol to oil molar ratio: The most effective variable affecting the methyl ester production yield during the transesterification reaction is the molar ratio of alcohol to waste vegetable oil. Since transesterification is an equilibrium reaction, a large excess of alcohol is required for the reaction to move forward and avoid the reversible reaction [16].

The biodiesel production has been investigated over a studied reactants molar ratios of methanol to oil from 6:1 up to 48:1. It was evident from Figure 3a that 6:1 reactants ratio recorded the lowest biodiesel conversion compared with 12:1 molar ration that attained the maximum biodiesel conversion. The increase in alcohol to oil molar ratio above 12:1 declines the biodiesel conversion. This is due to the reversibility behavior of transesterification reaction [17]. Figure 3b illustrates GC-MS of biodiesel produced at 24:1 methanol to oil molar ratio. It showed the appearance of a clear peak at 4.8 minutes retention time. This peak is verified to be glycerol. This result indicates the difficulty in separating the two produced layers of biodiesel and glycerol at high methanol to oil molar ratios regarding to the solubility of glycerol in excess methanol. This result proved that increasing the molar ratio higher than 12:1 was unflavored for the transesterification process using KM mixer.

Effect of catalyst concentration: The most common catalysts used for transesterification reaction are the alkali catalysts, like sodium and potassium hydroxides because they both react with the triglycerides to break them apart so that methanol can bond with the fatty acids and produce biodiesel. However, sodium hydroxide was selected to be utilized as catalyst in this investigation due to its low cost and availability [2].

A wide concentration range of sodium hydroxide from 0.5% to 2% (wt/wt of oil) has been tested as a catalyst that was premixed with methanol to form sodium methoxide (CH_3Na). The behavior of NaOH concentration regarding to the biodiesel production yield was shown in Figure 4a. Incomplete biodiesel conversion was indicated using 0.5% catalyst concentration. The highest biodiesel yield of 95%was achieved using 1% catalyst concentration. It was indicated from Figure 4a that as NaOH concentration increased above 1% significant decrease in the biodiesel yield was recorded. This result may be explained due to the soap formation owing to the excess of NaOH that reacts with oil fatty acids producing sodium oleate (soap) and water according to the following equation [18].

This prediction was confirmed from Figure 4b that investigates the GC-MS analysis for the biodiesel produced using 2% catalyst concentration. A clear peak of soap formation was indicated from this figure at six minutes retention time. Accordingly 1% wt/wt of NaOH was selected as the optimum catalyst concentration for high conversion biodiesel product.

Effect of reactant volumetric flow rate: Generally, the biodiesel production yield was enhanced through improvement the reactants residence time. The KM micro-mixer characterized by its rapid mixing property resulting from small micro-channel size. These small micro-channels provide fast and efficient mass transport rate versus short diffusion distance and also offer high surface to volume ratios, consequently the reaction residence time parameter at the KM micro-mixer is positively affected [10]. Figure 5 shows the behavior of changing the volumetric flow rates of reactants introduced into the KM micro-mixer. It was elucidated from this figure that the reactants flow rates of 20, 40 and 60 mL/h of two inlet reactants gave approximately equal biodiesel production yield around 96%. Further increase at the

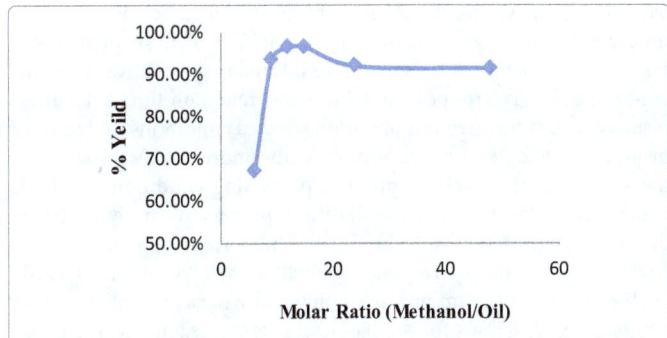

Figure 3a: Effect of Methanol to oil molar ratio on percentage biodiesel yield.

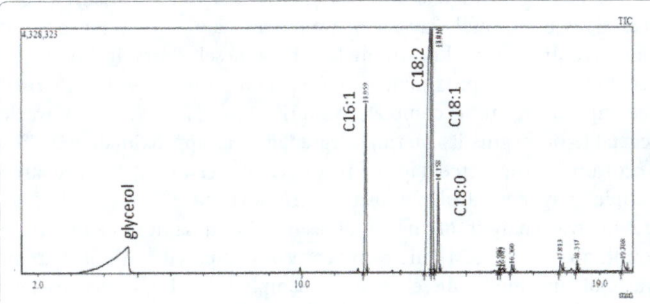

Figure 3b: GC-MS of biodiesel produced using 24:1 methanol:oil molar ratio.

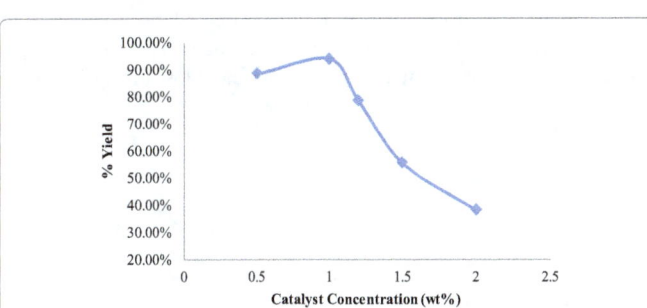

Figure 4a: Effect of NaOH concentration with respect to oil weight on percentage biodiesel yield.

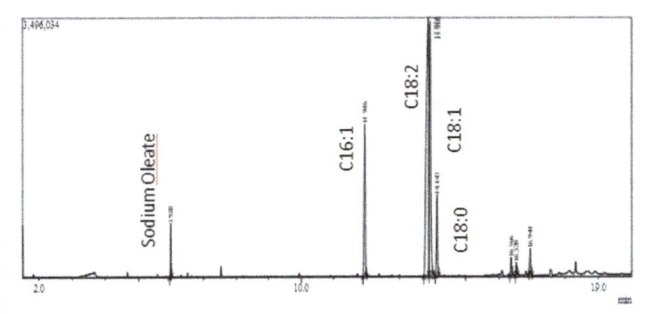

Figure 5: GC-MS of biodiesel produced using 2% NaOH catalyst concentration.

reactants flow rates above 60 mL/h the percentage biodiesel production yield showed obvious drop. This behavior may be explained by stating that for KM Mixer, higher flow rates will result in increasing the pressure drop inside the reactor due to the small mixing zone diameter of the

micro-mixer which affects the completion of the transesterification reaction [10]. Also very low flow rates were not durable for KM mixer causing it to lose its main privilege which is decreasing the reaction time. Accordingly, the flow rate of 60 mL/h is considered the optimum inlet flow rate of the reactants.

Effect of organic co-solvent presence: The main obstacle facing methanolysis of the waste vegetable oil is the presence of two immiscible phases that slows the reaction significantly. In order to conduct the transesterification reaction in a single phase and facilitate the diffusion of the two immiscible reactant fluids, an organic co-solvent has been suggested [4]. THF was preferable compared to other co-solvents because its boiling point is near to the boiling point of methanol that facilitates its separation from the excess methanol at the end of the reaction [4]. However, large amounts of THF are not favored to be used at the transesterification process regarding that the excess co-solvent may cause reagents dilution which declines the rate of transesterification process [19]. Moreover, using large THF amounts at the transesterification process increases the process cost aspects.

In this regards, the effect of presence co-solvent to methanol

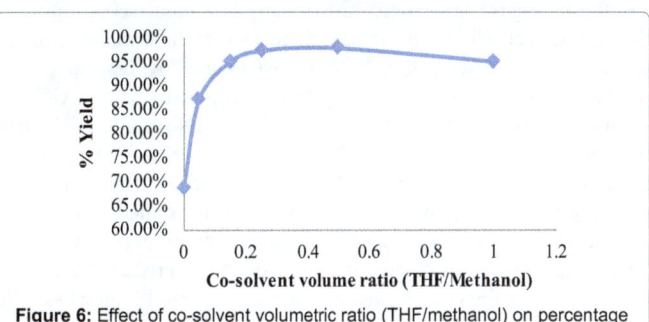

Figure 6: Effect of co-solvent volumetric ratio (THF/methanol) on percentage biodiesel yield.

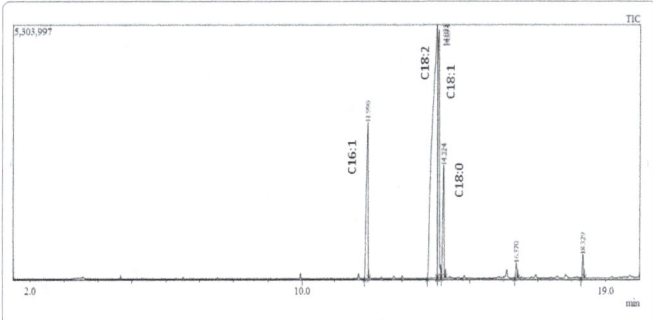

Figure 7a: Gas chromatography–mass spectrometry of biodiesel produced at optimum processing conditions using KM micro-mixer.

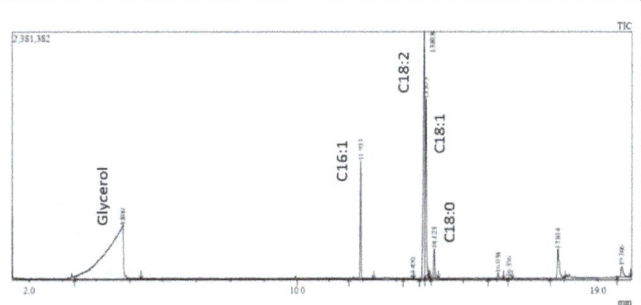

Figure 7b: Gas chromatography–mass spectrometry of biodiesel produced using 24:1 methanol: oil molar ratio.

volumetric ratio on biodiesel production yield was examined over the studied range from 0.2 to 1. It was indicated from Figure 6, noticeable improvement at the production yield using just small amount from THF to methanol ratio. Moreover, the biodiesel production yield was increased as the THF to methanol volumetric ratio increased. This behavior confirms the positive role of the co-solvent presence at the transesterification reaction. The optimum biodiesel production yield of 97.3% was recorded using THF to methanol volumetric ratio of 0.3:1. As the utilized THF to methanol volumetric ratio increased above the optimum selected value, there is no noticeable enhancement at the biodiesel production yield.

Characterization of produced biodiesel

In order to investigate the properties of the produced biodiesel at the predetermined optimum conditions using KM micro-mixer, it was compared with its parent waste vegetable oil using different characterization techniques.

Gas chromatography–mass spectrometry: GC-MS was used for determination of biodiesel methyl ester groups present at the produced biodiesel using KM micro-mixer to determine the optimum conditions. Figure 7a showed GC-MS analysis of biodiesel produced at the optimum conditions using KM micro-mixer. There were four main characteristics peaks of fatty acid methyl esters (FAMEs) appearing by the retention time and the fragmentation pattern data of GC-MS analysis. These four peaks identified FAMEs as 9-hexadecanoic acid methyl ester (C16:1), 9, 12-Octadecadienoic acid methyl ester (C18:2), 9-Octadecenoic acid methyl ester (C18:1) and Octadecanoic acid methyl ester (C18:0). The identified FAMEs were verified by retention time data and mass fragmentation pattern from previous studies [20]. As previously discussed, the glycerol characteristics peaks were only present at the GC-MS analysis of the prepared biodiesel sample

using excess methanol to oil molar ratio or at insufficient separation time conditions. Figure 7b investigates GC-MS analysis of biodiesel sample produced at 48:1 methanol to oil molar ratio. It was clear the appearance of glycerol peak at 4.7 minutes retention time [21]. These results confirm the previous optimum selected conditions for biodiesel production using KM micro-mixer. Furthermore, GC-MS analysis of the produced biodiesel at optimum processing conditions confirms completeness of the transesterification process of triglycerides in the waste vegetable oil into biodiesel. The percentage conversion of triglycerides to the corresponding methyl esters from the GC-MS analysis of the optimum produced biodiesel was calculated as 98% wt compared with 85%wt for the biodiesel prepared sample using excess methanol (48:1 molar ratio).

Thermal gravimetric analysis (TGA)7.3.1 Peak pressure: TGA analysis is one useful way for quantitative analysis for the produced biodiesel. It is well known that the biodiesel starts to thermally decompose at approximately 150°C and continues its thermal decomposition until complete vaporization. However, the waste vegetable oil begins its thermal degradation at approximately 350°C. Accordingly, the percentage of biodiesel conversion at the prepared sample may be calculated using TGA analysis [22]. The thermal gravimetric analysis of the biodiesel produced sample at optimum conditions using KM micro-mixer was compared with its parent waste oil and the biodiesel prepared sample using high inlet reactant flow rate of 200 mL/hr. Figure 8a clarifies that the parent vegetable oil starts its thermal degradation at approximately 300°C compared with 120°C for that other two biodiesel samples (Figure 8b and c). Biodiesel completes vaporization at around 330°C. The recorded over all percentage weight loss with in biodiesel degradation temperature range from 120°C to 330°C for the optimum biodiesel prepared sample (Figure 8b) was 96.5% compared with 60% weight loss for the biodiesel

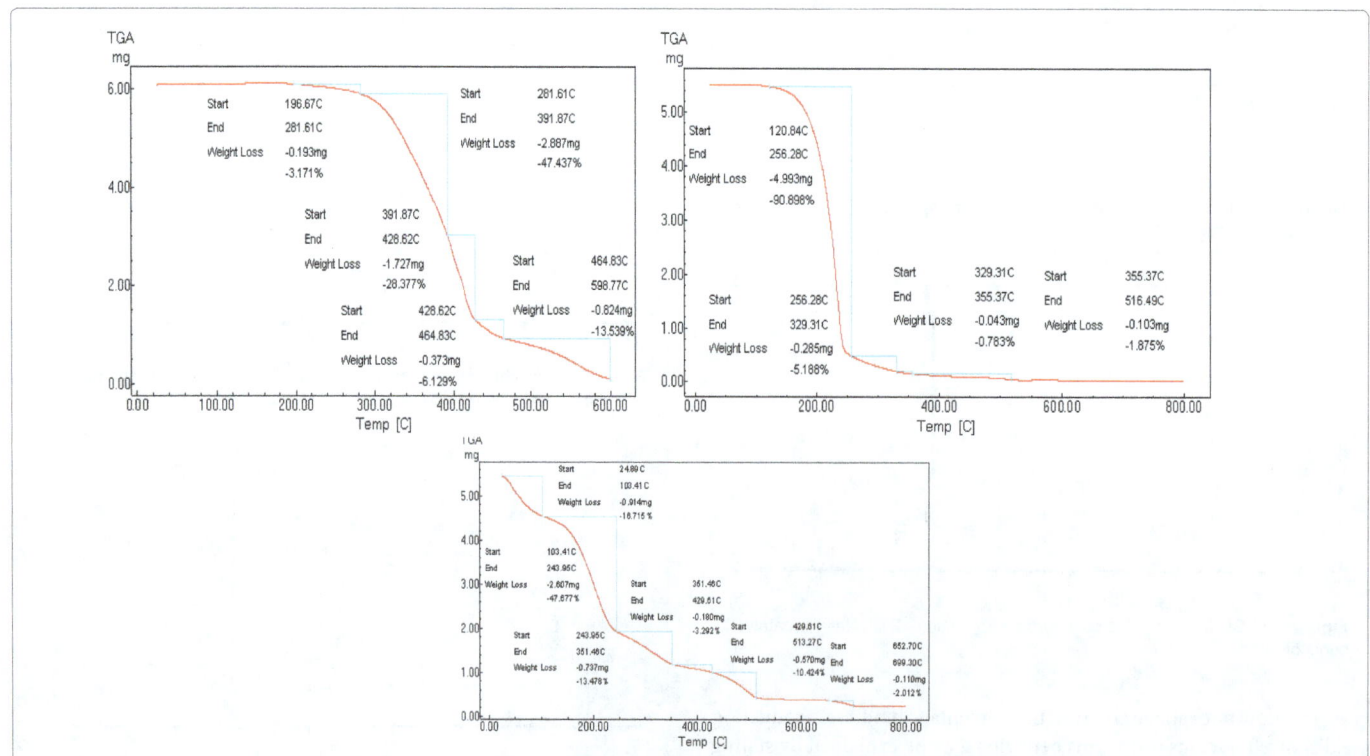

Figure 8: TGA diagram for waste vegetable oil. b. TGA diagram for Biodiesel produced at optimum conditions. c. TGA diagram for Biodiesel produced at high reactant flow rates.

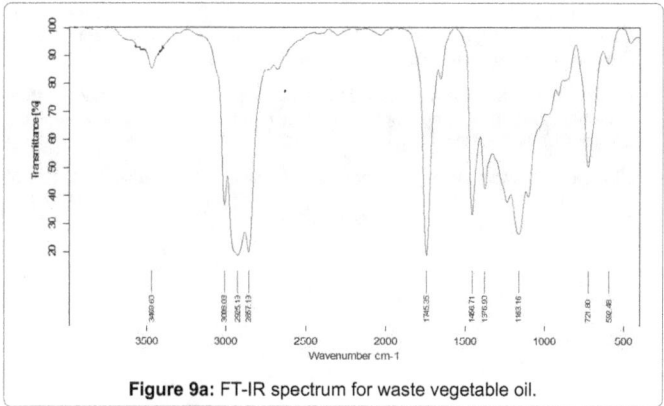

Figure 9a: FT-IR spectrum for waste vegetable oil.

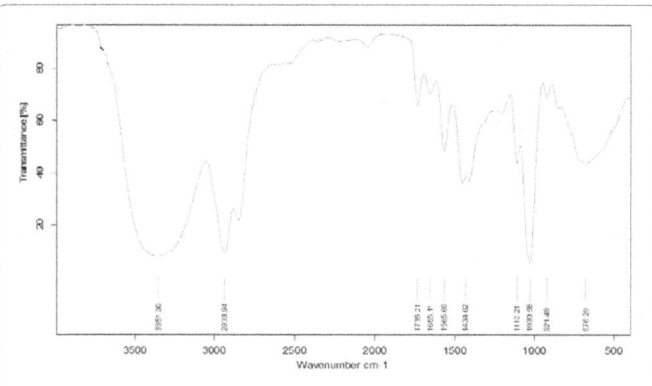

Figure 9b: FT-IR spectrum for biodiesel produced at optimum conditions.

prepared sample at the high reactant flow rate of 200 mL/h (Figure 8c). Moreover, it was indicated from Figure 8c that the biodiesel sample prepared at high reactant flow rates poses around 30% weight losses within the oil degradation temperature range, explained as unreacted oil [23]. These results confirm the successful biodiesel production in pure state at the optimum preparation conditions in contrast to biodiesel prepared at high reactant flow rates that contains remaining unreacted oil. Accordingly, the TGA analysis results confirm the previous GC-MS analysis results.

Fourier transform Infrared spectroscopy FT-IR: FT-IR spectrometry is a rapid and precise method for quantification of FAME. FT-IR spectrometry identifies the main functional groups presence at both the optimum produced biodiesel sample and its parent waste vegetable oil [24].

The most characteristics absorption peaks of the waste vegetable oil were indicated at figure (9a). The absorption peak appearing at 721 cm^{-1} is representative to -CH$_2$ rocking and the other one at 1745 cm^{-1} is representative to C=O ester stretch. Figure 9b showed the produced biodiesel absorption peaks appearing at 1434 cm^{-1} which is the methyl ester group (CO-O-CH$_3$) and the characterization peak at 1195 cm^{-1} corresponding to (C-O) ester peak. It was obvious the reduction of CH$_2$-O- groups in oil and the appearance of CH$_3$-O- vibrations in biodiesel. Also, the split of 1163 cm^{-1} in the oil sample into 1195 cm^{-1} and 1168 cm^{-1} in the biodiesel sample indicates the conversion of oil into biodiesel. The main difference between the two FTIR spectrums is related to the transformation of ester groups at the waste oil sample into methyl esters at the produced biodiesel [20].

Conclusions

This study investigated the use of KM micro-mixer in the production of biodiesel from waste vegetable oil. The effects of methanol/oil molar ratio, catalyst concentration, volumetric flow rates and the presence of a co-solvent on the transesterification reaction were examined. The study proved that the reaction can be completed giving higher percentage yield of biodiesel that reached 97%. In order to characterize the biodiesel product, its quality and quantity, GC-MS analysis has shown the characteristic peaks of FAMEs that ranged between C16-C18 methyl esters as main products with 98% yield. In addition, TGA and FTIR analysis were used to differentiate between the produced biodiesel and its parent oil. The results from both GC-MS and TGA methods were in good agreement regarding the quantity. The study confirms that the proposed KM micro-mixer designed with fourteen micro-channels was found effective for transesterification reaction completion. Thus, it can be employed in biodiesel production introducing many advantages over the batch reaction like time saving and higher yield and better conversion.

References

1. Alamu OJ, Waheed MA, Jekayinfa SO, Akintola TA (2007) Optimal transesterification duration for biodiesel production from Nigerian palm kernel oil. Agricultural engineering International CAIGE Journal 9: 1-11.

2. Demirbas A (2005) Biodiesel production from vegetable oils via catalytic and non-catalytic supercritical methanol transesterification methods. Progress in Energy and Combustion Science 31: 466-487.

3. Ali Y, Hanna M, Cuppett S (1995) Fuel properties of tallow and soybean oil esters. Journal of the American Chemical Society 72: 1557-1564.

4. Meher L, Vidyasagar D, Naik S (2006) Technical aspects of biodiesel production by transesterification-a review. Renewable and Sustainable Energy Reviews 10: 248-268.

5. Canakci M (2007) The potential of restaurant waste lipids as biodiesel feedstocks. Bioresource Technology 98: 183-190.

6. Matha MC, Kumarb SP, Chettyc SV (2010) Technologies for biodiesel production from used cooking oil -A review. Energy for Sustainable Development 14: 339-345.

7. Singh A, He B, Thompson J, van Gerpen J (2006) Process optimization of biodiesel production using different alkaline catalysts. Appl Eng Agr 22: 597-600.

8. Šalić A, Zelić B (2011) Microreactors portable factories for biodiesel fuel production. gorivaimaziva (fuels and lubricants journal) 2: 85-110.

9. Gude VG, Grant GE (2013) Biodiesel from waste cooking oils via direct sonication. Applied Energy 109 135-144.

10. Xie T, Zhang L, Xu N (2012) Biodiesel synthesis in microreactors. Green Process Synthesis 1: 61-70.

11. Noureddini H, Zhu D (1997) Kinetics of transesterification of soybean oil. Journal of the American Oil Chemists' Society 74: 1457-1463.

12. Kobayashi J, Mori Y, Kobayashi S (2006) Multiphase organic synthesis in microchannel reactors. Chem. Asian J 1: 22-35.

13. Wen Z, Yu X, Tu ST, Yan J, Dahlquist E (2009) Intensification of biodiesel synthesis using zigzag micro-channel reactors. Bioresource Technology 100: 3054-3060.

14. Nagasaw H, Aoki N, Mae K (2005) Design of a new micromixer for instant mixing based on the collision of micro segments. Chemical engineering & technology 28: 324-330.

15. Kulkarni MG, Dalai AK (2006) Waste cooking oil-an economical source for biodiesel: a review. Industrial and Engineering Chemistry Research 45: 2901-2913.

16. Bournay L, Casanave D, Delfort B, Hillion G, Chodorge JA (2005) New heterogeneous process for biodiesel production: A way to improve the quality and the value of the crude glycerin produced by biodiesel plants. Catalysis Today 106: 190-192.

17. Vyas AP (2011) Effects of Molar Ratio, Alkali Catalyst Concentration and Temperature on Transesterification of Jatropha Oil with Methanol under Ultrasonic Irradiation. Advances in Chemical Engineering and Science 1: 45-50.

18. Kumar D, Kumar G, Singh CPP (2010) Ultrasonic-Assisted Transesterification of Jatropha curcus Oil Using Solid Catalyst, Na/SiO$_2$. Ultrasonics Sonochemistry 17: 839-844.

19. Encinar JM, Gonzalez JF, Pardal A, Martinez G (2010) Transesterification of Rapeseed oil with methanol in the presence of carious co-solvents, Proceedings. Venice Third International Symposium on Energy from Biomass and Waste Venice 8: 839-844.

20. Tariq M, Ali S, Ahmad F, Ahmad M, Zafar M et al. (2011) Identification, FT-IR, NMR (^{1}H and ^{13}C) and GC/MS studies of fatty acid methyl esters in biodiesel from rocket seed oil. Fuel Processing Technology 92: 336-341.

21. McCurry JD (2011) Automation of a Complex, Multi-Step Sample Preparation Using the Standalone Agilent 7696A Work Bench. Agilent Technologies Publication Number 5990-7525EN.

22. Chand P, Reddy V, Verkade G, Wang T, Grewell D (2009) Thermogravimetric Quantification of Biodiesel Produced via Alkali Catalyzed Transesterification of Soybean oil. Energy and Fuels 23: 989-992.

23. Hamze H, Akia M, Yazdani F (2015) Optimization of biodiesel production from the waste cooking oil using response surface methodology. Process Safety and Environmental Protection 94: 1-10.

24. Oliveira JS, Montalvão R. Daher L, Suarez PAZ, Rubim JC (2006) Determination of methyl ester contents in biodiesel blends by FTIR-ATR and FTNIR spectroscopies. Talanta 69: 1278-1284.

17

Investigating the Effect of High Pressures and Temperatures on Corrosion Inhibition for Water-Based Drilling Fluids

Mahmood Amani*, Abdul Salam Abd, Abdulrahman Al-Hardan, Alireza Roustazadeh, Rommel Yrac

Texas A&M University, Qatar

Abstract

Corrosion is defined as the gradual degradation of materials as a result of reaction with their environment. In gas and oil sector and during the well life, equipment can corrode at any stage causing enormous losses in time and money. Inhibiting corrosion while drilling is considered to be one of the best solutions for corrosion, as chemical inhibition can be acquainted with the drilling fluid itself. The purpose of this paper is to study the effects of corrosion of steel pipes of different sizes (3.5', 4.5' and 5.5') and discuss the possible inhibition treatments.

In the lab, the change in thickness as well as weight was recorded. Then, the material properties were compared under the effect of diverse corrosion media conditions (temperature, base fluid, inhibitive fluid). Each sample was exposed to about 100 hours of corrosion. The final results showed that corrosion rate is the highest when only water based mud is present in the medium. However, corrosion rate is less severe under ambient temperature conditions contrary to High Pressure High Temperature (HPHT) medium were corrosion rate was severe (around 4.1 lbs/ft^2-year). It was also noted that the corrosion rate is inversely proportional to pipe thickness: as the diameter increases, the corrosion rate decreases accordingly.

When inhibitor (Conqor 404) is presented, it was observed that the rate of corrosion decreased drastically in the HPHT medium. Here, another relation can be established: as more inhibitor is injected into the mud, the corrosion rate reaches an economic margin where high concentration of the inhibitor is not feasible anymore.

Introducing inhibitor (OSL 1) to the mud instead of (Concor 404) will cause the corrosion rate to decrease to a low state, but higher than the rate achieved while using Concor 404 in same concentration. Mixing both inhibitors (OSL 1 + Concor 404) together will yield inhibition results better than using OSL 1 alone. Although Concor 404 was proven to be the best inhibitor when presented in considerably high concertation, it is recommended to use a combination of Concor 404 and OSL 1 as it has desirable results under HPHT conditions with feasible cost. The final decision depends merely on the metal type and limiting corrosion rate for that specific metal.

Keywords: Drilling; Corrosion; Degradation; Inhibitor; Efficiency

Introduction

Corrosion is defined as gradual degradation of metal caused by a chemical or electrochemical reaction with its environment. In oil and gas sector, components can corrode at any stage in the life of a field starting from drilling through to abandonment. Recent estimations showed that corrosion costs the oil industry in US yearly around $170 billion. In general, 50% of the operating expenditures in the drilling sector worldwide are for taming corrosion in drill pipe and down-hole equipment. On the other hand, "a corrosion inhibitor is a substance when added in a small concentration to an environment reduces the corrosion rate of a metal exposed to that environment. Inhibitors often play an important role in the oil extraction and processing industries where they have always been considered to be the first line of defence against corrosion" (SLB Glossary). Since the corrosion process is mostly due to the chemical reactions on the surface of the metal under HPHT condition, water-based mud properties used are hence of great effect. Mitigating corrosion is a very serious challenge for the oil and gas industry as it can't be totally eliminated. Because it is almost impossible to prevent corrosion, it is becoming more apparent that controlling the corrosion rate may be the most economical solution. Thus, the first step to tackle this problem is by determining the cause of the corrosion itself. This is vital as it helps understand the mechanism and the process behind corrosion to suggest more practical and helpful solutions. Nowadays, the urge to drill deeper to recover larger amounts of hydrocarbons exposes the drillers to High pressure/High Temperature (HPHT) zones. Wells with temperatures greater than 300F and pressures of 1000 psig are classified as HPHT

wells [1]. Moreover, using water based muds (WBM's) will increase the likelihood of a severe corrosion to happen under HPHT conditions.

Significance

This research is vital to the oil industry as it discusses a problem that has been ongoing for a long time. Corrosion is causing the oil companies a tremendous economic loss. In some cases, and in order to continue the drilling process, the tubing should be changed completely. There have been a lot of experiments on how to mitigate corrosion; however, the success rates are still low. Corrosion cannot be inhibited completely; however, the aim is to control it. Adding special additives to the drilling fluid or coating the tube with certain chemical are some ways to stop corrosion [2]. The aim of this research is to subject various metal samples of different grades to stress and strain similar to those caused by severe HPHT condition downhole, and compare the results of two main categories: treated samples and untreated samples.

*Corresponding author: Amani M, Texas A&M University, Qatar
E-mail: mahmood.amani@qatar.tamu.edu

The metals are expected to handle more stress when treatment is applied, proving the efficiency of the corrosion inhibitors compared to untreated samples. The challenge is to be able to manufacture an inhibitive chemical that can provide long term resistance as well as durable adherence on the steel [3].

Methodology

Corrosion needs 4 main elements to happen: anode, cathode, electrolyte (fluid) and external connection. In case any of these elements is absent, corrosion will not take place [4]. In our research, water based mud is the electrolyte of interest. In general, water helps in speeding the corrosion of metal where the steel itself serves as the external connection. The rate of corrosion depends purely on the grade of the metal and the generated potential due the dry cell effect. Oxygen (O_2) which plays an important role in corrosion is only present at the drilling stage and not in the producing formations. Water and Carbon dioxide (CO_2) injected at recovery operation can cause severe corrosion of completion string. Also, the presence of hydrogen sulfide (H_2S) gases at HPHT has a major role in the dynamics of corrosion. Thus, finding the effect of those elements (O_2, CO_2 and H_2S) in the corrosion process is very important to understand corrosion process.

First, the most common element that interfere in the corrosion process is the dissolved oxygen. The reaction of the iron to the oxygen contained in water will form iron rust (Figure 1). The equation below shows the reaction governing the process:

$$2\,Fe^{++} + \tfrac{1}{2}\,O_2 + H_2O = 2\,Fe^{+++} + 2\,OH^- \tag{1}$$

The formed rust is called ferric hydroxide which is characterized as insoluble.

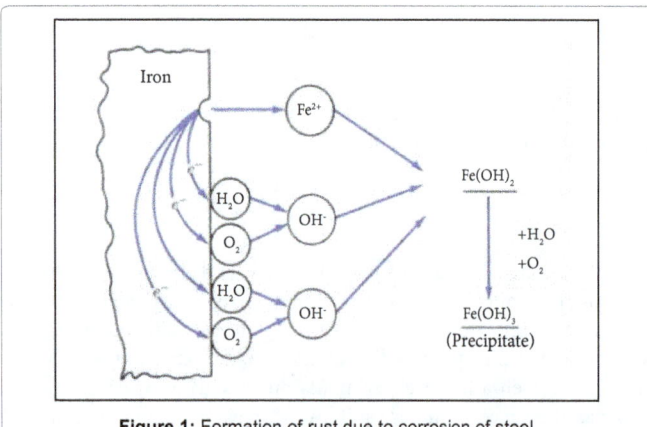

Figure 1: Formation of rust due to corrosion of steel.

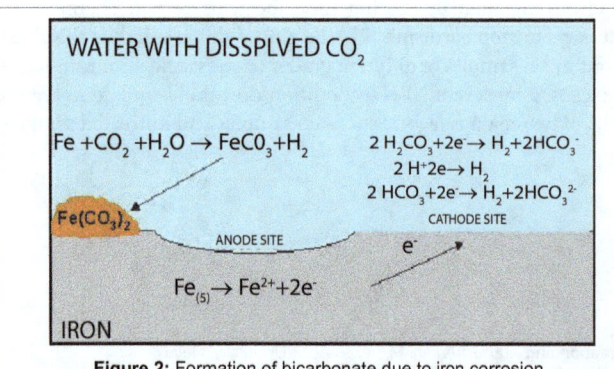

Figure 2: Formation of bicarbonate due to iron corrosion.

While drilling, we will have infinite oxygen as it is an open system operation, thus the corrosion will not cease. The corrosion rate is usually higher when the concentration of oxygen is low thus leading to rust that is impermeable to O_2 diffusion compared to that at high O_2 concentrations.

Second, the presence of dissolved CO_2 in water causes the steel to corrode where the rate of corrosion depends mainly on the quantity of CO_2 and O_2 present as well as temperature and composition of the material. This reaction is weaker than that induced by the presence of O_2 for equal quantities. In CO_2 based corrosion, carbon dioxide reacts with water to form bicarbonate (Figure 2). The following equation governs the reaction:

$$2CO_2 + 2H_2O + 2e{-} = 2HCO^{-3} + H_2 \tag{2}$$

This equation indicates that the CO_2, upon dissolving in water, acts like an acid. Thus, if we have dissolved CO_2 and O_2 combined in water, stronger corrosion rates will be observed.

Third, dissolved H_2S can be corrosive if dampness is present. The fact that H_2S is highly soluble in water creates a weak dibasic acid, which causes the degradation of iron because of the presence of oxygen. The reaction will be as follows:

$$H_2S + \tfrac{1}{2}\,O_2 = H_2O + 2S \tag{3}$$

The rate of corrosion is controlled by the concentration of the dissolved gas. If the dissolved H_2S is present in low quantities the corrosion will be severe. However, if the concentration of the dissolved H_2S is very, it might have reverse effect where it will act to inhibit the corrosion reaction.

When both CO_2 and H_2S are present, while having direct contact with O_2, there will create severe localized corrosion damage causing the material to crack and fail [5].

To go further with the influence of external factors on corrosion rate, we should consider the temperature of the medium. We should not only consider the fact that the reaction rate will increase simultaneously with temperature, but we should account for solubility and viscosity. The solubility of gases in water will decrease with temperature increase as well as the viscosity. However, this is scenario is not true in all cases. For example, when dissolved oxygen is present, the corrosion rate will increase with temperature till a critical point then it will start decreasing with oxygen solubility. If the system is open, the oxygen will escape. Otherwise, the oxygen will be trapped causing the rate of corrosion to increase at high temperatures.

As a result of corrosion, weight is expected to decrease and the weight loss is estimated according to the following equation:

$$\Delta m = W_{initial} - W_{finial} \tag{4}$$

Where $W_{initial}$ and W_{final} are the weight of each sample before and after corrosion in grams respectively

This allows for the estimation of a parameter: the corrosion rate according to equation (5) presented below:

$$CR = \frac{\Delta m}{\Delta t}\ (grams/day) \tag{5}$$

Where Δt is the time of the corrosion process in days.

The inhibitor efficiency is then estimated using the equation presented in the following:

$$Inhibitor\ Efficiency = \frac{CR\ uninhibited - CR\ inhibited}{CR\ uninhibited} \times 100 \tag{6}$$

Equipment and materials used

The equipment used in the experiments are:

a. (20x) 500 mL plastic bottles

b. Weight balance (1/1000g accuracy)

c. Plastic Trays

d. OFITE HTHP fluid loss double end- cap

e. Conducting wires

f. Steel modified-stem

g. Heating jacket

The materials used in the experiment are:

a. (1x) Corrosion rings of size 3.5"

b. 0.5%, 1%, 1.5% HCL solution

c. (1x) Corrosion rings of size 4.5

d. 0.5%, 1%, 1.5% Concor 404 solution

e. (3x) Corrosion rings of size 5.5

f. 0.5%, 1%, 1.5% OS1-L solution

g. Laboratory formulated Water Based Mud

Preliminary Procedure (Pre-experimentation)

Safety measurements

This experiment requires direct contact with synthetic chemicals which can hurt whoever is utilizing it. Inspecting the material safety data sheet (MSDS) of the chemicals utilized in this research is required. The MSDSs proved all the vital information on how to deal with the chemicals and how to act upon being exposed to those solutions. The MSDSs for the chemicals that will be used in this research will be provided in the Appendix [6].

Phase 1

Solution preparation: In the laboratory, a drilling water based fluid was prepared using mainly Drill water, Barite and Bentonite. Other additives such as NaCl, Flowzan, Soda Ash and Fine $CaCO_3$ were used as well. The properties of the formulated mud are summarized in Table 1 below. Different corroding solutions with varying composition were prepared and stored in plastic vessels as shown in Table 2.

Material preparation: The samples were prepared specific for each medium. The 3 corrosion rings were cut into 4 pieces in Texas A&M University – Qatar machine shop. The initial weight of each sample was

Rheology Temperature	120°F
Gel at 10 seconds	14 lbs/100 sq ft
Plastic Viscosity (PV)	24 cP
Yield Point (YP)	30 lbs/100 sq ft
YP/PV Ratio	1.25
Funnel Viscosity	59sec/qt
HTHP	230°F/500 psi Differential Pressure
API Fluid Loss	100 psi/30 minutes
Low gravity solids	8%
Mud Weight	12.00 ppg
pH	9.8
Chlorides	200 mg/L

Table 1: Initial test results for the water base mud properties.

Medium	Inhibitor Type	Inhibitor Concentration	Temperature
1	None – base fluid	-	Ambient
2	None – base fluid	-	Ambient
3	None – base fluid	-	Ambient
4	None – base fluid	-	HPHT - 230°F
5	None – base fluid	-	HPHT - 230°F
6	None – base fluid	-	HPHT - 230°F

Table 2: Different medium used in the reserach set-up for phase 1.

Medium	Inhibitor Type	Inhibitor Concentration	Temperature
1	Concor 404	0.5	HPHT - 230°F
2	Concor 404	1	HPHT - 230°F
3	Concor 404	1.5	HPHT - 230°F
4	OS1-L	0.5	HPHT - 230°F
5	OS1-L	1	HPHT - 230°F
6	OS1-L	1.5	HPHT - 230°F
6	Conqor 404+ OS1L	0.5	HPHT - 230°F
8	Conqor 404+ OS1L	1	HPHT - 230°F
9	Conqor 404+ OS1L	1.5	HPHT - 230°F

Table 3: Different medium used in the research set-up for phase 2.

measured using a high accuracy electronic balance. The purpose of the Initial weights is to determine the loss after all exposure and treatment operation. This will also help us understand how the condition of each set-up affects the corrosion rate. One sample (1/4 of a corrosion ring) was immersed in each of the mediums described in Table 2 [7].

Phase 2

Solution preparation: Table 3 below shows the solutions prepared for Phase 2 experimentation.

Material preparation: Only the corrosion ring of size 5.5" is used in this phase. Two corrosion rings were into 4 identical pieces in Texas A&M University – Qatar machine shop. Reasons for only using this size will be presented in the analysis.

Experiment conduction: Each sample of mud will contain each size of corrosion ring to determine corrosion accumulation and inhibition based on the size and type of corrosion ring. After approximately 100 hours of exposure, all corrosion rings will undergo the same procedures of inspection for the evaluation of the results.

Results and Discussion

Experiment 1

The weight of each sample was recorded before being immersed in the corrosion mediums and after their removal and cleaning. Tables 4-6 shows the data collected corresponding to corrosion rings of sizes 3.5", 4.5" and 5.5" respectively.

Experiment 2

Similarly, the weight of each sample was recorded before and after immersion and removal of each sample from the corrosion media. Tables 4-6 show the data collected corresponding to corrosion rings of sizes 3.5", 4.5" and 5.5" respectively (Tables 7-9).

Experiment 1: From the weight data collected, the weight was estimated using equation (4) and the corrosion rates and inhibitor efficiency were also estimated using equations (5) and (6) of the theory section respectively. The results for corrosion ring sizes 3.5", 4.5" and 5.5" are shown respectively in Tables 10-12.

Two samples of 3.5" corrosion rings were tested under ambient and HPHT conditions. Only water based mud was presented without any type of inhibition. The corrosion rate was severe in HPHT medium (4.1 lbs/ft 2-year) compared to ambient temperature conditions (2.2 lbs/ft 2-year). However, the corrosion rate was determined to be less

Corrosion Ring of Size 3.5"		
Medium	$W_{initial}$ (grams)	W_{final} (grams)
1	16.962	16.320
4	16.050	14.832

Table 4: Weight data for corrosion ring 3.5".

Corrosion Ring of Size 4.5"		
Medium	$W_{initial}$ (grams)	W_{final} (grams)
2	20.099	19.325
5	19.482	18.211

Table 5: Weight data for corrosion ring 4.5".

Corrosion Ring of Size 5.5"		
Medium	$W_{initial}$ (grams)	W_{final} (grams)
3	57.840	56.520
6	61.193	59.011

Table 6: Weight data for corrosion ring 5.5".

Corrosion Ring of Size 5.5" with Concor 404		
Medium	$W_{initial}$ (grams)	W_{final} (grams)
1	59.090	58.001
2	62.295	61.416
3	60.283	59.928

Table 7: Weight data for corrosion ring 5.5" with concor 404.

Corrosion Ring of Size 5.5" with OS1-L		
Medium	$W_{initial}$ (grams)	W_{final} (grams)
4	60.790	59.389
5	58.708	57.354
6	60.855	59.884

Table 8: Weight data for corrosion ring 5.5" with OS1-L.

Corrosion Ring of Size 5.5" with Concor 404 + OS1-L		
Medium	$W_{initial}$ (grams)	W_{final} (grams)
7	59.735	58.382
8	59.151	57.98
9	59.819	59

Table 9: Weight data for corrosion ring 5.5" with concor 404 + OS1-L.

Corrosion Ring of Size 3.5"			
Medium	Weight loss (grams)	Exposure Time (hours)	Corrosion rate (lbs/ft²-year)
1	0.642	99.08	2.2
4	1.218	99.87	4.1

Table 10: Corrosion rates for corrosion ring 3.5".

Corrosion Ring of Size 4.5"			
Medium	Weight loss (grams)	Exposure Time (hours)	Corrosion rate (lbs/ft²-year)
2	0.774	99.08	2
5	1.271	99.87	3.2

Table 11: Corrosion rates for corrosion ring 4.5".

Corrosion Ring of Size 5.5"			
Medium	Weight loss (grams)	Exposure Time (hours)	Corrosion rate (lbs/ft²-year)
3	1.32	99.08	1.8
6	2.182	99.87	2.9

Table 12: Corrosion rates for corrosion ring 5.5".

Corrosion Ring of Size 5.5" with Concor 404				
Medium	Weight loss (grams)	Exposure Time (hours)	Corrosion rate (lbs/ft²-year)	Inhibitor Efficiency (%)
1	1.089	99.9	1.5	48.2
2	0.879	99.78	1.2	58.6
3	0.355	100.67	0.5	82.7

Table 13: Corrosion rates for corrosion ring of size 5.5" with concor 404.

Corrosion Ring of Size 5.5" with OS1-L				
Medium	Weight loss (grams)	Exposure Time (hours)	Corrosion rate (lbs/ft²-year)	Inhibitor Efficiency (%)
4	1.401	100.57	1.9	34.4
5	1.354	100.43	1.8	37.9
6	0.971	100.22	1.3	55.1

Table 14: Corrosion rates for corrosion ring of size 5.5" with OS1-L.

Corrosion Ring of Size 5.5" Concor 404 + OS1-L				
Medium	Weight loss (grams)	Exposure Time (hours)	Corrosion rate (lbs/ft²-year)	Inhibitor Efficiency (%)
7	1.353	100.97	1.8	37.9
8	1.171	100.73	1.6	44.8
9	0.819	100.5	1.1	62

Table 15: Corrosion rates for corrosion ring of size 5.5" with concor 404+OS1-L.

for larger corrosion ring sizes for both mediums with higher rate in the HPHT medium.

The bargraph below allows us to draw a general conclusion relating corrosion rate with corrosion ring size. It is noticeable that the corrosion rate decreases with bigger size as it gets hard to corrode all the metal for same exposure time. Increasing the temperature will accelerate the corrosion process especially for small ring size (3.5") leading to a high and severe corrosion rate [8].

Experiment 2: Based on experiment 1 results, we assume that the corrosion rate decreases with size but increase with temperature. Thus, in this experiment we focused on the large ring size only under HPHT conditions to test different inhibition treatments while expecting to project the results onto the other sizes. Similar to the calculations done to obtain the results of experiment 1, weight loss, corrosion rate, and inhibition efficiency were obtained for the studied samples. Tables 13-15 show the impact of using different inhibitors combinations with different concentration on the corrosion rate (Figure 3).

In this experiment two different inhibitors were used. Inhibiting with concor 404 was the most efficient with increasing efficiency of inhibition up to 82 % for high concentrations. The corrosion rate without inhibition was around 2.9 lbs/ft 2-year; almost 6 times higher than when maximum inhibition is applied. On the other hand, using OS1-L, another type of inhibitor, reduced the corrosion rate to 1.3 lbs/ft 2-year for 1.5 ppb inhibitor concentration. This rate is almost three times higher than the rate when Concor 404 is applied. On the other hand, combining both inhibitor will yield and efficiency of 62% at maximum inhibitor concentration; a value that lies between Concor 404 and OS1-L efficiencies. A plot of corrosion rate versus different types of inhibitors and concentrations can be found below [9] (Figure 4).

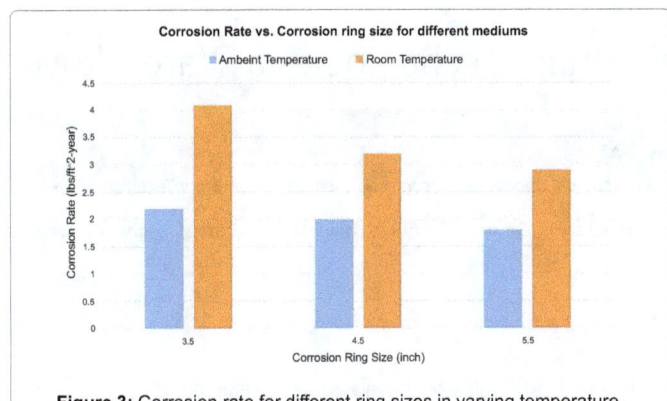

Figure 3: Corrosion rate for different ring sizes in varying temperature.

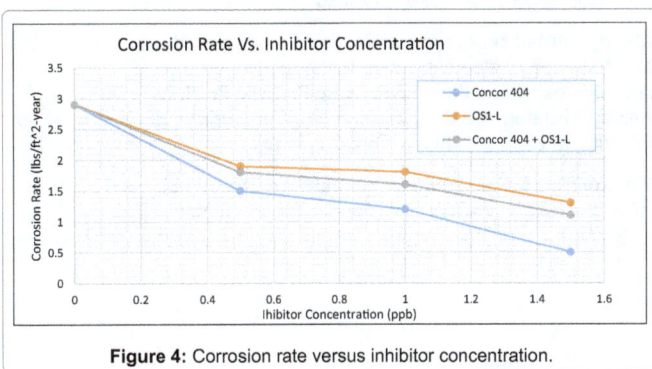

Figure 4: Corrosion rate versus inhibitor concentration.

In order to recommend an inhibition treatment, a balance between cost and efficiency should be achieved. While one of the treatments is the most cost-effective, others could be enhanced to be more cost-effective. The goal is to achieve the desired level of service at the least cost. In our case, the price of Concor 404 is estimated to be almost 10 time more than that of OS1-L. The acceptable rate of corrosion can range from 1 to 2 lbs/ft^2-year while Concor 404 achieves 0.5 bs/ft^2-year at 1.5 ppb concentration. Having a recommended concentrations of ppb, and combing Concor 404 and OS1-L will yield 1.6 2 lbs/ft^2-year corrosion rate which is tolerable. This combination saves money while maintaining integrity of the corroded metal.

Conclusions

This research focuses on corrosion inhibition treatments for various pressure and temperature conditions. Steel pipes are the most commonly used for oil and gas wells due to their low cost and high performance. During drilling, these steel materials are under very corrosive conditions due to elevated temperatures and pressures as well as exposure to CO_2 and O_2. Thus, an appropriate inhibitor should be used to prevent corrosion from affecting the integrity of the pipes. Various inhibitors were presented which are used as inhibitive chemicals for steel materials under HPHT conditions.

The following conclusions were made:

a. Efficiency is not the sole factor in treatment selection.

b. Detailed economic analysis including the cost effectiveness is required to successfully manage corrosion.

c. The most efficient and economic treatments amongst the different tested inhibitors was the chemical inhibition through using a mixture of Concor 404 and OS1-L.

Some of the limitations faced throughout this experiment are listed and particularized below:

a. Monitoring the values of thickness and weight periodically while performing the experiment was not possible due to time constraints. Cleaning each sample multiple times and measuring data in consistent time steps is a tedious process. However, if this was done, it would have allowed as to predict a correlation that can relate the corrosion rate with time under certain pressure and temperature conditions.

b. Introducing H_2S to the experiment was not allowed in Texas A&M Labs due to safety concerns. This would have allowed us to simulate downhole conditions of a reservoir in credible criteria.

On other note, some errors were also observed in the experiment mainly due to the interaction with the studied samples which might have altered the results a little bit. These errors, however, won't change the general trend of the observed results.

Recommendations

The following is recommended: The research should be further developed to study the effects of H_2S and CO_2 on corrosion rates along with temperature and pressure. A study for the composition of Concor 404 and OS1-L should be done to see why OS1-L is hindering the efficiency of Concor 404 compared to its presence alone.

Different treatment methods other than chemical inhibition should be evaluated and compared for efficiency and cost effectiveness.

Acknowledgment

"This publication was made possible by a UREP award [UREP No: 17-133-2-034] from the Qatar National Research Fund (a member of The Qatar Foundation). The statements made herein are solely the responsibility of the author."

References

1. Brownlee JK, Flesner KO, Riggs KR, Miglin BP (2005) Selection and qualification of materials for HPHT Wells. Society of Petroleum Engineers.

2. Godwin W, Ogbonna J, Boniface O (2011) Advances in mud design and challenges in HPHT Wells. Society of Petroleum Engineers.

3. Young K, Alexander C, Biel R, Shanks E (2005) Updated design methods for HPHT Equipment. Society of Petroleum Engineers.

4. Shadravan A, Amani M (2012) HPHT 101 - What petroleum engineers and geoscientists should Know about high pressure high temperature wells environment. Journal of Energy Science and Technology 4: 36-60.

5. Al-Tammar JI, Bonis M, Choi HJ, Al-Salim Y (2014) Saudi aramco downhole corrosion/scaling operational experience and challenges in HP/HT Gas Condensate Producers. Society of Petroleum Engineers.

6. Amani M, Hjeij DA (2015) Comprehensive review of corrosion inhibition methods in the oil and gas industry. Paper SPE-175337-MS presented at the 2015 SPE Kuwait Oil & Gas Show and Conference, Kuwait.

7. Bland RG, Mullen GA, Gonzalez YN, Harvey FE, Pless ML (2006) HPHT drilling fluid challenges. Society of Petroleum Engineers.

8. Lee J, Shadravan A, Young S (2012) Rheological properties of invert emulsion drilling fluid under extreme HPHT conditions. Presented at the IADC/SPE Drilling Conference and Exhibition, San Diego, California, U.S.A.

9. Weintritt DJ, Hughes RG (1965) Factors involved in high-temperature drilling fluids. Society of Petroleum Engineers.

18

Results of Introducing Innovative Thermal Mining Technologies at Yaregskoye Oilfield

Durkin SM[1], Moroziyuk OA[1], Ruzin LM[2], Polishvayko DV[2] and Abzaletdinov GA[3]

[1]Department of Exploitation and Development of Oil and Gas Fields and Subsurface Hydromechanics, Candidate of Technical Sciences, Ukhta State Technical University, Ukhta, Russia
[2]Department of Exploitation and Development of Oil and Gas Fields and Subsurface Hydromechanics, Doctor of Technical Sciences, Ukhta State Technical University, Ukhta, Russia
[3]Department of Petroleum Engineering, Louisiana State University, Baton Rouge, USA

Abstract

Presented in this paper are results of the introduction of a modernized single-horizon system of development with heat-insulated pipes at the inclined block "Северный" ("North") Oil Mine № 2 on the Substation Control «SC-2 bis» (ОПУ-2бис) and Substation Control «SC-3 bis» (ОПУ-3бис) pilot plots of Yaregskoye Oilfield.

In plots SC-2 bis and SC-3 bis, thermometry is systematically carried on the control wells to determine the temperature distribution in the reservoir, as well as the rational distribution of steam injection in the developed plot. A method for determining the effectiveness of injection wells on pilot plots was developed by staff at Ukhta State Technical University. In line with the method, studies which helped to determine the acceleration of underground injection wells necessary for calculating the volume of steam injection at each pilot plot were carried out.

Also, regular sampling of water extracted from wells for determining the presence of chlorides is carried out. Based on the analysis of these samples, wells in which have inflows of reservoir water from the aquifer are detected and develop measures to isolate them are then developed.

Keywords: Yaregskoye oilfield; Viscosity; Energy; Oil recovery

Introduction

Yaregskoye oilfield is located in Komi Republic 25 km south-west from the city of Ukhta which forms part of the Timan-Pechora oil and gas province. Crude Oil from Yaregskoye oilfield is heavy and has a high viscosity (oil density - 945 kg/m^3, oil viscosity at reservoir conditions - 16,000 mPa·s). The oilfield lies below ground level at a depth of 140 - 200 m. The geological reserves of the oilfield are 300 million tones [1].

Yaregskoye oilfield (Figure 1) consists of the following plots: Yaregskoye, Lyaelskoye and Vezhavozhskoye. Currently, only Yaregskoye plot is under industrial development while at the Lyaelskoye plot industrial pilot projects are being carried out for testing SAGD technology.

Yaregskoye oilfield was discovered in 1932. Since 1935 experimental development of it has been going on from the surface of vertical wells in a triangular grid on 2 plots of land with areas of 284,000 and 150,000 M^2, respectively. Experimental works continued until 1945, and for 10 years oil recovery didn't exceed 2%.

Mining method for the development of the Yaregskoye area began in late 1939. A dense mesh of polygon-penetrating production wells were drilled out of the fields drifts systems running along the cap of the production formation to reach the deposit. Oil production was possible due to the internal energy of the reservoir, as well as by the energy of the dissolved gas. In 1954, the technology of steadily inclined wells was also tested and used for drilling a grid of polygon-penetrating wells

Figure 1: Map of Yaregskoye oil-field.

*Corresponding author: Abzaletdinov German Albertovich, Department of Petroleum Engineering, Louisiana State University, Baton Rouge, USA, E-mail: gabzal1@lsu.edu

from the galleries located at the base of the formation in the area. The mining method's advantage was the close proximity of the oil reservoir which allowed the maximum use of its internal energy, but the final oil recovery using both technologies did not exceed 4% to 6%.

In 1972, thermal mining technology development started. Though innovative at the time, it is still unique till date. With this technology, steam is injected into the reservoir to heat the oil and then the heated oil with lower viscosity flows into the production wells drilled from the drilling galleries at the reservoir base. During that time several systems

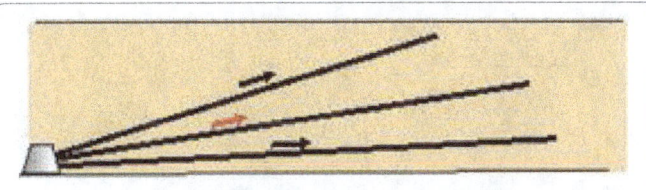

Figure 2: Single horizon system of development.

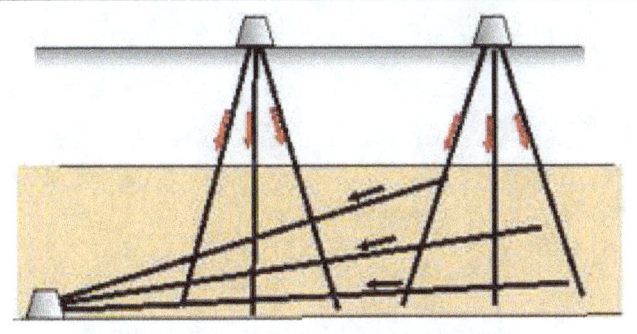

Figure 3: Double-horizon system of development.

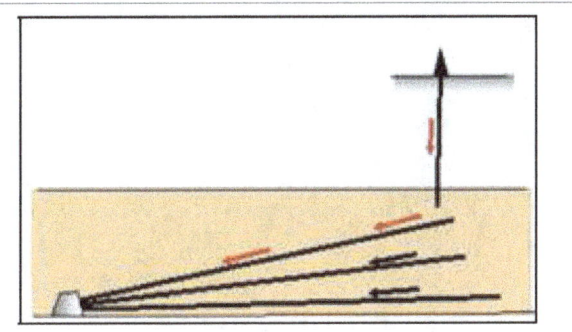

Figure 4: Underground-surface system of development.

of thermal mining development were tested and implemented. Using all the systems the area was developed in blocks, with each block having an inclination across the drilling gallery to the base of the formation from where the entire mesh of drilled polygon-penetrating production wells originates. The basic difference between the development systems is different arrangement of the injection wells. One of the first thermal mining technologies tested at the Yaregskoye oilfield was the single-horizon system (Figure 2). Its feature is that the production and injection wells can be drilled from a single drilling gallery, located at the base of the production formation. The wells are arranged in pairs, one of them in the upper portion of the formation through which steam is injected, and another at the bottom, through which heated oil flows.

The results of the application of this system proved its higher efficiency, but further use had to be abandoned. One of the key reasons for the refusal of the widespread introduction of steam was its breakthrough into the drilling gallery and that does not allow for the creation a safe working environment in the petroleum mine. The presence of two double-horizon systems on the field drifts passing over the site being developed allows for the drilling of injection wells using the system of polygon-penetration (Figure 3). The advantage of this system is a uniform coverage of the reservoir area by steam at a low injection pressure. This system was one of the most effective, but it was abandoned because of the high capital costs of tunneling field drifts.

To date, the main technology used on an industrial scale in the Yaregskoye oilfield is the underground-surface system (Figure 4). This system involves the drilling of injection wells from the surface. Its implementation allowed for the increase in the pace of development of the oilfield as well as raises the level of production at the mine, but the underground and surface system also is not without drawbacks. Its main drawback – the injection of steam into the reservoir under high pressure (10-12 atm.), which does not put into consideration the geological and physical characteristics of the deposit (high natural fracture). Included amongst the significant disadvantages is the high capital and operating costs for the drilling and maintenance of surface wells.

Given the experience of developing Yaregskoye oilfield and large supplies of heavy oil, the actual problem is to improve the systems of thermal mining development to increase oil recovery rates and reduce steam-oil relationship.

Material and Methods

The employees of Ukhta State Technical University developed a more effective technology for the development of Yaregskoye deposits that allows for an increase in the rate of steam injection and oil

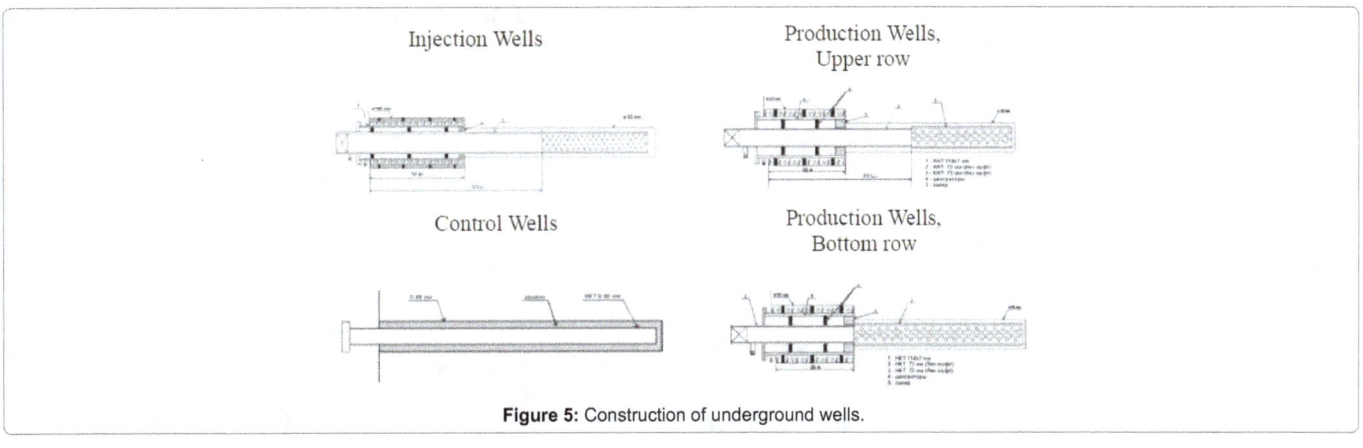

Figure 5: Construction of underground wells.

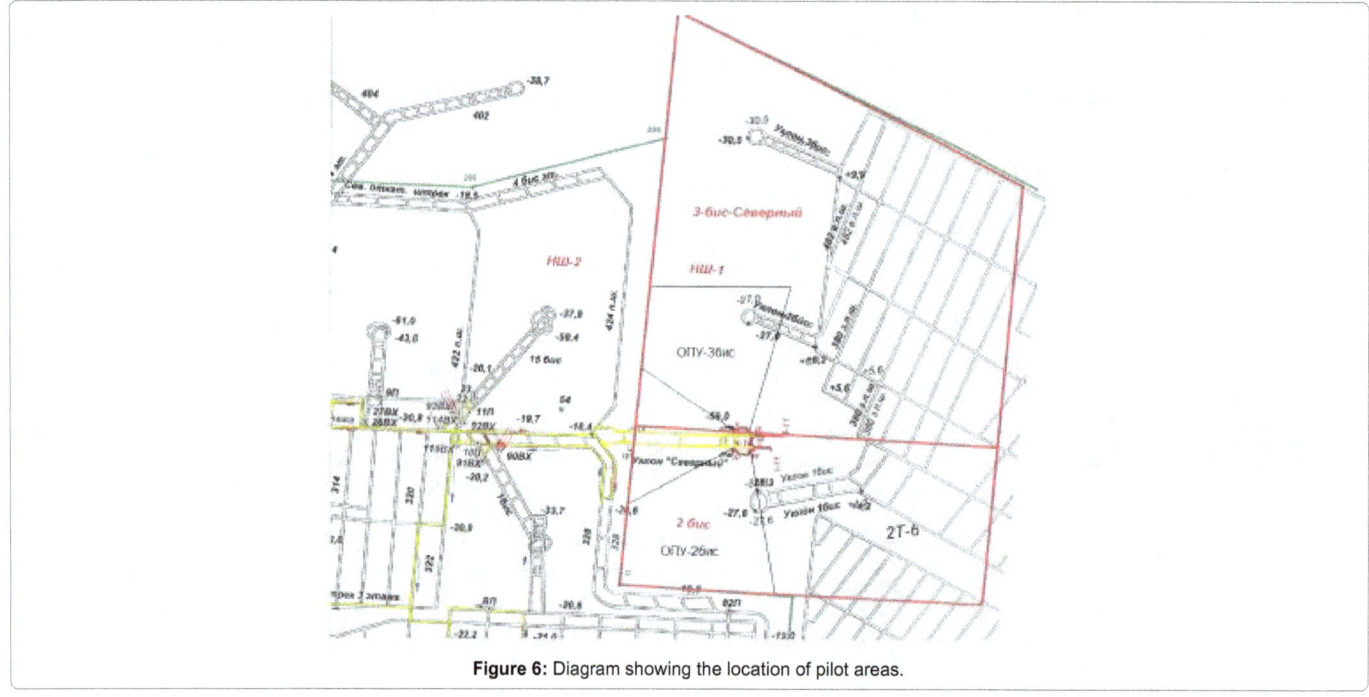

Figure 6: Diagram showing the location of pilot areas.

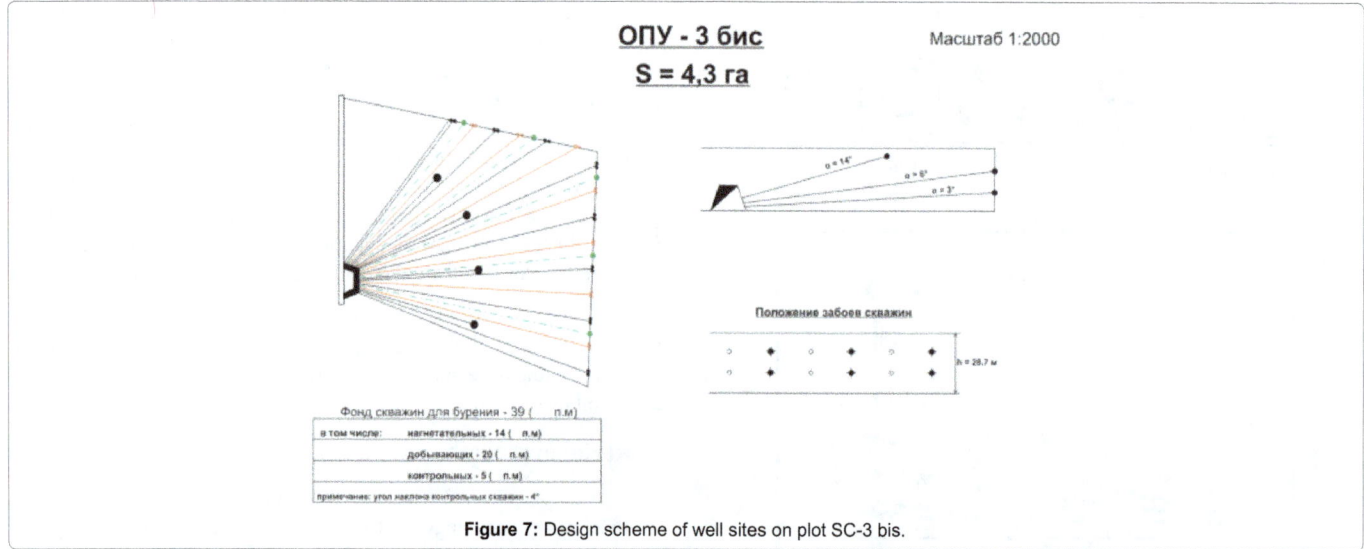

Figure 7: Design scheme of well sites on plot SC-3 bis.

recovery. A single-horizon system was adopted at the beginning as the technology is the best taking into account the geological features of deposits and reduced capital costs. Improvement to the technology was made possible due to new constructions of underground wells (Figure 5). To exclude steam leakages into the production gallery, production and injection wells are cased and cemented 50 meters from the top of specialized heat-insulated pipes.

Optimal parameters of the wells, their location and the length of the perforation were selected based on numerical simulation. According to the results of modeling, the optimal length of the perforation of injection wells was 2/5 the length of the holes, 3/5 the length for the top row of wells and the entire length of the lower production wells. Also, the project includes the development of control wells which were cased and cemented all through their entire lengths.

In 2010, "Lukoil-Komi" decided to conduct pilot projects for

testing the technology. On the recommendation of NSHU "Yareganeft" for testing the upgraded single-horizon development system, the inclination "Северный"/"North" (Figure 6) oil mine № 2 was chosen. It consists of 2 plots of Substation Control (SC)–2 bis/(ОПУ-2 бис) and Substation Control (SC)–3 bis/(ОПУ-3 бис). The plot areas are 4.7 hectares and 4.3 hectares respectively.

It should be noted that organization of control of the trajectory of wells in mine conditions is complicated by the following reasons: inability to use standard methods due to fire and explosion hazards and high costs of well logging using modern telemetry systems in explosion-proof instances (Figure 7).

In order to adequately monitor the trajectory condition of horizontal wells, a simple method for determining the zenith angle of wells during the process of drilling was proposed by Ukhta State Technical University (V.F. Buslaev) in 2006 which involves measuring the hydrostatic pressure of the liquid column in the wellbore using a manometer. This method is called gauge pressure method. Control of zenith angle using gauge pressure (manometer) method is carried out in most of the wells. The disadvantages of using gauge pressure (manometer) method include inability to accurately determining the azimuth of the trajectory and lack of information about the behaviour of the trajectory of wells after reaching the highest point.

Given the nature of the work accomplished, the project involved a directional survey of wells in the pilot area using modern inclinometers which aid in determining the position of the wells trajectories not only vertically but also in a horizontal plane i.e., azimuth. Table 1 shows the number and scope of well testing using an inclinometer.

As the table shows, 43.6% of well stock on plot SC-3 bis was tested, including 86% of injection and 25% of producing wells. Figure 8 shows a comparison of the design and trajectories studied using inclinometer in plot SC-3 bis. The solid lines show the trajectory of the tested wells using an inclinometer.

As can be seen from Figure 8 actual well trajectories deviate significantly from the design trajectories. Figures 9-10 show dynamics of technological parameters during development on plot SC-3 bis. Development of the pilot area began in October 2012 (Figures 9 and 10).

Current developmental indicators of plot SC-3 based bis site dates analysis was done are as follows: average daily oil production – 45-50 t/day, daily average of steam injection – 45.2 tons/day, the current steam-oil ratio – 1.0 t/t, average temperature of the produced fluid – 85°C to 89°C. In just 37 months (as at 01.12.15) 113,6 thousand tones of steam was pumped into the reservoir and at its expense 56,5 thousand tons

Plot №	Well category	Stock	Number of tested wells	Part of wells tested, %
SC-3 bis*	Injection	14	12	85.7
	Production	20	5	25.0
	Control	5	0	0.0
	Total	39	17	43.6
*Inclinometers for wells	Production № - 9, 11, 12, 17, 18		Injection № - All except 1, 2	

Table 1: Coverage of well tests using an inclinometer.

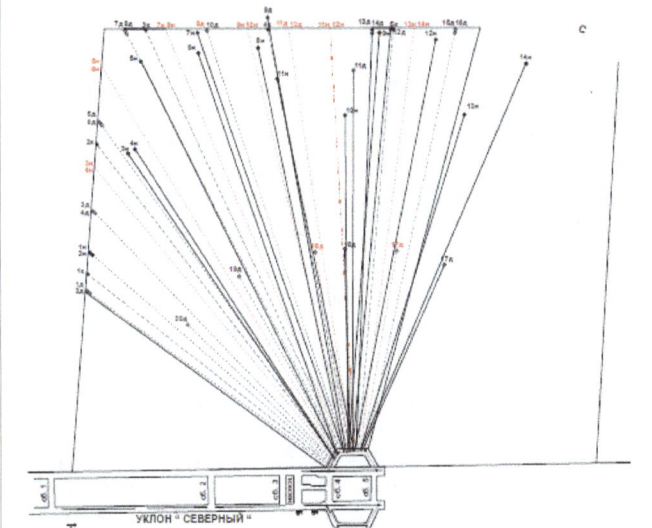

Figure 8: Comparison of the design and actual location of the trajectory of wells in plot SC-3 bis.

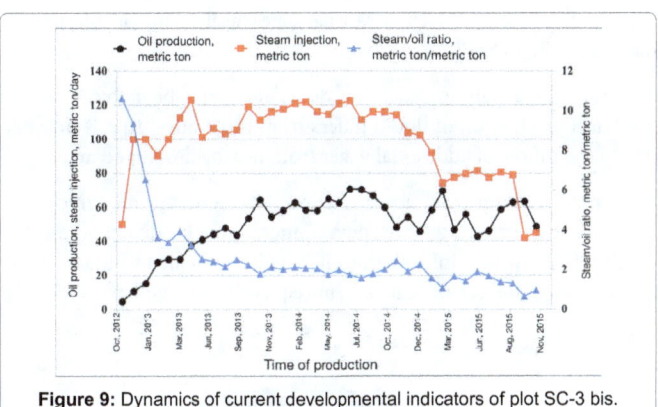

Figure 9: Dynamics of current developmental indicators of plot SC-3 bis.

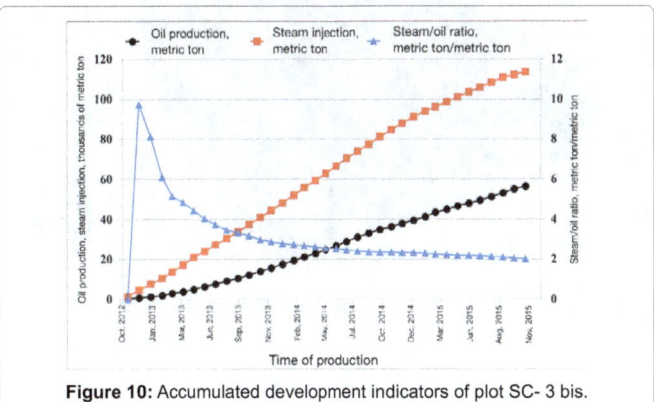

Figure 10: Accumulated development indicators of plot SC- 3 bis.

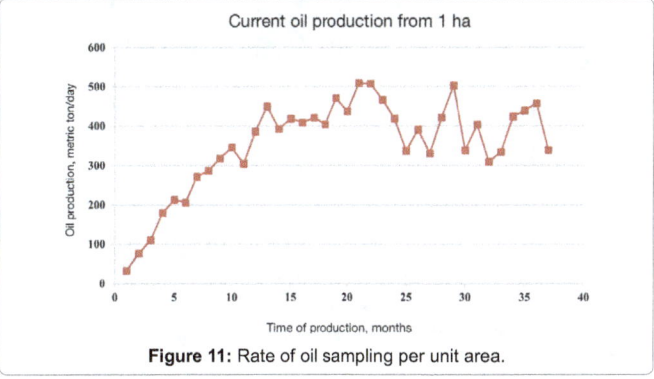

Figure 11: Rate of oil sampling per unit area.

of oil was produced. Oil recovery in the developed areas has reached 38.0%, including 27.4% due to heat exposure. Accumulated steam-oil ratio was 2,02 t/t.

In Figure 11 contains a graph of oil production rate per unit area is shown. Figure 11 shows that a year after steam injection, the pace of oil withdrawal pace from 1 hectare is almost identical and averages of 350-400 tons/month.

In 2010, as part of a mini-project, technological parameters for development of experimental plots were first predicted by numerical simulation [2]. Afterwards, industrial material was accumulated during the operation of the experimental plots. It formed basis for adaptation of the experimental plots models [3], which made it possible to clarify forward-looking information presented in the mini-project. Figures 12 and 13 shows a comparison of the design (mini-project), in view of adaptation (model) and the actual oil recovery factor for the experimental plots. CMG software package was used for calculating the indicators (Figures 12-14).

From the results of adaption done, it should be noted that the historical production of liquid differs from the model. That is possible due to the inflow of additional water from nearby developed areas.

As seen from the above figures, the quality of forecasting technological parameters of development greatly depends on the quality of the initial information. It is quite difficult to obtain at the initial design stage, and can be subsequently obtained by analyzing

Figure 12: Three-dimensional geological filtration model of plot SC-3 bis.

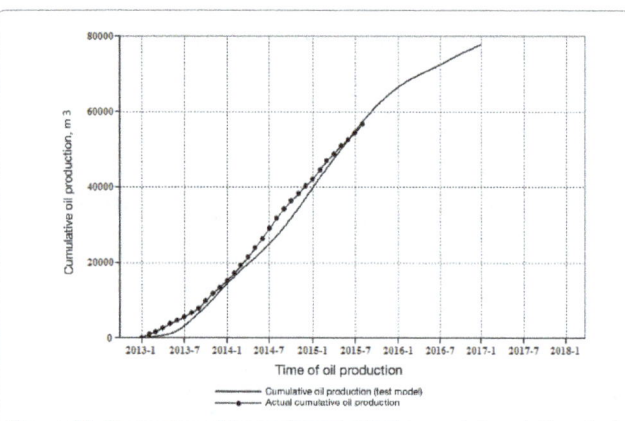

Figure 13: Comparison of the model and actual accumulation of oil production at plot SC–3 bis.

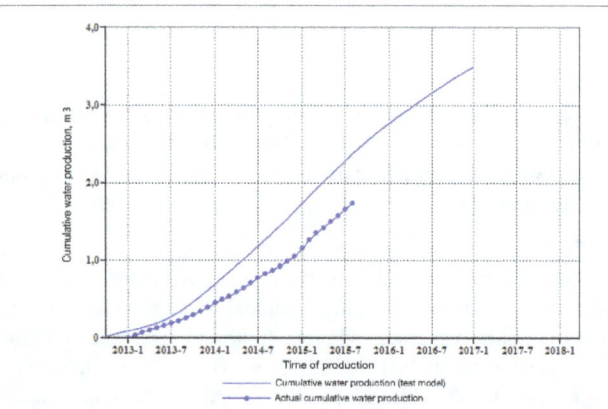

Figure 14: Comparison of the model and actual accumulated produced liquid across plot SC–3 bis.

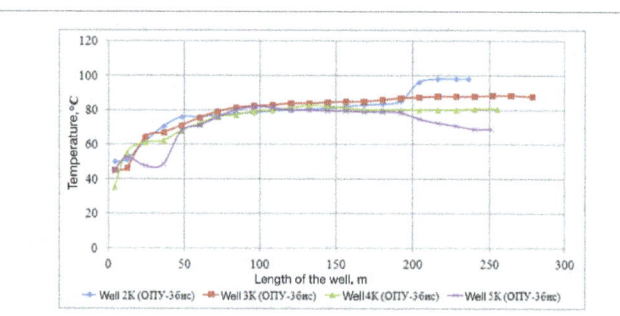

Figure 15: Temperature distribution along the wellbore (June 2015).

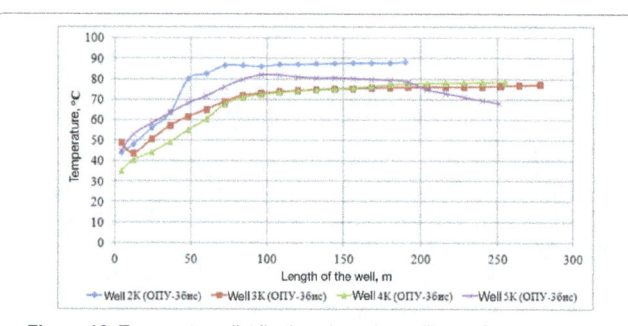

Figure 16: Temperature distribution along the wellbore (September 2015).

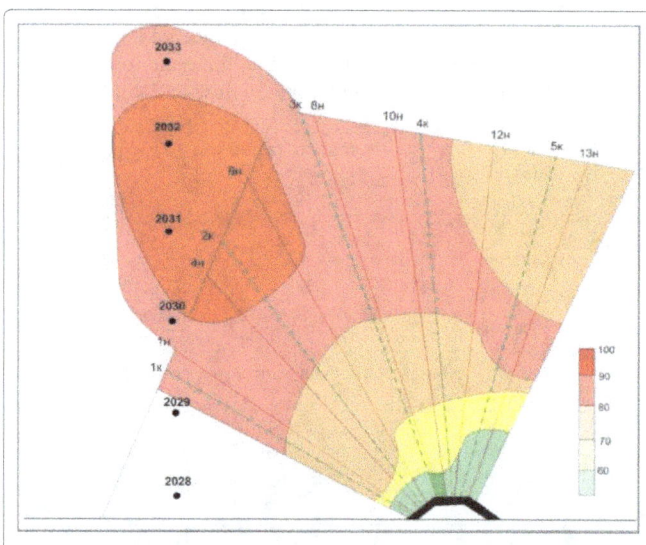

Figure 17: Map of thermal fields across plot SC–3 bis (June 2015).

field information using modern software applications which give more accurately predicted operational modes of the pilot sites [4,5].

In the fourth quarter of 2015, a study to determine collectability by injecting coolant into the reservoir was conducted on the pilot plot. Collectability at plot SC–3 bis which 18.69 t /(d•atm) was determined by the test results.

Also, temperature along the wellbore of control wells № 2K, 3K, 4K, 5K of plot SC–3 bis was measured in June and December of 2015 by the company «Argosy Analytics» LLC. The results of tests are shown in Figures 15 and 16, which shows that at the wellhead zone where production wells are cased to a depth of 50 m using heat-insulated columns, the formation temperature ranges from 35°C to 65°C (well № 2K-80°C) and 55°C on average. That is significantly lower than in the more remote areas at the top of the reservoir. The rest of the formation

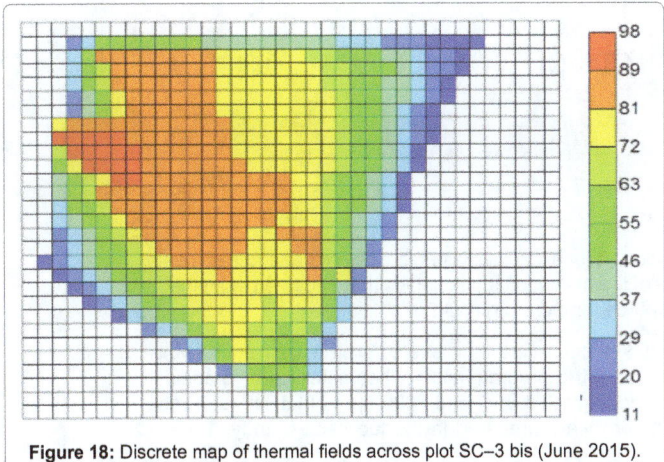

Figure 18: Discrete map of thermal fields across plot SC–3 bis (June 2015).

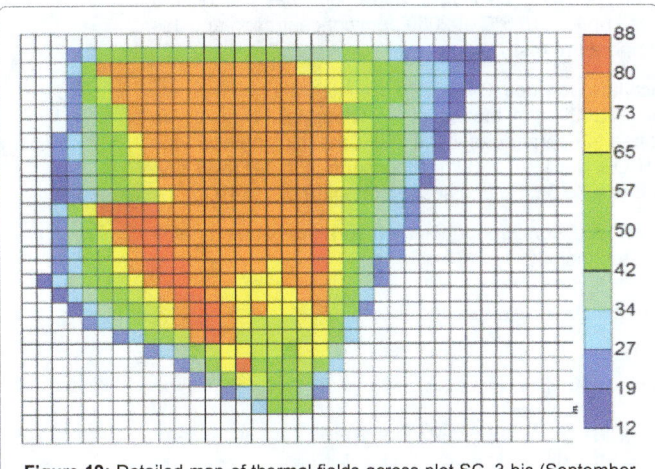

Figure 19: Detailed map of thermal fields across plot SC–3 bis (September 2015).

to the bottom of the well is uniformly warmed to a temperature of up to 80°C on average, with the temperature at the bottom of well № 2K up to 90°C to 100°C [6].

The tests made mapping of the thermal fields of plot SC-3 bis (Figures 17-19) possible for the first time. When creating the map, thermometric test results of surface wells № 2028-2033 drilled along the contour of the pilot plots were also used. Testing the wells likewise produced the following results. Average temperature in the formation interval: in well № 2030 - about 80°C, well № 2031 - about 100°C, well № 2033 - about 90°C.

Results

These data fully confirm test results obtained from underground wells. The most and least heated areas can be seen on the maps shown in Figures 17-19 (June 2015) will continue to develop recommendations for a more rational distribution of vapor on the developed area of the site. The maps shown in Figures 17-19 (June 2015) can be seen the most and least warmed areas. That makes it possible for future development recommendations relating to a more rational distribution of vapor on the developed area of the site. Following the events carried out in September using well temperature measurement data, an updated map of the thermal field was being created (Figure 19). The figure shows that the implementation of measures for redistribution of steam into the reservoir, helped to reduce the temperature in the left side of the reservoir (from 100°C to 90°C) and slightly increased (align) it into the right side of the reservoir.

Conclusion

1. Application of a thermal mining method is a very effective technology for development of shallow heavy oil fields. Yaregskoye deposit is the first oil field, where steam assisted gravity drainage in mining form is used since 1968.

2. Currently, various mining systems are used for development of heavy oil fields. For example, one of the very effective systems is a single horizon system (analog SAGD). The efficiency of this system has been tested upon experimental grounds.

3. For prediction of technological parameters CMG software (Stars module) has been used.

4. Nowadays, specialists in mining oil fields utilize modern methods and equipment for control and monitoring of technological parameters and safety.

References

1. Ruzin LM, Chuprov IF, Moroziyuk OA, Durkin SM (2015) Technological principles of developing abnormally viscous oil and bitumen deposits. M. Izhevsk: Institute of Computer Research, Moscow, Russia.

2. Durkin SM, Ruzin LM, Moroziyuk OA, Volik AI (2015) Evaluating the effectiveness of steam injection in horizontal wells of greater lengths based on numerical modelling.

3. Ruzin LM, Moroziyuk OA, Durkin SM (2013) Mechanism of oil recovery in heterogeneous reservoirs containing highly viscous oil 8: 54–57.

4. Ruzin LM, Moroziyuk OA, Durkin SM (2013) Features and innovative directions for developing high-viscosity oil resources.

5. Ruzin LM, Moroziyuk OA, Durkin SM (2013) New thermal mining technologies and assessment of their effectiveness by method of numerical simulation.

6. Durkin SM, Moroziyuk OA (2014) Evaluating the effectiveness of developing Yaregskoye oilfield by method of block and panel arrangement of wells on the basis of numerical modelling: Materials from the All-Russian Scientific-Practical Conference. Oil and Gas complex: Education, science and industry - Almatevsk: Almatevsky State Oil Institute.

19

Experimental Study on the Effect of Inhibitors on Wax Deposition

Muhammad Ali Theyab* and Pedro Diaz

London South Bank University, 103 Borough Road, SE1 0AA, London, UK

Abstract

A challenge facing offshore oil production is wax deposition. It leads to increases in operational and remedial costs while suppressing oil production. Wax inhibitors are one of the mitigation technologies that had been examined its influence on crude oil viscosity, pour point and wax appearance temperature (WAT).

The performance of some of wax inhibitors was evaluated to determine their effects on the pour point, wax appearance temperature and the viscosity of the crude oil using the programmable Rheometer rig at gradient temperatures (55°C) and shear rate 120 1/s before and after adding 1000 ppm and 2000 ppm of inhibitors to the crude oil. Three different inhibitors which were not tested before were prepared in the lab of this study. These inhibitors works well compared with its original components.

The first inhibitor was coded Mix01 by mixing polyacrylate polymer (C16-C22), and copolymer + acrylated monomers. The reduction of pour point of the waxy crude oil was up to a 16.6°C at 2000 ppm concentration and this reduces the crude oil viscosity to about 61.9% at a seabed temperature of 4°C.

The second inhibitor was coded Mix02, by mixing polyacrylate polymer (C16-C22), alkylated phenol in heavy aromatic naphtha, and copolymer dissolved in solvent naphtha. At 2000 ppm, the reduction of pour point of the crude oil up to a 15.9°C and decreases the viscosity to 57% at a seabed temperature of 4°C. Finally, the third inhibitor was Mix03, by mixing polyacrylate polymer (C16-C22), and brine (H_2O + NaCl). At 1000 ppm concentration, the reduction of pour point of the oil was up to a 14.4°C and reduced the viscosity to 52.5% at a seabed temperature of 4°C.

This unique blend of the inhibitory properties and significant reduction in pour point temperatures and crude oil viscosity is providing an original development in wax mitigation technology.

Keywords: Inhibitors; Wax; Oil; Phenol; Viscosity; Crystallization

Introduction

Paraffin wax deposition is a phenomenon that plagues the oil industry. It can choke the production lines thereby reducing the oil production to uneconomic levels. Wax inhibitor, alternatively known as pour point depressant/ wax crystal modifier, can reduce the growth of the wax crystal and form smaller crystals allowing large free space of the liquid fraction of the crude oil to flow freely [1].

The chemical addition is one of the inhibitors that have been used to reduce or prevent wax deposition in crude oil production. These inhibitors can be divided into four types: pour point depressants (PPD), crystal modifiers, dispersants, and solvents. PPD hinders the formation and growth of wax crystals by modifying the crystal structure (by merging with the edge of a growing wax crystal). Although it reduces the viscosity, yield stress, and pour point of oil, it cannot reduce the wax deposition rate [1].

The crystal modifier has a similar molecular structure to wax. It co-precipitates or co-crystallizes with a wax crystal by replacing wax molecules on the crystal lattices. It imposes steric hindrance on paraffin crystals that interfere with the proper alignment of the new incoming paraffin molecules such that growth terminates. Typical crystal modifiers are polyethylene, copolymer esters, ethylene/vinyl acetate copolymers, olefin/ester copolymers, ester/vinyl acetate copolymers, polyacrylates, polymethacrylates, and alkyl phenol resins. Dispersants are similar to surfactants in their molecular structure. Dispersants are breaking wax crystals up into much smaller particles and reduce the rate of wax deposition and prevent it by minimizing wax adhesion to the pipe wall [1,2].

Solvents increase the solubility of wax in oil and so dissolve already deposited wax. The solvents most commonly used today include aromatic compounds (toluene, and xylene), white or unleaded gasoline, and pine-derived terpenes [1].

The advantage of the wax inhibitor addition to the crude oil sample is the deposition can be mitigated without stopping production. Even though many wax inhibitors have been developed, there is currently no universal type can be used for all types of crude oils due to the varying properties of the crude oils [1,3].

Hoffmann and Amundsen [4] found that about 60% to 90% of wax thickness was reduced by applying different inhibitor concentration during experimental work investigation. The presence of the small amount of inhibitor concentration such as poly (ethylene-co-vinyl acetate (EVA)) and poly (maleic anhydride-alt-1-octadecene (MA)), can coalesce with wax crystals and interfere the crystal growth of the crystals [5].

Polymers have been used successfully as crystal modifiers in some areas; their use should expand as more effective polymers are developed. The polymers molecular weight also has influence on the pour point depression. Short or lower molecular weight polymers may

***Corresponding author:** Muhammad Ali Theyab, London South Bank University, 103 Borough Road, SE1 0AA, London, UK, E-mail: theyabm@lsbu.ac.uk

cause little disruption to the wax crystal agglomeration and growth, while very long and high molecular weight polymers can interact within the molecule itself instead of with the wax structures. This interaction reduces the rate of wax formation, leading to formation of softer wax that is easy to transport [6,7]. The reduction in the pour point and the crude oil viscosity had been making the transportation of the crude oil easier [8,9].

All the inhibitors have been used in the current work are based on polymers which are normally used as pour point depressants/crystal modifiers such as Alkylated phenol in Heavy Aromatic Naptha (HAN), Polyacrylate based polymer (C16-C22), Copolymer + acrylated monomers, Co-polymer dissolved in solvent naphtha and three inhibitors, which were not tested before were prepared by mixing some of the previous inhibitors as will illustrated in the methodology.

Methodology

Regarding to study the effects of inhibitors on wax deposition that have been used in this study, the crude oil was heated to 60°C using hot bath water for one hour to dissolve all wax crystals present in the oil and to bring the crude up to a suitable temperature for treatment. A Bohlin Gemini II Rheometer (Figure 1) was used to measure the viscosity, pour point and wax appearance temperature of the crude oil.

Treated samples

- The crude oil should be maintained at a temperature approximately 20°C above the wax appearance temperature in a hot water bath for 1 hour before commencing the test work.

- Measure 40 ml of crude oil into a test tube then dose with the required inhibitor at concentration 1000 ppm and 2000 ppm or leave undosed to act as a blank.

- Stirring the mixture for few minutes at 60°C to make sure of the reaction and to increase the influence of the inhibitor on wax crystals.

Viscosity/temperature profile

- A Bohlin Gemini II Rheometer was used to measure the viscosity of the crude oil with and without inhibitors at a cooling range from 55°C down to 0°C at a rate of 5°C/min and shear rate of 120 1/s.

- The data for the dosed samples should then be compared to the blank results, and the chemical that maintains the flattest, smoothest trace for the longest time is considered the most effective wax inhibitor.

Pour point

A Bohlin Gemini II Rheometer was used to measure the pour point of the crude oil with and without inhibitors. It was determined from the elbow point in the viscosity curve of the crude oil, where at which the liquid converted from non-Newtonian to a Newtonian liquid as shown in Figure 2.

Wax appearance temperature (WAT)

WAT is the temperature at which the first wax crystals start to form and precipitate, from crude oil, on the cold surface of the pipe [10]. The rheometer was used to measure the viscosity of the crude oil at shear rate 120 1/s and different temperatures. The WAT was determined from the converted point in the viscosity curve from the straight line to

the incline line, at which the viscosity start to increase gradually when the temperature is decreased, as shown in Figure 3.

Preparing the new inhibitors

Three wax inhibitors, which were not tested before were prepared in the lab of this work to study its effects on wax deposition and compare the results with original components. The first one coded Mix01 by mixing 33% of each of W802, W804 and W805 (Table 1), at 70°C to increase the reaction between the mixtures (Table 1).

The second inhibitor Mix02 was developed by mixing 33% of each of W802, W302 and W510. Finally, the third inhibitor settled in this subject was Mix03 by mixing 50% of W802 and 50% of brine (H$_2$O + NaCl).

Results and Discussion

Effect of inhibitors on viscosity

The performance of some of wax inhibitors was evaluated to determine their effects on the wax precipitation. The effect was on the pour point, wax appearance temperature and the viscosity of the crude oil.

The analysis of the crude oil viscosity with the inhibitors shows that the new mixtures Mix01 and Mix02 produced the greatest reduction

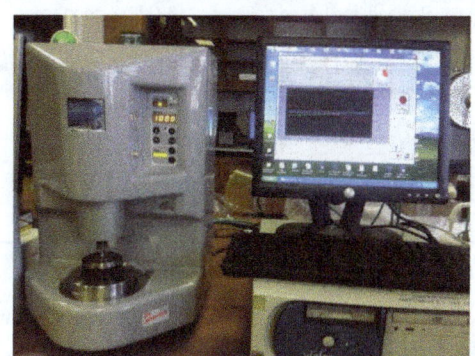

Figure 1: A Bohlin Gemini II Rheometer was used to determine pour point, viscosity and WAT of the crude oil before and after adding the inhibitors.

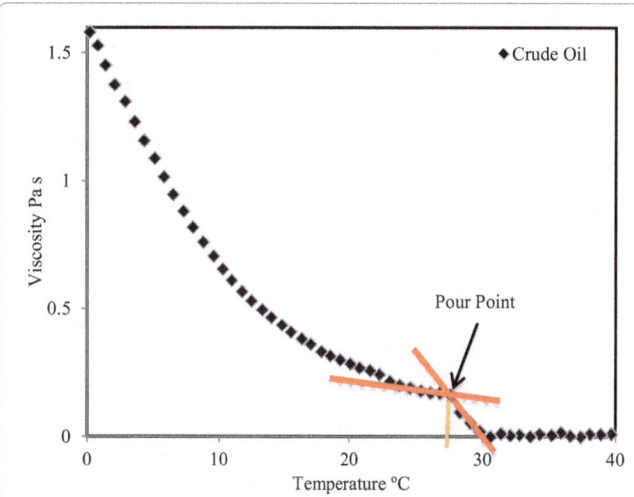

Figure 2: Determine the pour point from the viscosity curve of the crude oil using the rheometer.

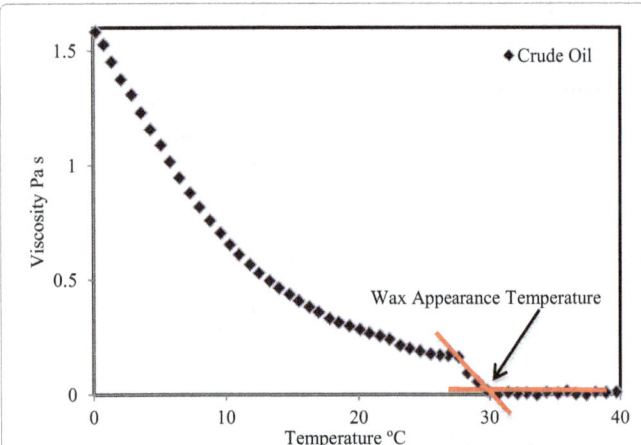

Figure 3: Determine the wax appearance temperature from the viscosity curve of the crude oil using the Rheometer.

Inhibitor Code	Inhibitor Chemistry
W302	Alkylated phenol in Heavy Aromatic Naptha (HAN).
W802	Polyacrylate based polymer (C16-C22).
W804	Copolymer + acrylated monomers.
W805	Copolymer + acrylated monomers.
W510	Co-polymer dissolved in solvent naphtha.
Mix01 (New)	Polyacrylate based polymer (C16-C22) + copolymer + acrylated monomers.
Mix02 (New)	Polyacrylate based polymer (C16-C22) + Alkylated phenol in Heavy Aromatic Naptha (HAN) + Co-polymer dissolved in solvent naptha.
Mix03 (New)	Polyacrylate based polymer (C16-C22) + Brine (H_2O + NaCl).

Table 1: The chemistry of wax inhibitors has been used during this study.

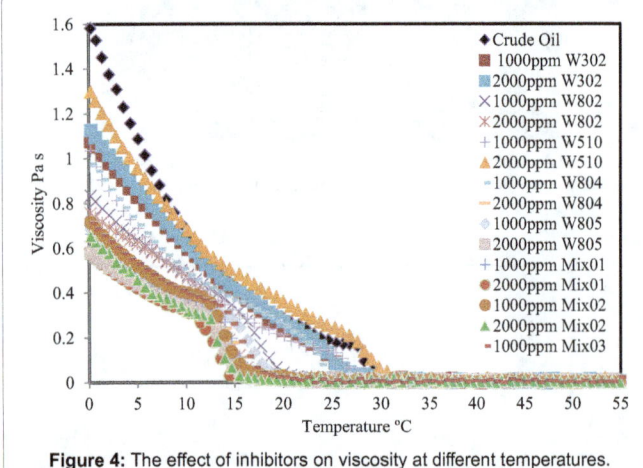

Figure 4: The effect of inhibitors on viscosity at different temperatures.

in viscosity at concentration 2000 ppm comparing with its original components at the same concentration, see Figure 4. The products were all tested at 1000 ppm and 2000 ppm.

The prepared mixtures were produced better result compared with its original components, due to increase the monomers in the mixture and that means increase the ability to prevent wax crystal formation. The first inhibitor Mix01 at 2000 ppm was reduced the crude oil viscosity up to 61.9% at seabed temperature of 4ºC, as shown in Figure 5. The second inhibitor Mix02 at 2000 ppm concentration was decreased the

viscosity up to 57% at seabed temperature of 4ºC, as shown in Figure 6. Finally, the third inhibitor Mix03 at 1000 ppm concentration was reduced the viscosity up to 52.5% at seabed temperature of 4ºC, as shown in Figure 7.

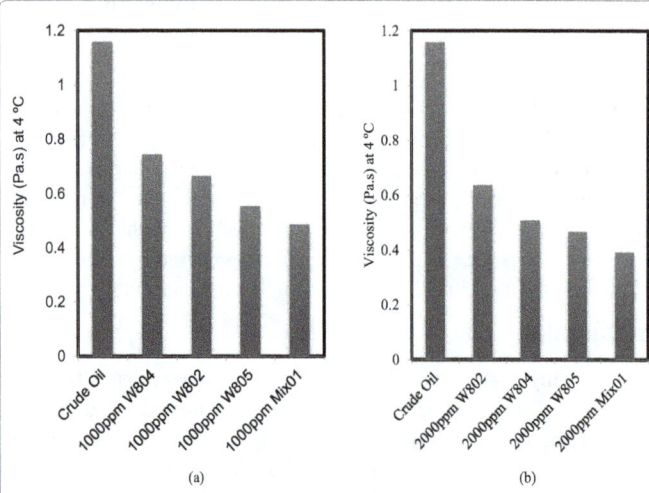

(a)   (b)

Figure 5: The effect of Mix01 and its components on crude oil viscosity at 4ºC, (a) 1000 ppm (b) 2000 ppm.

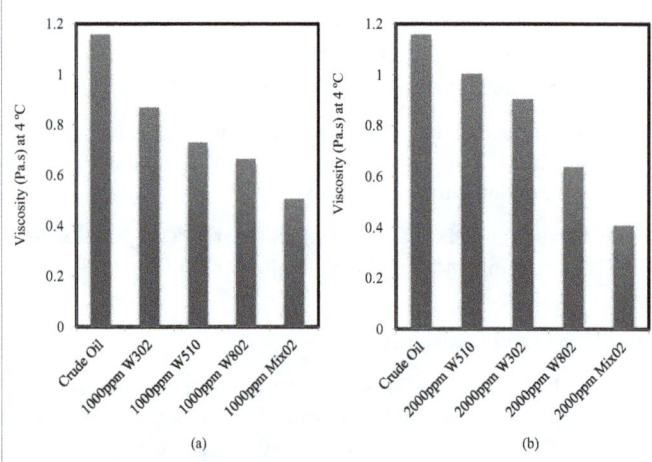

(a)   (b)

Figure 6: The effect of Mix02 and its components on crude oil viscosity at 4ºC, (a) 1000 ppm (b) 2000 ppm.

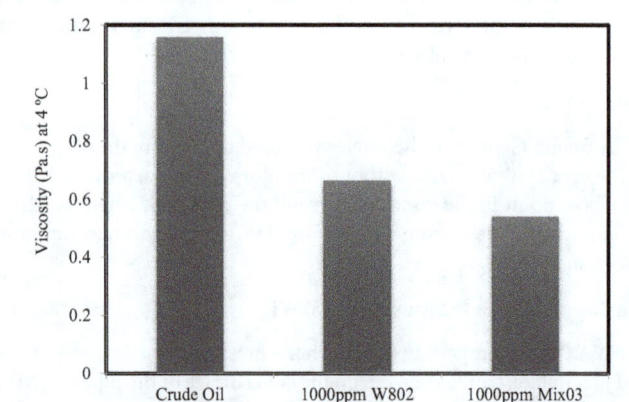

Figure 7: The effect of Mix03 and its components on crude oil viscosity at 4ºC and concentration 1000 ppm.

Effect of inhibitors on pour point

The pour point is the temperature at which it becomes semi solid and loses its flow characteristics. The first inhibitor Mix01 was reduced the pour point of the waxy crude oil from 27.6°C to 11°C at 2000 ppm concentration, as shown in Figures 8 and 9. The second inhibitor Mix02 at 2000 ppm concentration was produced better result compared with its components, where it decreased the pour point of the crude oil from 27.6°C to 11.7°C, as shown in Figure 10. Finally, the third inhibitor Mix03 at 1000 ppm concentration was reduced the pour point of the oil from 27.6°C to 13.2°C, as shown in Figure 11.

Effect of inhibitors on wax appearance temperature

The wax appearance temperature of the crude oil has been reduced by adding the new inhibitors of this study, where it is decreased up to 52% by adding the new inhibitor Mix01 and up to 48.3% by adding Mix02 at concentration 2000 ppm respectively; and up to 41% by adding the inhibitor Mix03 at concentration 1000 ppm as shown in Figures 12, 13 and 14 respectively.

This can be interpreted as by increasing the concentration of Mix01 from 1000 to 2000 ppm the quantity of the polyacrylate polymer and the acrylated monomers will be increased, providing more structures to interfere and merge with the edge of a growing wax crystal.

The reduction in WAT and viscosity due to add Mix02 can be explained as by increasing the concentration of polyacrylate polymer, alkylated phenol and co-polymer dissolved in naphtha, the wax crystals will be decreased due to increase the molecules that prevent wax crystal formation and preserve it in smaller particles.

The third Mix03 contain of mixing polyacrylate polymer and brine

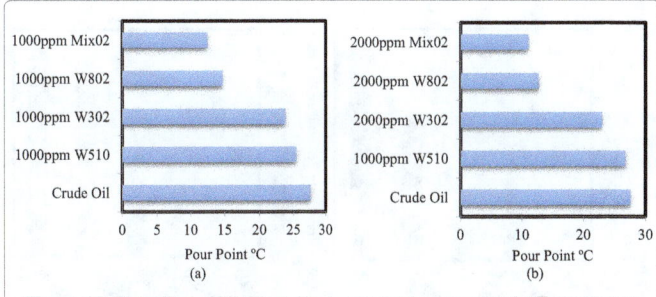

Figure 10: The effect of Mix02 and its components on pour point Temperature (a) 1000 ppm and (b) 2000 ppm.

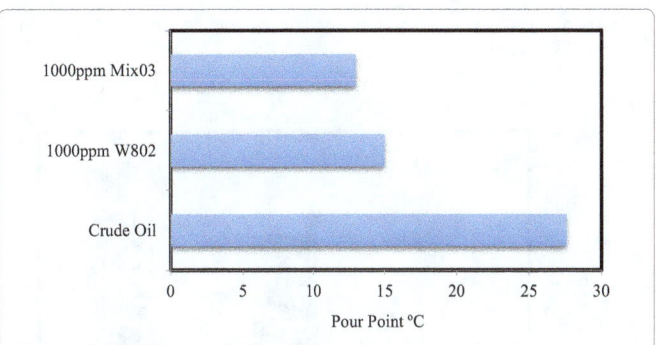

Figure 11: The effect of Mix03 and its components on pour point temperature at concentration 1000 ppm.

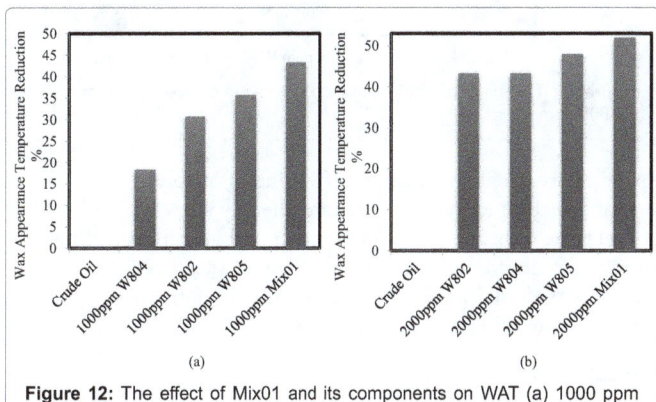

Figure 12: The effect of Mix01 and its components on WAT (a) 1000 ppm and (b) 2000 ppm.

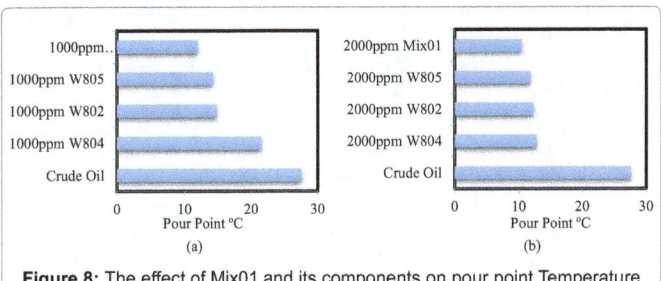

Figure 8: The effect of Mix01 and its components on pour point Temperature (a) 1000 ppm and (b) 2000 ppm.

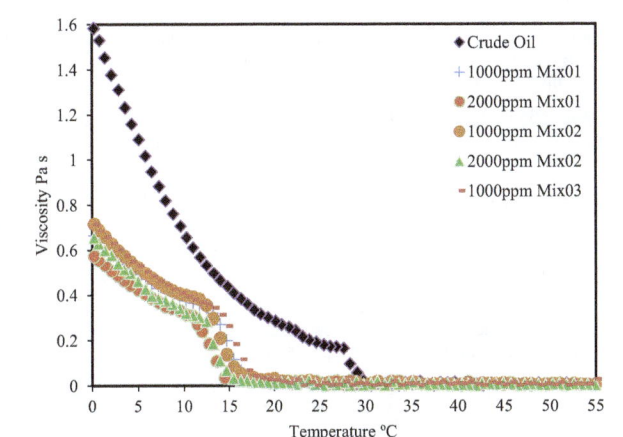

Figure 9: Comparison between the new mixtures Mix01, Mix02 at 1000 ppm and 2000 ppm, and Mix03 at 1000 ppm as shown in Figure 8.

(H_2O + NaCl) and this will lead to producing sodium polyacrylate and this will absorb and merge with the wax crystals and prevent it to combine together.

Comparison between the prepared mixtures

From Figures 8 and 9, it was noticed that at concentration 1000 ppm that Mix02 was produced better results, compared with Mix01 and Mix03, in pour point and wax appearance temperature 12.5°C and 16.3°C, respectively. At 2000 ppm Mix01 was produced best results compared with the prepared mixtures and the original inhibitors, where it was the pour point and the wax appearance temperature 11°C and 14.5°C, respectively.

These inhibitors (Mixtures) at concentration 1000 ppm and 2000 ppm improved the reduction in wax crystal formation by interfering with wax crystallization and prevent growth process. However, this interfering mechanism has not yet been fully understood [11]. The

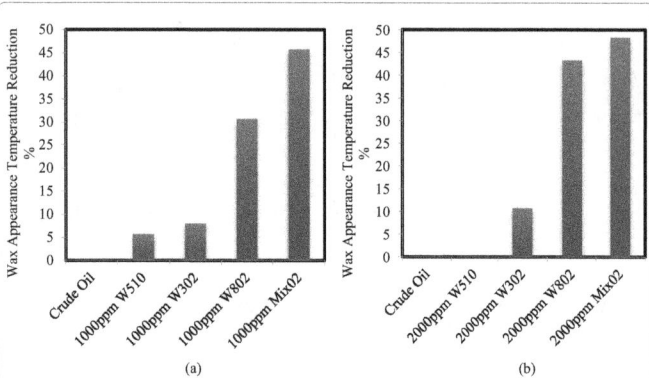

Figure 13: The effect of Mix02 and its components on WAT (a) 1000 ppm and (b) 2000 ppm.

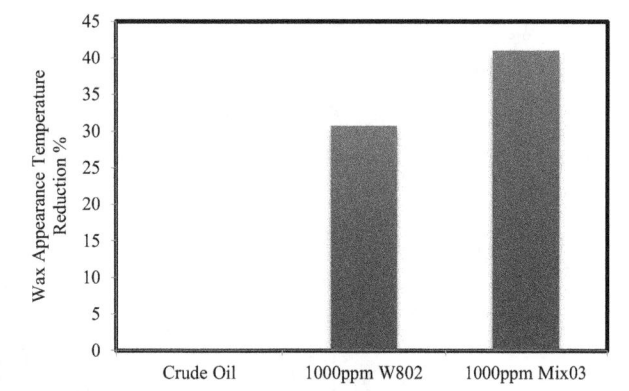

Figure 14: The effect of Mix03 and its components on WAT of the crude oil at 1000 ppm.

major theory stated the possibility of wax inhibitor polymers containing similar structure to the wax structure, thereby allowing the inhibitor crystal to be incorporated into the wax crystal growth. Sometimes the structural part of the polymer covers the wax site, thereby preventing further wax crystal growth and promoting the formation of smaller wax aggregates [9,11].

Conclusion

Wax deposition in offshore pipelines and other production equipment can pose significant flow problems requiring remediation. Wax inhibitors are considered one of the suitable mitigation technologies in the deep water because it does not need to stop production.

The performance of some of wax inhibitors was evaluated to determine their effects on the pour point, wax appearance temperature and the viscosity of the crude oil using the programmable Rheometer rig at gradient temperatures (55°C) and shear rate 120 1/s before and after adding 1000 ppm and 2000 ppm of inhibitors to the crude oil.

During this research, three different inhibitors were prepared, these inhibitors works more efficient compared with its original components. The first inhibitor was coded Mix01 by mixing 33% of each of polyacrylate polymer (C16-C22), and copolymer + acrylated monomers. The reduction of pour point of the waxy crude oil was from 27.6°C to 11°C at 2000 ppm concentration of Mix01 and this reduces the crude oil viscosity to about 61.9% at a seabed temperature of 4°C.

The second inhibitor was coded Mix02, by mixing 33% of each of polyacrylate polymer (C16-C22), alkylated phenol in heavy aromatic naphtha, and copolymer dissolved in solvent naphtha. At 2000 ppm, the reduction of pour point of the crude oil from 27.6°C to 11.7°C and decreases the viscosity to 57% at a seabed temperature of 4°C. Finally, the third inhibitor was Mix03, by mixing 50% of each of polyacrylate polymer (C16-C22), and brine (H₂O+NaCl). At 1000 ppm concentration, the pour point of crude oil was reduced from 27.6°C to 13.2°C and reduced the viscosity to 52.5% at a seabed temperature of 4°C.

A comparison between the prepared mixtures was completed, where it was noticed that at concentration 1000 ppm that Mix02 was produced better results, compared with Mix01 and Mix03, in pour point and wax appearance temperature 12.5°C and 16.3°C, respectively. At 2000 ppm Mix01 was produced best results compared with the prepared mixtures and the original inhibitors, where it was the pour point and the wax appearance temperature 11°C and 14.5°C respectively.

This unique blend of the inhibitory properties and significant reduction in pour point temperatures, wax appearance temperatures and crude oil viscosity is providing a forward step in wax mitigation technology to be study.

References

1. Kang PS, Dong-Gun Lee, Jong-Se Lim (2014) Status of wax mitigation technologies in offshore oil production. Proceedings of the Twenty-Fourth International Ocean and Polar Engineering Conference Busan, Korea.

2. Dobbs JB (1999) A unique method of paraffin control in production operations. SPE 55647 presented at SPE Rocky Mountain Regional Meeting, Gillette.

3. Ridzuan N, Adam F, Yaacob Z (2014) Molecular Recognition of wax inhibitor through pour point depressant type inhibitor. International Petroleum Technology Conference.

4. Hoffman R, Amundsen L (2010) Single-phase wax deposition experiments. Energy & Fuels 24: 1069-1080.

5. Jafari Ansaroudi HR, Vafaie-Sefti M, Masoudi S, Behbahani TJ, Jafari H (2013) Study of the morphology of wax crystals in the presence of ethylene-co-vinyl acetate copolymer. Pet. Sci. Technol. 31: 643-651.

6. Jang YH, Blanco M, Creek J, Tang Y, Goddard WA (2007) Wax inhibition by comb-like polymers: Support of the incorporation-perturbation mechanism from molecular dynamics simulations. Journal of Physical Chemistry 111: 13173-13179.

7. Han S, Song Y, Ren T (2009) Impact of alkyl methacrylate-maleic anhydride copolymers as pour point depressant on crystallization behaviour of diesel fuel. Energy and Fuel 23: 2576-2580.

8. Pedersen KS, Ronningsen HP (2003) Influence of wax inhibitors on wax appearance temperature, pour point, and viscosity of waxy crude oils. Energy and Fuels 17: 321-328.

9. Adeyanju OA, Oyekunle LO (2014) Influence of long chain acrylate ester polymers as wax inhibitors in crude oil pipelines. Society of Petroleum Engineers.

10. Dantas Neto AA, Gomes EAS, Barros Neto EL, Dantas TNC, Moura CPAM (2009) Determination of wax apperance temperature (WAT) in paraffin/solvent systems by photoelectric signal and viscosimetery. Brazilian Journal of Petroleum and Gas 3: 149-157.

11. Jennings DW, Newberry ME (2008) Paraffin inhibitor applications in deepwater offshore development, Paper Presented at the International Petroleum Technology Conference, Kuala Lumpur, Malaysia.

20

Damage Assessment of Bitumen Refineries Using Simapro (LCA) Inventory Data

Saeed Morsali*

Faculty of Applied Sciences, Department of Environmental Science, Gazi University, Ankara, Turkey

Abstract

Oil refineries are complex facilities. Several processes, such as distillation, vacuum distillation, or steam reforming are required to produce a large variety of oil products such as gasoline, light fuel oil or bitumen. The environmental impacts of oil refineries are assessed using the technique of life cycle assessment (LCA). In this paper, only the material production phase of the bitumen life cycle is considered. To improve the quality of the LCA, a regionalized life cycle inventory (LCI) database for the Oil refineries and commercial LCI databases are used to validate and model unit processes with LCA software.

Keywords: Bitumen refinery; LCA: Life Cycle Assessment; Damage assessment of bitumen; Simapro

Introduction

The importance of quantifying the impact of products and services on the environment is growing due to the recent changes in patterns of climate, living and ecosystem quality. Consumers and governments are increasingly demanding information about the sustainability of products and interest in comparing potential solutions based upon scientific data is necessary in order to do this.

Petroleum refining is a unique and critical link in the petroleum supply chain, from the wellhead to the pump. The other links add value to petroleum mainly by moving and storing it (e.g., lifting crude oil to the surface; moving crude oil from oil fields to storage facilities and then to refineries; moving refined products from refinery to terminals and end-use locations, etc.). Refining adds value by converting crude oil (which in itself has little end-use value) into a range of refined products, including transportation fuels. The primary economic objective in refining is to maximize the value added in converting crude oil into finished products. Petroleum refineries are large, capital-intensive manufacturing facilities with extremely complex processing schemes. They convert crude oils and other input streams into dozens of refined (co-products) including:

a. Liquefied petroleum gases (LPG).

b. Gasoline.

c. Jet fuel.

d. Kerosene (for lighting and heating).

e. Diesel fuel.

f. Petrochemical feedstocks.

g. Lubricating oils and waxes.

h. Home heating oil.

i. Fuel oil (for power generation, marine fuel, industrial and district heating).

j. Asphalt (for paving and roofing uses).

Of these, the transportation fuels have the highest value; fuel oils and asphalt the lowest value. Many refined products, such as gasoline, are produced in multiple grades, to meet different specifications and standards (e.g., octane levels, sulfur content). More than 660 refineries, in 116 countries, are currently in operation, producing more than 85 million barrels of refined products per day (1: 2011: 2). Each refinery has a unique physical configuration, as well as unique operating characteristics and economics. A refinery's configuration and performance characteristics are determined primarily by the refinery's location, vintage, availability of funds for capital investment, available crude oils, product demand (from local and/or export markets), product quality requirements, environmental regulations and standards, and market specifications and requirements for refined products.

In published studies about petroleum industry life cycle assessments the most focused subjects are energy consumption and some emissions such as CO_2, CO, SO_2, SO. In this study the analysis lead to product specific allocation factors for energy, airborne and waterborne pollutants. Furthermore working material consumption, additive requirements, production waste, and infrastructure are included [1-3].

Typical refineries

Petroleum refining first began in earnest as a value-added process in 1856 near the site where the Killing Holm/Humber refinery sits today. Today's refineries are decidedly more sophisticated than in 1856 and rely on a fixed configuration that produces fixed output depending on the quality of the crude inputs and the capacity of the refinery. There are several processes involved in processing crude inputs to make them useable and marketable fuel outputs. The main refining processes can be described in terms of the order in which they occur. The most common form of petroleum refining is known as fractional or atmospheric distillation, which involves pumping the crude petroleum into the bottom of a heated column and then separating the fuels via different temperature levels (Energy Institute). All fuels go through the initial distillation process to separate from crude oil on the way to further processing. The residue of the distillation column, much heavier than crude oil is then sent to a second distillation unit while the other fuel products are sent to other processes. The lightest

*Corresponding author: Saeed Morsali, Faculty of Applied Sciences, Department of Environmental Science, Gazi University, Ankara, Turkey
E-mail: morsali.saeed@gmail.com

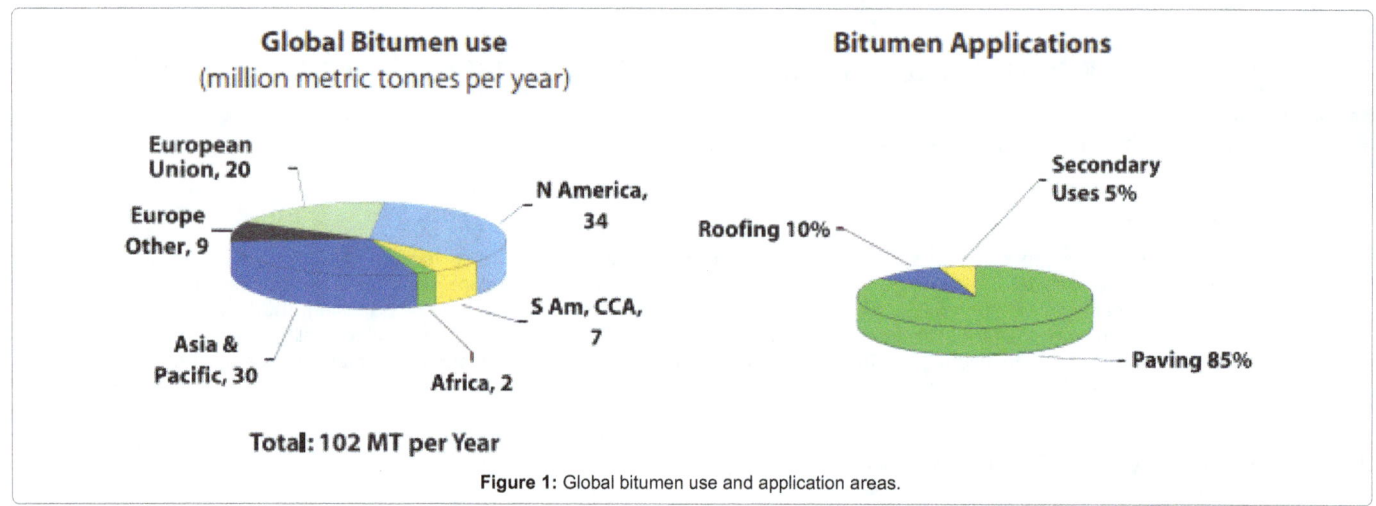

Figure 1: Global bitumen use and application areas.

products, liquefied petroleum gases, mostly butane, propane and naphtha require little to no further processing in order to be sold to market. However, other products require more processing in order to become marketable. The main fuel products can be classified in the same way as the distillation column for simplicity's sake. The lighter products within the column rise while the heavier products sink [4-8]. The additional processing that occurs can be summarized by fuel type and carbon structure. Liquid petroleum gases (LPG) are most commonly in the form of naphtha, butane and propane. LPGs typically require little to no further processing except for sulfur removal. Sulfur removal (desulfurization) is also entirely dependent on the source of the crude and how much sulfur it contains. Petrol is generally removed from the distillation unit and cleaned in what is called a unifiner. A unifiner removes sulfur and nitrogen compounds in the fuel and creates hydrogen sulfide and ammonia as wastes. Then the molecular structure is modified to increase the octane levels of the fuel so that it is suitable for combustion in motor vehicles and other petrol burning engines. Sulfur is a by-product of this process and is recycled in other processes or sent to waste processing. Petrol can also be separated from the heavy distillate residues through a process called catalytic cracking. Generally the more complex plants have catalytic crackers and are capable of refining heavier fuels. Catalytic cracking is an additional process and while adding value, also adds cost and emissions. The last step in petrol processing is fuel blending as required by national fuel specifications guided by the European Fuel Quality Directive. Jet fuel and kerosene are generally grouped together because they have a similar carbon structure. They emerge from the distillation process requiring desulfurization. This is done through what is known as a merox unit, which washes the fuel with sodium hydroxide (caustic washing) and other additives which also help to reduce the impurities in the fuel. Diesel and gas oil are used for combustion engines and heating purposes mostly. They require post-distillation processing in a unit known as a hydrotreater. The hydrotreater removes sulphur and other impurities using hydrogen recycled from other processes as a catalyst. The diesel and gas oil is typically ready for market after this process. Fuel oils are generally used for heating and ship transport. These fuels require additional distillation through a process known as vacuum distillation. Vacuum distillation is a similar process to the primary distillation process except that the pressure within the distillation column is greatly reduced so that additional lighter fuels can be separated and captured for further processing. The lighter fuels that come out of the vacuum distillation unit are sent to a catalytic cracking unit and separated by fuel type to go through the remaining refining processes. Finally, Bitumen is obtained

by vacuum distillation or vacuum flashing of atmospheric residue from the vacuum distillation column. This is "straight run bitumen". This process is called bitumen production by straight run vacuum distillation [9-12].

An alternative method of bitumen production is by precipitation from residual fractions by propane or butane-solvent deasphalting. The bitumen thus obtained has properties which derive from the type of crude oil processed and from the mode of operation in the vacuum unit or in the solvent deasphalting unit. The grade of the bitumen depends on the amount of volatile material that remains in the product: the smaller the amount of volatiles, the harder the residual bitumen.

It is estimated that the current world use of bitumen is approximately 102 million tons per year. The primary use of bitumen is for paving and roofing applications; 85% of all the bitumen is used as the binder in various kind of asphalt pavements: Pavements for roads, airports, parking lots, etc.

About 10% of the bitumen is used for roofing. The rest of the bitumen, approximately 5% of the total, is used for variety of purposes, each very small in volume. This sector is referred to as "secondary uses" (2: 2011: 2).

This figure shows estimated yearly bitumen production worldwide by different area, it also represents bitumen applications by sector (Figure 1).

Inventory Analysis

In this study database for emission factors for the Swiss and for the average Western European refinery are used. Airborne emissions comprise CO, CO_2, SO_2, NO_x, particulate matter, hydrocarbons (specified), acids and heavy metals (specified). Waterborne pollutants comprise hydrocarbons (specified), and inorganic substances (sulfates, phosphates and nitrate). Different production waste and their further treatment are distinguished. The environmental impacts modeled include energy consumption and greenhouse gas (GHG) emissions from oil refineries.

System boundaries

The model describes the production of oil products for energetic and partly non energetic uses and the production of thermal energy and electricity in Switzerland and Western Europe. The inventory tables for oil products include oil field exploration, crude oil production, long

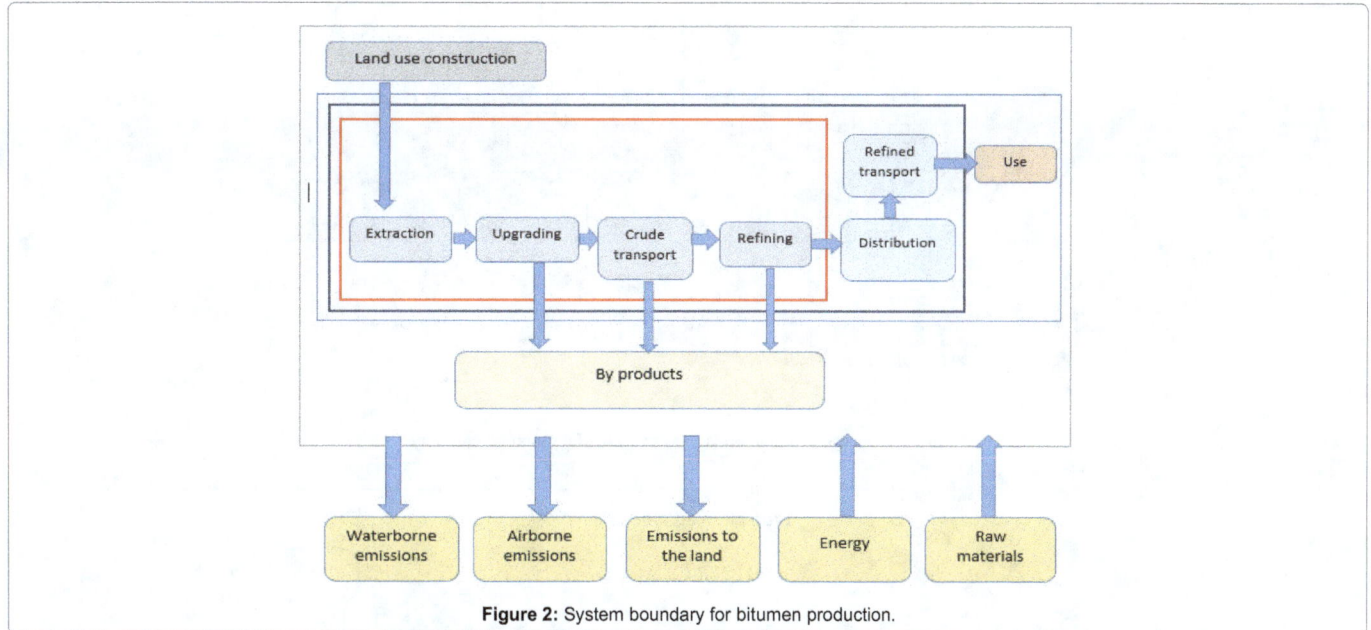

Figure 2: System boundary for bitumen production.

distance transportation, oil refining, regional distribution and the use of oil products in domestic and industrial boilers, in power plants and in spark ignition engines (of trucks, personal cars, excavator, locomotives and ships).

For all these steps air and waterborne pollutants as well as energy and working material requirements, production waste, and the production of the equipment are considered [13,14].

This study covers the bitumen production chain, starting from raw material extraction and ending with a bitumen product ready for delivery to a customer. The process is divided into four stages: crude oil extraction, transport, production and storage. A schematic description of the system boundary is given in Figure 2.

Inventory data

Inventory data for this study is taken from commercial Simapro 7.1 program database and the oil fuel chain in particular is divided into the following process steps:

Oil field exploration: Include emissions caused by drilling activities, barite and bentonite consumption and the emissions of oily drilling fluids into the sea (emission data for North Sea exploration is used).

Crude oil production: The variation in drilling efforts and energy consumption per barrel oil extracted between different regions is modeled.

Long distance transportation: Distance is used according to the specific supply situation of Switzerland and Western Europe.

Oil refining: *Regional distribution*: Regional distribution includes storage in large stock and the supply to the costumer. The requirements and emissions during the regional distribution are considered. The infrastructure and the energy consumption for the movement of goods, production waste and hydrocarbon emissions are included.

Fuel oil boilers: Three different sizes of boilers are considered, namely 10 KW, 100 KW and 1 MW and also manufacturing of boilers including tank room and chimney is considered.

Impacts assessment of bitumen refineries

For this paper the used method is ECO INDICATOR 99 which has three main impact categories; human health, ecosystem quality and resources. Simapro uses Pt unit to show these impacts. The Pt unit used in eco indicator method defined as a dimensionless value. The value of 1 Pt means one thousandth of the yearly environmental load of one average European inhabitant. The environmental impact scores of life cycle assessments are often presented in units that are difficult to grasp, such as kg CO_2 equivalents or CTUh. One way to make interpreting such scores easier is to normalize them: dividing your scores by a reference situation's scores. This reference situation could be one person's – Average Joe's – share of all emission and resource use in the world during one year. Normalization converts complicated units into fractions of Joe's scores per impact category (3: 2000: 9) (Figure 3 and Graph 1) [14]. The scores for climate change, human toxicity, and many other impact categories are all compared to the annual impact Average Joe has, and expressed in fractions. For 1 ton bitumen production the following values are analyzed.

Resources: Mankind will always extract the best resources first, leaving the lower quality resources for future extraction. The damage of resources will be experienced by future generations, as they will have to use more effort to extract remaining resources. This extra effort is expressed as "surplus energy".

As the Table 1 shows the highest impact accrues in resources category and the Figure 4 represents a tree analysis for resource category, a tree analysis shows all processes which cause any impact on resources as well as their share in whole resource depletion process. Figure 4 also shows the priority of the process.

Under this category Eco-Indicator 99 considers two main impacts, namely fossils fuels and minerals:

Minerals: Surplus energy per kg mineral or ore, as a result of decreasing ore grades.

Fossil fuels: Surplus energy per extracted MJ, kg or m³ fossil fuel, as a result of lower quality resources.

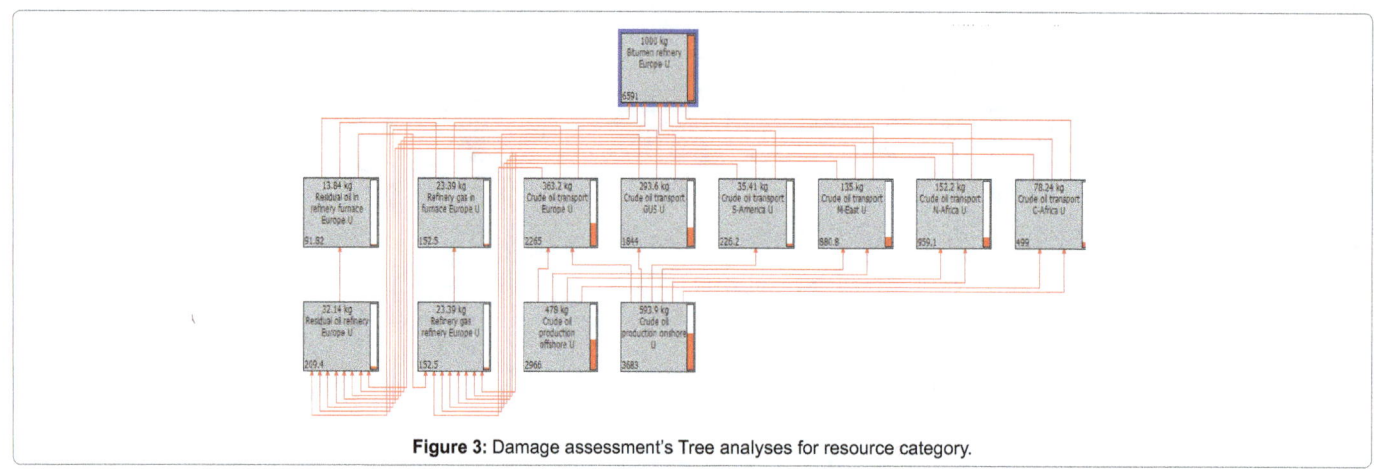

Figure 3: Damage assessment's Tree analyses for resource category.

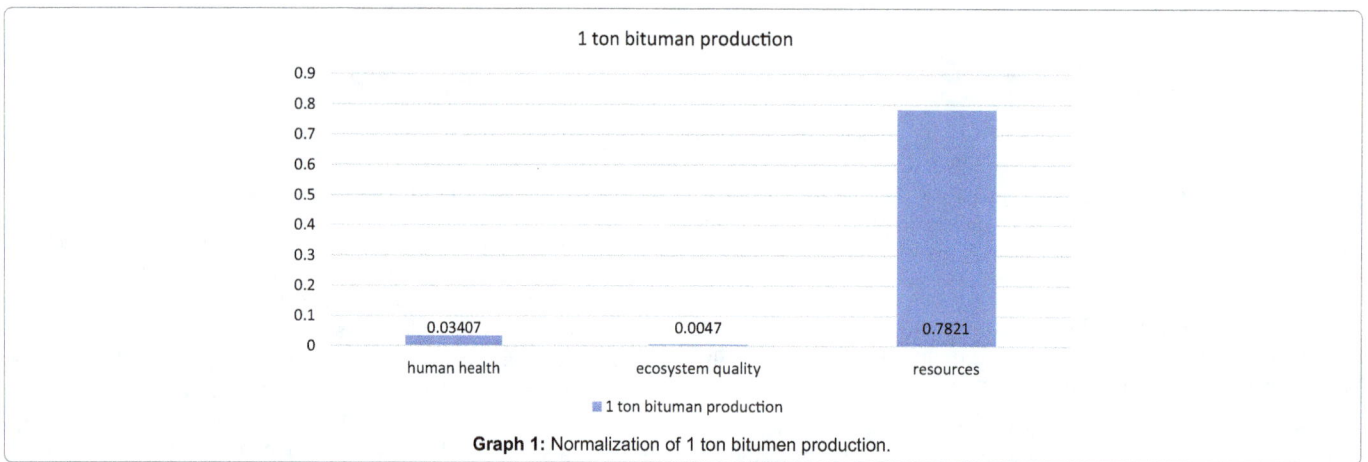

Graph 1: Normalization of 1 ton bitumen production.

No	Substance	Compartment	Sub-Compartment	Unit	Bitumen refinery Europ
	Total of all compartments			MJ surplus	6591
1	Oil, crude, 42.6 MJ per kg, in ground	Raw	in ground	MJ surplus	6572
2	Gas, natural, 35 MJ per m3, in ground	Raw	in ground	MJ surplus	12.84
3	Coal, 18 MJ per kg, in ground	Raw	in ground	MJ surplus	2.889
4	Tin, in ground	Raw	in ground	MJ surplus	1.127
5	Gas, mine, off-gas, process, coal miningil<g	Raw	in ground	MJ surplus	0.8796
6	Copper, in ground	Raw	in ground	MJ surplus	0.7595
7	Iron, in ground	Raw	in ground	MJ surplus	0.2661
8	Nickel, in ground	Raw	in ground	MJ surplus	0.07882
9	Bauxite, in ground	Raw	in ground	MJ surplus	0.02575
10	Lead, in ground	Raw	in ground	MJ surplus	0.02292
11	Chromium, in ground	Raw	in ground	MJ surplus	0.004774

Table 1: Damage assessment of Substances which are effect on resources.

Table 1 shows substances which effect on resource during bitumen production, the unit used for this table is MJ surplus which explained earlier. Also Table 1 shows the nature of these substances in case of how they released in environment and where they release like ground or as gases in the air.

Human health: Damage to human health, expressed as the number of year life lost and the number of years lived disabled. These are combined as Disability Adjusted Life Years (DALYs), an index that is also used by the World Bank and the WHO (4: 2016: 53). This category includes; climate change, ozone layer depletion, carcinogenic effects, respiratory effects and ionising radiation.

The Figure 4 is a tree analysis for human health category; all processes which have any effect on human health are shown as a tree with their impact share and their relevance which each other. The unit used for Figure 4 is DALY unit.

Human health category includes six subcategories which shown in the Tables 2 and 3; Carcinogenic affects due to emissions of carcinogenic substances to air, water and soil, Respiratory organics effects resulting from summer smog, due to emissions of organic substances to air, causing respiratory effects. Respiratory inorganics effects resulting from winter smog caused by emissions of dust, sulphur and nitrogen oxides to air. Climate change Damage, resulting from an increase of diseases and death caused by climate change. Radiation Damage,

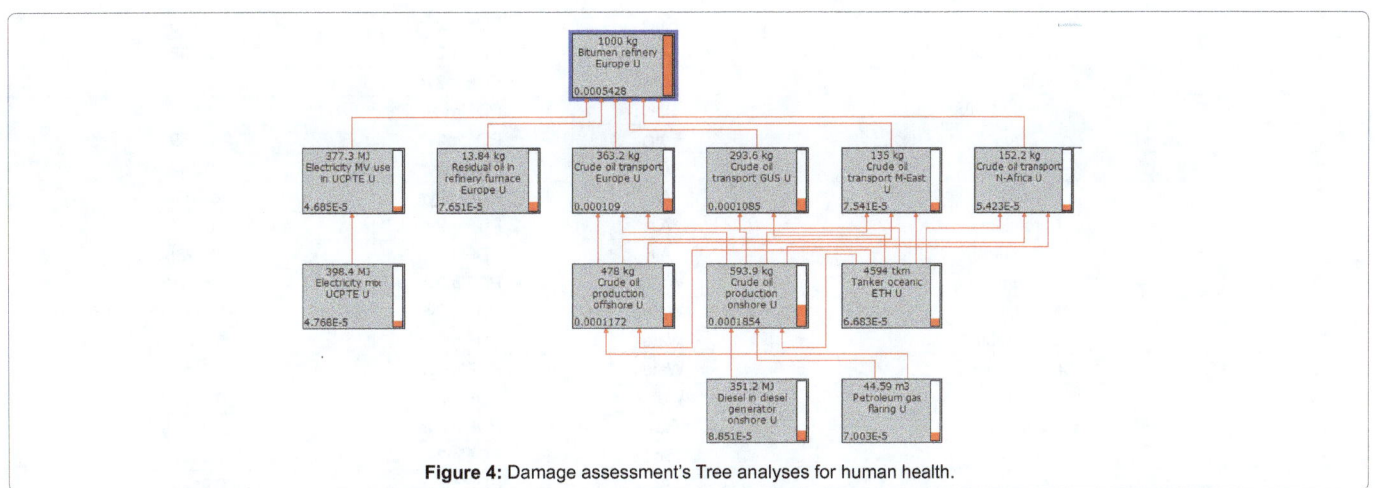

Figure 4: Damage assessment's Tree analyses for human health.

No	Substance	Compartment	Unit	Bitumen refinery Europ
1	Nitrogen oxides	Air	DALY	0.0001783
2	Sulfur oxides	Air	DALY	0.0001415
3	Carbon dioxide	Air	DALY	0.00009124
4	Particulates,< 10 um (stationary)	Air	DALY	0.00006937
5	Methane	Air	DALY	0.00001894
6	NMVOC, non-methane volatile organic compounds, unspecified origin	Air	DALY	0.00001028
7	Particulates, < 10 um (mobile)	Air	DALY	0.000007473
8	Arsenic, ion	Water	DALY	0.00000714
9	Cadmium	Air	DALY	0.000005344
10	Cadmium, ion	Water	DALY	0.000004691
11	Methane, bromotrifluoro-, Halon 1301	Air	DALY	0.000002287
12	PAH, polycyclic aromatic hydrocarbons	Water	DALY	0.000001817
13	Radon-222	Air	DALY	0.000001528
14	Dinitrogen monoxide	Air	DALY	7.796E-07
15	Arsenic	Air	DALY	6.176E-07
16	Arsenic	Soil	DALY	3.253E-07
17	Carbon monoxide	Air	DALY	2.046E-07
18	Carbon-14	Air	DALY	1.522E-07
19	Butane	Air	DALY	1.129E-07
20	Cesium-137	Water	DALY	9.505E-08
21	Pentane	Air	DALY	8.727E-08
22	Propane	Air	DALY	8.152E-08
23	Nickel	Air	DALY	4.684E-08
24	Hexane	Air	DALY	4.137E-08
25	Ethane, 1,2-dichloro-1, 1,2,2-tetrafluoro-, CFC-114	Air	DALY	3.011E-08
26	Benzene	Water	DALY	2.881E-08
27	Benzene	Air	DALY	2.47E-08
28	Heptane	Air	DALY	2.144E-08
29	Ammonia	Air	DALY	2.057E-08
30	Xylene	Air	DALY	1.785E-08
31	Radon-222	Air	DALY	1.679E-08
32	Toluene	Air	DALY	1.624E-08
33	Hydrocarbons, aliphatic, alkanes, unspecified	Air	DALY	1.588E-08
34	Cobalt-60	Water	DALY	1.156E-08
35	Ethene	Air	DALY	1.078E-08
36	Propene	Air	DALY	9.234E-09
37	Cesium-134	Water	DALY	8.503E-09

Table 2: Damage assessment of 1 ton bitumen production on human health.

S. No.	Substance	Compartment	Unit	Bitumen refinery Europe
1	Nitrogen oxides	Air	PDF'm2yr	11.49
2	Nickel	Air	PDF'm2yr	7.751

3	Land use 11-111	Raw	PDF'm2yr	2.736
4	Sulfur oxides	Air	PDF'm2yr	2.698
5	Land use I I-IV	Raw	PDF'm2yr	1.611
6	Land use 111-IV	Raw	PDF'm2yr	1.297
7	Zinc	Air	PDF'm2yr	0.789
8	Cadmium	Air	PDF'm2yr	0.382
9	Lead	Air	PDF'm2yr	0.3082
10	Chromium	Air	PDF'm2yr	0.134
11	Copper	Air	PDF'm2yr	0.1288
12	Chromium, ion	Water	PDF'm2yr	0.05413
13	Nickel, ion	Water	PDF'm2yr	0.04436
14	Copper, ion	Water	PDF'm2yr	0.03762
15	Cadmium, ion	Water	PDF'm2yr	0.03163
16	Zinc, ion	Water	PDF'm2yr	0.02708
17	Land use IV-IV	Raw	PDF'm2yr	0.02189
18	Chromium	Soil	PDF'm2yr	0.01688
19	Arsenic	Air	PDF'm2yr	0.01486
20	Cadmium	Soil	PDF'm2yr	0.01056
21	Mercury	Air	PDF'm2yr	0.005566
22	Ammonia	Air	PDF'm2yr	0.003769
23	Lead	Water	PDF'm2yr	0.002451
24	Zinc	Soil	PDF'm2yr	0.002314
25	Arsenic, ion	Water	PDF'm2yr	0.001239
26	Toluene	Water	PDF'm2yr	0.001005
27	Benzene	Water	PDF'm2yr	0.0003357

Table 3: Damage assessment of 1 ton bitumen production and substances which effect on ecosystem quality.

expressed in DALY/kg emission, resulting from radioactive radiation Ozone layer Damage, expressed in DALY/kg emission, due to increased UV radiation as a result of emission of ozone depleting substances to air. These effects expressed in DALY/kg emission.

Table 2 shows materials which have negative effects on human health during the 1 ton bitumen production, all included processes were described in introduction section. Compartment column shows where the substances release.

Ecosystem quality: Under this category two subcategories, Ecotoxicity and Acidification/ Eutrophication are considered. Ecotoxicity Damage to ecosystem quality, as a result of emission of ecotoxic substances to air, water and soil. Damage is expressed in Potentially Affected Fraction (PAF)*m^2 *year/kg emission and Acidification/ Eutrophication Damage to ecosystem quality, as a result of emission of acidifying substances to air, Damage is expressed in Potentially Disappeared Fraction (PDF)* m^2 *year/kg emission.

Table 3 shows all substances that have negative effects on ecosystem quality, these materials are producing from 1 ton bitumen production in different stages, the used unit also described earlier in this paper.

Conclusion

According to this paper in typical oil refineries the most damage occurs in resources category, in case of Pt unit resources category has 156.9 Pt units per 1 ton bitumen production, it means for obtaining 1 kg crude oil it will take 42.6 MJ energy. From extraction step to distribution bitumen in markets the most released emissions to the air are; nitrogen oxides, sulfur oxides, carbon dioxide, nickel and methane which all have negative effects on human health category, in this category respiratory inorganics subcategory has the highest effect. In respiratory inorganic subcategory using of diesel in diesel generators has the most negative impact on environment which produces nitrogen oxides to the air.

References

1. ICCT (2011) An introduction to petroleum refining and the production of ultra-low sulfur gasoline and diesel fuel. International Council of Clean Transportation, Bethesda, Maryland.

2. The Bitumen Industry a global perspective (2011) Asphalt institute Inc and European bitumen association-eurobitume, second edition, USA.

3. Eco Indicator 99 Manual for Designers (2000) Ministry of housing spatial planning and the environment, The Netherlands.

4. PRé (2016) Putting the metrics behind the sustainability: SimaPro Database Manual Methods Library. Netherlands.

5. Kennepohl G (2008) Ashalt pavements and the environment. ISAP International Symposium, Zürich.

6. Lattanzio RK (2014) Canadian Oil Sands: Life-cycle assessments of greenhouse gas emissions. Congressional Research Service, R42537.

7. Azhar Butt A (2014) Life cycle assessment of asphalt roads. Doctoral Thesis, KTH Royal Institute of Technology, Stockholm, Sweden.

8. Galatioto F, Huang Y, Parry T, Bird R, Bell M (2015) Traffic modelling in system boundary expansion of road pavement life cycle assessment. Transportation Research Part D: Transport and Environment 36: 65-75.

9. Han J, Forman GS, Elgowainy A, Cai H, Wang M, et al. (2015) A comparative assessment of resource efficiency in petroleum refining. Fuel Journal 157: 292-298.

10. Smith S, Durham SA (2016) A cradle to gate LCA framework for emissions and energy reduction in concrete pavement mixture design. International Journal of Sustainable Built Environment 5: 23-33.

11. Thomson H, Corbett J, Winebrake J (2015) Natural gas as a marine fuel. Journal of Energy Policy 87: 153–167.

12. Banar M, Özdemir A (2015) An evaluation of railway passenger transport in Turkey using life cycle assessment and life cycle cost methods. Transportation Research Part D: Transport and Environment 41: 88–105.

13. Environmental Impact of the Petroleum Industry (2003) Published by the Hazardous Substance Research Centers/South & Southwest Outreach Program.

14. Normalization (2015) New developments in normalization sets

21

Improving Recovery Through Surfactant Desorption on An Oil Wet Limestone Reservoir

Prince MJA*

Department of Petroleum Engineering, AMET University, Chennai - 603112, Tamil Nadu, India

Abstract

The current paper investigates on reducing surfactant adsorption after its application onto limestone surface by Ethelene Oxide (EO) conjugated with Sodium Dodecyl Sulphonate (SDS) to improve oil recovery. SDS has been treated with EO as a nonionic surfactant with low critical micelle concentration to increase its hydrophilic nature that leads to desorb itself from oil wet surfaces like limestone reservoirs after altering wettability.

Although, surfactant has a great impact on Oil recovery, adsorption makes them ineffective. It's been a great concern for petroleum industry during enhanced oil recovery operations. This study focuses on finding a mechanism to reduce SDS adsorption on an oil wet limestone core sample and altering its wettability through EO at different concentrations.

Critical Micelle Concentration (CMC) of SDS was found at 500 ppm by conductivity test was chosen to reduce interfacial tension between oil and brine composition. Due to surface charge variation SDS was observed to adsorb onto limestone surface through core analysis. It has been flooded after water under core flooding operations, which shows near to field observations. For reducing this effect, EO was introduced with different concentrations to alter the hydrophilic properties of SDS. Being limestone oil wet surface, which leads SDS to adsorb onto its surface.

Since, SDS would adhere onto the inner layers of core lead to alter wettability by recovering crude. The recovery of crude from a limestone core has been carried in two ways. Firstly, SDS was treated to get adsorb onto the core surface and secondly desorption of SDS by enhancing its hydrophilic nature through EO. By its application, the recovery of oil has been improved by reduction in adsorption of SDS successfully have been reported.

Keywords: Wettability alteration; Surfactant adsorption; Core flooding; Critical micelle concentration

Introduction

Globally, there is more than 50% of known oil reserves are in carbonate structures. The majority being oil wet primary and secondary recoveries are not sufficient to extract complete oil. Selection of a proper EOR method is required to alter the complex nature of carbonate reservoirs, lead more challenges in chemical flooding operations. Almost an average of 60% OOIP is left behind primary or secondary operations, majority at deep Oil wells. The concern area for surfactant application onto carbonate reservoirs has been limited to laboratories [1]. The unique structure and ability to alter surface properties makes surfactant more reliable for enhancing recovery [2].

Chemical flooding operations are less satisfied due to adsorption of surfactants on reservoir rocks and precipitation [3]. Adsorption and wettability depends on oil composition, structure of surfactants, blending mechanism and surface properties of rock. The mineral composition of rock plays a major role in adsorption and wettability alterations, which acts at solid liquid interface [4].

The application of surfactants on fields is limited because of fluctuations in oil prices. Even though some laboratory results are promising, the major concern is the large-scale availability of surfactants [5].

Adsorption of surfactant is the adherence of nonpolar molecules of surfactants which are organic in nature onto the carbonate surface by ion exchange and lipophillic bonding [6]. The adsorption depends on the availability of divalent ions, salinity and HLB ratio. Adsorption of surfactants means the loss in altering the surface properties which is uneconomical for chemical enhanced oil recovery [7].

Surfactant adsorption onto the surface depends on double layer at interface where there is polarity in charges [8]. At low concentrations, the adsorption is dependent on electric double layer. But at high concentrations near to CMC it depends on salinity, HLB and ionic strength. The adsorption of anionic surfactants has been observed to increase by increasing salt concentration (salinity), temperature and pH by addition of alkali [9].

In the current study, adsorption of surfactants has been reported at different temperatures with salt concentrations. SDS has been chosen to alter wettability for carbonate samples at CMC level. Due to high adsorption, the recovery of oil was observed to be low. The application of EO at different concentrations to reduce adsorption and enhancing oil recovery additional to the recovery obtained through SDS has been reported.

Experimental Methodology

Core flooding apparatus

The apparatus consists of a $3^{11} \times 5^{11}$ core holder which holds cores

---

'Corresponding author: Prince MJA, Department of Petroleum Engineering, AMET University, Chennai - 603112, Tamil Nadu, India
E-mail: prince466@gmail.com

Figure 1: Capillary pressure curves.

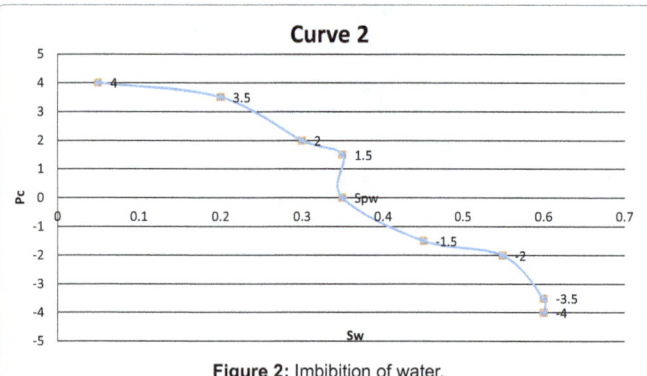

Figure 2: Imbibition of water.

of diameter less than its inner diameter of 3 inch as shown in Figure 1. Cores of $3^{11} \times 3^{11}$ has been kept inside core holder and mounted by an inlet tube. To make the core static it has been cemented between the inner surface of core holder and inlet tube. At the end of core holder, a porous plate has been mounted to bypass fluids from the core.

Cores can be flooded with low pressures based on column height to extract fluids from core. Wettability of a core sample can be calculated by constructing capillary pressure curves through core flooding. Capillary pressures are measured as the pressure difference between nonwetting oil phase and wetting water phase.

Curve 1: First drainage of water

Clean and dry limestone Cores have been placed inside core holder. Initially core has been flooded by water injected through inlet column. The pore volume of cores will be saturated by water has been displaced by oil injection. The pressure exerted by water inside core is wetting phase pressure (Pw) and the pressure by oil is non-wetting phase pressure (Po). Po has to reach higher than Pw will result in displacing water by oil. The resultant Pc will be positive and reaches maximum until the oil breakthrough at outlet. This indicates complete drainage of water by oil and the core is left with connate water saturation (Sw).

Curve 2: First imbibition of oil

The core of complete oil saturation with Sw would be displaced by water injection. Water has to be injected slowly to displace oil until the curve reaches Zero. At Pc zero the water saturation will be recorded as Spontaneous water saturation (Spw) which indicated water have been

saturated by itself or spontaneously. From here a little more pressure has to be applied on water to displace remaining oil in place until it reached residual state recorded as (Sor). The Pc would move towards extreme negative.

Curve 3: Second drainage of water

The process of oil injection will be repeated like curve 1. Slow injection of oil is preferable to displace water and to reach Pc at zero. At this level oil have been saturated in core spontaneously by itself and recorded as Spontaneous oil saturation(Spo) [10].

Estimation of wettability

Wettability is the nature of a reservoir to have partial attraction towards a fluid. Limestone samples have been aged with oil at reservoir conditions in a core oven. Then it has been cleaned by soxhlet apparatus with the treatment of heptane. Through core analysis with water and oil simultaneous flooding saturation exponents can be observed by constructing capillary pressure curves. The core has been found to be oil wet by amott wettability index. According to Amott wettability Index

I_w is imbibation of water and I_o is imbibation of oil

$$I_w = \frac{S_{spw} - S_{cw}}{1 - S_{cw} - S_{or}} \quad \text{and} \quad I_o = \frac{S_{spo} - S_{or}}{1 - S_{cw} - S_{or}}$$

If the difference between imbibation of water and oil is negative then core is Oil wet and positive for water wet. The core is intermediate wet at zero [11].

Critical micelle concentration test

Micelle is a form of droplet appears at the interface of oil and water by addition of surfactants. The formation of droplet will increase by increasing surfactant concentration [12]. At specific concentration, the micelle will appear with its lowest size leads to lower IFT at optimum. The concentration, where IFT is minimum is considered to be critical micelle concentration of that surfactant can be analyzed by conductivity [13].

In this test conductivity rises with increasing concentration of surfactants until the formation of micelle is completed [14]. Beyond addition of surfactants will increase the number of micelles, which has no effect on conductivity [15]. CMC can be observed by a peak variation on a graph between conductivity and surfactant concentration shown in Figure 2.

Emulsion tests

In this test, the concentration which has been chosen for core flooding operation should be suited for dissolution [16]. The CMC concentration from conductivity test will be tested with different proportions of brine and alkali for complete de emulsification. The suited proportion will be chosen by observing three clear layers in an emulsion after treatment with surfactants [17].

In the second stage, the selected proportion from first stage of three layers has been treated with EO at different concentration to increase hydrophilic nature by increasing HLB.

HLB calculation

EO mol. wt = 44 g/mol hydrophilic nonionic surfactant,

SDS mol. wt= 288.44 g/mol,

Curve 1			Curve 2			Curve 3		
Capillary Pressures (Po-Pw) psi	S_o	S_w	Capillary Pressures (Po-Pw) psi	S_o	S_w	Capillary Pressures (Po-Pw) psi	S_o	S_w
1	0	1	3.5	0.8	0.2	-4	0.4	0.6
1.5	0.2	0.8	2	0.7	0.3	-3.5	0.45	0.55
2	0.6	0.4	1.5	0.65	0.35	-2	0.5	0.5
3	0.95	0.05	0	0.65	0.35 Spw	-1.5	0.65	0.35
4	0.95	0.05 Scw	-1.5	0.55	0.45	0	0.75 Spo	0.25
4	0.95	0.05	-2	0.45	0.55	0	0.75	0.25
4	0.95	0.05	-3.5	0.4	0.6	0	0.75	0.25
4	0.95	0.05	-4	0.4 Sor	0.6	0	0.75	0.25

Table 1: Capillary pressures vs. saturations.

Number of emulsions	Brine Concentration in moles	SDS Concentration in ppm	Conductivity mS/cm
1	0.5	200	10.5
2	0.5	300	17.5
3	0.5	400	26.2
4	0.5	500	30.1
5	0.5	600	30.2
6	0.5	700	30.2

Table 2: SDS concentrations for conductivity.

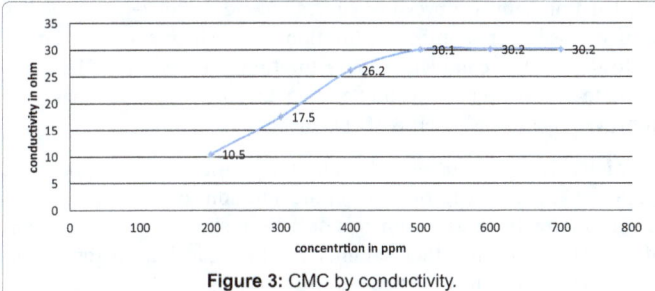

Figure 3: CMC by conductivity.

Hlb calculation for mixture of 10mole SDS and 20 mole EO will be

HLB= (20*44) / ((20*44) + (10*288.44)) = 0.23

0.23*100 = 23

HLB = 23/5 = 4.6

Six different emulsions with EO has been prepared and treated separately in core flooding operation. The increase in HLB will raise water solubility by addition of EO and reduces adsorption on an oil wet surface reservoirs.

Results

The limestone core has been saturated in oil for seven days to make it oil wet before core flooding. Then during core flooding capillary pressure curves were constructed to observe the level of wettability shown in Table 1.

According to amott wettability Index formulae Iw = 0.55 and Io = 0.64

Iw-Io = -0.09, which indicated oil wet.

After wettability, CMC was estimated by preparing six concentrations of Sodium dodecyle sulphonate (SDS) surfactants from 200 ppm to 700 ppm as shown in Table 2. The CMC has been observed at 500 ppm due to sharp deviation observed on graph between conductivity vs. concentration shown in Figure 3. After selecting 500

ppm of SDS as suitable concentration, six emulsions were prepared with different proportions shown in Table 3.

From the emulsion test 500 ppm of SDS with 1% wt alkali was observed to appear three clear layers as shown in Figure 4.

The pore volume (PV) of core sample has been calculated to be 30.5cc by Ruska porometer. During core analysis, core has been injected by 2 PV of water for 2 days with Pc of 0.09psi, where 1.35 PV was collected at outlet. Then 2 PV of oil has been injected into core to displace water. Upto 0.6 PV out of 0.65 PV saturated water was collected at outlet. The remaining water was considered as connate water saturation Scw of 0.05 PV at Pc of 4 psi shown in Figure 5. At outlet, 0.85 PV out of 2 PVoil has been collected. It shows core have been saturated and adsorbed with 1.15 PV of oil and 0.05 PV of connate water. Since, the core has a space limit of 1 PV with consideration of 0.05 PV of Scw the total absorbed amount of oil was observed to be 0.2 PV apart from saturated 1 PV.

Number of emulsions	SDS ppm in 15ml	NaCl wt%	Na_2CO_3 wt%	Appearance in layers	Inference
1	500	0.0	0.0	1 phase	w/o emulsion
2	500	0.5	0.0	2 phases	slightly w/o emulsion
3	500	1.0	0.0	2 phases	Light w/o emulsion
4	500	0.0	0.5	2 phases	Light w/o emulsion
5	500	0.0	1.0	3 clear phases	De emulsification
6	500	0.5	0.5	2 phases	Light o/w emulsion

Table 3: SDS concentrations with Nacl and alkali.

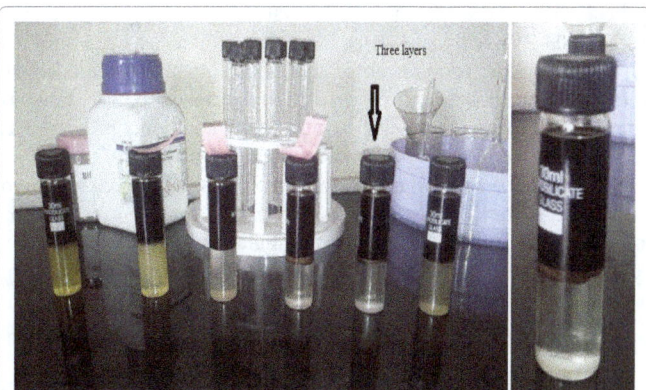

Figure 4: Six emulsion were prepared indicating three-layer middle micro emulsion.

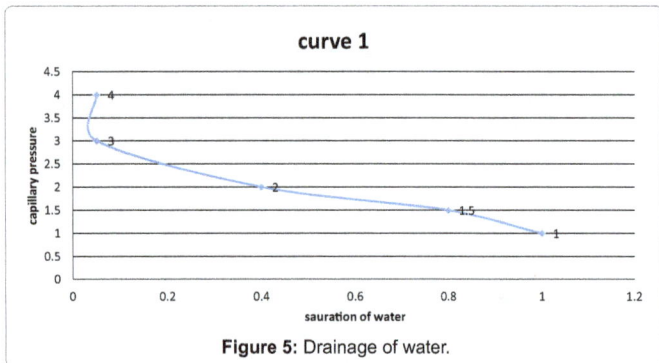

Figure 5: Drainage of water.

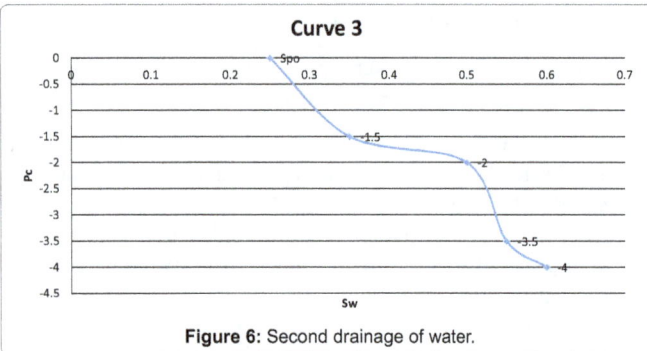

Figure 6: Second drainage of water.

ppm	Brine wt %.	concentration in moles	SDS ppm	HLB	Oil recovery in PV	Surfactant recovery SDS PV
				EO		
1000	0.5	10	500	2.6	0.05	0.33
2000	0.5	20	500	4.6	0.12	0.35
3000	0.5	30	500	6.2	0.13	0.44
4000	0.5	40	500	7.5	0.12	0.36
5000	0.5	50	500	8.6	0.11	0.36

Table 4: SDS concentration vs HLB.

The core sample was treated by injecting 5 PV of water to displace oil at Pc of 4 psi. It leads to collect 0.55 PV of oil out of 1.2 PV at outlet until breakthrough. Water has been imbibed by displacing oil spontaneously upto both pressures were equal. The saturation of water at this level is considered to be spontaneous saturation of water Spw of 0.35 PV beyond where additional pressure has been applied by injecting more water. Spw and Sor have been recorded at 0.35 PV and 0.4 PV respectively shown in Figure 3. In that 0.2 PV of oil is considered to be adsorbed. The left-out oil after water flooding is 0.4 PV as Sor in that 0.05 PV is connate water saturation. Water has been imbibed by displacing oil spontaneously upto both pressures were equal. The saturation of water at this level is considered to be spontaneous saturation of water of 0.35 PV beyond where additional pressure has been applied by injecting more water. The residual oil saturation of oil was observed at 0.4 PV, where the water saturation was maximum upto 0.6 PV as shown in Figure 3. The same process has been repeated by injecting oil to displace water until the capillary pressure becomes zero. Spontaneous oil saturation was recorded at this level to be 0.55 PV of oil as shown in Figure 6.

Discussion

Surfactant flooding through CMC and Emulsion tests were initiated after water flooding. 5 PV of diluted Emulsions has sent and 0.35 PV out of 0.55 PV oil was collected until breakthrough. The surfactants

were collected by inlet of 4.4 PV and lost 0.6 PV was observed to be absorbed by core. Remaining connate water 0.05 PV and residual oil 0.2 PV were left behind.

In this test the loss of surfactant emulsion is the resultant of adsorption due to electrostatic charge polarity between carbonate minerals and SDS. This has been reduced by treating the same emulsion with EO has a HLB enhancer. EO leads to increase hydrophilic nature of surfactants contrary to core nature makes it desorbed and enhances recovery.

Before treating with EO, five different concentrations have been chosen for miscibility with the emulsions already sent into core by Table 4. These five concentrations have yield different recoveries of oil and surfactants. From the Table 4 it has been observed that 3000 ppm of EO at HLB 6.2 is the effective combination that could recover oil of 0.13 PV out of 2.0 PV of 65% and SDS of 0.44 PV of 0.6 PV upto 73%.

Conclusion

The application of surfactants onto carbonate reservoirs has been effective under chemical EOR process. During core flooding analysis, the loss of surfactants was observed due to opposite ion interaction with the surface. Capillary pressure curves are considered to be one of the effective methods for estimating wettability of a core sample. While constructing capillary pressure curves second drainage of water has been stopped at spontaneous saturation of oil which can be extended up to the level of complete water saturation. The recovery of oil was found to be less after surfactant flooding due to adsorption. It has been improved by increasing HLB of SDS by EO.

Methods for reducing adsorption of surfactants have a great scope for enhancing recovery of crude addition to its flooding. HLB is one among the most parameters have been altered by the treatment of EO. There may be other parameters which can reduce adsorption of surfactants has to investigated. This process can be extended for reducing adsorption onto dolomite and sandstone reservoirs.

References

1. Orivri DU, Taiwo OA, Olafuyi OA (2014) Characterizing wettability Effect on Recovery from Surfactant Flooding in a lighty oil Porous Media. The Journal of Nig. Institution of Prod. Engineers 17: 143-152.

2. Atsenuwa JB, Taiwo OA, Mohammed IU, Dala A, Olafuyi OA (2014) Effect of viscosity of heavy oil (Class-A) on oil recovery in SP flooding using lauryl sulphate and gum Arabic. SPE 172401, Presented at SPE-NAICE, Annual Meeting, Lagos.

3. Avwioroko JE, Taiwo OA, Mohammed IU, Dala JA, Olafuyi OA (2014) A laboratory study of ASP flooding on mixed wettability for heavy oil recovery using gum arabic as a polymer. SPE 172401, Presented at SPE-NAICE, Annual Meeting, Lagos.

4. Raffa P, Wever DAZ, Picchioni F, Broekhuis AA (2015) Polymeric surfactants: Synthesis, properties, and links to applications. Chem. Rev. 115: 8504–8563.

5. Olajire AA (2014) Review of ASP EOR (alkaline surfactant polymer enhanced oil recovery) technology in the petroleum industry: prospects and challenges. Energy 77: 963–982.

6. Raffa P, Brandenburg P, Wever DAZ, Broekhuis AA, Picchioni F (2013) Polystyrene-poly (sodium methacrylate) amphiphilic block copolymers by ATRP: effect of structure, pH, and ionic strength on rheology of aqueous solutions. Macromolecules 46: 7106–7111.

7. Onuoha SO, Olafuyi OA (2013) Alkali/Surfactant/Polymer flooding using Gum Arabic; A comparative analysis. This paper was presented at the Nigeria Annual International Conference and Exhibition held in Lagos, Nigeria.

8. Xu F, Guo X, Wang W, Zhang N, Jia S, et al. (2011) Case study: Numerical simulation of surfactant flooding in low permeability oil field. A paper presented at presented at SPE Enhanced Oil recovery. Conference held in Kuala Lumpur, Malaysia.

9. Samanta A, Ojha K, Sarkar A, Mandal A (2011) Surfactant and surfactant-polymer flooding for enhanced oil recovery. Advances in Petroleum Exploration and Development 2: 13-18.

10. Wever DAZ, Ramalho G, Picchioni F, Broekhuis AA (2013) Acrylamide-b-N-Isopropylacrylamide block copolymers: synthesis by atomic transfer radical polymerization in water and the effect of the hydrophilic–hydrophobic ratio on the solution properties. J. Appl. Polym. Sci. 131: 39785.

11. Hongya W, Xulong C, Jichao Z, Aimei Z (2009) Development and Application of Dilute Surfactant-Polymer Flooding System for Shengli Oilfield. J. Pet. Sci. Eng. 65: 45-50.

12. Julius P (2015) Analysis on capillary pressure curves by wettability modification through surfactants. Indian Journal of science and technology.

13. Ahmadi MA, Shadizadeh SR (2012) Adsorption of novel nonionic surfactant and particles mixture in carbonates: Enhanced oil recovery implication. 26: 4655–4663.

14. Nasralla RA (2012) Double-layer expansion: Is it a primary mechanism of improved oil recovery by low-salinity waterflooding? Society of Petroleum Engineers.

15. Bo Gao MMS (2012) A New Family of Anionic Surfactants for EOR Applications. Society of Petroleum Engineers, USA.

16. Salari Z, Ahmadi MA, Ahmadi R, Kharrat R, Shahri AA (2011) Experimental studies of cationic surfactant adsorption onto carbonate rocks. Australian Journal of Basic and Applied Sciences 5: 808-813.

17. Austad T (2010) Chemical mechanism of low salinity water flooding in sandstone reservoirs. Society of Petroleum Engineers. USA.

22

Theoretical Application of Decision Support System in Petroleum Contaminated Ogoniland in South-Southern Nigeria

Arinze Emmanuel Emeka*

Department of Civil Engineering, Michael Okpara University of Agriculture, Umudike, Abia State, Nigeria

Abstract

Land is an indispensable natural resource made up of soil and groundwater, both of which have many functions for which we depend on, support of agricultural activities, engineering structures, portable water for domestic and industrial use as well as sustenance of flora and fauna in order to maintain favorable ecosystem. Contaminants in land pose a number of threats to public health and the environment; other natural resources; and have detrimental effects on property such as buildings, crops and live stocks. The most effective method of dealing with these contaminants is to cleaning up and returning the sites to beneficial use. The cleanup process involves making a choice from amongst competing remediation methods, where the wrong choice may have disastrous social, economic and environmental impact. This work presents the development of a Decision Support System via thorough literature survey. The Developed DSS was applied to petroleum contaminated site in Ogoniland South-East Nigeria. Finally, Air sparging, phyto-remediation and soil vapour extraction methods were systematically recommended.

Keywords: Decision support system; Petroleum; Contaminated land; Remediation

Introduction

Land is a limited resource that is increasingly getting polluted as a result of land contamination. Land is made up of soil and groundwater both of which have numerous functions for which humanity depend on, including provision of food and water, supporting shelter, natural flood defence, waste containment, maintaining natural cycles etc. Contaminants in land pose a number of threats to public health and ecosystem and have detrimental effects on lives and properties [1,2].

Petroleum hydrocarbon contaminants are amongst the most commonly occurring at contaminated sites. According to the European Environmental Agency (EEA) approximately 14.1 percent of identified contaminated lands in its countries are cause by the oil industry, with heavy metals, mineral oils and hydrocarbon contaminants constituting approximately 90% percent of the total contaminants found on sites [3]. According to available statistics, in the last 30 years more than 400,000 tons of oil have spilled into the creeks and soils of southern Nigeria. About 70% of the oil has not been cleaned basically because of inefficient cleaning or remediation method, ERA 2010. Figure 1 shows some contaminated sites in Ogoniland South Nigeria. The clean-up process involves making a choice from amongst competing remediation methods, where the wrong choice may aggravate the problem resulting to disastrous social, economic and/or environmental negative impact [1-8].

Contaminated land management is therefore much complex than the selection and implementation of removal solutions, and requires extensive data collection and analysis at huge cost and effort [2]. The need for decision support in contaminated land management decision making has long been widely recognised [4], and in recent years a large number of Decision Support System (DSS) have been developed. Nigeria is ploughed by numerous oil spills. The United Nations Environmental Programme (UNEP) announced that oil firms contaminated a 1,000 sq km (386 sq miles) area of Ogoniland, in the Niger delta with disastrous consequences for human health and wild life. Ogoniland is just a representative of many southern part of Nigeria that are contaminated [5]. The three-year investigation of UNEP discovered that;

i. Heavy contamination of land and underground water causes, sometime more than 40 years after oil was spilled.

ii. Soil contamination is located more than five metres deep in many areas studied.

iii. Most of the spill site oil firms claimed to have cleaned are still highly contaminated.

iv. The UNEP report also cited lack of adequate remediation procedure or institutional framework and lack of trained manpower in government supervisory agencies as another

Figure 1a: The cumulative impact of artisanal refining puts significant environmental pressure on Ogoniland.

*Corresponding author: Arinze Emmanuel Emeka, Department of Civil Engineering, Michael Okpara University of Agriculture, Umudike, Abia State, Nigeria, Tel: +2348188388313; E-mail: emmanuel.arinze@mouau.edu.ng

Figure 1b: Aerial view of artisanal refining site (Bodo West, Bonny LGA).

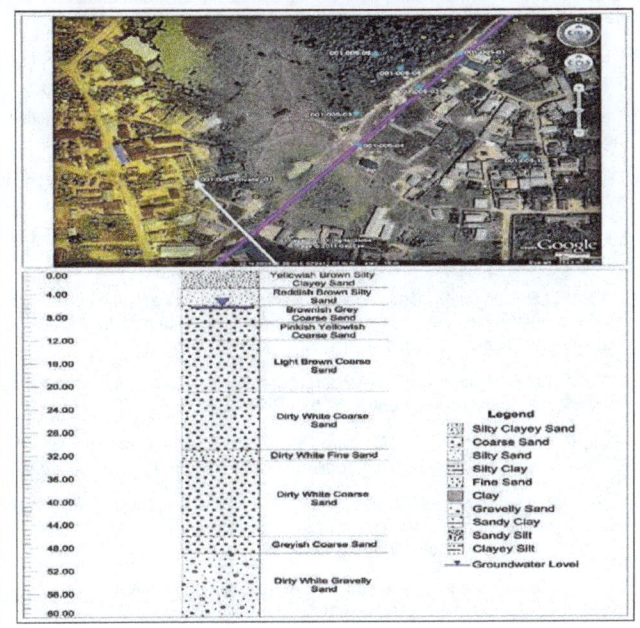

Figure 2: Soil logs from Nsioken Agbi Ogale, Eleme L.G.A.

reason why it is very difficult to embark on, and enforce appropriate remediation procedures.

The structure of Practise Guide for investigation and remediation of contaminated land (PG) follows the stages of the contamination assessment and remediation process as summarized inflow chart in Figure 2 [6-15].

Methodology (Decision Support System)

The Decision Support System is designed by ranking the contaminants, soil type, cost, efficiency and duration of operation as shown in Table 1 [16-23]. The comparative evaluation of soil remediation techniques using decision support system comprising of contaminant type, cost, soil type (Table 2).

Theoretical Application, Result and Discussion (A case study of Nsioken Agbi Ogale, Eleme L.G.A Ogoniland)

The result of soil logs from Nsioken Agbi Ogale, Eleme is shown in Figure 2; Considering that the oil spills took place between 1986 and 1990, natural attenuation, or biodegradation of contaminants has not proven effective in reducing contaminant concentrations to safe levels in the affected area [5,24-33]. Another reason why natural attenuation has failed is the presence of silty clay or pure clay soil in some areas. Natural attenuation does not work well in fine grained soils [34-40].

Critically analysing Table 2 in relation to the soil log result of Nsioken Agbi Ogale between 0-4 m, the most suitable method is the combination of phyto-remediation and soil vapour extraction. Phyto-remediation has to be applied first between 2 to 3 years. First, because of its effectiveness in all types of soil type, secondly, because of its cost effectiveness (Figure 2) [41-48].

After this period, the most effective, the relative cost effective method, that is soil vapour extraction should be used for at least one year to give the final recovery touch. SVE is emerging as the most frequently used technology. SVE not only promises good result in short time, but it is also cost effective [40,49-55].

Furthermore, between 4-6 m depth, methods effective in groundwater such as Air Sparging should be used in combination with SVE. SVE will take care of deficiencies evident in Air sparging such as inability to work where soil permeability is low and little lag in efficiency. Air sparging should be applied for six months before the application of SVE for one year [56-59].

Conclusion and Recommendations

The methods recommended in section 4 above should be applied in the study area. In summary, Air Sparging should be applied for six months between 4 to 6 metres whereas phyto-remediation should be applied for 2 years between 0-4 m. After this period soil vapour extraction should be used for at least one year over the entire length of six metres. After the remediation, TPH value should be checked to make sure it is within the acceptable limit, if not the SVE should continue for another one year. Finally, Nigeria government should put

Contamination		Soil Type		Cost		Efficiency		Duration	
Type	Representation	Type	Representation	Range	US/Tonnes	Range	Rank	Range	Rank
VOCS	A	Fine clay	A	>150	A	>90%	A	1-6 months	A
SVOCS	B	Medium clay	B	75-150	B	75-90%	B	6-12 months	B
Medium to heavy hydrocarbons	C	Silty clay	C	50-75	C	50-75%	C	1-2 years	C
		Clay silt	D	25-50	D	<50%	D	2-5 years	D
		Silty sand	E	25	E			>5 years	E
		Silt	F	<10	F				
		Sandy clay	G						
		Sandy Silt	H						
		Sand	i						

Table 1: Decision support system for petroleum containment land.

Technique	Contaminant	Soil type	Cost	Efficiency	Duration
Enhanced Bioremediation	B & C	A-I	C-D	B	D-E
Bioventing	B & C	D-F	C-E	A	A-B
Natural Attenuation	A & B	F-I	Variable	A-B	E
Phylo-remediation	A, B & C	Independent	D-E	C-D	D-E
Air Sparing	A & B	F-I	C	B	A
Soil Vapour Extraction	A & B	F-I	C	A	B-C
Thermal Treatment	A & B	A-I	E-F	B	A-B
Soil Washing	B & C	F-I	A-B	A	A
Incineration	A, B & C	A-I	E-F	B	A-B
Thermal Desorption	A & B	A-F except C	C-E	A	A-B
Excavation and Disposal	A, B & C	A-I	A-B	A	A-B

Table 2: Comparative evaluation of soil remediation techniques.

up a strong legal and institutional framework to enforce remediation of petroleum contaminated land. Government should train adequate manpower that will take charge of the aforementioned institutions as well as pay them what is obtainable in multinational oil companies in order not to lose their services.

References

1. Sanchez-Narre M, Gilbert K, Sodja RS, Skeyer JP, Struss P, et al (2008) Intelligent environmental decision support system. Development in Integrated Environmental Assessment 3: 11-44.

2. Vegter JJ (2001) Sustainable contaminated land management: A risk-based land management approach. Land Contamination & Reclamation 9: 95-100.

3. EEA (2007) The unseen threat to water quality: Diffused water pollution in England and wales report. European Environmental Agency.

4. Clarinet Association (2003) Sustainable management of contaminated land-An Overview.

5. UNEP (2012) Assessment of contaminated soil and groundwater in Ogoni south-south Nigeria. UNEP report.

6. Adams JA, Reddy KR (2003) Extent of benzene biodegradation in saturated soil column during air sparging. Ground Water Monitoring and Remediation 23: 85-94.

7. Alkorta I, Garbisu C (2001) Phytoremediation of organic contaminants in soils. See comment in PubMed Commons below Bioresour Technol 79: 273-276.

8. Alpaslan B, Yukselen MA (2002) Remediation of lead contaminated soils by stabilization/ solidification- water. Air and soil pollution 133: 253-263.

9. Asente-Dual DK (1996) Managing contaminated sites. Problem diagnosis and development of site remediation. Wiley, New York, NY.

10. Baker RS, Moore AT (2000) Optimizing the effectiveness of in-situ bioventing. Pollution Engineering 32: 44-47.

11. Barter MA (1999) Phytoremediation-an overview. Journal of New England Water Environmental Association 33: 158-164.

12. Bass DH, Hastings NA, Brown RA (2000) Performance of air sparging systems: a review of case studies. J Hazard Mater 72: 101-119.

13. Dambatta B, Javadi AA (2009) Risk based assessment and management of total petroleum hydrocarbon contamination in soil. 23rd European Conference on Operational research, Bonn, Germany.

14. Benner ML, Mohtar RH, Lee LS (2002) Factors affecting air sparging remediation system using field data and numerical simulations. Journal of Hazardous Material 95: 305-329.

15. Boire P (1998) Air sparging bioremediation of petroleum hydrocarbon contaminated soils. Biorem Technologies Inc Water100.

16. Chu W (2003) Remediation of contaminated soils by surfactant-aided soil washing. Practice Periosical of Hazardous. Toxic and Radioactive Waste Management 7: 19-24.

17. Chu W, Chan KH (2003) The mechanism of the surfactant-aided soil washing system for hydrophobic and partial hydrophobic organics. Science of the Total Environment 307: 83-92.

18. CPEO (1998) Thermal desorption. Center for Public Environmental Oversight 425, Market Street San Francisco, CA.

19. DENIX (1995) Natural attenuation for petroleum contaminated sites at Federal Facilities. Defense Environmental Network and Information Exchange.

20. Dermatas D, Mang X (2003) Utilization of flyash for stabilization/solidication of heavy metal contaminated soils. Engineering geology 70: 377-394.

21. Diele F, Notarnicola F, Sgura I (2002) Uniform air velocity field for a bioventing system design: Some numerical result. Int J Eng Sci 40: 1199-1210.

22. EPD (2011) Environmental Protection Department. The government of the Hong Kong Special Admin Region.

23. Erickson LE, Banks MK, Davis LC, Schivab AP, Muralidharan N, et al. (1999) Using vegetation to enhance In-Situ Bioremediation. Centre for Hazardous Substances Research, Kansas University. Manhattan KA.

24. ESEPA (1998a) Bioventing of the underground Storage Tank, US Environmental Protection Agency. Publication# EPA 510-B-95-007.

25. Feng D, Lorenzon L, Aldrich C, Mare PW (2001) Ex-situ diesel contaminated soil washing with mechanical methods. Mineral Engineering 14: 1093-1100.

26. Filler DM, Lindstorm JE, Braddock JF, Johnson RA, Nickalaski R (2001) Integral biopile components for successful bioremediation in the Arctic. Cold regions science and Technology 32: 143-156.

27. FRTR (1999d) Thermal desorption federal remediation technology roundtable. USEPA 401 M Street SW Washington DC.

28. FRTR (1999) Slury phase biological treatment. Federal Remediation Technologies Round Table. USEPA 401 M Street SW Washington DC.

29. FRTR (1999b) Bioventing federal remediation technologies roundtable. USEPA 401 M Street SW Washington DC.

30. FRTR (1999c) Biopiles federal remediation technologies roundtable. USEPA 401M street SW Washington DC.

31. GWRTAC (1996a) Phytoremediation groundwater remediation technologies analysis center six Avenue regional Enterprise Tower Pittsburgh PA.

32. GWRTAC (1996b) Technical documents-technical Overview reports. Groundwater Remediation Technologies Analysis Center. 425 6th Avenue. Regional Enterprise Tower Pittsburgh PA.

33. Halmemies S, Grondahl S, Arffman M, Nenonen K, Tuhkamen T (2003) Vacuum extraction based response equipment for recovery of Fresh Fuel spills from soil. J Hazardous Material 97: 127-143.

34. Harper BM, Stiver WH, Zytner RG (2003) Non-equilibrium nonaqueous phase liquid mass transfer model for soil vapour extraction system. J Environmental Engineering 129: 745-754.

35. Jørgensen KS, Puustinen J, Suortti AM (2000) Bioremediation of petroleum hydrocarbon-contaminated soil by composting in biopiles. Environ Pollut 107: 245-254.

36. Khan FI, Husain T (2003) Evaluation of a petroleum hydrocarbon contaminated site for natural attenuation using 'RBMNA' methodology. Environmental Modeling and Software 18: 179-194.

37. Khan FI, Husain T (2002) Evaluation of contaminated sites using risk based

monitored natural attenuation. Chemical engineering progress AICHE USA 34-44.

38. Khan FI, Husain T, Hejazi R (2004) An overview and analysis of site remediation technologies. See comment in PubMed Commons below J Environ Manage 71: 95-122.

39. Li P, Sun T, Stagnitti F, Zhang C, Zhang H, et al. (2003) Field-scale bioremediation of soil contaminated with crude oil. Environmental Engineering Science 19: 277-289.

40. Mihopoulos PG, Suidan MT, Sayles GD, Kaskassian S (2002) Numerical modeling of oxygen exclusion experiments of anaerobic bioventing. J Contam Hydrol 58: 209-220.

41. Mihopoulos PG, Suidan MT, Sayles GD (2001) Complete remediation of PCE contaminated unsaturated soils by sequential anaerobic-aerobic bioventing. Water Sci Technol 43: 365-372.

42. Nyer EK (1996) In situ treatment technology. Lewis Publishers, Boca Raton, FL.

43. Pulford ID, Watson C (2003) Phytoremediation of heavy metal-contaminated land by trees-a review. Environ Int 29: 529-540.

44. RAAG (2000) Evaluation of risk based corrective action model remediation alternative assessment group, Memorial University of newfoundland St. John's NF, Canada.

45. Rai JPN, Singhal V (2003) Biogas production from water hyacinth and channel grass used from phytoremediation of industrial effluents. Bioresource Technology 86: 221-225.

46. Riser-Roberts E (1998) Bioremediation of petroleum contaminated sites CRC Press Raton FL.

47. Suffersan SS (1997) Remediation Engineering: Design concepts. Lewis Publishers, Boca Raton, FL.

48. Urum P, Pekdemir T, Gopur M (2003) Optimum conditions for washing of crude oil-contaminated soil with biosurfactant solutions. Process Safety and Environmental Protection: Transactions of the Institution of Chemical Engineers, part B 81: 203-209.

49. USEPA (1995a) How to evaluate alternative clean-up technologies for underground storage tank sites. Office of solid Waste and Emergency responses US Environmental Protection Agency Washington D.C.

50. USEPA (1996a) A citizens guide to natural attenuation. Office of solid waste and emergency response. US environmental protection agency publication # EPA 542-F-96-015 Washington DC.

51. USEPA (1996b) A citizen's guide to phytoremediation office of solid waste and emergency response, US Environmental Protection Agency. Washington DC.

52. USEPA (1998c) Soil vapour extraction (SVE) Office of the Underground Storage Tank, US Environmental Protection Agency.

53. USEPA (1998b) Biopiles, Office of the Underground Storage Tank, US Environmental Protection Agency.

54. USEPA (1996e) A citizen's guide to in-situ thermal desorption, Office of solid Waste and Emergency Response. US Environmental Protection-Publication # EPA 542-F-96-005, Washington, DC.

55. USEPA (1996d) A citizen's guide to soil washing. Office of Solid Waste and Emergency Response, US Environmental Protection Agency Publication #EPA 542-F-96-002, Washington D.C.

56. Vouillamoz J, Milke MW (2001) Effect of compost in phytoremediation of diesel-contaminated soils. Water Sci Technol 43: 291-295.

57. Wait ST, Thomas D (2003) The characterization of base oil recovered from the low temperature thermal desorption of drill cuttings. SPE/EPA Exploration and Production, Environmental Conference, Mar 10-12, San Antonio, TX – pp 151-158.

58. Wiedemeier TH, Newell CJ, Rifai HS, Wilson JI (1999) Natural attenuation of fuels and chlorinated solvents in the sub forces. Wiley, New York.

59. Zhan H, Park E (2002) Vapour flow to horizontal well in unsaturated zones. Soil Science Society of American Journal 66: 710-721.

23

An Experimental Study on the Application of Ultrasonic Technology for Demulsifying Crude Oil and Water Emulsions

Mahmood Amani*, Idris M, Abdul Ghani M, Dela Rosa N, Carvero A and Yrac R

Texas A&M University at Qatar, Qatar

Abstract

An emulsion is the mixture of two immiscible fluids, where one fluid appears as droplets within another. In the oil and gas industry, produced crude oil generally comes with an appreciable amount of water within it in an emulsified form. Before produced crude oil can be prepared for purchase, the water associated with it must be removed. A process known as demulsification is required in order to separate an emulsion into its two phases. In the industry, a number of demulsification techniques are already present; these include thermal, mechanical, chemical, and electrical techniques.

Crude oil and gas produced from wells originally come with water, salts, and volatile gases such as oxygen, carbon dioxide, and sometimes hydrogen sulfide, etc. Hence, the petroleum mixture needs to be refined-water, salt, and non-hydrocarbon gases to be separated from the mixture, in order to meet certain oil and gas specifications (which state the maximum concentrations of such contaminants) and make it ready for purchase and transportation.

Sonication provides a cheap, simple, and harmless (as it involves mainly the propagation of sound waves) way of separating crude oils from water droplets via demulsification. In addition, if needed, it can be used for emulsification processes as well. Hence, a study of sonification as a way for crude refinement or chemical mixing has important implications for the oil and gas. This investigation proposes the use of ultrasonication as a new and cost-effective technique to aid in the demulsification of crude oil emulsion. The effectiveness of this technique was gauged through its comparison to the already present methods in the industry. Based on the investigation it was found that centrifuge served as the best demulsification method for it reduced the turbidity by 86%. In addition, the reduced turbidity achieved with proposed ultrasonication method ranges from 20%-60%.

Keywords: Ultrasonic; Crude oil; Emulsions

Significance

In the production of crude oil, usually a significant amount of water is also produced. Many times, this water and oil mixture is in the form of emulsions. Emulsion separation is very time consuming, requires additional surface facilities and can be very costly. Any new method that can increase the efficiency of this process and or provide a cheaper method would be very much helpful to the petroleum industry.

Objective

The objective of this research is to investigate the potential use of ultrasonication as a new and cost-effective technique to aid in the demulsification of crude oil emulsion. This research is looking at various existing techniques and will focus on the use of ultrasonic waves as a potential technique for emulsion separation.

Introduction

More than thirty percent of all crude oil produced in the world comes to the surface with an appreciable amount of water in an emulsified form [1]. A recurring issue in the oil industry is the separation of water from produced oil, its significance as an issue "is shown by an estimated 15-20 million dollars expended for chemicals each year to treat the world's oil production" [2]. An emulsion is defined as the dispersion of one liquid as droplets in another immiscible liquid. The water-in-oil emulsion (W/O) that occurs because of crude oil production is the most common oil emulsion that is discussed in the industry and is the focus of this investigation; however, oil emulsions can come in many forms. The second most common form of oil emulsions is oil-in-water, sometimes referred to as reverse emulsions (Figure 1). Emulsions can also occur in more complex manners such water-in-oil-in-water, illustrated in Figure 2, where the droplets themselves house a third immiscible liquid.

In order to make the produced crude oil ready for purchase and transportation, the crude oil needs to be refined – water, other

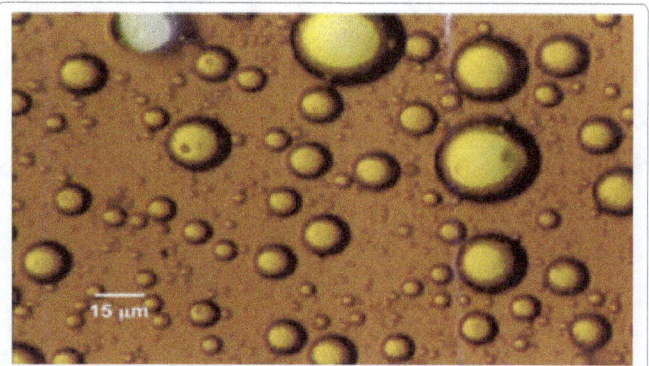

Figure 1: Photomicrograph of a W/O emulsion and a W/O/W emulsion.

*Corresponding author: Mahmood Amani, Texas A&M University at Qatar, Qatar
E-mail: mahmood.amani@qatar.tamu.edu

nonpetroleum fluids, and solids to be separated from the mixture. Sonication is a cost-effective, simple, and safe (since it only involves sound wave propagation) means of separating water droplets from crude oil (demulsification). It may even be used for emulsification purposes as well. Therefore, sonication has important implications for the oil and gas industry.

We begin by discussing the characteristics and stabilizing components of emulsions. Second, we describe demulsification and the factors that affect such process. Third, we present the current demulsification methods used by the oil and gas industry. Fourth, we cover basic theory behind waves and sounds, and sonication. Finally, we examine sonication as means of demulsification.

Emulsion

Emulsions can be characterized by their kinetic stability, their stability over a period. In other words, emulsions are grouped based on their separation rates. Based on kinetic stability, emulsions can be divided into three groups: loose, medium and tight emulsions. Loose emulsions can be thought of as highly unstable with a low separation time while tight emulsions are very stable with relatively high separation times. Kinetic stability of an oil emulsion is a result of two parameters: droplet size and the interfacial film surrounding the droplets. Droplet size is quite intuitive as smaller droplet sizes are a clear indication of a better size fluid and therefore a more stable emulsion. Likewise, the interfacial films that surround water droplets protect water droplets from combining with each other and thus increasing the emulsion stability. More viscous interfacial films lead to more stable oil emulsions. The interfacial films around the water droplets play the largest role in creating a stable oil emulsion. Interfacial films with higher viscosity work the best in preventing the emulsion in separating. These films can be enhanced with the addition of an emulsifying agent. Emulsifying agents are generally naturally occurring and come mixed with the produced crude oil. They can be divided into two main categories: surfactants and fine solids. Surfactants work by attaching to the oil water interface, due to their unique affinity to both oil and water, where they form an interfacial film. Examples of surfactants are asphaltenes and resins. Fine solids work to stabilize emulsions by providing a mechanical barrier in between the droplets in addition to the interfacial film. These solids are most effective when they are smaller than the dispersed droplets. Examples of fine solids found in crude oil emulsions include clay, sand, and silt.

Given a mixture of two immiscible fluids, mechanical agitation, or disturbance, causes dispersion of one phase as tiny droplets (dispersants) throughout the other continuous liquid phase, and

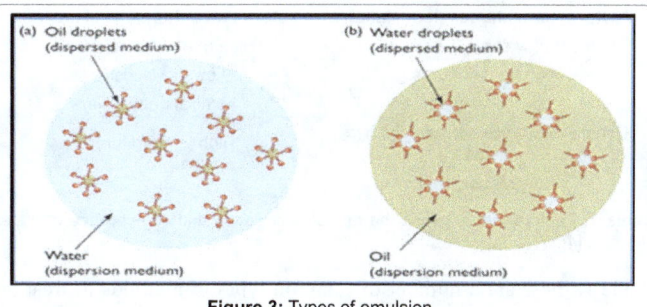

Figure 3: Types of emulsion.

subsequently the formation of the interfacial films of surfactants around the dispersed droplets for stabilizing an emulsion. Hence, sound wave propagation, which causes agitation via pressure fluctuations, can be used for emulsification purposes such as mixing viscosifier and oil into water-based drilling muds to obtain appropriate rheological properties (Figure 3).

Demulsification of crude oil (water-in-oil emulsion)

Demulsification is the separation of the emulsion into its separate components (water and oil). Demulsification as a treatment process is defined by the:

1. Rate of separation

2. Remaining water/salt content in oil.

3. Remaining oil content in water.

Ideally, an optimum demulsification method would separate the crude oil while maximizing on these three parameters. Reduction of the rate of separation will aid in maximizing profits by increasing the efficiency of the treatment process. Moreover, the goal of any demulsification is to eliminate the water in oil emulsion that occurs as part of the desalting process (intentional mixing of crude and fresh water) as well as the emulsion with the water content that comes naturally with produced crude. The separation of the water droplets containing dissolved salt is ideal in avoiding production problems including "corrosion, scale accumulation, lowering of activity analysts and plugging or fouling in pipeline [3]. Moreover, demulsification is critical in achieving a marketable standard of crude oil. Finally, ideal demulsification would see that the separated water phase is composed of low oil content so that its disposal or reuse satisfies the environmental standards established by the industry.

Demulsification can be defined as breaking emulsion kinetic stability. As a result, demulsification can be said to occur in a two-step process consisting of flocculation and coalescence. Flocculation involves the gathering of water particles to near proximity of each other, while coalescence occurs when the individual films holding the water droplets break to allow the droplets to combine and culminate into bigger drops. Neither of these processes occurs faster than the other; however flocculation general must occur first before coalescence. As a result, in designing a demulsification program one must subject the emulsion to a combination of treatments that enhance both the flocculation and coalescence of the water droplets. Table 1 provides a list of the factors that affect both of these processes'. It is important to note that a high flocculation will aid in enhancing coalescence.

Curent demulsification techniques

Demulsification programs are very specific to the oil field and the crude oil that is produced. However, based on the factors presented in

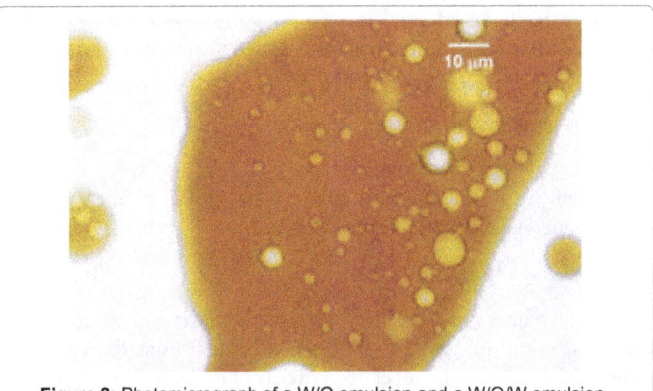

Figure 2: Photomicrograph of a W/O emulsion and a W/O/W emulsion.

Factors that enhance flocculation	Factors that enhance coalescence
Amount of water in the emulsion	High rate of flocculation
High temperatures	Low oil viscosity
Low oil viscosity	Chemcial demulsifiers
Density difference between the two fluids	High temperatures
Electrostatic field	

Table 1: Factors that enhance the rate of flocculation and coalescence of water droplets [4].

Table 1, there are a number of common types of methods to treat an emulsion. Emulsion treatment methods can be grouped into four main categories; these are thermal, mechanical, electrical, and chemical.

Thermal: The application of thermal energy is probably the simplest method of demulsification used in the application of thermal energy is probably the simplest method of demulsification used in the industry. Thermal energy reduces the viscosity of the oil and increases the water settling rates." [4]. In addition, the increased temperatures "also result in the destabilization of the rigid films caused by reduced interfacial viscosity" and in some cases even rupturing of the films which in turn aids in the coalescence of the water particles [4]. The increased temperature also leads to an overall increase of kinetic energy of the particles leading to a higher coalescence frequency of the water droplets because of their increased mobility. In other words, heat accelerates the demulsification process in a variety of ways, however it is important to note that heat alone is not known to completely separate an emulsion on its own especially when dealing with stable emulsions. In fact, the application of thermal energy to an oil emulsion can lead to some undesirable effects, such as a reduction of the API gravity, and an "increased tendency for corrosion and scale deposition in treating vessels" [4]. In light of this, it's always important to weigh the costs of heating an oil emulsion against the adverse effects of the treatment in deciding whether thermal energy is an economic demulsification technique for the given emulsion case.

Electrical: An electrical demulsification method that is commonly used in the industry involves the application of an electric field perpendicular to the direction of flow. The associated charge of the water droplets along with the electric field can result in one of three main phenomena in occurring, all of which aid in destabilizing the emulsion. These phenomena include:

• The electric field aligns the water droplets based on polarity, thus positive ends of droplets are brought next to negative ends of other droplets. Through electrostatic attraction these droplets come closer and closer together eventually leading them to coalesce

• The polarity caused by an electric field can also make water particles attracted to an electrode. This can lead to large amount of water particles gathering and collecting in one-area allowing forming larger water droplets and eventually settling and separating from the oil.

• When an AC field is applied to an oil emulsion that effect is a weakening or even rupturing of the film surrounding the water droplets. This occurs because of the cyclic nature of an AC electrostatic grid. During the high voltage, phase of this cycle the water droplets are elongated along either end of the droplet, when the cycle returns back to the low voltage phase the droplets snap back to a spherical shape.

Regardless of which of the three phenomena occur when an electrostatic grid is applied to an oil emulsion the result is an acceleration of the demulsification process. An umbrella term for all three phenomena is known as Electrostatic Dehydration. This method is best used in combination with chemical or thermal techniques [5].

Chemical: The most common of the four techniques is the use of chemicals, known as demulsifiers, in treating an oil emulsion. Demulsifiers work by breaking the interfacial film protecting the trapped water droplets. In order to optimize the use of demulsifiers a number of factors are considered before the addition of a demulsifier into a crude oil emulsion including, proper selection of demulsifier accurate dosage, and adequate mixing time.

Demulsifier selection

In general, demulsifiers can be broken down to 3 critical components; solvents, surface-active ingredients, and flocculants. Over the years the selection of chemicals used as demulsifiers in the industry has vastly increased. With such a large number of options available it's very important that the right chemical combinations are selected in order to ensure the best results. Demulsifier selection begins first with characterization of the crude oil emulsion in question; breaking down the type of crude oil/brine and all the contaminants/solids that make up the emulsion. Next, the production conditions of the crude oil emulsion should be considered including operating temperatures, production rates, treating vessel characteristics etc. Finally, it's important to consider any past record on the performance of the demulsifier with other crude oil emulsions. Demulsifier selection is always a unique experience and can vary considerably between different well case studies.

Demulsifier dosage and mixing time

Similar to demulsifier selection, demulsifier dosage is unique and depends entirely on the chosen demulsifier and the properties of the emulsion. Proper dosage is highly important in ensuring an optimized treatment of the emulsion. In fact, while too small of a demulsifier dosage will leave an emulsion as is, too much demulsifier can actually result in a more stable emulsion. Again, due to the large number of factors that can vary significantly between scenarios it's important to conduct accurate testing with the case in question in determining the optimum dosage. Equally as important is ensuring the chosen demulsifier is mixed adequately into the emulsion. Demulsifiers work by attaching to the oil and water interface and therefore the mixture has to be subjected to enough agitation in order to allow the demulsifier to reach the interfacial film. However, an unnecessary amount of mixing can reverse the effects of the demulsifier and stabilize the emulsion. Therefore, it's important to monitor the mixing of the demulsifier into the emulsion and decrease agitation significantly once the demulsifier begins to break the emulsion.

Mechanical

There are a variety of emulsions treating equipment that can be used to emulsify an oil emulsion. Some of the common ones used in the industry include: free knock out water drums; two/three phase separators, desalters, and settling tanks.

Free knock out water drums

Aside from taking out water from crude oil, free-water knockout drums can also separate associated gases. This equipment usually supplements main demulsification equipment.

Three-phase separators

As its name suggests, the three-phase separator separates the produced crude oil into three phases existing within the crude-oil, water, and gas. They can be set up vertically or laterally. The separator is appropriately assigned set retention time for sufficient separation at any required throughput rate. The separator can have components of heat section, wash water, filter section, stabilizing section, and electrostatic grids. It is no surprise then that large varieties of separators are in used today. What specific separator design would a demulsification task require is a complicated engineering task, which must consider many factors?

A flexible separator design, which allows for modification, is most preferable. Pressures and temperature conditions, water cuts, and oil and brine compositions change over the life of the field; hence, a well-designed separator is one that allows for adaptations to these fluctuating operation conditions.

Coalescer packs are a good way to increase separator efficiency. These packs increase the amount of fluid traversing through the separator. Water droplets coalesce as the emulsion rolls or wipes through the packing. Installation of spreaders, which increase droplet collision frequency, are also viable choice for improving separation efficiency.

Desalters

Separators usually are not enough to reduce the amount of water in the crude oil to marketable standards. In addition, some salt must be taken out as well. Hence, a desalter must be employed for removing the remaining water and salts. Desalters usually employ combinations of settling, chemical addition, and electrostatic treating. The time for settling depends on oil specification. With chemical addition, fresh water is also added for decreasing salt concentration (i.e. dilution) in the effluent water and crude. A one-stage desalter schematic is shown in Figure 4.

These are some of the most common techniques used in the industry to achieve demulsification. It is interesting to note that some techniques have overlapping components. For instance, desalters and three-phase separators may both employ a settling component and an electrostatic treating. Generally, a combination of these techniques is used in conjunction with each other to develop a complete demulsification program.

Basic wave theory, sound, and sonication

Harmonic motion: Harmonic motion is any repeating and oscillating motions. For instance, the back and forth motion of a mass attached to a fixed spring which is stretched, the swinging of a pendulum, and the motions of particles in a vibrating mass are all harmonic motions. An object in harmonic motion is an oscillator. A harmonic motion results from two things: A restoring force and inertia. A restoring force acts on the object in the direction towards equilibrium or decreasing potential energy. Inertia is the object's resistance to motion, or more specifically, the resistance to return to equilibrium. Hence these two-act opposite to each other in harmonic motion.

Period and frequency are the important parameters for describing the harmonic motions. Period T is the amount of time given in seconds (s). For one cycle of the motion to occur, and frequency f is the number of cycles per unit time and is given in Hertz (Hz=1/s). Also, they are related as:

$T=1/f$

When the restoring force is directly proportional to the object's displacement, the motion is called simple harmonic motion (SHM). This can be stated as.

$F=-kx$

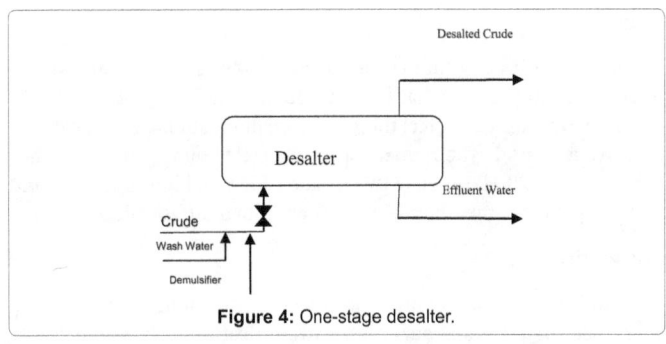

Figure 4: One-stage desalter.

The minus sign is due to the restoring force acting opposite the displacement. The examples of harmonic motion given previously are all simple harmonic motions, although not necessarily for the vibrating mass case. Simple harmonic motions are described mathematically as sinusoidal functions:

$x\ (t)=A\cos\ (\omega t+\varphi)$

Where x, t, A, ω, and φ are the displacement of the object, time, angular frequency, the constant called amplitude, and the constant called phase angle, respectively. The angular frequency is defined to be:

$\Omega=2\pi/T$

Amplitude is the maximum displacement from equilibrium position of the object. For example, the amplitude for a swinging pendulum is the pendulum's maximum vertical displacement from its position when it is not in motion.

Wave

Waves are disturbances or oscillations carrying energy, and propagates through space (and are then electromagnetic waves), light for instance or mass (and are then called mechanical).

Mechanical waves

Mechanical waves are disturbances that propagate through matter, called the medium. As mechanical waves propagate continuously through the medium, the particles move roughly in a simple harmonic motion. Mechanical waves are generated for industrial or experimental purposes using simple harmonic oscillators.

The propagation speed of mechanical waves, or wave speed v, increases with higher restoring forces F, and decreases with higher inertia μ. These relationships are more precisely given by:

$V=F/\mu$

Transverse waves

When the particles move perpendicular to the direction of wave propagation, the wave is transverse. Consider a string is fixed at one end to a wall and tied to a continuously moving oscillator in the other. Then as the wave generated by the oscillator propagates through the string, each of the string's particle move up and down, perpendicular to the wave direction. The collective motion of the particles results in a sinusoidal waving string. This is appropriately called a sinusoidal motion.

Longitudinal waves

When the particles move parallel to the direction of wave propagation, the wave is longitudinal. When a longitudinal wave propagates through a medium, particles move in or opposite the wave direction, and causing alternating regions of unidirectional rarefaction, low pressure, and compression, high pressure.

Sound

Sounds are longitudinal mechanical waves. These are micro fluctuations in pressure that travel through a medium, commonly air. Our sensitive ears can detect these pressure fluctuations, which we then perceive as sound. The human ears can detect sound with frequency ranging from 20 Hz to 20,000 Hz, called the audible range. A sound with frequency higher than 20,000 Hz are called ultrasonic.

Sonication

Sonication is simply the propagation of sound waves into a substance for agitation purposes. Ultrasonication uses sound waves of frequency above 20 KHz (ultrasonic) and is commonly used. A sonication equipment may comprise of an electricity generator that generates AC electricity, a transducer that converts the AC electricity into mechanical vibrations, a transformer that amplifies this vibration, and a tool tip which makes the amplified vibration energy available for application (hence, tool tip is the oscillator to be applied) [1].

Demulsification using ultrasonification

Demulsification can be achieved using ultrasonification. Currently, there is not one unanimously agreed mechanism for which ultrasonification causes demulsification, although reasonable explanations have been proposed. The following is such an explanation [1]:

1. Ultrasound waves of high intensity produce cavitation or regions of free fluid spaces in the liquid.

2. Cavitation leads to rapid fluctuation of pressure.

3. Relatively large masses of vacuous, small bubbles form.

4. Bubbles grow until it reaches critical size.

5. Bubbles implode and generate intense shockwaves leading to high temperature and microstreaming of the liquid.

6. High shear gradient is produced throughout the mixture which weakens the interfacial films.

7. Emulsion is destabilized.

Demulsifiers are commonly employed for demulsification purposes, and ultrasonification can be used to complement them. Demulsification using demulsifiers is made more efficient through ultrasonification since the agitation due to ultrasonic waves aids in mixing the demulsifiers through the emulsion mixture better. This way, the demulsifiers may reach more of the interfacial oil-water film, drain the film of surfactants, and decrease the film's strength. This induces the coagulation of dispersed water blobs, and consequently the destabilizing of emulsion.

Methodology

As this project attempts to compare the efficiency of ultrasonication to other methods of oil demulsification, each of those techniques were performed separately. Before any experimentation was done

% Solution, oil in Water	Oil to Solvesso ratio	Highest turbidity, NTU
0.1	0.05:0.05	975
0.08	0.04:0.04	713
0.06	0.03:0.03	541
0.04	0.02:0.02	252
0.02	0.01:0.01	152

Table 2: The base-turbidity of each emulsion.

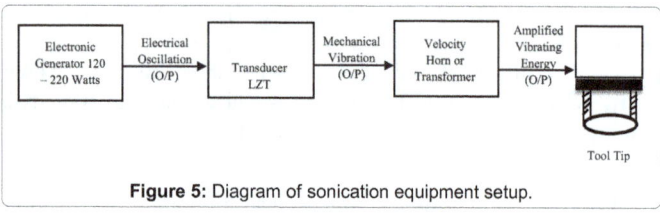

Figure 5: Diagram of sonication equipment setup.

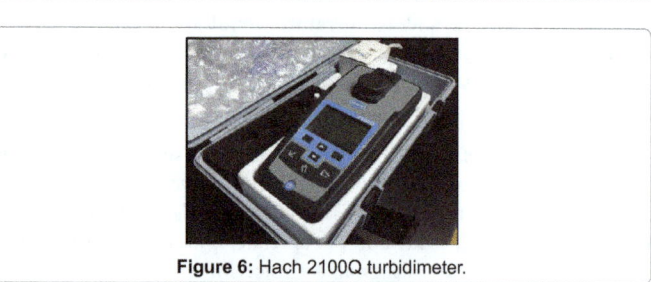

Figure 6: Hach 2100Q turbidimeter.

Figure 7: Solvesso 150.

to determine the demulsificating effect of any mechanism, turbidity for each oil-water concentration was obtained. This initial turbidity was then compared to the turbidity of the solution after any specific test was performed. From basic analysis, the test that results in the most turbidity decrease can be said to be the most effective test for demulsifying an emulsion.

Base-turbidity determination

To determine the base-turbidity, different concentrations of emulsions were created. To create each emulsion, studied concentrations of oil were added to a relevant amount of water. To form a stable emulsion, an emulsifier (Solvesso 150) was added alongside the oil after which rigorous mixing was done for 70 seconds with an electric solution mixer. This procedure was carried on for every emulsion created in this project. After mixing, the resulting solution was distributed into 5 beakers right after mixing and was allowed to set for 7 min before a turbidity reading was taken.

Many trials of this experiment were repeated and the concluded base turbidity for each mixture of different oil concentration is shown in Table 2. Turbidity reading was done on a turbidimeter from Hach Instruments Company (Figures 5-7).

Emulsion breaker

The concentration of the demulsifying agent which serves as a catalytic agent through time and external conditions such as temperature and pressure can satisfy and bridge to the comparison of all subjected samples.

As with the base turbidity test, emulsions were created by mixing different concentrations of oil and Solvesso 150 with water. In this test the following procedure was followed:

1. For a specific concentration of oil, an emulsion of around 1000

| Concentration | | Turbidity, NTU | | | | | | | | | |
| Oil in water | Emulsion breaker | Trial 1 | | | Trial 2 | | | Trial 3 | | |
%	ppm	Initial	Final	%Reduction	Initial	Final	%Reduction	Initial	Final	%Reduction
0.08	80	905	747	17	865	629	27	717	675	6
	60	924	691	25	890	659	26	890	762	14
	40	950	756	23	908	618	32	686	592	14
	20	951	742	22	902	695	23	644	564	12
	15	963	795	17	928	666	28	987	808	18
0.06	80	618	614	1	492	477	3	656	642	2
	60	571	578	-1	489	462	6	629	696	-11
	40	579	551	5	337	327	3	599	592	1
	20	555	525	5	336	300	11	611	599	2
	15	527	493	6	337	293	13	594	546	8
0.04	80	327	304	7	275	286	-4	320	329	-3
	60	312	299	4	274	276	-1	273	282	-3
	40	309	285	8	262	261	0	270	269	0
	20	292	274	6	263	253	4	260	260	0
	15	286	261	9	271	264	3	263	254	3
0.02	100	118	138	-17	117	126	-8	202	214	-6
	80	123	123	0	172	179	-4	190	197	-4
	60	125	115	8	172	166	3	192	187	3
	40	125	106	15	167	156	7	192	182	5
	20	118	99	16	168	160	5	178	174	2

Note: Shaded cells highlight results with the highest turbidity reduction.

Table 3: The effect of different concentrations of emulsion breaker on different concentrations of emulsion.

| Concentration | | Turbidity, NTU | | |
| Oil in water | Emulsion breaker | Trial 1 | | |
%	ppm	Initial	Final	%Reduction
0.08	80	632	604	4.4
	60	674	615	8.8
	40	571	617	-8.1
	20	701	621	11.4
	15	670	657	1.9
0.06	80	420	376	10.5
	60	463	396	14.5
	40	422	389	7.8
	20	411	399	2.9
	15	445	428	3.8
0.04	80	520	491	5.6
	60	480	446	7.1
	40	500	383	23.4
	20	508	409	19.5
	15	508	412	18.9
0.02	80	92.2	91.2	1.1
	60	85.5	92.1	-7.3
	40	88.7	90.6	-2.1
	20	92.4	88.7	4
	15	98.8	94.3	4.6

Table 4: The Effect of different concentrations of emulsion breaker on different concentrations of emulsion (continuation).

mL was created then immediately distributed in 5 beakers after mixing and left to set for 7 min.

2. After setting for 7 min, an initial turbidity reading was recorded and a noted amount of emulsion breaker was added.

3. The mixture of emulsion and emulsion breaker was mixed through a magnetic stirrer on 60 rpms for 2 min, 40 rpms for 2 min then 20 rpms for the last 2 min. A final turbidity was then recorded

after mixing. This process was repeated for 3 trials. The results of all those trials are presented in Table 3.

Water bath

Since the density and viscosity depends on the temperature of the samples, this test aims to demulsify the samples by the function of time, temperature, viscosity and density.

To perform this section of the project, two things need to be optimized: the time and the temperature of exposure. To do that, the regular emulsion previously discussed was created and separated in different beakers. After taking and initial turbidity reading, these beakers were placed in a water bath at a certain temperature and each of them was left inside for a certain amount of time. After removal of the beakers from the water bath, the final turbidity reading was measured. This was repeated for different temperatures. The results of this are presented in Table 4.

This technique has been done in an SDM water bath device with a maximum heating temperature of 70°C. This device is shown below on Figure 8.

Centrifuge

To test the demulsification efficiency of the centrifugal technique, the test needed to be optimized. For a fixed speed of 5000 rpm and a

Figure 8: Water bath.

Oil in water, %	EB concentration	% Reduction
0.08	20	11
0.06	60	15
0.04	40	23
0.02	15	5

Table 5: The optimum emulsion breaker concentration for each oil in water concentration.

Figure 9: Centurion centrifuge.

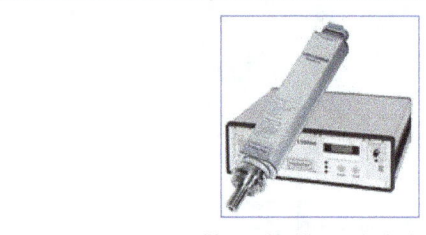

Figure 10: Ultrasonic device.

concentration of oil of 0.08, the test was run for different durations. After finding the optimum duration for the test, emulsions of different concentrations were created then placed in centrifugal tubes and run for the optimum time at different rotation speeds. The reduction of turbidity of each trial was recorded and the results are presented in Table 5. This technique has been conducted on a Centurion Centrifuge with a rotation speed of 1000 rpm to 5000 rpm (Figure 9).

Ultrasonication

This test proposed several stages of time duration, temperature, amplitude and power to systematically determine the optimum level. This will provide a full analysis of the direct effect of ultrasonic waves on the emulsions and will determine the sonication durations and intensities that result in the most efficient separation.

This technique, an ultrasonic device from Hielscher Ultrasonics Model UIP1500hd has been used. This ultrasonic device has a maximum power of 1500 Watts and at 20 kHz. This device is also able to deliver up to 170 microns of Amplitude. Please refer to Figure 10.

Results and Discussion

For each of the four demulsification techniques performed the turbidity before and after were recorded for different operating conditions. The percentage reduction of turbidity for each test was computed to find the method resulting in the greatest reduction. This section Presents those collected results and the direct computations that stem from them.

Base-turbidity determination

The value of turbidity for each solution was found by taking the highest turbidity reading for each specific concentration during any of the trials performed. This follows the observation that as the oil, emulsifier, and water are mixed, splashes of water carry with them

droplets of the oil that cause the emulsion to be of less concentration than the calculated value resulting in a value of turbidity less than the supposed value. The highest turbidity reading at any trial is then considered the reading for which the least splashing was done thus the most accurate. The value of this highest reading considered as the base turbidity for each concentration is presented in the following table.

These values of base turbidity were, however, disregarded as the experiment progressed. As it was observed, the turbidity of any solution changes with each trial so assigning a concentration a specific value of turbidity for the reduction to be compared with is not viable. Instead, a % decrease notation was followed in which the turbidity of each mixture is measured before any demulsification technique was applied to it. After the demulsification the turbidity was measured again and the % reduction in turbidity was noted. Some of the samples are shown on the photos below.

Emulsion breaker

The method involves using a specific amount of emulsion breaker or demulsifier to the predetermined oil in water emulsion concentrations. The emulsion breaker (EB-8956) used in this project is coming from a world leader emulsion breaker supplier in the industry MI Swaco. The first step taken was with determination of the optimum amount of emulsion breaker to be used for each concentration of oil in water emulsion. Please refer to Table 3.

Oil in water concentration was tested for 5 different concentrations of emulsion breaker ranging from 15 to 80 ppm to find the optimum concentration for each concentration. This was first done over 3 trials which reported varying results that can be seen in Table 3. The experiment was repeated one last time for a fourth trial to confirm the variations in the data. This last trial is represented in Table 4.

After finding the optimum concentration of emulsion breaker for oil in water concentration, Table 5 was created to summarize the results. In the table, the optimum emulsion breaker concentration alongside the % reduction caused by it is presented. Kindly refer to the graphical representation Plot 1 of Table 5 below.

From these plots, we can deduce that as the concentration of the oil

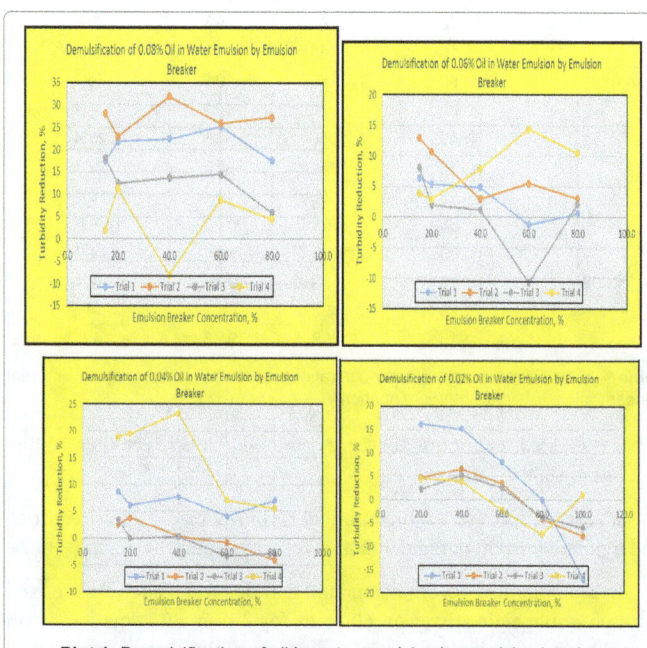

Plot 1: Demulsification of oil in water emulsion by emulsion breaker.

Oil in water concentration	Water bath temperature	Turbidity, NTU								
		Trial 1			Trial 2			Trial 3		
%	Degree C	Initial	Final	%Reduction	Initial	Final	%Reduction	Initial	Final	%Reduction
0.08	70	750	556	26	897	592	34	956	630	34
	60	703	475	32	703	490	30	703	445	37
	50	991	987	0	991	990	0	991	997	-1
0.06	70	739	514	30	739	512	31	739	504	32
	60	483	323	33	483	313	35	483	341	29
	50	799	793	1	799	804	-1	799	789	1
0.04	70	339	188	45	339	197	42	339	203	40
	60	332	226	32	332	227	32	332	238	28
	50	379	302	20	379	295	22	379	283	25
0.02	70	140	91	35	140	90	36	140	93	34
	60	151	110	27	151	109	28	151	116	23
	50	95.6	74.1	22	95.6	69.6	27	95.6	70.8	26

Note: Shaded cells highlight results with the highest turbidity reduction.

Table 6: The temperature optimization of the water bath.

Time, mins	Turbidity, NTU											
	Trial 1			Trial 2			Trial 3			Trial 4		
	Initial	Final	%Reduction	Initial	Final	%Reduction	Initial	Final	%Reduction	Initial	Final	%Reduction
10	750	598	20	956	709	26	897	655	27			
15	750	576	23	956	702	27	897	628	30			
20	750	546	27	956	689	28	897	624	30			
25	750	536	29	956	614	36	897	616	31			
30	750	556	26	956	630	34	897	592	34			
35	750	510	32	956	605	37	897	575	36			
40	750	489	35	956	610	36	897	564	37			
45				956	597	38	897	545	39			
50										817	498	39
55										817	461	44
60										817	461	44
65										817	474	42
70										817	440	46

Note: Empty cells do not have values due to sample limitations; shaded cells highlight highest turbidity reduction.

Table 7: The reduction in turbidity for 0.08% oil-in-water emulsion with time at 70°C.

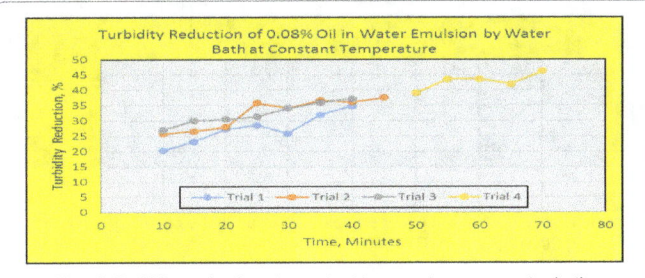

Plot 2: Turbidity reduction at constant temperature on a water bath.

in water emulsion sample reduces and as the concentration of emulsion breaker increases, sample turbidity increases.

Water bath

To test the reduction of turbidity used a water bath, two parameters needed to optimize: temperature and time. Firstly, four different concentrations of oil in water were mixed and placed in the water bath at different temperatures at 10°C increments from 40°C to 70°C. The turbidity readings were taken before and after the treatment allowing for the calculation of the % reduction in the turbidity. No reductions were found for 40°C water temperature, and the reductions at other temperatures are presented in Table 6.

As seen in the above table, the most reduction happened at a

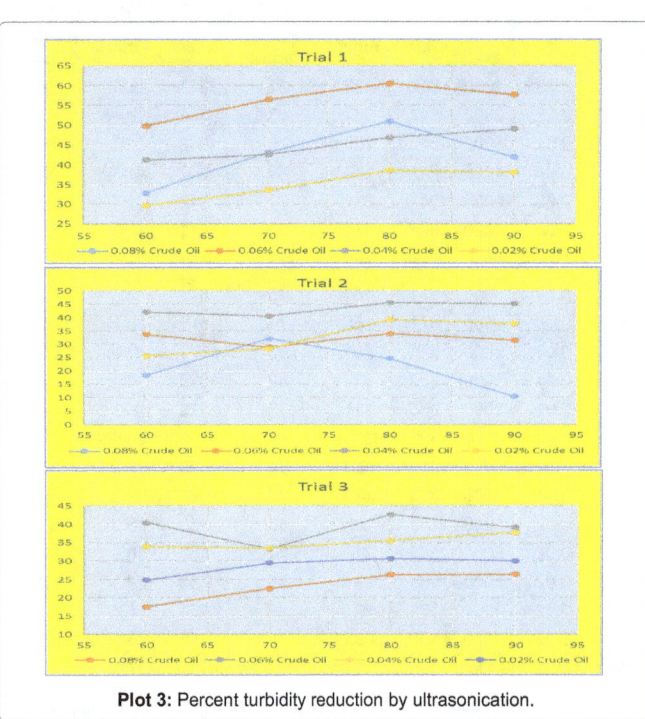

Plot 3: Percent turbidity reduction by ultrasonication.

temperature of 70°C. With that said, the temperature of the water bath was set at 70°C and a solution of 0.08% of oil-in-water was prepared and placed in the water bath to optimize the time of treatment this is presented in Table 7 (Plots 2 and 3).

From the table above, it is found that the longer the time the sample was subject to the water bath the higher (results highlighted in read) is the turbidity reduction of the sample at a constant temperature of 70°C. That is, demulsification with water bath is directly proportional to time at constant temperature. The plot above shows the increasing turbidity reduction with respect to increasing time submerges in the water bath at constant temperature of 70°C.

Centrifuge

As with the water bath, both an optimization of the rotational

Time	Concen-tration, %	RPM	Turbidity, NTU					
			Trial 1			Trial 2		
			Initial	Final	%Reduc-tion	Initial	Final	%Reduc-tion
5	0.08	5000	992	195	80	992	193	81
8	0.08	5000	992	157	84	992	164	83
10	0.08	5000	992	196	80	992	167	83
15	0.08	500	992	177	82	992	143	86
Note: Shaded cells highlight results with the highest turbidity reduction.								

Table 8: Time optimization of the centrifugal demulsification technique.

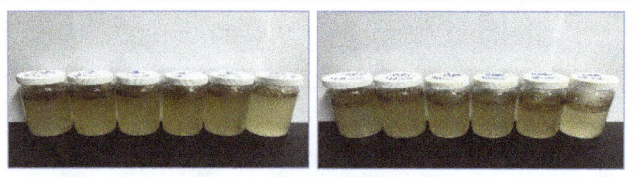

Figure 11: A 0.08% oil in water sample subjected to water bath at constant temperature at varying time (from 10 min to 65 min).

Time	Concen-tration, %	RPM	Turbidity, NTU					
			Trial 1			Trial 2		
			Initial	Final	%Reduc-tion	Initial	Final	%Reduc-tion
8	0.08	1000	942	456	52	942	395	58
8	0.08	2000	942	352	63	942	314	67
8	0.08	3000	942	230	76	942	210	78
8	0.08	4000	942	144	85	942	183	81
8	0.08	5000	942	210	78	942	197	79
8	0.06	1000	671	274	59	671	290	57
8	0.06	2000	671	147	78	671	184	73
8	0.06	3000	671	161	76	671	112	83
8	0.06	4000	671	153	77	671	99.2	85
8	0.06	5000	671	181	73	671	134	80
8	0.04	1000	325	125	62	325	139	57
8	0.04	2000	325	114	65	325	110	66
8	0.04	3000	325	121.1	63	325	68.6	79
8	0.04	4000	325	82.9	74	325	78.2	76
8	0.04	5000	325	75.7	77	325	57.3	82
8	0.02	1000	153	64	58	153	68.9	55
8	0.02	2000	153	39.8	74	153	48.9	67
8	0.02	3000	153	23.5	85	153	28.3	82
8	0.02	4000	153	19.2	87	153	43	72
8	0.02	5000	153	40.2	74	153	37.6	75
Note: Shaded cells highlight results with the highest turbidity reduction.								

Table 9: RPM optimization of the centrifugal demulsification technique.

Figure 12: Oil in water emulsion samples.

Figure 13a: A 0.08 % oil in water emulsion sample before (left sample) and after (right sample) treatment.

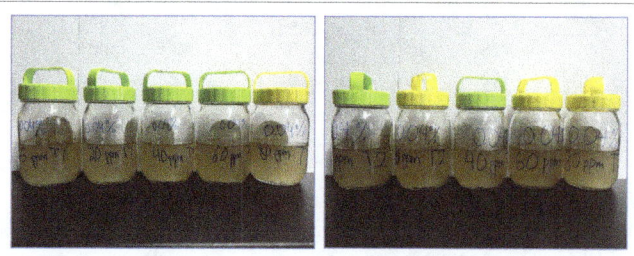

Figure 13b: A 0.04% oil in water emulsion sample before (left sample) and after (right sample) treatment.

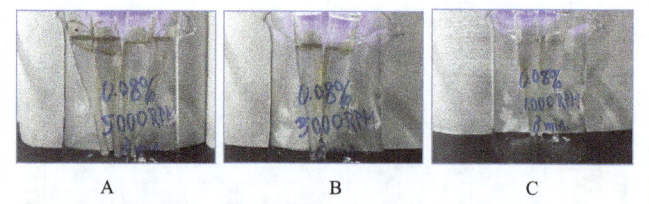

Figure 14a: A 0.08% oil in water sample subjected to the centrifuge at (A) 5000 RPM, (B) 3000 RPM and (C) 1000 RPM.

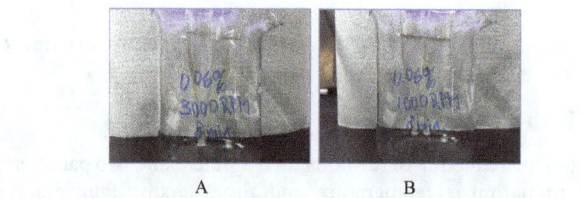

Figure 14b: A 0.06% oil in water sample subjected to the centrifuge, (A) 3000 RPM and (B) 1000 RPM.

speed and the treatment time need to be performed for this technique. To start, a solution of 0.08% oil in water was prepared and run at 5000 rpm at different times as shown in Table 8 (Figure 11). Through that, the optimum treatment time was found to be 8 min. From there an optimization of the rotational speed was done. This was through preparing solutions of different oil-in-water concentrations that were placed in the centrifuge for 8 min at different rpms. The results of this optimization are shown in Table 9 where it can be seen that the optimum speed is 4000 rpm (Figures 12-18).

Ultrasonication

As the efficiency of each of the demulsification techniques has been tested, it is necessary to compare them to that of Ultrasonication. As before, both the time and amplitude of the ultrasonication treatment need to be optimized. Initially the test was run at amplitude of 100% while varying the time of treatment between 2 min and 14 min for a sample of 0.08% oil-in-water. The results of this initial run are presented in Table 10 where it can be seen that running the treatment for a time

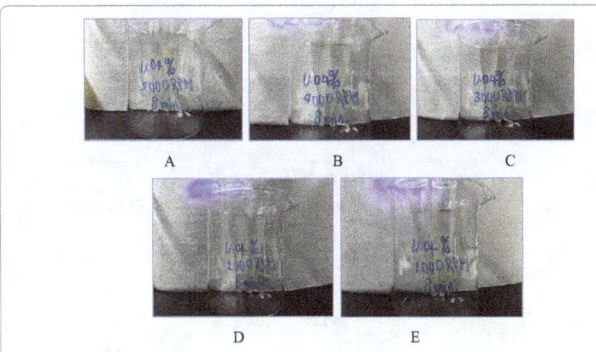

Figure 14c: A 0.04% oil in water sample subjected to the centrifuge at (A) 5000 RPM, (B) 4000 RPM, (C) 3000 RPM, (D) 2000 RPM and (E) 1000 RPM.

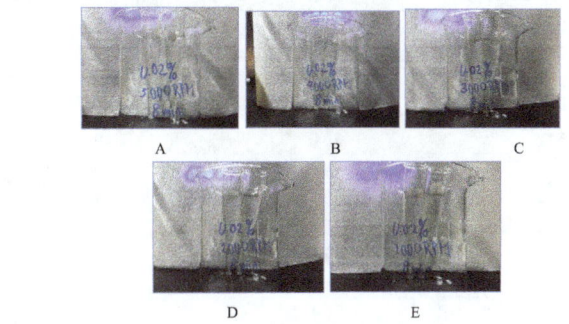

Figure 14d: A 0.02% oil in water sample subjected to the centrifuge at (A) 5000 RPM, (B) 4000 RPM, (C) 3000 RPM, (D) 2000 RPM and (E) 1000 RPM.

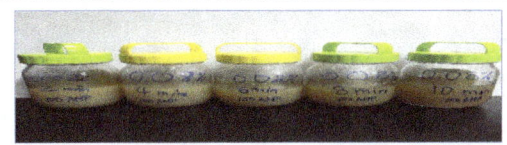

Figure 15: Time optimization of 0.08% oil in water sample by ultrasonication.

Figure 16: Amplitude optimization of 0.08% oil in water sample by ultrasonication.

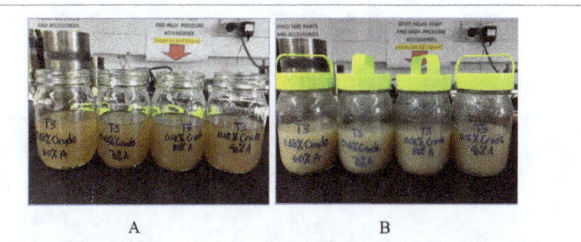

Figure 17a: (A) 0.08% water in oil sample before and (B) after ultrasonication.

Figure 17b: (A) 0.06% water in oil sample before and (B) after ultrasonication.

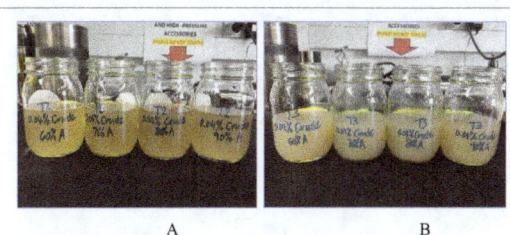

Figure 17c: (A) 0.04% water in oil sample before and (B) after ultrasonication.

Figure 17d: (A) 0.02% water in oil sample before and (B) after ultrasonication.

Figure 18: Hielscher ultrasonic device model UIP1500hd in sound proof cabinet.

Time, mins	Amplitude, %	Concentration, %	Initial turbidity, NTU	Final turbidity, NTU	% Turbidity reduction
2	100	0.08	986	OR	-
4	100	0.08	989	950	4
6	100	0.08	961	787	18
8	100	0.08	943	551	42
10	100	0.08	923	460	50
12	100	0.08	734	602	18
14	100	0.08	699	587	16

Note: Shaded cells highlight results with the highest turbidity reduction.

Table 10: Time optimization of ultrasonication.

Time, mins	Amplitude, %	Concentration, %	Initial turbidity, NTU	Final turbidity, NTU	% Turbidity reduction
5	60	0.08	948	OR	-
5	70	0.08	948	908	4
5	80	0.08	948	790	17
5	90	0.08	948	638	33
5	100	0.08	948	594	37

Note: Shaded cells highlight results with the highest turbidity reduction.

Table 11: Amplitude optimization of ultrasonication.

Concentration	Amplitude	Time	Turbidity, NTU		
%	A	min	Initial	Final	% Reduction
Trial 1					
0.08	60	5	991	664	33
0.08	70	5	991	563	43
0.08	80	5	991	486	51
0.08	90	5	991	575	42
0.06	60	5	887	445	50
0.06	70	5	887	386	56
0.06	80	5	887	350	61
0.06	90	5	887	375	58
0.04	60	5	512	301	41
0.04	70	5	512	294	43
0.04	80	5	512	272	47
0.04	90	5	512	261	49
0.02	60	5	202	142	30
0.02	70	5	202	134	34
0.02	80	5	202	124	39
0.02	90	5	202	125	38
Trial 2					
0.08	60	5	931	759	18
0.08	70	5	931	632	32
0.08	80	5	931	702	25
0.08	90	5	931	833	11
0.06	60	5	777	514	34
0.06	70	5	777	533	29
0.06	80	5	777	512	34
0.06	90	5	777	532	32
0.04	60	5	496	287	42
0.04	70	5	496	295	41
0.04	80	5	496	270	46
0.04	90	5	496	272	45
0.02	60	5	191	142	26
0.02	70	5	191	137	28
0.02	80	5	191	116	39
0.02	90	5	191	119	38
Trial 3					
0.08	60	5	995	819	18
0.08	70	5	995	770	23
0.08	80	5	995	733	26

0.08	90	5	995	732	26
0.06	60	5	777	463	40
0.06	70	5	777	518	33
0.06	80	5	777	446	43
0.06	90	5	777	473	39
0.04	60	5	483	319	34
0.04	70	5	483	321	34
0.04	80	5	483	311	36
0.04	90	5	483	301	38
0.02	60	5	176	132	25
0.02	70	5	176	124	30
0.02	80	5	176	122	31
0.02	90	5	176	123	30

Note: Shaded cells highlight results with the highest turbidity reduction.

Table 12: Turbidity reduction with varying amplitude for different oil in water emulsion concentrations.

of 10 min causes the most reduction. Although the optimum condition was found to be 10 min, the actual test was performed for 5 min due to the threat to safety if performed for longer times.

From the table above it was observed that the highest turbidity reduction was at the 100% amplitude. But no further test has been done for 100% amplitude due to the hazards that arise from such magnified amplitude and for safety purposes of the ultrasonic device. To proceed with, the ultrasonic test was performed for the specified time at three different amplitudes ranging from 70 to 90. The results of the varying reduction in turbidity with amplitude are presented in Table 11. Optimum amplitude can be noted at a value of 80%. With the ultrasonication demulsification technique turbidity reduction ranges between 26% to 61% which can be seen from the plots above.

Conclusions

In this project, we determined the effectiveness of the thermal (water bath), mechanical (centrifuge), chemical (demulsifiers), and ultrasonication as means of demulsification by applying it on various oil-in-water mixtures, and measuring the turbidities of each before and after application.

1. Based on our attempt to establish base turbidity values for each oil-in-water mixture, it was found that the turbidity of these mixtures was difficult to keep sufficiently constant. This is because the turbidity is affected by some uncontrollable factors. These factors include the imprecision of our liquid transferring equipment and instrument to measure turbidity.

2. In finding an optimum emulsion breaker concentration, consistent trend cannot be identified. However, out of the three oil-in-water mixtures tested, the optimum emulsion breaker concentration decreased as percent oil in water decreased. In other words, the lower the oil to water ratio, the lower the amount of demulsifier should be added.

3. The most reduction in turbidity for water bath was attained at the highest temperature (70°C) and longest duration (70 min). Hence, the higher the temperature and longer the duration for thermal demulsification, the more effective.

4. The optimum time for centrifuge method was found to be 8 min. Hence the optimum duration for such method is not always the longest. The optimum centrifuge speed was found to be 4000 rpm. Again, this is not the highest rotational speed. So, the optimum time and rotational speed must be determined for specific oil and water emulsions.

5. Optimum time for ultrasonication is 10 min, which again isn't the longest duration (14 min). But max reduction in turbidity is achieved at maximum amplitude, in this case at 100%.

6. As data shows, there was a larger turbidity reduction on the centrifuge technique than any other technique.

7. Ultrasonication does reduce the turbidity of the sample ranging from 20% up to 60%.

8. 100% amplitude was not done in the study for the reason that the machine was having a history of being damaged at 100% amplitude.

9. Turbidity reading variation was observed on different samples of the same oil-in-water solution concentration.

Nomenclature

T: Period, seconds,

f : Frequency, Hertz

F: Force, Newton

k: Proportionality constant for Hooke's law, Newton/meters

x: Displacement, meters

ω : Amplitude, 1/seconds

φ: Phase angle, dimensionless

t: Time, seconds

π: Irrational constant pi

v: Wave speed, m/s

μ: Inertial quantity, unit varies

Acknowledgment

"This report was made possible by a UREP award [UREP No: 18-102-2-042] from the Qatar National Research Fund (a member of The Qatar Foundation). The statements made herein are solely the responsibility of the authors".

References

1. Singh B, Padney B (1992) Ultrasonication for breaking water-in-oil emulsions. Proc Indian natn Sci Acad 58: 181-194.

2. Schoeppel RJ, Howard AW (1966) Effect of ultrasonic irradiation on coalescense and separation of crude oil-water emulsions. Society of Petroleum Engineers, SPE-1507-MS.

3. Gholam R (2014) Two-stage ultrasonic irradiation for dehydration and desalting of crude oil: A novel method. Chemical Engineering and Processing.

4. Kokal S (2008) Crude oil emulsions: Everything you wanted to know but were afraid to ask. Society of Petroleum Engineers.

5. Islam MR, Genyk R, Malik Q (2000) Experimental and mathematical modelling of ultrasonic treatments for breaking oil-water emulsions. Petroleum Society of Canada.

24

An Experimental Study on the Influence of Ethanol and Automotive Gasoline Blends

Tarek M Aboul Fotouh[1,2]*, Omar A Mazen[2] and Ibrahim Ashour[3,4]

[1]Department of Mining and Petroleum Engineering, Faculty of Engineering, Al-Azhar University, Egypt
[2]Department of Chemical Engineering, The British University in Egypt, Egypt
[3]Department of Chemical Engineering, Faculty of Engineering, Minia University, Egypt
[4]Zewail City University of Science and Technology, Sheikh Zayed 12588, 6th October, Egypt

Abstract

The objective of this work is to investigate the production possibility of high octane environmental ethanol gasoline blends based on Euro specifications. The environmental gasoline is the key element to keep the environment safe and clean. Moreover, it reduces gas emissions after combustion of gasoline. One of the main methods to produce the environmental gasoline is blending gasoline with oxygenated compounds such as ethanol. Ethanol is chosen among other oxygenated compounds as it has a high influence on physico-chemical characteristics of gasoline rather than other oxygenated compounds. In addition, it has a high octane number as well as it is not polluting the environment and clean additive. In the experimental study, the choice of environmental gasolines are based on Euro-3 specifications for samples without ethanol blend and Euro-5 specifications for samples with ethanol blend; after upgrading. Various blend stocks have been prepared which have reformate, isomerate, full refinery naphtha (FRN), heavy straight run naphtha (HSRN), hydrocracked naphtha, heavy hydrocracked naphtha, coker naphtha and heavy coker naphtha. In this study, ASTM standard methods are performed for spark ignition fuels to characterize its physical and chemical properties. The results show that one has exhibited the optimum specifications of Euro-3 and thus its physico-chemical characteristics are 755.11 kg/m^3 of density, 55.88 of °API and 95 of RON, 88 of MON, 40% by volume of aromatic content and 0.66% by volume of benzene content. Moreover, ASTM distillation curve shows that the volume percentage at 150°C is 83. At the same time, the final boiling point (FBP) and recovery volume percent are 198°C and 96% respectively. While another sample has the poorest physical as well as chemical properties so that it is blended with ethanol to upgrade its characteristics. Therefore, the target is determining the optimum ethanol volume percent to be blended with poorest sample to yield the highest properties of gasoline. These blends are namely as E0, E5, E10, E15, E20. The results indicate that E5 is the optimum one for Euro-5 specifications after upgrading and thus its physico-chemical characteristics are 745.55 kg/m^3 of density, 58 of °API, 101 of RON, 98 of MON, 32.65% by volume of aromatic content and 0.47% by volume of benzene content. Moreover, ASTM distillation curve illustrates that the volume percentage at 150°C is 75. At the same time, the final boiling point (FBP) and recovery volume percent are 190°C and 97% respectively. In addition, its Reid vapor pressure equals 8.1 psi and the heat of combustion equals 35 MJ/L. In the final, Blending gasoline with ethanol is an essential issue concerning the production of environmental gasolines.

Keywords: Automotive; Gasoline; Fuel; Energy

Introduction

Nowadays the whole world has witnessed an industrial revolution in all fields, huge population growth and increase in energy consumption which have brought about an unsustainable situation. As a result, the pollution volume and impact have increased dramatically and thus reflect a bad impact on human's health. The increase in energy consumption has a direct influence to change in fuel prices in addition to a global environment as CO_2 and CO emissions increased significantly. The European Union stated that sector of transportation consumes about one third of total energy consumption [1].

This sector is mainly dependent on petroleum products (gasoline, diesel) as source of fuel. These fuels are considered a major contributor to greenhouse phenomena as they produce a high amount of greenhouse gases. Thus it becomes a necessity to find an alternative fuel instead of gasoline and diesel. This problematic situation is nearly the same in most developed and advance countries. That's why some countries have modified their fuel regulation in order to reduce CO_2 and CO emissions and keep environment clean as much as possible [2,3].

This fuel should be renewable and clean one to minimize greenhouse gases and produce an environmental fuel. Biofuels characterized by their production from renewable sources and considered as a clean fuel. It is biodegradable, and produces significantly less air pollution than fossil fuel. The fossil fuel exhaust is a potential carcinogen, since the use of bio-fuel has been found to reduce risks of cancer because it reduces the production of cancer-causing compounds, such as aromatics. Bio-fuel also produces less greenhouse gases such as CO_2. When either bio-fuel or petroleum is burned, the carbon content of the fuel returns to the atmosphere as CO_2. Plants grown to make ethanol for bio-fuel draw CO_2 out of the atmosphere for photosynthesis, causing a recycling process that result in less accumulation of CO_2 in the atmosphere. Thus, bio-fuel does not contribute to global warming in the same way that petroleum does [3-5].

In addition, Bio-fuel leads to produce exhaust NOx, hydrocarbons

---

*Corresponding author: Tarek M Aboul Fotouh, Department of Mining and Petroleum Engineering, Faculty of Engineering, Al-Azhar University, Egypt
E-mail: tarekfetouh@yahoo.com

and smokes lower than a regular gasoline fuel. The gasoline fuel replacement is regulated by the amount of ethanol in the blend. Problems arise, however, due to the presence of water in the blend because commercially available ethanol is seldom found in an anhydrous state [6].

These biofuels can be used either as fuel or blending it with petroleum products. Using biofuels as an additive is much better as most of vehicles and other means of transportation are operated with petroleum products. The main standard fuel additives for gasoline are ethanol and ethyl tert-butyl ether (ETBE). The majority of advanced countries use both of them. Blending gasoline with ethanol has a great advantage as it decreases greenhouse gases significantly compared to pure gasoline and produces an environmental gasoline. Many studies have been done on ethanol gasoline blends to find the appropriate amount of ethanol. Various blending have been made to determine the optimum amount of ethanol that should be used [7-10]. A research study at Southern Illinois University has found that with bio-fuel blends engine power and specific fuel consumption slightly increase [11-14]. The aim of this study is to determine the optimum gasoline sample without any blending of oxygenated compounds based on Euro 3 specifications. Furthermore, the optimum sample is then blended with ethanol to find the optimum volume percent of ethanol based on Euro 5 specifications.

Methodology

The Experimental work is mainly a quantitative analysis to find out the optimum volume percent of ethanol added to gasoline to enhance its physical properties. This can fulfill by preparing various samples of gasoline with different volume percent from each cut (light naphtha, heavy naphtha, coker naphtha, reformate, isomerate and hydrocracker naphtha) and then applying experimental tests for the various gasoline samples (ASTM distillation, API, Reid vapor pressure, octane numbers, GC analysis). Furthermore, the optimum sample of gasoline based on Euro-3 specifications has been chosen as an environmental gasoline sample without any percentage of ethanol. Adding different volume percentages of ethanol to the gasoline sample No 5 to upgrade its physico-chemical characteristics have been achieved. Then experimental tests for the various gasoline-ethanol samples are applied (ASTM distillation, API, Reid vapor pressure, octane numbers, GC analysis). Moreover, the optimum oxygenated sample based on Euro-5 specifications has been chosen as an environmental gasoline sample with the optimum percentage of ethanol.

Experimental Work

This section is mainly showing a brief overview for the experimental work and procedures of each test conducted.

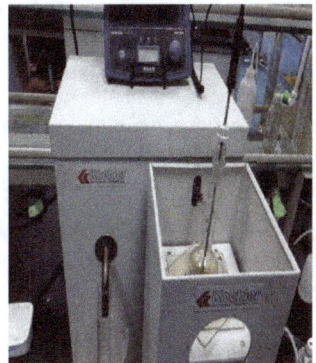

Figure 1: ASTM distillation apparatus.

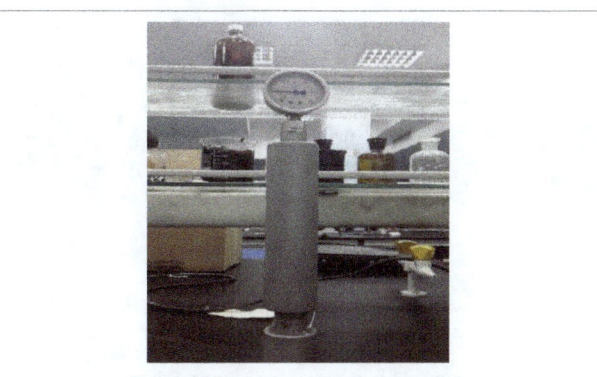

Figure 2: Reid vapor pressure apparatus.

Materials

This section is concerned with materials used in the experimental work. These materials are ethanol, isomerate, reformate, hydrocracker naphtha, heavy hydrocracker naphtha, coker naphtha, heavy coker naphtha, light straight run naphtha (LSRN), heavy straight run naphtha (HSRN), 50 ml volumetric flask and 50 ml Pycnometer.

ASTM distillation test

ASTM distillation is one of the main standard tests applied for crude oil and its products as shown in Figure 1 [16]. This test measures the temperature corresponding to volume percent evaporated from sample. The distillation of petroleum products is done in simple distillation equipment without fractionation column. The test is performed under atmospheric pressure for light products like gasoline, kerosene, diesel and heating oil. On the other hand for heavy cuts, reduced pressure is employed to decrease the boiling point of this cut. The results obtained from this test may be used to determine other physical properties like flash point.

Relative density or density

Specific gravity and API of crude oil and its products are considered two major properties which reflect the kind and price of products. As the API of the fraction is high, this indicates that the cut is light and contain valuable products so its price will be high and vice versa. The specific gravity is related with API by this relation:

$$API = \frac{141.5}{specific\ gravirty} - 131.5$$

Where specific gravity is defined as the density of the liquid cut over the density of water at specified temperature which is 60/60°F. As the specific gravity decreases, the value of the product increases as contents of light products increase [17].

Reid vapour pressure

Reid vapour pressure (RVP) of a cut or product is defined as the vapour pressure measured in a volume of air four time volume of liquid at 37.8°C which demonstrates in Figure 2 [18]. This property measures the vapour lock tendency occurring in a spark ignition engine operated with gasoline fuel in which excess vapours generated in line cause interruption in liquid fuel supplying to engine. Also it is important tool to estimate the hazard degree of fuel.

Heat of combustion using bomb calorimeter

Heat of combustion is one of the main characteristic that should be identified for hydrocarbons fuels. Combustion of fuels is either complete or incomplete combustion depending on the amount of oxygen. Always complete combustion is more desired than incomplete because it does not affect environment and cleaner than incomplete. Complete combustion of hydrocarbon fuels is an exothermic reaction and yields are carbon dioxide, water vapour and large amount of heat. The main aim of this test is to determine net heating value produced after combustion of fuel. A standard test is performed by ASTM using a bomb calorimeter to measure heat of combustion (Figure 3).

Octane numbers (RON and MON)

An octane number is considered one of the major properties in gasoline that must be determined accurately for motor fuels like gasoline [19,20]. This property has a great impact on motor performance and its life time so improving it is done continuously to keep it as high as possible. As the ability of motor or spark ignition engines increases to resist auto ignition the octane number of gasoline increases. The octane number of gasoline can be determined by measuring knocking value of this fuel and compared it to a mixture of iso octane (2, 2, 4, tri-methyl pentane) and normal heptane. The octane of pure n-heptane is assigned to be zero while octane of iso octane is assigned to be 100. For example a mixture of 80% iso octane and 20% n- heptane will have octane number 80. Hence if knocking value of gasoline fuel is corresponding to a mixture of 90% iso-octane and 10% n-heptane it means that gasoline fuel has octane number equal to 90. Octane number is classified into two types MON and RON. MON indicating the performance of engine at severe conditions and high speed may reach to 900 rpm (Figure 4). While RON is measuring performance of engine at smooth conditions and low speed may reach 600 rpm. AKI is the arithmetic average of RON and MON.

Detailed hydrocarbon analysis by gas chromatography (PIONA)

Detailed hydrocarbon test is one of the most main tests that should be performed for petroleum fractions. This test can be implemented on petroleum gases and naphtha cuts using a gas chromatography unit. This device consists of various units which are ordered in a certain sequence for the purpose of test. This test gives a detailed analysis for hydrocarbon family contents (paraffins, iso paraffins, olefins, naphthenes and aromatics) in petroleum fractions (Figure 5).

Results and Discussion

This section is mainly elaborating the results of experiments and giving an interpretation for what obtained in the experimental section.

ASTM distillation

According to Figure 6 which shows the volume % collected versus temperature, the following curves represent distillation curves for prepared gasoline samples as presented in Table 1. The distillation curves show the volume percent of volatile cuts with its corresponding temperatures. The curves represent mainly the initial, mid and final boiling points. The recovery of each sample reaches to 96% or 97% of total volume of samples and about 3% or 4% losses. The difference in distillation curves is due to the difference in each blend stock.

According to Figure 7, the following curves represent distillation curves for ethanol –gasoline blends as demonstrated in Table 2. The distillation curves show the volume percent evaporated with its corresponding temperature. The curves illustrate the initial, mid, and final boiling points. The recovery of each sample reaches to 96% or 97% of total volume of samples and about 3% or 4% loss. The difference in the behaviour of distillation curves is due to the volume percentages of ethanol.

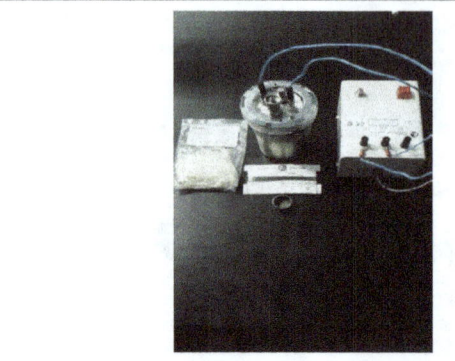

Figure 3: Oxygen bomb calorimeter.

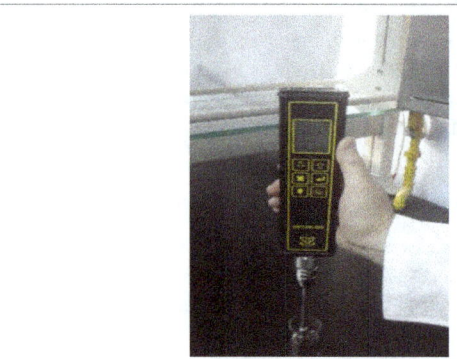

Figure 4: Octane-meter appartus.

Figure 5: Gas chromatography device.

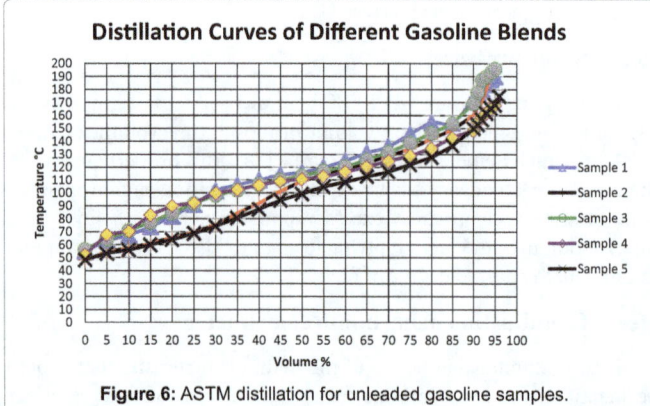

Figure 6: ASTM distillation for unleaded gasoline samples.

Density

The density values of unleaded gasoline samples are measured at 15.5°C. Table 3 shows density values of unleaded gasoline samples that presented in Table 1. Table 3 elaborates that the density of unleaded gasoline samples varies from (755 Kg/m³ to 772.5 Kg/m³). The density

Blend-stocks, vol.%	Sample 1	Sample 2	Sample 3	Sample 4	Sample 5
Reformate	52	60	51	51	36
Isomerate	12	9	8	8	17
FRN	17	10	13	-	17
HSRN	-	-	-	13	-
Hydrocracker naphtha	17	20	25	-	-
Heavy hydrocracker naphtha	-	-	-	25	25
Coker naphtha	2	1	3	-	5
Heavy coker naphtha	-	-	-	3	-

Table 1: Blend-stocks volume percent for unleaded gasoline samples.

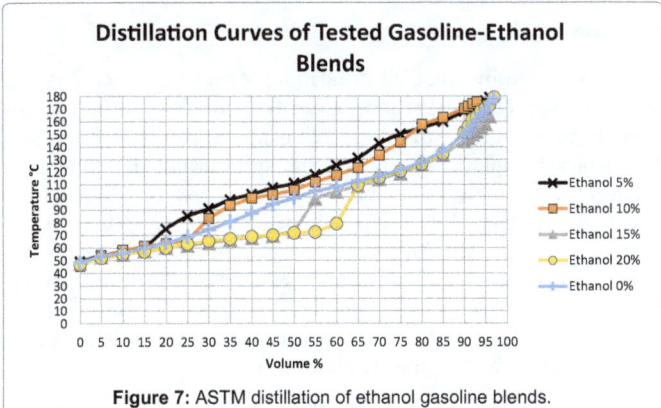

Distillation Curves of Tested Gasoline-Ethanol Blends

Legend: Ethanol 5%, Ethanol 10%, Ethanol 15%, Ethanol 20%, Ethanol 0%

Figure 7: ASTM distillation of ethanol gasoline blends.

Blend-stocks, vol. %	E0	E5	E10	E15	E 20
FRN	17	16	15.5	14	14
Reformate	36	34	31.5	31	28
Isomerate	17	16	15.5	14	14
Hydrocracker naphtha	25	24	23	22	20
Coker naphtha	5	5	4.5	4	4
Ethanol	0	5	10	15	20

Table 2: Blend-stocks volume percent of ethanol gasoline samples.

shows a variance as each sample differs from the other in the amount of blend-stocks used. As the lighter components increase in the sample, consequently its density decreases and vice versa. Sample 3 exhibits the optimum one according to Euro 3 specifications.

Table 4 represents the results obtained for ethanol-gasoline blends (E0, E5, E10, E15 and E20). It appears from the results that density range from 739 Kg/m³ to 754 Kg/m³. The densities of E5 and E 10 are slightly smaller than the base gasoline sample by 0.66% and 1.4% respectively, while E15 and E 20 are slightly greater than base gasoline sample by 0.2% and 0.4% respectively. This indicates that as volumes of ethanol increase more than 10%, densities of samples increase and that does not sound good, while as volume percent of ethanol is 10 or less, the sample become lighter and that sounds good. So densities of E5 and E 10 are more desirable than E 15 and E 20. Eventually E5 will be the optimum blend based on Euro 5 specifications.

Reid vapour pressure

Reid vapour pressure is applied to all samples to obtain its Reid vapour pressure at 37.8°C. The experiment proves that E5 sample has a Reid vapour pressure equal 8.1 psi (Table 4). Euro standard for gasoline should be in the range from 8 to 9 psi. It is considered a physical property that should be taken into consideration especially in cold and hot countries. In cold countries RVP should have a high value to avoid operational problems in spark ignition engines. While as hot countries, RVP should be low to avoid excess vapours produced which may cause lock and interrupt liquid supply to engine.

Heat of combustion

According to Table 4, it means that everyone litre of sample E5 combusted gives calorific value equal to 35 MJ. This property is very essential to show that the thermal efficiency of the fuel is qualified or not. Also it is important for determining whether this fuel will be able to operate equipment, engine or not. Finally this test proves that sample E5 is qualified according to Euro 5 specifications.

Octane number (MON, RON)

Table 3 represents MON and RON for unleaded gasoline samples that are prepared according to proposed blend. The five samples show a high RON and MON except the sample 5. Sample 2 as well as sample 1 shows the highest octane number, but that does not mean sample

Test	Method	Unit	Sample 1	Sample 2	Sample 3	Sample 4	Sample 5
Density @ 15.5°C	ASTM D1217-16	kg/m³	768.1400	772.4280	755.1124 (748-762) Euro-3	769.1600	750.5480
RVP	ASTM D323-15a	Psi	7	7.5	8.4	7.2	8.6
RON	ASTM D2699-15a		95.6	98.2	95	90	88
MON	ASTM D2700-16		85.8	91.1	88	86	81.7
Aromatic	ASTM D6839-16	Vol. %	42.8420	46.6960	40 (29-42) Euro-3	40	32.6540
Paraffins	ASTM D6839-16	Vol. %	18.4680	16.3602	18.6639	18.6639	21.6228
Isoparaffins	ASTM D6839-16	Vol. %	25.2160	24.8070	26.3960	26.3960	27.8200
Naphthenes	ASTM D6839-16	Vol. %	11.8335	10.8500	12.9820	12.9820	15.1632
Olefins	ASTM D6839-16	Vol. %	1.6405	1.2868	1.9581	1.9581	2.7400
Benzene	ASTM D6839-16	Vol. %	0.68	0.78	0.66 <1 Euro-3	0.66	0.47
IBP	ASTM D86-04b	°C	55	48.1	56.3	52.4	48.5
T_{10}	ASTM D86-04b	°C	67.5	57.5	71	70.6	56
T_{50}	ASTM D86-04b	°C	116	109	113.3	110.7	99.2
FBP@ 96 Vol.%	ASTM D86-04b	°C	195	197	198 (190-215) Euro-3	170	174.8
Dist. @ 100 °C	ASTM D86-04b	Vol. %	30	45	34	30	50
Dist. @ 150 °C	ASTM D86-04b	Vol. %	75	85	83 (81-87) Euro-3	90	90

Table 3: Physico-chemical characteristics for unleaded gasoline samples.

Test	Method	Unit	E0	E5	E10	E15	E20
Density at 15.5°C	ASTM D1217-16	kg/m³	750.5480	745.5528 (743-756) Euro-5	739.3120	752.5500	754.1000
RVP	ASTM D323-15a	Psi	8.6	8.7 (8.1-8.7) Euro-5	8.8	7.9	7.4
RON	ASTM D2699-15a		88	101	106	103	97.6
MON	ASTM2700-16		81.7	98	105	102	89.5
Aromatic	ASTM D6839-16	Vol. %	32.6540	31.0910 (29-35) Euro-5	29.6855	28.1948	26.1013
Paraffins	ASTM D6839-16	Vol. %	21.6228	20.5910	19.6571	18.2024	18.0120
Isoparaffins	ASTM D6839-16	Vol. %	27.8200	26.4840	25.2909	24.0913	23.0232
Naphthenes	ASTM D6839-16	Vol. %	15.1632	14.4211	13.7847	13.1254	12.1240
Olefins	ASTM D6839-16	Vol. %	2.7400	2.5081	2.4909	2.0826	2.0121
Benzene	ASTM D6839-16	Vol. %	0.47	0.47 <1 Euro-5	0.46	0.46	0.45
IBP	ASTM D86-04b	°C	48.5	49.3	45.6	48.2	47
T_{10}	ASTM D86-04b	°C	56	57	58.2	55	55
T_{50}	ASTM D86-04b	°C	99.2	111	105.6	73	71.8
FBP@ 97 Vol.%	ASTM D86-04b	°C	178	190 (190-210) Euro-5	188	166	179
Dist. @ 100°C	ASTM D86-04b	Vol. %	50	38	40	55	64
Dist. @ 150°C	ASTM D86-04b	Vol. %	90	75	77.5	92.5	90
Heat of Combustion	ASTM D 4809-13	MJ/L	-	35	-	-	-

Table 4: Physico-chemical characteristics for ethanol gasoline blends.

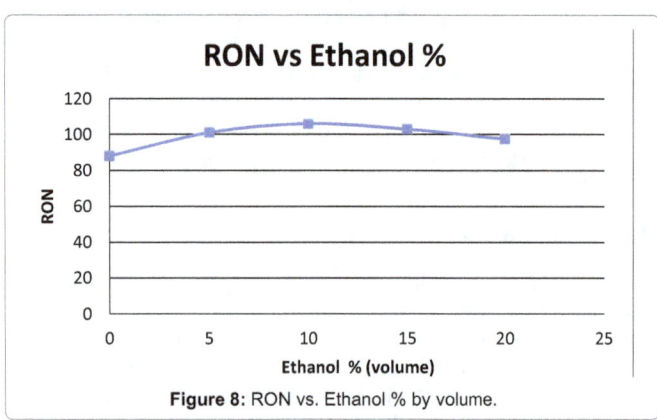

Figure 8: RON vs. Ethanol % by volume.

is matching Euro specifications. Further tests will determine the optimum sample.

According to Table 4, the following results show RON and MON of ethanol gasoline blend samples measured by octane-meter. RON of samples varies from 88 to 97.6 while MON varies from 81.75 to 105. The RON and MON shows a major variance in its value depending on volume % of ethanol added to the sample. RON and MON show an increase from the base gasoline for E5 and E10 while as it shows a decrease for E15 and E 20. The RON and MON for E5 and E10 are slightly higher than the base gasoline by 14%, 20.4%, 19.8%, and 28.4% respectively.

Figure 8 represents RON of ethanol gasoline samples with percentage of ethanol added. The curve indicates that RON is increasing until E 10 (a positive impact) and then RON begins to decrease. This proves that as the volume percent of ethanol exceed 10% by vol., RON begins to decrease. Therefore, there is no need to increase ethanol content above 10% as it has a negative impact on both RON and MON. Eventually, it has a main role in improving a spark ignition fuel if it is added within limited range.

Conclusions and Recommendations

This paper is discussing one of the main petroleum refinery issues which is the production of an environmental gasoline based on specifications of Euro-3 as well as Euro-5 using oxygenated compounds. The base gasolines in this work are mainly consists of various blend stocks produced by crude oil distillation, conversion process as well as upgrading processes. In more explanation, the Gasoline pool in this paper is mainly consists of straight run naphtha, isomerate, reformate, hydrocracker naphtha, coker naphtha, and ethanol. By comparing results of each sample, an optimum sample is chosen matching Euro-3 specifications. Also poorest sample in octane number is chosen to be blended with ethanol to enhance its physico-chemical characteristics. In addition, the choice of an optimum ethanol gasoline blend sample is based on Euro-5 specifications. Finally, it can be concluded the following points, by performing experimental results for both of base gasoline samples and ethanol gasoline blends:

a. The Production of environmental, clean and high octane number gasoline blends are the best solution for our environment.

b. The optimum unleaded gasoline sample matching Euro-3 specifications is the sample 3.

c. The optimum ethanol gasoline blend matching Euro-5 specifications is the sample E5.

d. Ethanol-gasoline-blends can be used as an alternative fuel for a variable speed spark-ignition up to 5 vol. % blends.

e. The high yield of gasoline production is based on different blend stocks not only straight run naphtha and reformate.

f. Using oxygenated compounds lead to reduce the aromatic content and consequently reduce carcinogenic compounds as well as improve octane numbers.

g. Maximizing the quality and quantity of an environmental gasoline according to standard European regulations (Euro-5).

h. An Environmental gasoline provides a great potential benefit to the refinery in view of minimizing operating costs, product quality improvement, safe and healthy living environment.

The following recommendations could be put for future work:

a. This research should be applied in the industry to prevent the hazards of air pollution.

b. The optimum composition of refinery gasoline blend should be applied for maximizing its quantity and quality with ethanol percentages.

References

1. Fahim M, Al-Sahhaf T, Elkilani A (2010) Fundamentals of petroleum refining. Oxford: Elsevier.

2. Anderson JE, DiCicco DM (2011) High octane number ethanol–gasoline blends: Quantifying the potential benefits.

3. Gary JH, Handwerk GE (2001) Petroleum refining technology and economics. CRC Press, New York: Marcel Dekker.

4. Rodríguez AU, Hernández CM, Sanz FP (2012) Experimental determination of some physical properties of gasoline, ethanol and ETBE blends.

5. Majid ABD, Hadi A (2010) Optical study of ehanol gasoline blends with or without heating.

6. Riazi MR (2010) Characterization and properties of petroleum fractions.

7. De Melo C, Tadeu C, Machado, Guilherme B (2012) Hydrous ethanol–gasoline blends – Combustion and emissions investigations.

8. Aboul-Fotouh TM, Hussein MS (2016) Experimental determination of physico-chemical characteristics of new environmental gasoline, ethanol and isopropanol blends. The 2016 Spring Meeting and 12th Global Congress on Process Safety of AIChE, Houston, TX, USA.

9. Abdellatief TMM, El-Bassiouny AMA, Aboul-Fotouh TM (2015) An environmental gasoline (Enhancing the properties of the gasoline through modified blending operations). Lab Lambert Academic Publishing, Germany.

10. EL-Bassiouny A, Aboul-Fotouh TM, Abdellatief YMM (2015) Maximize the production of environmental, clean and high octane number gasoline-ethanol blends by using refinery products. International Journal of Scientific and Engineering Research 6: 1792-1803.

11. Roger W (2005) Blending of ethanol in gasoline for spark ignition engines. Stockholm.

12. David SJ, Pujado PR (2006) Handbook of Petroleum Processing. USA, Springer.

13. Kheiralla AF, El-Awad M, Hassan MM, Mathani Y (2012) Experimental determination of fuel properties of ethanol gasoline blends. ICMAR, Malaysia.

14. Alvydas P, Saugirdas P, Juozas G (2003) Influence of composition of gasoline – Ethanol blends on paremeters of internal combustion engines. Journal of KONES Internal Combustion Engines 10: 3-4.

15. EL-Bassiouny A, Aboul-Fotouh TM, Abdellatief TMM (2015) Upgrading the commercial gasoline A80 by using Ethanol and refinery products. International Journal of Scientific and Engineering Research 6: 405- 417.

16. ATSM (2006) Standard test method for distillation of petroleum products at atmospheric pressure. D86-04b.

17. American Society for Testing and Materials (2015) Standard test method for density and relative density (specific gravity) of liquids by Bingham pycnometer. D1217-15.

18. American Society for Testing and Materials (2015) Standard test method for vapor pressure of petroleum products. Reid Method. D323-15a.

19. American Society for Testing and Materials (2016) Standard test method for research octane number of spark-ignition engine fuel. D2699-16.

20. American Society for Testing and Materials (2016) Standard test method for motor octane number of spark-ignition engine fuel. D2700-16.

25

Research and Practice of the Early Stage Polymer Flooding on LD Offshore Oilfield

Kuiqian Ma, Yanlai Li, Ting Sun*

China National Offshore Oil Corporation (CNOOC) Ltd., Tianjin, China University of Petroleum-Beijing, P.R. China

Abstract

Literature survey shows that polymer flooding was generally conducted during high water-cut stage (WCT>80% to 90%). Even the first China Offshore polymer flooding project was carried out in SZ when water cut was 60%. By then, conduction of polymer flooding in early phase (WCT<10%) was just discussed in theory. For offshore oilfield, the treatment of water could be costly. Because polymer improves mobility ratio of replacement fluid over oil and sweep efficiency, less water is injected and less water is produced. So, we did enormous research about the polymer flooding on early stage by theoretical analysis, series of experiments and chemical flooding simulation. Based on these researches, we carried out the first field test of polymer flooding on early stage in LD. Single well polymer injection test was started in Mar 2006 when the water cut in the pattern was lower than 10%. After the trial, there were other 5 water injectors being converted to polymer injectors from 2007 to 2009. The polymer flooding controlled reserve was about 25,250,000 m³. For the early stage polymer flooding, the characteristics of the responses on producers were different from the case in which polymer flooding was conducted during high water cut stage. The water producing of the producers continued to rise up after polymer flooding, but the simulation research showed that the water cut increasing rate was lower than the rate during merely water flooding. In addition, we observed the drop-down on the water cut in some wells, such as A11, A12, A13, A15, etc. For the well A11, the highest water cut reduction reached 41% after the injectors (A5/A10) profiles controlled, and net incremental oil for A11 even reached 154,510 m³. By Dec 2014, the total incremental oil by polymer flooding was about 754,650 m³, and the stage oil recovery efficiency was enhanced by 3.0%. The polymer flooding is still effective, and we will get more oil from the polymer flooding.

Keywords: EOR; Polymer Flooding; Offshore oil field

Abbreviations: RRF: Residual Resistance Factor; RF: Resistance Factor; PORFT: Accessible Pore Volume; FREQFAC: Reaction Frequency Factor; ADMAXT: Maximum Adsorption Capacity; ADRT: Residual Adsorption Level; PI: Pressure Index; WCT: Water Cut.

Introduction

With the improvement of offshore exploration extent, there are more and more heavy oil found. The oil recovery was only 18% to 25% by the conventional oil production method of water injection. So more challenges rise up, such as how can we get more oil? How can we improve the oil recovery? For the offshore oilfield, a method which can be used to develop oilfield more efficiently must be found due to the life time of offshore platform is limited (with the life of 25 years to 30 years).

Polymer flooding was already a mature set of technology to improve oil recovery on the onshore oilfields in China. Literature survey shows that polymer flooding was generally carried out during high water-cut stage (WCT > 80% to 90%). Even the first China Offshore polymer flooding project was carried out in SZ when water cut was 60%. By then, conduction of polymer flooding in early phase (WCT<10%) was just discussed in theory. So, we did enormous research about the polymer flooding on early stage by theoretical analysis, series of experiments and chemical flooding simulation, as shown in (Figures 1 and 2). Gel

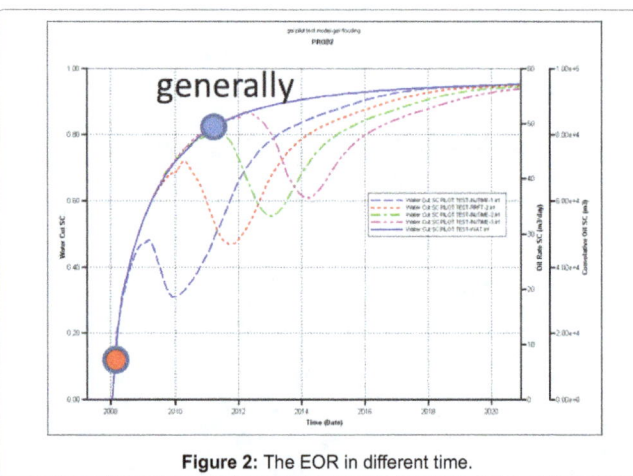

Figure 2: The EOR in different time.

flooding technology combines the function of improvement of mobility ratio by polymer flooding and injection profile control by cross linked gel injection.

LD oilfield is located in Bohai Bay, characterized by huge thickness,

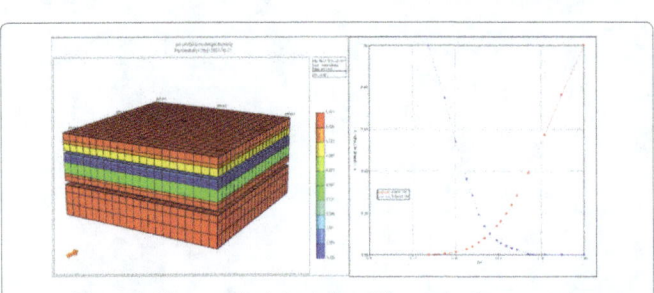

Figure 1: The numerical simulation of the mechanism research.

***Corresponding author:** Ting sun, Institute for Ocean Engineering, China University of Petroleum-Beijing, China
E-mail: ting.sun@cup.edu.cn

Method of displacement	The water cut start to the polymer injection (%)	Viscosity of the oil (cp)	Polymer concentration (mg/l)	Core size	Other lab test conditions
Polymer	0	70	1750	L: 30 cm W: 4.5 cm H: 4.5 cm	Temperature: 65°C Rate of displacement: 1 m/d Polymer injection PV: 0.25 PV
	28				
	53				
	97				
Water	--				

Table 1: The geometric parameters of the model.

Method of displacement	The water cut start to the polymer injection (%)	The water flood recovery before polymer flood (%)	The ultimate recovery factor (%)	Improve oil recovery by polymer flooding (%)
Polymer	0	0	51.35	6.23
	28	22.2	51.08	5.96
	53	23.1	50.75	5.63
	97	45.1	50.35	5.23

Table 2: The results of lab test experiments.

The water cut start to the polymer injection (%)	0	20	40	60	80
Rw (%)	39.86				
Rp (%)	49.28	48.98	48.73	48.3	46.71
ΔR (%)	9.39	9.09	8.84	8.42	6.82

Table 3: Statistical recovery efficiency in the development of 30 years.

The polymer					
Polymer concentration (mg/L)	500	700	1000	1200	1500
Polymer viscosity (cp)	3.5	4.5	6	8	9.5
The cross-linked polymer (P: Cr³=60:1)					
Polymer concentration (mg/L)	500	700	1000	1200	1500
Polymer viscosity (cp)	3	4	5.5	7	8.5

Table 4: The relationship between viscosity and concentration of gel and polymer.

Polymer concentration (P: Cr³)	Viscosity of the crosslinked polymer (cp)				
	Initial	After 2 days	After 5 days	After 10 days	After 15 days
1200 mg/L, (P: Cr³=40:1)	8.2	420	582.3	613	602.3
1200 mg/L, (P: Cr³=60:1)	8.1	312	452.3	523	572.3
1000 mg/L, (P: Cr³=60:1)	5.9	62.9	87.5	137	210.5
1000 mg/L, (P: Cr³=40:1)	6.2	154	231.4	336	341.2

Table 5: The viscosity of the cross-linked polymer.

Core number	Kg ('10⁻³ mm²)	Displacement system	RF	RRF
R3-9	1350	The crosslinked polymer Cp=1200 mg/l (P: Cr³⁺=20:1)	45.6	58.2
12-4-5	1355	The crosslinked polymer Cp=1200 mg/l (P: Cr³⁺=40:1)	39.9	48.4
12-3-5	1360	The crosslinked polymer Cp=1200 mg/l (P: Cr³⁺=60:1)	34.9	39.3
L12-1-1	1345	The crosslinked polymer Cp=1200 mg/l (P: Cr³⁺=80:1)	32.9	35.6
12-4-2	1320	The crosslinked polymer Cp=800 mg/l (P: Cr³⁺=20:1)	38.9	46.7
12-3-4	1380	The crosslinked polymer Cp=800 mg/l (P: Cr³⁺=40:1)	33.9	42.9
L12-1-2	1346	The crosslinked polymer Cp=800 mg/l (P: Cr³⁺=60:1)	29.9	36.3
12-4-10	1368	Polymer Cp=800 mg/l	12.9	1.8
12-4-10	1349	Polymer Cp=1000 mg/l	16.2	2.5
12-4-13	1342	Polymer Cp=1200 mg/l	21.3	3.2

Table 6: The RF and RRF of the cross-linked polymer.

high permeability, severe heterogeneity, high crude oil density (0.947 g/cm³) and medium oil viscosity (7.2 cp to 19.4 cp). LD oilfield was put online in January of 2005, and started to inject water on September of 2005. Based on well understanding of the mechanism and effect of the early polymer flooding, we carried out the single well polymer injection pilot test from 2006 when the water cut in the pattern was lower than 10%. After the trial, there were other 5 water injectors were converted to polymer injectors from 2007 to 2009. The polymer flooding controlled reserve was about 25,250,000 m³. For the early stage polymer flooding, the characteristics of the responses on producers were different from the case in which polymer flooding was conducted during high water cut stage. The water producing of the producers continued to rise up after polymer flooding, but the simulation research showed that the water cut increasing rate was lower than the rate during merely water flooding. Of course, we also observed the drop down on the water cut in some wells, such as A11, A12, A13, A15, etc. For the well A11, the highest water cut reduction reached 41% after the injectors (A5/A10) profiles controlled, and net incremental oil for A11 even reached 154,510 m³. By December 2014, the total incremental oil by polymer flooding was about 754,650 m³, and the stage oil recovery efficiency was enhanced by 3.0%. The polymer flooding is still effective, and we will get more oil from the polymer flooding.

Experiment Tests and Polymer Flooding Scheme Design

Mechanism study on early stage polymer flooding

Geometric size of the core model is 30 cm × 4.5 cm × 4.5 cm, and the parameters of experiments were shown in Tables 1 and 2). The results of lab test experiments were shown in Table 2. From the results of the lab test, the difference of the EOR between cases in which polymer flooding conducted at different time was not obvious. But for offshore oilfield, the platform life was about 25 years to 30 years. We must produce more oil in limited time. From numerical simulation of the mechanism research, we know that the larger enhanced recovery value can be obtained, if polymer can be injected earlier (Table 3) [1-3].

Experiment tests and mechanistic models of gel

The relationship between viscosity and concentration of gel and polymer in early stage was shown in Table 4. The experiment about the crosslinked polymer viscosity in the case of different polymer concentrations and ratios of the polymer to Cr³⁺ was done. As is shown in the Table 5, the viscosity also increased with time going by. The results of RF and RRF test were shown in Table 6. RRF of the polymer is 1.8 to 3.2 from the experiment, and the gel is 39.3. According to the experiment results and the SZ oilfield mature experience, RRF of LD gel is 5, and the polymer is 2.5. The static adsorption and dynamic adsorption tests were shown in Tables 7 and 8.

Injection scheme design

The size of injection slug was 0.163 PV, and the period of the slug injection was 5 years as shown in Table 9. Injection system was the polymer and Chromium ion (Cr^{3+}) crosslinking agent. The concentration of the polymer was 1200 mg/L (Polymer: Cr^{3+}=600:1-60:1). The polymer injection rate was 0.033 PV per year. The predicted improve recovery factor was 6.1%.

The Result of the Field Test

From March of 2006, the polymer injection trial has been conducted on the A23 well when the water cut of the pattern was lower than 10%. After the trial, there were other 5 injectors (A1\A5\A10\A14\A18M) were converted to polymer injectors from 2007 to 2009. The polymer flooding controlled reserve was about 25,250,000 m³.

Characteristic of gel-injection well effectiveness is shown in Figure 3 and Table 10. From the table we can conclude: injecting pressure rise, while water injectivity index decline and the values of RF and RRF remain above 1.

Number	Polymer concentration (mg/L)	Static adsorption (mg/g)
1	600	2.51
2	800	2.82
3	1000	3.13
4	1200	4.31
5	1400	5.16

Table 7: The static adsorption experiment result.

Displacement solution	Core permeability (10⁻³mm²)	Polymer concentration (mg/l)	Dynamic adsorption (μg/g)
DQKY polymer	1096	1400	126
	1103	1200	95
DQGF gel	1133	1200	332
	1108	1000	226
DQGF polymer	1202	1200	104
	1100	1000	72

Table 8: The dynamic adsorption experiment result.

For the early stage polymer flooding, the characteristics of the responses on producers were different from the case in which polymer flooding was conducted during high water cut stage. The water producing of the producers continued to rise up after polymer flooding, but the simulation research showed that the water cut increasing rate was lower than the rate during merely water flooding (Figures 4-6). Of course, we also observed the drop down on the water cut in some wells, such as A11, A12, A13, A15, etc. For the well A11 (Figure 5), the highest water cut reduction reached 41% after the injectors (A5/A10) profiles controlled, and net incremental oil for A11 even reached 154,510 m³.

By December 2014, the total incremental oil by polymer flooding was about 754,650 m³, and the stage oil recovery efficiency was enhanced by 3.0%. And the polymer flooding is still effective now, and the polymer will be injecting until 2017. So we will get more oil from the polymer flooding. And the reasonable expectation of recovery beyond 2014 will be 2.8%. Totally we can get the oil recovery enhanced by polymer in LD oil field will be 5.8%. The actual EOR from polymer will be less than the plan of 6.1% EOR. The reason was that there were some wells blocked by polymer and sand. And we are doing lots works on solving the sand and polymer blockage [4,5].

Discussion and Conclusion

a. For the offshore oilfield, the platform life was about 25 years to 30 years. From lab test and numerical simulation research, we both know that the larger enhanced recovery value can be obtained, if polymer can be injected earlier.

b. We designed the polymer flooding on early stage, and the field test was carried out on the LD heavy oilfield of China offshore when the water cut was less than 10%.

c. By December 2014, the total incremental oil of polymer flooding was about 754,650 m³, and the stage recovery was improved by 3.0%. The polymer flooding is still effective, and more oil will be produced by the polymer flooding.

Acknowledgment

The data presented in this paper was obtained from many other departments of CNOOC Ltd.-Tianjin, such as the production department and offshore department etc. And the polymer and gel suppliers also provided some laboratory test data. We would like to thank them all for their contributions to this work.

Injecting slug	Displacement pattern	Time (month)	Displacement agent	Slug concentration (mg/L)	Mother liquor concentration (mg/L)	Injecting pressure (Mpa)	Implementation date
First slug (0.0275 PV)	Gel	10	DQGF+crosslinking agents (P: Cr³=600:1~60:1)	1200	5000 to 6200	<10	The polymer injection trial was conducted on the A23 well in March of 2006. After the trial, there were other 5 injectors were converted to polymer injectors from 2007 to 2009.
Second slug (0.11 PV)	Polymer	40	DQKY	1600			
Third slug (0.0275 PV)	Gel	10	DQGF+crosslinking agents (P: Cr³=600:1-60:1)	1200			

Table 9: The dynamic adsorption experiment result.

Comparision before and after gel injection		A01	A05	A10	A14	A23	A35
Injection pressure (MPa)	Before	2.4	81	6.5	7.7	5.5	6.8
	After	12.2	12	12.2	12.4	12.5	6
	Increase value	9.8	3.9	5.8	4.7	7	-
Apparent injectivity index (m³/d/Mpa)	Before	11.4	102.8	104.5	90.7	107.5	122.2
	After	46.2	65	89.5	62.6	50.6	127
	Decrease percent	59.5	36.8	14.4	31	52.9	--
RF		2.3	1.5	1.3	1.5	3.6	Acidification

Table 10: Characteristic of gel-injection well effectiveness.

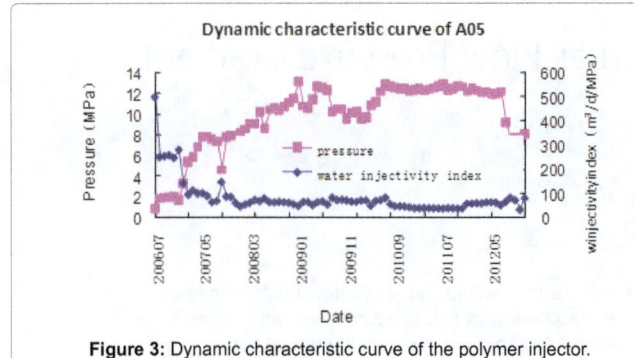

Figure 3: Dynamic characteristic curve of the polymer injector.

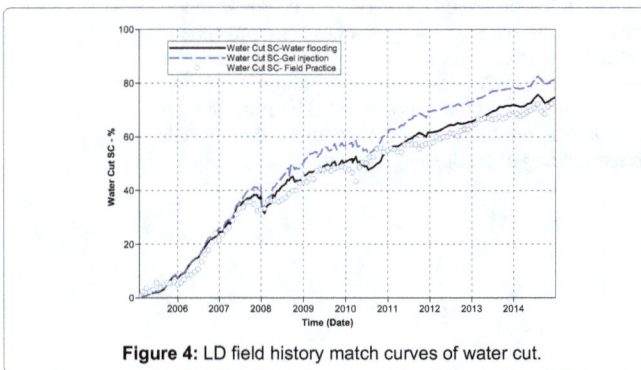

Figure 4: LD field history match curves of water cut.

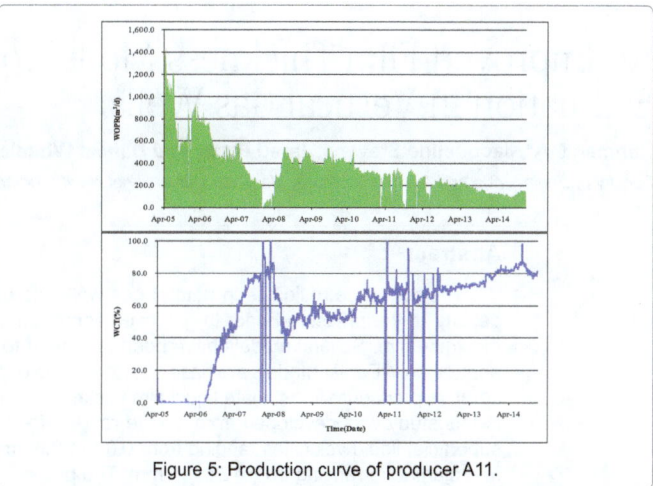

Figure 5: Production curve of producer A11.

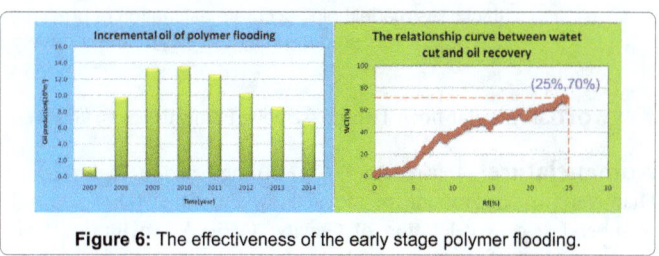

Figure 6: The effectiveness of the early stage polymer flooding.

References

1. Yanlai L, Yanchun S, Kuiqian M, Qizeng L, Xiaofei J (2011) Study of Gel Flooding Pilot Test and Evaluation Method for Conventional Heavy Oil Reservoir in Bohai Bay. SPE 146617.

2. Xiaodong K, Guozhi F, Xiansongv Z, Fujie S (2007) A method of hall derivative curves to evaluate effects of polymer flooding. China Offshore Oil and Gas 19: 173-175.

3. Yanlai L, Yanchun S, Kuiqian M The practice of chemical treatments for heavy oil reservoirs of China offshore oil field WHO C14-196, China.

4. Kuiqian M, Xiaofei J, Jing Y, Yingxian L, Zongbin L (2013) Nitrogen foam flooding test for controlling water cut and enhance oil recovery in conventional oil reservoirs of China offshore oilfield. SPE 165808.

5. Yanchun S, Yanlai L, Lixin T, Kuiqian M, Lilei W (2012) Gel treatments pilot test in conventional heavy oil reservoirs of China offshore oil field. SPE 157762.

26

An Improved Film Thickness Model for Annular Flow Pressure Gradient Estimation in Vertical Gas Wells

Rahman MA*, Jacqueline Stevens, Jared Pardy and Danika Wheeler

Faculty of Engineering and Applied Sciences, Memorial University of Newfoundland, St. John's, NL, Canada

Abstract

The presence of liquids in natural gas wells increases the pressure loss within the well due to differences in density of the pressure head. In gas, well annular flow, liquid may be present in entrained droplets as well as in the liquid film. Several models have been proposed to predict liquid film thickness in pipes with vertical two-phase annular flow. Earlier models are based limited range of experimental data. The earlier models also require exhaustive iterative procedure to estimate liquid film thickness. On the other hand, the proposed modified film thickness model in this study was developed from a wide range of experimental data. The experimental data covers conditions of superficial liquid velocities ranging from 0.6 to 38.8 cm/s; superficial gas velocities ranging from 13.4 to 110.6 m/s; and diameters ranging from 12 to 51 mm. The proposed model is compared with the available experimental data in the literature. Model predictions are in good agreement with the available experimental data set. The modified film thickness model helps accurate estimation of pressure gradient in vertical annular flow, which in turn is beneficial to the natural gas production industry as it further develops the understanding of production mechanics.

Keywords: Annular flow; Two-phase flow; Film thickness; Gas well

Nomenclature: *A:* Area (m²); *Cf:* Fanning Friction Factor (-); *D:* Diameter (m); *E:* Fraction of Liquid Flow Entrained (-); *Fr:* Froude Number (-); *g:* Acceleration of Gravity; *INTw:* Wave Intermittency (-); *Lwave:* Size of Disturbance Waves; (m); *LF:* Fraction of Film With Linear Velocity Profile (-); (kg s⁻¹): Mass Flow Rate; *Nμ:* Viscosity number (-); *RD:* Droplet Deposition Flux; *Re:* Reynolds Number (-); *U:* Velocity (m s⁻¹); *u:* Friction velocity (m s⁻¹); *We:* Weber Number (-); *x:* Flow Quality (-); *y:* Radial Coordinate (m); *δ:* Film Thickness (m); *ϵ:* Roughness (m);·: Non-dimensional Roughness (-); *μ:* Viscosity (kg m-1 s-1); *v:* Kinematic Viscosity (m² s⁻¹); *ρ:* Density (kg m⁻³); *σ:* Surface Tension (N m⁻¹); *τ:* Shear (Pa).

Subscripts: *Base:* Base Film; *Core:* Core; *Crit:* Critical; *Ent:* Entrainment; *Film:* Film; *G:* Gas; *I:* Gas-Liquid Interface; *L:* Liquid; *Max:* Maximum (E.G., Entrainment); *S:* Superficial (E.G., Velocity); *Rough:* Roughness; *Trans:* Transitions Between Base Film And Waves; *Wave:* Waves.

Introduction

The presence of liquids in natural gas wells increases the pressure drop within the well due to differences in density of the pressure head. Two-phase flow is occurred when gas and liquid flow together in a pipe and this can take several forms of flow patterns. A common type of two-phase flow is annular flow, which is characterized by a slow-moving liquid film along the pipe walls and a fast-moving gas in the center of the pipe. In annular flow, liquid may be present in entrained droplets as well as in the aforementioned film. Understanding the pressure drop is paramount important for equipment sizing, operation, and production. To model the pressure, drop, the density of the two-phase fluid must be determined. To determine this density, the amount of liquid, gas, and entrained liquid must be accurately computed. The density is a function of the void fraction between liquid and gas, which in turn is dependent on the film thickness and the entrained droplets within the gas.

The liquid film is characterized in two parts: a base film and a wavy disturbance layer, as shown in Figure 1. The amount of entrained liquid is dependent on the behaviour of this wavy disturbance layer and its

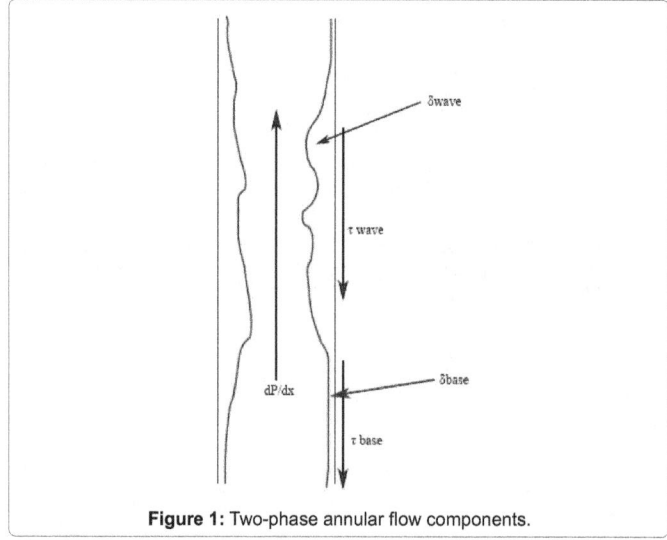

Figure 1: Two-phase annular flow components.

interaction with the gas component. In addition to density of the fluid, an understanding of the forces acting on the system is also important in understanding the pressure drop. Gravity, shear stresses, and drag forces all contribute to the velocity profiles, which in turn affect the characteristics of the liquid film and entrained droplets. The base film and wave film forces must be calculated separately in order to accurately determine the pressure drop of the system. A number of models have

*Corresponding author:** Aziz Rahman, Adjunct Assistant Professor, Faculty of Engineering and Applied Science, Memorial University of Newfoundland, St John's, NL, Canada, A1B 3X5, E-mail: marahman@mun.ca

been proposed over the past 50 years to predict fluid behaviour and characteristics of two-phase annular flow. Many of the methods are applicable only to the experimental data set upon which they were based, while others are simply modifications of previous models. The following literature review discusses some of the significant proposed models and correlations related to film thickness, liquid entrained fraction, and pressure drop.

Film thickness models

Past attempts to quantify liquid base film thickness (δ) in a two-phase system have resulted in significant progress in modeling natural gas transportation behaviour. There are numerous models to determine film thickness and entrainment. Schubring [1] completed a series of experiments using optical techniques, which measured base film thickness in annular two-phase flow. In these experiments horizontal and vertical conditions were examined and a correlation for each case was determined; the first being an empirical fit of the experimental data, and the second using a critical friction factor alongside an empirical fit. From the experiments, Schubring made a series of observations for the vertical case, such as: average base film thickness is inversely related to gas flow rate (superficial velocity) at constant liquid flow rate; for high gas flow rates, the dependence of average base film thickness on liquid flow rate vanishes; the average base film thickness is larger for larger tube diameters, however the dependence is less than linear; the film becomes more symmetric as the gas flow rate increases; vertical and horizontal average base film thicknesses have similar trends, however significant dependence on liquid flow rates remains at higher flow rates for the vertical geometry. The experiments included measuring superficial gas velocity against film thickness at fixed superficial liquid velocity values. This experiment included 3 horizontal cases using inside tube diameters of 8.8 mm, 15.1 mm, and 26.3 mm and a vertical case having an inside diameter of 23.4 mm.

Schubring were able to develop a model using an empirical correlation between film thickness roughness and tube length, gas and liquid density, and a Reynolds number. However, when incorporating the critical friction factor to develop the model an iterative process was developed. As the authors state, this process is most reliable when realistic values for film thickness and the base film velocity at the gas-liquid interface are used to calculate flow rates and the Reynolds number. This can be seen in the following set of equations.

$$\frac{\delta}{D} = 4.7 \frac{1}{x} \left(\frac{p_g}{p_l} \right)^{\frac{1}{3}} R_G^{\frac{2}{3}} \tag{1}$$

Where $R_G = \dfrac{G}{\mathrm{i}_l}$, $x = \dfrac{m_g}{m_g + m_i}$, $G = \dfrac{m_g + m_i}{A}$

Based on the findings, Schubring concluded that base film thickness is most strongly connected to gas flow and the effect of liquid flow rate is more significant at lower gas flows and in the vertical geometry. They also noted that uncertainties varied with the film thickness, tube thickness, and surface roughness. Typical base film thickness uncertainties for the horizontal tube were around 2% for high gas flows and 5% for low gas flows. Uncertainties were higher in the vertical tube (approximately 15%) due to the decreased wall thickness. Expanding upon their past research, Schubring [2] conducted another series of experiments using planar laser-induced fluorescence (PLIF) in a vertical tube with an annular flow regime of an air-water mixture in order to quantify the liquid film thickness. By introducing fluorescent dye into the water, the thin film layer could be measured after a series of imaging processing. By detecting the presence of the dye at the surface

of the film the thickness could then be determined. This experimental method proved advantageous over past methods as it allowed for a visual of the film after processing, which showed the differences between the base film and the waves that are observed during annular flow. The data obtained was applied to the Wallis [3] singlezone interfacial shear correlation, the Owen and Hewitt [4] correlation, and finally, the two-zone model by Hurlburt. For the Wallis [3] correlation, results at low liquid flow rates are consistent with the experimental data. However, for high liquid flow rates the Wallis model is not suitable for annular flow film thickness predictions due to its inconsistencies with the experimental data. Typically, the Owen and Hewitt [4] correlation, essentially a more complex version of the Wallis correlation, uses the entrainment fraction as an input to output film height and pressure loss, however, by using film height as an input, the entrainment fraction and pressure loss are found. It was observed that the Owen and Hewitt [4] correlation performed similarly to the Wallis [3] correlation in that for low liquid flow rates the predictions for pressure loss are consistent with experimental data, however, at high liquid flow rates the model diverges. The analysis noted that the Owen and Hewitt [4] model was no better than other correlations.

Finally, a two-zone model presented originally by Hurlburt [5] was applied to the experimental data. This model uses a roughness-modified log law to model interfacial shear and was found to be inferior to the inaccurate aforementioned models. More specifically, the effect of liquid flow on interfacial shear is underestimated while the effect of gas flow is overestimated. Furthermore, for this model the entrainment fraction is not a strong function of gas flow, which is not consistent with the experimental data presented. In its entirety, this model provided an acceptable prediction, particularly at a low superficial liquid velocity, however it is bested by the Wallis [3] correlation, and the Owen and Hewitt [4] correlation.

From the analysis, Schubring [6] concluded that average film thickness is found to be an increasing function of liquid flow and a decreasing function of gas flow. For the present experimental data sets, the models discussed here all under-predict the importance of increasing liquid flow on pressure loss and interfacial shear. Finally, since high liquid flow rates in annular flow induce disturbance wave and entrainment activity, further modeling in these areas is needed. This paper proposes a new film thickness model that has been developed by correlating to a wide range of experimental data. This model is based on Reynolds, Weber, and Froude dimensionless numbers and requires only diameter, fluid properties, and flow rates as inputs.

Entrained liquid fraction models

In addition to accurate predictions of film thickness, determining the fraction of entrained liquid in the gas core is critical in correctly calculating core density and pressure drop. A number of experimental data sets and models have been proposed to attempt to estimate entrainment. Most models however are limited to the experimental data sets from which they were correlated. Entrainment is the fraction of liquid entrained as droplets in the gas core – it has a range between 0 and 1, with 0 being perfect gas/ liquid separation and 1 being the transition from annular to dispersed mist flow. One of the earliest models for predicting entrained liquid fraction is the work of Steen and Wallis [6]. This model was developed graphically and is highly limited to low liquid and gas flow rates. A number of pressure drop models used this model to estimate entrainment due to its simplicity. Recent work on studying entrainment and associated models has been done by Cioncolini and Thome [7,8] who compared a number of proposed entrainment models against a broad range of experimental data and proposed another model that incorporates a refined Weber number. The data set includes 8 different gas-liquid combinations and over 19 different tube diameters. Inlet effects are not considered in this study, as the experimental setups have been designed to remove any dependence on the inlet conditions.

Cioncolini and Thome [8] observed that the best predictions of entrainment for the entire data set are given by the Sawant [9] correlation, followed by Ishii and Mishima and Pan and Hanratty [10,11]. If the entrained liquid fraction is above 0.5, the best predictions are given by Oliemans followed by Sawant [9,12]. The correlation presented by Oliemans requires an elaborate equation for entrainment:

$$\frac{e}{1-e} = 10^{b_0} \rho_l^{b_1} \rho_g^{b_2} \mu_l^{b_3} \mu_g^{b_4} \sigma^{b_5} d^{b_6} J_l^{b_7} J^{b_8} g^{b_9} \tag{2}$$

In this equation, entrainment is a function of many different parameters. The exponents b0 - b9 are dependent on the liquid film Reynolds number, and are given in a tabular form in the literature. The definition of the Reynolds number for this model depends on the fraction of entrained liquid. This results in a tedious iterative procedure to calculate entrainment. This model does not include density and viscosity of the droplet laden core but uses density and viscosity of the gas and liquid separately and therefore fails to properly capture the effect of the entrainment process on the core flow properties Cioncolini and Thome [8].

Ishii and Mishima [9] developed a correlation to predict the liquid entrainment fraction in the quasi-equilibrium annular flow region. The equilibrium entrainment fraction is when the rate of droplet entrainment equals the rate of droplet deposition. Their correlation is non-dimensional, explicit and was developed based on low pressure air-water data. The correlation was developed using an entrainment fraction of 1.0 as the upper boundary limit. More recent experimental data has shown that it is not possible to entrain all of the liquid from the liquid film even at very high gas velocities therefore their model fails to predict the observed trends at higher gas velocities. Sawant [9] proposed a correlation that uses dimensionless numbers for ease of use. Up until the introduction of this model, other correlations have proved to only be accurate for low flow, low pressure conditions. The model of Sawant [9] was developed to be able to accurately predict entrainment behaviour at high flow and high pressure conditions.

$$\frac{E}{E_m} = \tanh\left[2.31\times10^{-4} R_l^{-0.3} \left(W - W_c\right)^{1.2}\right] \tag{3}$$

The calculation requires the prediction of maximum entrained liquid fraction, *Emax*:

$$E_m = 1 - \frac{13N_i^{-0.5} + 0.3\left(R_l - 13N_i^{-0.5}\right)^{0.9}}{R_l} \tag{4}$$

Sawant [9] stated that the prediction of the maximum entrainment parameter requires further refinement and is therefore the constraint to their model. Cioncolini and Thome [8] determined that the core flow Weber number is the dominant dimensionless group in predicting the entrained liquid fraction. They developed a correlation based on this parameter:

$$E = \left(1 + 13.18W_c^{-0.6}\right)^{-1.7} \tag{5}$$

$W_{e_{core}}$ is the core flow Weber number, given as:

$$W_c = \frac{\rho c V_c^2 d_c}{\acute{o}} \tag{6}$$

They note, however, that accurate prediction of the core Weber number requires an accurate prediction of the average liquid film thickness, which results in an iterative calculation procedure. Their recommendations for further work include more experimental investigation concentrated on obtaining accurate data on the entrained liquid fraction. They recommend focusing on annular flows with high entrained liquid fractions as this appears to be where most models fail to accurately predict experimental data. This paper incorporates the Sawant [9] entrainment model into a modified pressure drop model. The choice of this model is validated by comparing its results with experimental data and by comparing it with results from other widely used entrainment models proposed by Steen and Wallis, Ishii and Mishima [7,10].

Pressure-drop models

Theoretical analysis of vertical flow began in the 1930s however it was not until the 1950s that correlations between all the various factors involved began to be documented. Gilbert [13] proposed complex empirical correlations for pressure drop in gas-liquid vertical flow but his method is based on limited production rates and conditions. Poettman and Carpenter [14] used an energy-balance method, a simplified Reynolds number and a limited amount of low flow rate measured field data as the basis of their theory. This method treats the system as a single-phase, which fails to take into account all the components of the energy balance and does not consider liquid/gas void fraction in the calculation of the overall mixture density. The discrepancies between measured data and these early correlations justified the need for large-scale laboratory experimental data to be used to find more suitable models and correlations to understand two-phase annular vertical flow. Duns and Ros [15] proposed the use of the pressure balance equation rather than an energy balance equation and also included the liquid holdup/void fraction in the density calculation. The pressure balance equation includes static, friction, and acceleration gradients which allows for distinguishing between the vertical two-phase flow patterns. Ultimately, this introduction to differentiating between flow patterns led to the refinement of modeling vertical two-phase annular flow. All the early work was done in short-tube laboratory apparatuses making it difficult to apply to actual field conditions where well tubing lengths are much longer. Hagedorn and Brown [16] ran experiments in longer tubes and used the data to develop more generalized correlations for pressure drop in vertical two-phase flow. Orkiszewski [16] developed pressure drop prediction correlations that were based on identifying the specific two-phase flow pattern. His method roved to be complex and is not fully tested for annular flow patterns. The models presented in the early work from the 1960s are valid only for the conditions under which the experimental data was taken. Aziz and Ansari [17,18] developed models that can be applied to a wider range of operating conditions. Their correlations elaborated on the early work and proposed models that predict annular flow behaviour using physical characteristics. Up until this date, all the early models were characterized by the rigorous steps involved to describe the physical nature of the annular flow which relied on determination of liquid film thickness, liquid entrainment, and interfacial shear stress. Hasan and Kabir [19] present a simplified model that assumes the liquid-film thickness is too small to have any significant contribution to computing accurate pressure drops. This method simplifies the mechanical energy balance by assuming the liquid film thickness is zero such that all of the liquid is moving through the core at the same velocity as the liquid. The core velocity equals the sum of the superficial liquid velocity and the superficial gas velocity. The Chen [20] correlation is used to determine the friction factor used in the pressure drop equation. The roughness value used in the correlation is that for the pipe wall, ignoring the wavy-liquid film. When compared to other models, this approach proved to match experimental data just as well. A recent model proposed by Schubring and Shedd [2] ignores the simplified approach and presents a model that is predominantly based on mass balance principles and by dividing the model into a base film zone and a wave zone. The model does not assume an average liquid film thickness but requires a base film thickness input and then assumes an average wave film thickness of two times that of the base film. Interfacial shear of the base film, wave roughness, and wave transitions are all estimated. While this method requires a number of parameters to be calculated, the inputs are only flow rate, fluid properties, and geometry. A time-averaged pressure gradient can be calculated based on the intermittency of the liquid film waves. The proposed modified model is compared against the Schubring [2], Ansari [18], and Hasan and Kabir models [19].

Theory

Liquid entrainment model

Using experimental data in the literature Sawant [9] proposed an entrainment model as follows:

$$E = E_m \times \tan h \left[2.31 \times 10^{-4} R_l^{-0.3} \left(W_e - W_c \right)^{1.2} \right] \tag{7}$$

$$E_m = 1 - \frac{13N_\mu^{-0.5} + 0.3 \left(R_l - 13N_\mu^{-0.5} \right)^{0.9}}{R_l} \tag{8}$$

Superficial liquid Reynolds number Rel and viscosity number $N\mu$ is defined as follows:

$$R_l = \frac{\rho_l U_s D}{\mu_l} \tag{9}$$

$$N_\mu = \mu_l \left(\rho_l \sigma \sqrt{\frac{\sigma}{g \left(\rho_l - \rho_g \right)}} \right)^{-0.5} \tag{10}$$

Weber number and critical Weber number can be calculated as follows:

$$W = \frac{\rho_g U_s^2}{\sigma} \left(\frac{\rho_l - \rho_g}{\rho_g} \right)^{0.2} \tag{11}$$

$$W_c = \frac{\rho_g U_s^2 D}{\sigma} \left(\frac{\rho_l - \rho_g}{\rho_g} \right)^{0.2} \tag{12}$$

Critical Superficial Gas Velocity ($Re \leq 1635$)

$$U_s = \frac{11.78 Rl^{-0.3} N_\mu^{0.8}}{\frac{\mu_l}{\sigma} \sqrt{\frac{\rho_g}{\rho_l}}} \tag{13}$$

New film thickness model

Based on available experimental data, a new film thickness model was developed from a range of experimental data sets. These data sets come from the works of Schubring [1,21-23], Bai and Newell, Alamu, Azzopardi, Paz, Shoham and Butterwoth [24-27]. The experimental data sets cover conditions of superficial liquid velocities ranging from

0.6 to 38.8 cm/s; superficial gas velocities ranging from 13.4 to 110.6 m/s; and diameters ranging from 12 to 51 mm. The new film thickness model is based on Reynolds, Weber, and Froude dimensionless numbers and thus requires only diameter, fluid properties, and flow rates as inputs.

$$\delta_{mod} = 1.93 \times 10^{-3} (Re)^{-0.246} (We)^{-0.161} (m_L / m_G)^{0.546} (Fr)^{0.15} \tag{14}$$

Where, $Re = \rho g Usg D / \mu l$, $We = \rho g Usg^2 D / \sigma$, $Fr = Usl^2 / gD$

Previously proposed models rely on correlated methods and empirical data from experiments that cover a narrow range of data to determine base film thickness as well as wave thickness. The new film thickness model has been developed from past experimental results, which collectively cover a broad range of diameters, fluid properties, and flow rates. The new model therefore provides a better model than those previously proposed.

Pressure drop model

The newly proposed film thickness model and Sawant [9] proposed entrainment model is used to calculate the pressure loss in an annular flow using Schubring and Shedd model as follows:

$$\frac{d}{d} = (1 - \pi_w) \frac{d}{d_b} + \pi_w \frac{d}{d_w} \tag{15}$$

Results and Discussion

Validation of new model with experimental data

Results from the new model were compared with results presented by Schubring [1,22], Alamu and Azzopardi and Butterwoth [26,27]. Each author has developed their respective film thickness models based on a series of experiments that use diameters ranging from 19 mm to 31.75 mm. As such it is not acceptable to directly compare the values for experimental film thickness with the values calculated by the new model. In order to circumvent the inability to directly compare these values, both experimental film thickness and modeled film thickness have been normalized with their respective diameters and presented in the form of δ/D. The Figure 2 represents a comparison between normalized values of experimental film thickness values and film thickness values calculated by the new model.

For complete agreement between experimental data and model results, the values from Figure 3 should lay on the line y = x, however it can be seen that the majority of the points in this figure lie within

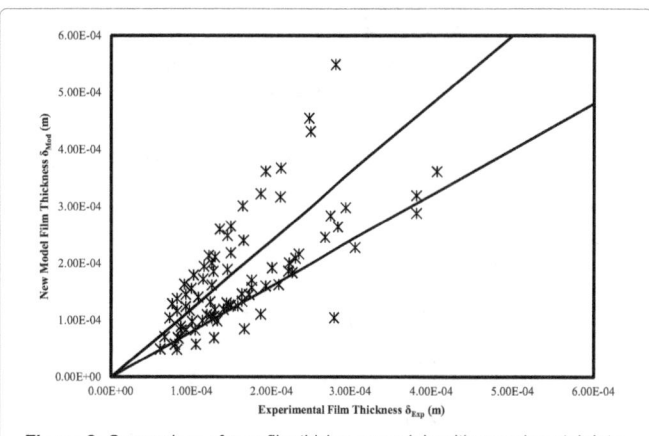

Figure 2: Comparison of new film thickness models with experimental data.

the +/- 20% tolerances. Therefore, the comparison between the new film thickness model and experimental film thickness data reveals that the model is indeed an appropriate approximation of experimental film thickness values for wider range of experimental data.

Further analysis of the new model and past models developed by Elvis, and Alamu with experimental measurements of film thickness reveal that the new model does indeed provide better predictions [26]. Figure 4 shows the comparison of film thickness models with experimental data and other models. The comparison of film thickness models with experimental data and other models clearly shows that while the values for the new model mostly reside within the acceptable area of +/- 20% of experimental data, values from Elvis and Alamu [22] reside outside this area. It can be noted that the models presented by Elvis and by Ansari [19] show a tendency to over predict film thickness, while the model presented by Elamu, greatly under predicts experimental film thickness values.

The inability to predict experimental results from models presented

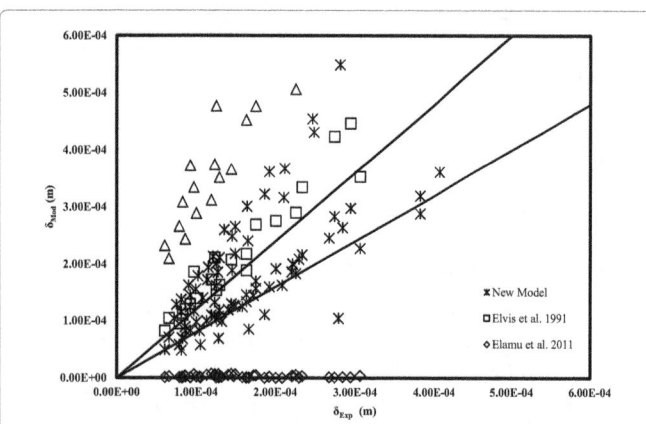

Figure 3: Comparison of film thickness models with experimental data.

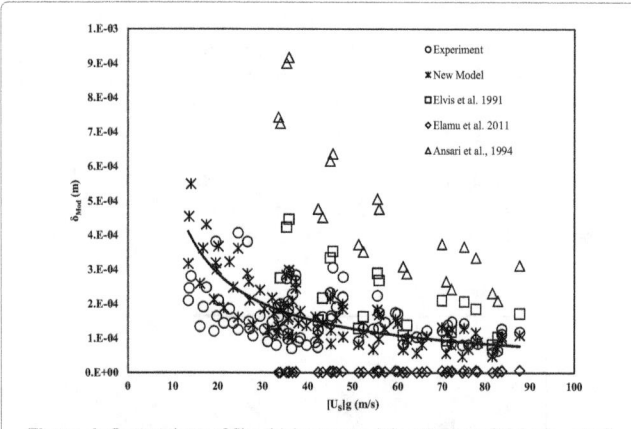

Figure 4: Comparison of film thickness models with superficial gas velocity.

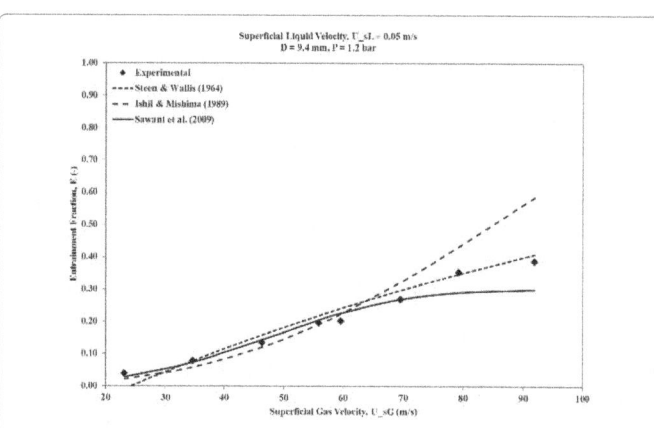

Figure 5: Entrainment model results vs. Sawant [1] data, UsL = 0.05 m/s, P = 1.2 bar.

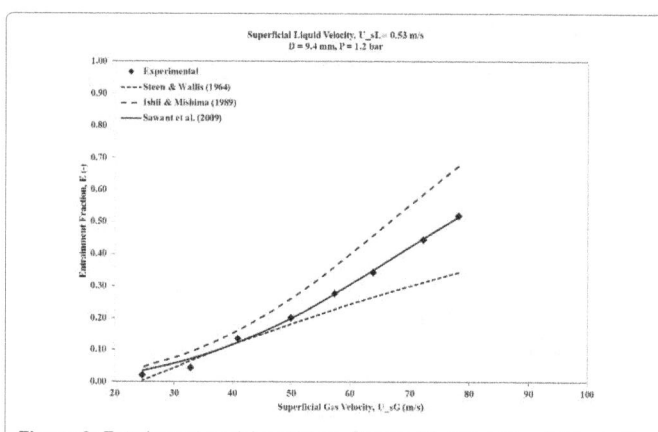

Figure 6: Entrainment model results vs. Sawant [1] data, UsL = 0.53 m/s, P = 1.2 bar.

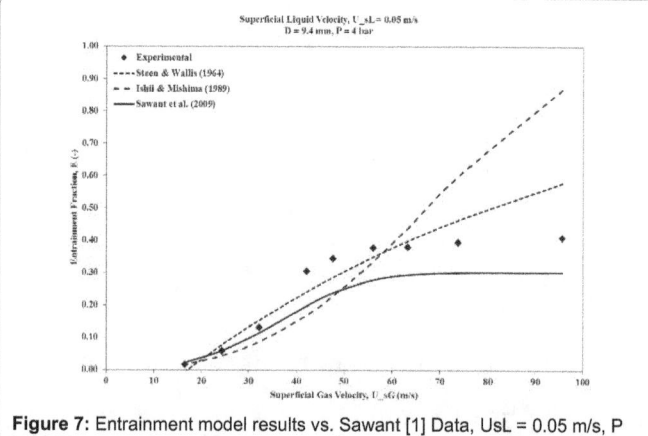

Figure 7: Entrainment model results vs. Sawant [1] Data, UsL = 0.05 m/s, P = 4 bar.

by Elvis and Elamu can also be observed in Figure 4 by the interaction between film thickness and varying superficial gas velocity. The new model remains the more accurate model, as it does not appear to deviate from the film thickness trends captured by the experimental data therefore validating the use of this new model developed in this study.

Entrainment model selection

To simplify the model presented by Schubring and Shedd [2],

	Diameter (mm)	U_{sg}	U_{sl}	Gas Density (kg/m³)	Liquid Density (kg/m³)	Inlet Pressure (kPa)
Sawant et al. (2008)	9.4	21.1 – 99.4	0.05 – 0.53	2	1000	120
		15.7 – 97.8		4		400
Owen et al. (1989)	31.8	97.4 – 470	0.05 – 0.11	2	1000	100

Table 1: Liquid entrainment fraction experimental data conditions.

a model to estimate liquid entrainment fraction was chosen from existing models. Steen, Wallis, Ishii, Mishima and Sawant [7,9,10] were compared against experimental data to determine a suitable choice. Experimental data was extracted from literature from Sawant and Owen [4]. Table 1 presents the conditions under which the experimental data sets were generated.

Validation with Sawant experimental data [1]: Figures 5-8 compare the Sawant [9] entrainment model with the Steen and Walls and Ishii and Mishima [7,10] models against the Sawant [12] experimental data. The graphs show the trend of entrainment fraction with increasing superficial gas velocity. While each model's predictions agree well with experimental data at low pressure and low liquid flow rates, only the Sawant [12] model predicts the trends at higher liquid flow rates and higher pressures. Ishii and Mishima [9] over predict at higher

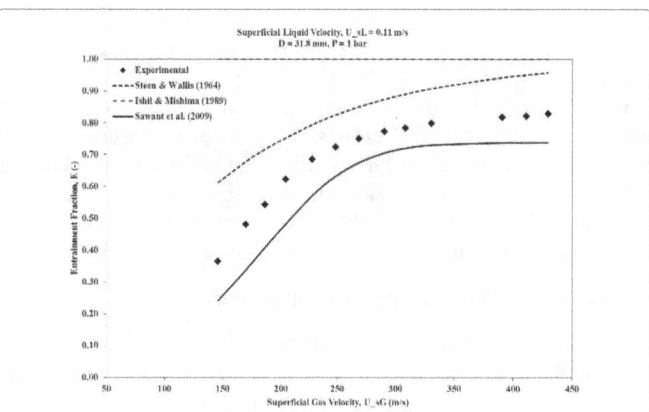

Figure 10: Entrainment model results vs. Owen [4] Data, UsL = 0.11 m/s, P = 1 bar.

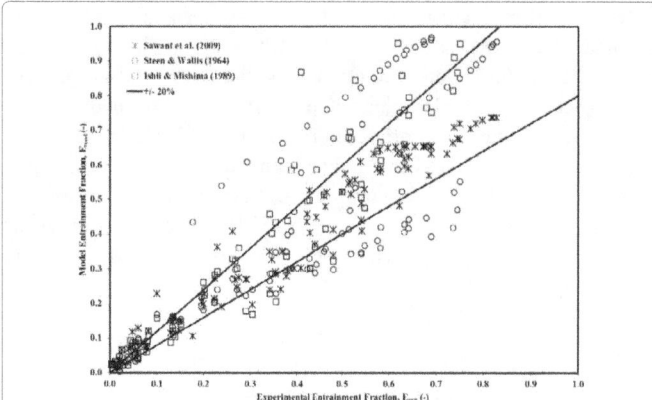

Figure 11: Comparison of entrainment model results with experimental data.

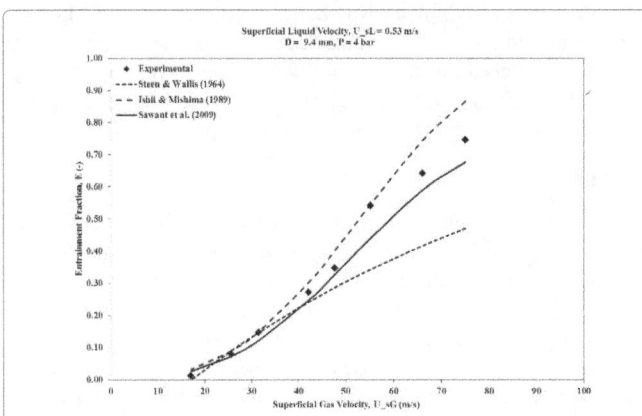

Figure 8: Entrainment model results vs. Sawant [1] Data, UsL = 0.05 m/s, P = 4 bar.

	Average % Difference = (Model – Exp)/Exp*100			
	Usl = 0.05m/s	Usl = 0.53m/s	Usl = 0.05m/s	Usl = 0.53m/s
	P = 1.2bar	P = 1.2bar	P = 4bar	P = 4bar
Sawant et al. (2009)	-6%	16%	-15%	5%
Ishii and Mishima (1989)	3%	52%	3%	29%
Steen and Wallis (1964)	-7%	-19%	-7%	-29%

Table 2: % Difference of entrainment model results with Sawant et al. (2008) experimental data.

	Diameter (mm)	Usg (m/s)	Usl (m/s)	Gas Density (kg/m3)	Liquid Density (kg/m3)	Inlet Pressure (kPa)
Schubring et al. (2010c)	23.4	34 - 76	0.04 – 0.35	1.2	1000	100

Table 3: Pressure drop experimental data conditions.

gas flow rates because their model was developed using an entrainment fraction of 1.0 as the upper boundary limit. In reality, as presented by the experimental data, the entrainment fraction approaches the limit much lower than one at high gas velocities. Table 2 summarizes the average % difference of each model with experimental data.

Validation with Owen experimental data [4]: Figures 9 and 10 compare the Sawant [9], Steen and Walls [6], and Ishii and Mishima [9] entrainment models against the Owen [4] experimental data. Again, due to the assumption inherent in the Ishii and Mishima [9] model that all the liquid can be entrained at high gas flow rates, their model fails to predict experimental data for this set. Their model predicts entrainment of ~1.0. Steen and Wallis [6] predicts the overall trend but at the low liquid flow rate the predictions are ~63% greater than experimental. Similarly, at the high liquid flow rate the predictions are ~22% greater than experimental. The Sawant [9] model matches the observed trends in close agreement, within +/-20% for both data sets. At the low liquid flow rate the predictions are on average ~0.4% less than experimental. Similarly, at the high liquid flow rate the predictions are ~16% less than experimental. For both experimental data sets of Sawant and Owen,

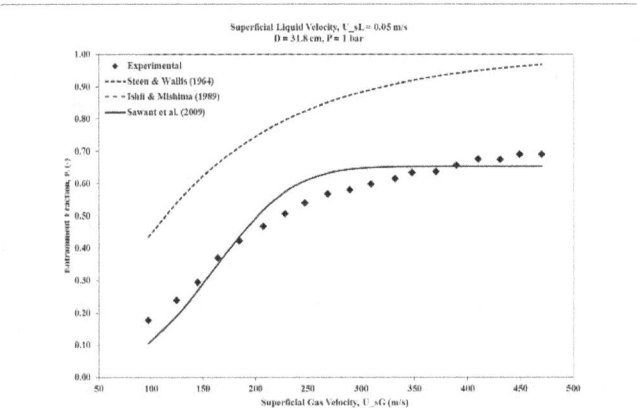

Figure 9: Entrainment model results vs. Owen [4] Data, UsL = 0.05 m/s, P = 1 bar.

Petroleum Technology and the Environment

the Sawant [4,6,9] model agrees within +/-20%. Figure 11 summarizes the comparison of all three models with experimental data. Steen and Wallis and Ishii and Mishima [7,10] both fail to agree with experimental data within suitable limits and fail to predict observed trends at high gas flow rates and higher pressures. Ability to predict entrainment at these conditions is critical for modelling annular flow in natural gas wells as high gas flow rates and high pressures are typical operating conditions. Sawant [9] is a suitable choice of entrainment model due to its simplicity and validation with experimental data.

Validation of modified pressure drop model

The modified pressure drop model presented in the previous section was compared against experimental data and models of Ansari and Hasan and Kabir [19,20] to validate that the model, though simplified, agrees with Schubring and Shedd's [2] results and predicts better than the other models studied. The modified model is a simplification of the Schubring and Shedd [2] mass balance / two zone approach. Ansari [19] is a semi-mechanistic approach that uses Steen and Wallis [6] to estimate entrainment and uses Wallis [3] to estimate the core friction factor which requires determining a wall friction factor and a superficial liquid friction factor. Hasan and Kabir [20] is a homogeneous model that is less rigorous than both Schubring and Shedd and Ansari [2,19]. It assumes that the liquid film may be ignored and that all the liquid therefore moves through the core with the gas phase at equal velocities. The model uses the correlations proposed by Chen to compute the two-phase friction factor [21]. Experimental data was extracted from

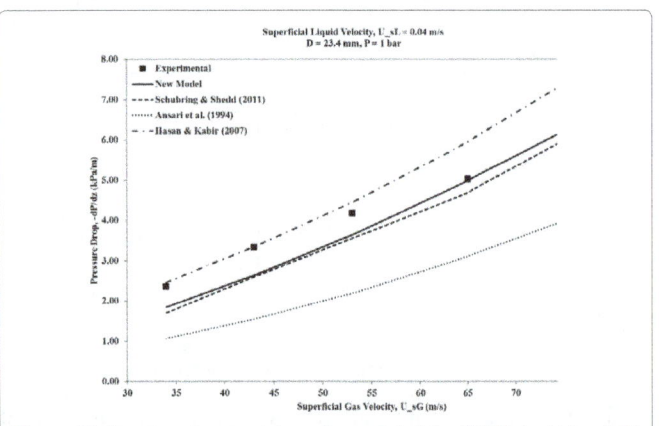

Figure 12: Pressure drop model results vs. Schubring [22] Data, UsL = 0.04 m/s, P = 1 bar.

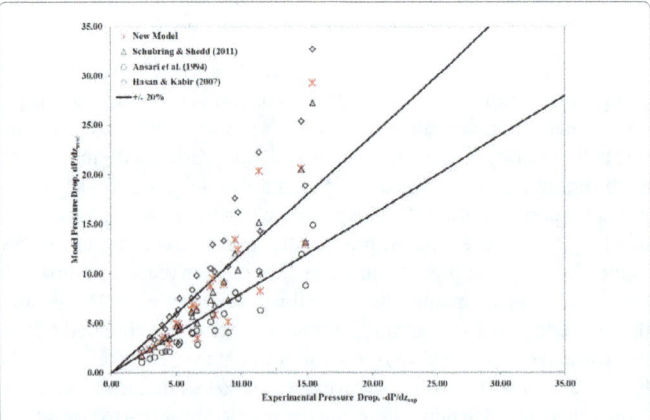

Figure 13: Comparison of pressure drop models with experimental data.

literature from Schubring [23]. Table 3 presents the conditions under which the experimental data sets were generated.

Validation with Schubring experimental data

Figure 12 compares the modified pressure drop model with the Schubring and Shedd [2], Ansari [19], and Hasan and Kabir [20] models against the experimental data of Schubring [23]. All models capture the trend that with increased flow rates, pressure drop will also increase. Hasan and Kabir [20] consistently over-predict while Ansari [19] under-predict. The modified model proves to agree with the predictions of the Schubring and Shedd [2] model. As presented in Figure 13 at high gas flow rates (65 and 76 m/s) and high liquid flow rate (0.35 m/s), experimental data shows that pressure drop does not continue to increase. All models fail to capture this trend. Some authors have seen that at these high rates the entrained droplets cluster together within the core and exhibit a wispy-annular type flow regime. The two-zone model, and therefore the modified model, are not able to predict this behaviour accurately. The modified model, while a simplified version of Schubring and Shedd [2], still agrees with experimental data and agrees with Schubring and Shedd's [2] results.

Conclusion

A modified two-phase vertical annular flow pressure drop model has been proposed based on simplification of Schubring and Shedd's [2] model. A new film thickness model has been proposed and incorporated into the model for simplification purposes. The new film thickness model is a simply model that requires only flow rates and fluid properties and is based on dimensionless terms. Further simplification was achieved by incorporating the entrainment model of Sawant [9] and by simplifying the characteristic gas velocity calculation to eliminate an integral term. The results of the simplified modified model agree well with experimental data and agree with Schubring and Shedd's [2] results and predict better than other earlier pressure drop models of Ansari, Hasan and Kabir [19,20].

The conclusions drawn from this study are as follows:

a) A new model was developed to calculate the film thickness for annular flow.

b) Hydrodynamic force balance was applied to develop the new model.

c) The newly developed model can reliably predict experimental data compared to other models.

d) The new model will help to reliably predict gas well pressure gradient for a wide range of operating conditions.

Acknowledgement

The author would like to acknowledge Mr. Jared Pardy, Ms. Danika Wheeler and Ms. Jacqueline Stevens for a contribution to this paper through their project work. The author also would like to acknowledge Mr. Xiao Xiong for editorial contribution.

References

1. Schubring D, Ashwood A, Hurlburt E, Shedd T (2008) Optical measurement of base film thickness in annular two-phase flow. FEDSM2008. Jacksonville, Florida: ASME Fluids Engineering Conference.

2. Schubring D, Shedd T (2011) A model for pressure loss, film thickness, and entrained fraction for gas-liquid annular flow. International Journal of Heat and Fluid Flow.

3. Wallis G (1969) One-dimensional two-phase flow. New York: McGraw-Hill.

4. Owen D, Hewitt G, Bott T (1985) Equilibrium annular flows at high mass fluxes; data and interpretation. PCH Physiochem. Hydrodynamics.

5. Schubring D, Ashwood A, Shedd T, Hurlburt E (2010) Planar laser-induced fluorescence (PLIF) measurements of liquid film thickness in annular flow. International Journal of Multiphase flow 36: 815-835.

6. Steen D, Wallis G (1964) AEC report, No. NYO-31142-2.

7. Cioncolini A, Thome J (2010) Prediction of the entrained liquid fraction in vertical annular gas-liquid two-phase flow. International Journal of Multiphase Flow 36: 293-302.

8. Sawant P, Ishii M, Mori M (2009) Prediction of amount of entrained droplets in vertical annular two-phase flow. International Journal of Heat and Fluid Flow 30: 715-728.

9. Ishii M, Mishima K (1989) Droplet entrainment correlation in annular two-phase flow. International Journal of Heat and Mass Transfer. p: 1835-1846.

10. Pan L, Hanratty T (2002) Correlation of entrainment for annular flow in vertical pipes. . Int J Multiph Flow.

11. Sawant P, Ishii M, Mori M (2008) Droplet entrainment correlation in vertical upward co-current annular two-phase flow. Nuclear Engineering and Design 238: 1342- 1352.

12. Gilbert W (1954) Flowing and gaslift well performance. Drill. and Prod. Pract, API, 126.

13. Poettman F, Carpenter P (1961) Multiphase flow of gas, oil and water through vertical flow strings. Drill. and Prod. Pract, API, 257.23

14. Duns H, Ros N (1963) Vertical flow of gas and liquid mixtures in wells. Netherlands: Section II-Paper 22-PO6.

15. Hagedorn A, Brown K (1965) Experimental study of pressure gradients occuring during continuous two-phase flow in small-diameter vertical conduits. J Pet Technol 17: 475-484.

16. Orkiszewski J (1967) Predicting two-phase pressure drops in vertical pipe. J Pet Technol 19: 329-338.

17. Aziz K, Grovier G, Fogarasi M (1972) Pressure drop in wells producing oil and gas. J Can Petrol Technol 11: 38-48.

18. Ansari A, Sylvester N, Sarica O, Shoham O, Brill J (1994) A comprehensive mechanistic model for upward two-phase flow in wellbores. SPE Production & Facilities SPE.

19. Hasan A, Kabir C (2005) A simple model for annular two-phase flow in wellbores. SPE Production and Operations.

20. Chen N (1979) An explicit equation for friction factor in pipe. Ind Eng Chem Fundamentals 18: 296-297.

21. Schubring D, Ashwood A, Shedd T, Hurlburt E (2010a) Planar laser-induced fluorescence (PLIF) measurements of liquid film thickness in annular flow. Part I: Methods and Data. Int J Multiph Flow.

22. Schubring D, Shedd T, Hurlburt E (2010c) Studying disturbance waves in vertical annular flow with high-speed video. Int J Multiph Flow.

23. Schubring D, Shedd T, Ashwood A, Hurlburt E (2010b) Planar laser-induced fluorescence (PLIF) measurements of liquid film thickness in annular flow. Part II: Analysis and comparison to models. International Journal of Multiphase Flow.

24. Bai X, Newell T (2002) Investgation of two-phase viscous liquid flow. International Refrigeration and Air Conditioning Conference. Lafayette, Indiana: Paper 590.

25. Alamu M, Azzopardi B (2011) Simultaneous investigation of entrained liquid fraction, liquid film thickness and pressure drop in vertical annular flow. J Energ Resour Tech.

26. Paz R, Shoham O (1999) Film-thickness distribution for annular flow in directional wells: Horizontal to vertical. Society of Petroleum Engineers.

27. Butterworth D (1972) Air-water annular flow in a horizontal tube. Progress in Heat and Mass Transfer.

Lignin Black Liquor in Chemical Enhanced Oil Recovery

Sulaiman WRW, Abbas AH*, Jaafar MZ and Idris AK

Faculty of Chemical and Energy Engineering, University of Technology Malaysia, Malaysia

Abstract

Enhance oil recovery (EOR) technique has been long used in residual oil recovery. Chemical EOR is emerging as a vital technology to recover additional residual oil through several mechanisms. The selection of high performance chemical for EOR is a challenging and time consuming task. Chemical EOR has been under spotlight for decades with enormous research efforts for specific reservoir applications. The research efforts reached its pinnacle with the development of surfactant families and formulation rules as well as the use of lignin from black liquor waste as surfactant or sacrificial agent. This complex chain of hydrocarbon and heavy molecular weight lignin, attribute to its high performance as surfactant. Moreover, lignin is a natural product; thus makes it as a low cost chemical. This review paper emphasizes on the importance of lignin in chemical EOR. The current trends and issues on its utilisation and its importance as an environmental sustainable biodegradable polymer matrix are discussed thoroughly. It is hope that this study will increase the research-based culture in lignin-based polymer composites; thus lead to generation of new ideas in the domain.

Keywords: EOR; Lignin; Adsorption; Biodegradable

Introduction

There are three stages in recovering oil from reservoir. In the primary stage, less than 20% oil were recovered from the original oil in place. Secondary recovery is used when there is insufficient underground pressure for left over oil [1,2]. Most of a reservoir's oil remains in place after the natural energy pressurizing the reservoir has been dissipated. Several techniques for injecting fluids into an oil reservoir to augment the natural forces have been widely used for many years. Such fluid injection is generally recognized as secondary recovery. Nevertheless, many reservoirs still jump to their last stage in less than fifteen years, which requires the application of tertiary recovery techniques after secondary oil recovery attempts [3-5]. The chemical flooding or chemical EOR is applied to recover the residual oil after water flooding. The process involves the use of surfactant/polymer, polymer, and alkaline flooding. In chemical flooding, surfactants are used to increase the capillary number by lowering the interfacial tension between the aqueous and oil phases. A suitable surfactant should lower the interfacial tension to (10^{-5} or 10^{-4}), which is accomplished by the disturbance of the oil and water molecules [1,6-8]. Chemical EOR has been under the spotlight for decades and has been attracting academic research attention particularly in the area of specific reservoir applications. Several studies have attempted to address the chemical flooding challenges using chemical adsorption to preserve the concentration of effective groups during flooding [9-11]. Empirical findings suggest that surfactants have the ability to perform in multi salinity reservoirs with less adsorption and work in high temperatures [12-14]. However, it is not always possible to find the suitable surfactant for the reservoir condition to work. Thus, in a bid to minimize surfactant adsorption, the term "sacrificial agent" appeared. Sacrificial agents are substances that mitigate the adsorption of surfactants into a formation. It also helps to reduce the retention of the surfactant inside the formation [12,15]. The sacrificial agent acts to modify the surface formation or be kept in the formation to reduce or eliminate the adsorption of the surfactant [6,10,16]. Compares to surfactant, the sacrificial agents are cheaper and thus cost-effective. Alkali, lignosulfonates, cellulose and cellulose derivatives, starch and starch derivatives, low cost surfactant and polybasic carboxylic acids

[15,17,18] are among the sacrificial agents used in the chemical EOR. Although each has their advantages over the other, recent studies suggest lignin as the most sought after sacrificial agent for a host of reasons including its renewable and sustainable property, environmental friendly, readily available in commercial quantity and very affordable [19,20]. This review paper therefore highlights the various derivatives and different applications of lignin as a sacrificial agent for chemical EOR. It summarizes the recent advances and issues involved in utilising lignin for new polymer composite materials development. It begins by classifying lignin based on its different derivatives, functions, applications, extraction methods and finally highlights why it is the ideal candidate as the most environmental sustainable and renewable form of biodegradable polymer matrices.

Derivatives of Lignin

Lignin or lignum meant wood in Latin. In the plant world, lignin is the second abundant and important element. Lignin is a renewable material that can replace the petrochemical in energy and raw fabric industries [21]. Research on the feasibility of lignin as polymer materials has been over 30 years and many properties about lignin were discovered throughout the years. It can be obtained easily, abundance, contain aromatic structure and can be modified in various way makes lignin attractive materials to polymer scientist. However, its complex structure may pose some challenge in its application. Furthermore, lignin is highly dependent on both the botanic origin and the isolation techniques.

*Corresponding author: Abbas AH, Faculty of Chemical and Energy Engineering, University of Technology Malaysia, Malaysia, E-mail: azzahashim2008@gmail.com

Wood generally comprises of cellulose, hemicelluloses and lignin. The quantities differ depending on the type of wood. Phenyl propane units (coniferyl alcohol and sinapyl alcohol) made up the hardwood, while substantial amount of p-coumaryl alcohol units together with coniferyl alcohol made up the softwoods. Free phenolic group on the other hand is the functional group for lignin (Figure 1) [22].

The structure of lignin is complex, but today, the structural elements are quite well known. Due to its hydrophilic nature, the lignin system is sensitive to high clay content. Clay content represents a key risk factor or failure mechanism for this system. The lignin phenol is manufactured from black liquor, a by-product of pulp mills. There are many steps involve in lignin processing: isolation, liquefaction and sulfonating. Despite the attempts made by several researchers to define lignin in terms of its chemical structure, there is still no acceptable definition. The conversion of lignin into valuable products is one way to unleash lignin's potential. For example, technical lignins such as Kraft lignin, lignosulphonates, and soda lignin, which are readily available in large amounts, are considered as a potentially interesting raw material because they are produced in processes dealing with the treatment of lingo-cellulosic materials [23,24].

Lignin is an undefinable compound. It can be used in surfactant system without the need to modify it. The modification process is costly; thus currently, the need to find a way to avoid modification step has led to the use of unmodified Kraft lignin, water soluble sulfonate, and the oil-soluble organic amine blending for chemical EOR.

Lignin by Kraft Process

Pulping process in paper mills is one of the mostly used methods to extract lignin in the form of black liquor. This black liquor is used for power generation, chemical recovery and process steam in the pulp mill [25]. The black liquor can be gassified to create some few value added products in the pulping mills. The gasification can also generate power in large scale. Thanks to the technology, lignin now has transit into a major stream process in bio-refinery. It is no longer a fuel, but is a green fuel, it can be converted into mixed alcohols, ethanol, and other syngas products. Now, the application and research on lignin has extends to high-molecular weight applications and research is being made for its utilization as aromatic products [26,27].

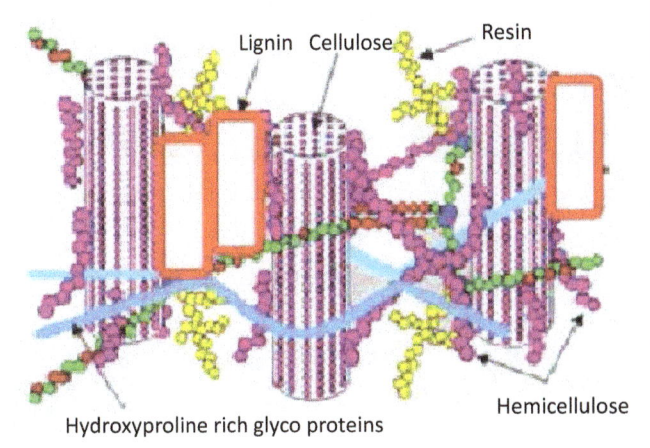

Figure 1: Wood microscopic structure showing the lignin position within lingo-cellulosic matrix [24].

Disposal of Lignin in Black Liquor

One of the waste products generated by the alkaline Kraft process in wood pulping is known as black liquor. It is a highly viscous aqueous waste consisting of organics constituents extracted from wood as well as inorganic pulping chemicals [27]. The solid composition of black liquor varies between 15% and 40% in weight out of which 30-40% is lignin [28,29]. Though the annual amount of lignin generated from black liquor is high [19], only about 10% of the lignin is recovered [20]. The rest are burnt during the extraction. The black liquor-lignin helps in increasing the Kraft- recovery system and increases its pulping capacity. Therefore, if lignin is left untouched without a proper treatment, it can harm the environment and human because lignin can increase the biological oxygen demand (BOD) and chemical oxygen demand (COD) when introduced to wastewater streams [30]. This can be averted as well as used as a great economic resource by extracting lignin from black liquor, and apply it in various areas such as combustible fuel gases or as a raw material for the production of activated carbons, phenols, etc. [31].

Lignin as a Sacrificial Agent

The importance of lignin was first highlighted in 1977, introducing salt lignosulfonate as a sacrificial agent. A laboratory study found that surfactant loss could be reduced significantly by pre-treatment of rock with a lignosulfonate preflush [17]. Subsequently, lignosulfonate was field tested in conjunction with the Glenn Pool surfactant flood expansion project [32,33]. Due to the success of lignosulfonate, another trend appeared to use raw source to replace it, this trend showed the importance of lignin as cheap raw source [34,35].

Johnson Jr and Westmoreland [36] extend the work of sacrificial agent by a finding that showed sacrificial adsorbates can outperformed the use of surfactant and adsorb on potential adsorption sites in the formation matrix, and consequently reduce the overall surfactant adsorption as the emulsion progresses. They also found that preparation of ligosulfonates is expensive. Thus, they introduced sacrificial agent consisting of effluent derived from the caustic extraction stage or weak black liquor.

Dardis [37] studied the soda-anthraquinone lignin as a sacrificial agent for surfactant. He investigated its use as a pre-flood agent in a mixture with the miceller dispersion, and found that the soda-anthraquinone lignins are strongly adsorbed by the adsorption sites in the formation.

Howard and Stirling [38] found the E-stage bleach plant as an active agent that can prevent the adsorption of anionic surfactant (sulphonated oil type surface active agents) on clays and sandstones. However, the main disadvantage of the use of E-stage bleach plant effluent, is the low concentration of active ingredients in the liquid.

Naae and DeBons [39] create a more effective surfactant from lignin by poisoning hydrogenation catalyst with sulphur. The lignin phenols that are produced from the reduction reaction can be recovered from the reaction mixture with a suitable organic solvent such as benzene, toluene, ether or diethyl ether. Later on the same year, they discovered the importance of surface-active compound chemicals. Useful surfactants have been produced by investigating the ways to increase oil recovery by reducing the interfacial tension to improve the displacement ability of water floods to act between oil and water in the reservoir [40].

Morrow [41] stated that lignin derivatives can be prepared by dissolving brines in fresh water followed by an addition of other compounds. This is because divalent brines cannot solubilized lignin, thus it must be prepared in fresh water prior to use with lignin. However, this may become a disadvantage to the process since large volume of fresh water is needed.

Kieke [42] suggested that the surfactant mobilizes the oil remaining in place after conventional production and allows it to be swept into production wells. Lignin is a waste by-product that the pulping industry produces in prodigious amounts. As a result, a large research effort has been undertaken over the last 40 years in attempts to find uses for the large volume of lignin by-product. Lignin itself is a major non carbohydrate constituent of wood and woody plants. It functions as a natural plastic binder for the cellulose fibers and permeates the membranes.

Findings and Conclusion

The role of lignin in industrial applications is negative because it is considered a waste material from the production of polysaccharide component from plants. This creates a lot of disposal problems given the huge amount of lignin generated from several industries such as paper industries. However, the inherent properties of lignin being biodegradable, CO_2 neutral, available abundantly in industrial waste, low in cost, and environmentally friendly, and having antioxidant, antimicrobial, and stabilizer properties makes it potentially attractive as a polymer composite.

The depletion of petroleum resources together with the increase in environmental awareness and demand toward sustainability and renewable resource has created concern among researchers to look for the best alternative renewable materials. Lignin then were discovered as potential component for polymer including surfactants and sacrificial agents.

Lignin has variety of use. It is not merely function as polymer composite, but also in other industrial sectors. However, the complexity of lignin structure may cause limitation in the application of lignin in industries. Thus, in order to fully utilised or manipulate the use of lignin, some studies began to propose techniques for the surface modification of lignins to make it compatible with other polymer matrices. The research are still new, thus is very expensive even at industrial scale. Also, careful consideration should be paid to gaining insight into the chemical and physical properties of lignin molecule.

In this review paper, the feasibility of using black liquor lignin as an environmentally friendly, low cost polymer composite, especially in the chemical EOR, is highlighted. Overall characteristics of lignin; derivatives, functions, applications, structure, extraction methods and bio-renewable properties in polymer composites matrices are the main focus. The latest trend and prospects in the utilization of lignin-reinforced polymer composites were also highlighted.

References

1. Sheng J (2010) Modern chemical enhanced oil recovery: theory and practice: Gulf Professional Publishing, USA.

2. Amyx JW, Bass DM, Whiting RL (1960) Petroleum reservoir engineering: physical properties: McGraw-Hill College.

3. Lake LW (1989) Enhanced oil recovery.

4. Alvarado V, Manrique E (2010) Enhanced oil recovery: An update review. Energies 3: 1529-1575.

5. Robel R (1978) Enhanced oil recovery potential in the United States. Interstate Oil Compact Comm. Com. Bull. USA.

6. Donaldson EC, Chilingarian GV, Yen TF (1985) Enhanced oil recovery, I: fundamentals and analyses: Elsevier.

7. Pope GA (2007) Overview of chemical EOR. In: Presentation Casper Eor workshop.

8. Watkins C (2009) Chemically enhanced oil recovery stages a comeback. Inform 20: 682-685.

9. Thomas S (2006) Chemical EOR: The past-does it have a future? (Russian).

10. Bera A, Kumar T, Ojha K, Mandal A (2013) Adsorption of surfactants on sand surface in enhanced oil recovery: isotherms, kinetics and thermodynamic studies. Applied Surface Science 284: 87-99.

11. Dąbrowski A (2001) Adsorption-from theory to practice. Advances in colloid and Interface Science 93: 135-224.

12. Southwick JG, Buijse MM, Van Batenburg DW, Van Rijn CHT (2014) Enhanced oil recovery fluid containing a sacrificial agent. Google Patents.

13. AlQuraishi AA, Alsewailem FD (2011) Adsorption of guar, xanthan and xanthan-guar mixtures on high salinity, high temperature reservoirs. In: Offshore Mediterranean Conference and Exhibition, Italy.

14. Flaaten A, Nguyen QP, Pope GA, Zhang J (2008) A systematic laboratory approach to low-cost, high-performance chemical flooding. SPE Symposium on Improved Oil Recovery, USA.

15. Shamsijazeyi H, Hirasaki G, Verduzco R (2013) Sacrificial agent for reducing adsorption of anionic surfactants. SPE International Symposium on Oilfield Chemistry.

16. Gogoi SB (2011) Adsorption–desorption of surfactant for enhanced oil recovery. Transport in Porous Media 90: 589-604.

17. Kalfoglou G (1977) Lignosulfonates as sacrificial agents in oil recovery processes. Google Patents.

18. Muherei MA, Junin R (2009) Equilibrium adsorption isotherms of anionic, nonionic surfactants and their mixtures to shale and sandstone. Modern Applied Science 3: 158.

19. Tejado A, Pena C, Labidi J, Echeverria J, Mondragon I (2007) Physico-chemical characterization of lignins from different sources for use in phenol–formaldehyde resin synthesis. Bioresource Technology 98: 1655-1663.

20. Calvo-Flores FG, Dobado JA (2010) Lignin as renewable raw material. Chem Sus Chem 3: 1227-1235.

21. Carrott P, Carrott MR (2007) Lignin–from natural adsorbent to activated carbon: a review. Bioresource technology 98: 2301-2312.

22. Ek M, Gellerstedt G, Henriksson G (2007) Ljungberg Textbook: Pulp and Paper Chemistry and Technology: Fibre and Polymer Technology, KTH.

23. Vishtal A, Kraslawski A (2011) Challenges in industrial applications of technical lignins. BioResources 6: 3547-3568.

24. Agrawal A, Kaushik N, Biswas S (2014) Derivatives and applications of lignin–an insight. The Sci-Tech Journal 1: 30-36.

25. Zaied M, Bellakhal N (2009) Electrocoagulation treatment of black liquor from paper industry. Journal of Hazardous Materials 163: 995-1000.

26. Jin W, Tolba R, Wen J, Li K, Chen A (2013) Efficient extraction of lignin from black liquor via a novel membrane-assisted electrochemical approach. Electrochimica Acta 107: 611-618.

27. Demirbas A (2008) Recovery of oily products from organic fraction of black liquor via pyrolysis. Energy Sources, Part A 30: 1849-1855.

28. Cardoso M, De Oliveira ÉD, Passos ML (2009) Chemical composition and physical properties of black liquors and their effects on liquor recovery operation in Brazilian pulp mills. Fuel 88: 756-763.

29. Zhao XY, Cao JP, Morishita K, Ozaki J, Takarada T (2010) Electric double-layer capacitors from activated carbon derived from black liquor. Energy & Fuels 24: 1889-1893.

30. Lataye DH, Mishra IM, Mall ID (2006) Removal of pyridine from aqueous solution by adsorption on bagasse fly ash. Industrial & engineering chemistry research 45: 3934-3943.

31. Kohl AL (1986) Black liquor gasification. The Canadian Journal of Chemical Engineering 64: 299-304.

32. Angert P, Leventhal S (1985) Preflush analysis of the Glenn Pool Surfactant Flood Expansion Project. SPE Annual Technical Conference and Exhibition.

33. Bae J, Petrick C (1986) Glenn pool surfactant flood pilot test: Comparison of laboratory and observation-well data. SPE Reservoir Engineering 1: 593-603.

34. Wang H, Zou J, Shen Y, Fei G, Mou J (2013) Preparation and colloidal properties of an aqueous acetic acid lignin containing polyurethane surfactant. Journal of Applied Polymer Science 130: 1855-1862.

35. Cerrutti B, De Souza C, Castellan A, Ruggiero R, Frollini E (2012) Carboxymethyl lignin as stabilizing agent in aqueous ceramic suspensions. Industrial Crops and Products 36: 108-115.

36. Johnson JS, Westmoreland CG (1982) Sacrificial adsorbate for surfactants utilized in chemical floods of enhanced oil recovery operations. Google Patents.

37. Dardis RE (1985) Soda-anthraquinone lignin sacrificial agents in oil recovery. Google Patents.

38. Howard J, Stirling M (1987) Sacrificial agents for enhanced oil recovery. Google Patents.

39. Naae DG, DeBons FE (1988) Recovering hydrocarbons with water soluble alkylphenol lignin surfactants. Google Patents.

40. Naae DG, Whittington LE, Ledoux WA, DeBons FE (1988) Recovering hydrocarbons with surfactants from lignin. Google Patents.

41. Morrow LR (1992) Enhanced oil recovery using alkylated, sulfonated, oxidized lignin surfactants. Google Patents.

42. Kieke DE (1999) Use of unmodified kraft lignin, an amine and a water-soluble sulfonate composition in enhanced oil recovery. Google Patents.

28

Preparation of Sorbitol Palmitate by Organic Catalysis and Its Application for Base Oil Stabilization

Noura El-Mehbad*

Faculty of Science, Najran University, Saudi Arabia

Abstract

Esters are excellent lubricants and high performance industrial fluids, but they are often costly. We prepared the ester sorbitol palmitate via an inexpensive phase-transfer catalysis method as an additive for the retardation of oil oxidation. The effects of the sorbitan palmitate content on the lubricant properties and oxidation stability of a base oil were determined. The addition of sorbitan palmitate to the oil retarded oxidation and enhanced the pour point depression. A novel method for inhibiting oxidation through the action of micellar cores was suggested. This micellar inhibition offers a new concept for the protection of lubricants against oxidative degradation.

Keywords: Hydrocarbons; Oxidation; Alkylation; Esterification; Wear

Introduction

Esters are widely used as lubricants and high performance industrial fluids. They are characterized by good biodegradability, low volatility, good lubricity, good thermal stability, and low pour points [1]. Ester oils are now used in many applications, including as automotive engine oils, hydraulic fluids, and compressor oils [2]. The antiwear and antifriction characteristics of alkyl octadecceoates increase with an increase in the number of polar linkages in the alkyl octadecenoate backbone. This is primarily due to the increase in number of sites amenable to chemisorption on the surface and, consequently, their reactivity. The antiwear and antifriction characteristics of derivatives of ethyl octadecenoates are inferior to the corresponding methyl 12-hydroxyoctadecenoate derivatives. This can be attributed to the increased reactivity of the latter at surfaces due to the additional hydroxyl moiety [3,4].

The properties of a non-ionic surfactant are related to its chemical structure. In particular, the structure of its hydrophobic and hydrophilic groups and their interactions are of great importance. The oxidative stability of a lubricating oil has a critical influence on its performance during service. In this paper, the non-ionic surfactant sorbitan palmitate (SPT) was prepared by phase-transfer catalysis, and its physicochemical (e.g. cloud point, critical micelle concentration (CMC) and performance properties as an antioxidant were investigated [5-7].

Ester oils are typically prepared by alcoholysis reactions catalysed by simple inorganic compounds [7]. Transesterification can be catalysed by both acids and bases, with the latter usually proceeding at much faster rates [7]. Aqueous solutions of sodium alkanoates will react with alkyl halides in a second phase provided an amine is added as a catalyst [8,9]. Hennis [9] showed that the catalyst must have at least one moderately long alkyl group to function well. The poor nucleophilicity of acetate ion toward various substrates in condensed systems has been attributed to a combination of polarizability, basicity, and solvation factors. Liotta [10] reported that the acetate ion, solubilized as the potassium salt in acetonitrile or benzene containing 18-crown-6, becomes sufficiently nucleophilic to react smoothly and quantitatively, even at room temperature, with a wide variety of organic substrates. Because of the cost of some esterification methods, we explored a phase-transfer catalysis approach for the preparation of SPT.

The aim of this work was to study the preparation and performance of SPT as an additive which can act as an antioxidant and affect the pour and cloud points for a base oil. SPT was added to the base oil in different concentrations, and its antioxidant activity was evaluated as a function of time. The degradation of the oil was monitored by total acid formation. A mechanism of action based on the formation of micelles and micellar inhibition of radical propagation was suggested. The oxidation stability of the lubricating oil was largely affected by the sulphur and aromatic hydrocarbon concentration in the oil, with an increased sulfur content leading to increased oxidative stability. The prepared compounds gave higher oxidation stability than imported compound (IRGANOX® L 135, Ciba) [11].

Experimental Procedure

Base oil sample

The physicochemical properties of the base oil are listed in Table 1.

Preparation of the additive by phase-transfer catalysis

Method 1: Anhydrous aluminium chloride (1 mol) was add with stirring over 1 h to 1-chloropalmitoyl chloride (1 mol) in CCl_4 (100 mL). The reaction temperature was kept at 20°C and stirring was continued for another hour to form the aluminium chloride complex. Sorbitan (0.1 mol) was condensed with palmitoyl chloride (0.6 mol) in a three-necked flask in the presence of tetraethylammonium bromide (0.01 g) as a phase-transfer catalyst. The reaction mixture was heated with continuous stirring until the theoretical amount of water was collected. The product was purified by washing with a hot solution of 5% sodium carbonate, and was then dissolved in petroleum ether (b.p. 40°C to 60°C). The sorbitol palmitate was completely characterized by IR, ^1H NMR, and mass spectroscopy, as discussed in the results and discussion section below.

*Corresponding author: Noura El-Mehbad, Faculty of Science, Najran University, Saudi Arabia, E-mail: dr.n.almehbad@hotmail.com

Property	Base oil	Test
Density (g/mL) at 15.5°C	0.8918	D. 1298
Refractive index (nD^{20})	1.4945	D. 1218
ASTM colour	4.5	D. 1500
Kinematic viscosity (cSt) at 40°C at 100°C	18.56 27.15	D. 445 D. 455
Pour point (°C)	12	ASTM D 97
Molecular weight (g/mol)	450	GPC
Total paraffinic content (wt %)	61.353	Urea adduction (7)
Carbon residue content (wt %)	1.5	ASTM D524
Ash content (wt %)	0.0311	ASTM D482
Naphthenes (wt %)	24.49	ASTM 3238/85
Aromatics (wt %)	9.51	ASTM 3238/85

Table 1: Physicochemical properties of the base oil.

The esters can be also prepared from haloalkanes by phase-transfer catalysis (ptc), as follows:

A mixture of 1-bromohexadecane (0.25 mol), sodium acetate trihydrate (0.2 mol), and tetraethyl-ammonium bromide (1 g) was heated at 105°C with vigorous stirring for 1 h. Then, water (300 mL) was added, the organic layer was separated and dried over anhydrous Na$_2$SO$_4$, and 1-hexadecyl acetate was recovered in 97% yield.

Method 2: The transesterification of methyl palmitate with 1-sorbiton was catalysed by calcium oxide. The ester was charged with 1-sorbiton complex in a glass reactor equipped with a Dean-Stark trap. Methanol was removed by azeotropic distillation with isooctane, as per the following equation:

Methyl palmitate + 0.1 Sorbitan → Sorbitan palmitate (C$_{22}$H$_{42}$O$_6$) (SPT) + CH$_3$OH

Oxidation stability study

The oxidation tests were carried out at 120°C according to the ASTM D 943 standard method. The base stock sample was subjected to oxidation with pure oxygen at a flow rate of 0.1 L/h for a maximum of 70 h. The SPT additive was added in different concentrations (from 1×10^{-6} to 1×10^{-3} mol/L). The viscosity, pour point depression, and total acid number were determined.

Results and Discussion

The successful preparation of SPT was confirmed by IR and NMR spectroscopy (Figures 1 and 2). In the IR spectrum shown in Figure 2, characteristic CH$_2$ and CH$_3$ stretching bands at 2916–2860 cm^{-1} are observed, as is a band at 1750 cm^{-1} typical for a carbonyl group. The molecular weight was confirmed by mass spectroscopy (Figure 3), with the product affording a parent ion at m/z 402.5. The melting point is 47°C.

The synthesis conditions for SPT were probed by studying the rate of product formation as a function of the catalyst loading. The preparation of SPT from alkyl carboxylates depends on the rate of quaternization relative halides and aqueous sodium (Figure 4).

According to Hennis [9], the rate can be determined according to the following:

N(Et)$_4$ Br + R-X   -------------^{K}--- (Et)$_3$N$^+$RX$^-$

(Et)$_3$N$^+$RX$^-$/dt = K[(Et)$_3$N/RX]

Evaluation of SPT as a pour and cloud point depressant and a flow improver for the base oil

Figures 5 and 6 present the data for the changes in the pour and cloud

points as a function of the SPT additive concentration. The additive clearly depresses both the pour and cloud points. This indicates that SPT prevents aggregation of the wax nuclei, but the value is nearly equal to the critical micelle concentration (CMC) in the oil phase (5×10^{-5} mol/L). Thus, the additive disperses the wax molecules and disrupts the formation of aggregates, in accordance with the findings in a previous

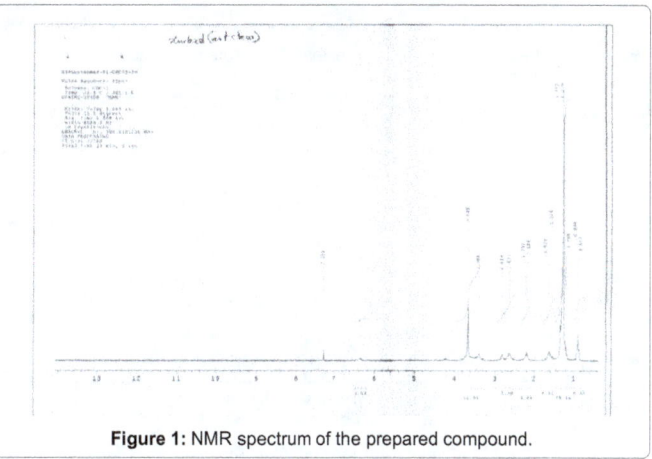

Figure 1: NMR spectrum of the prepared compound.

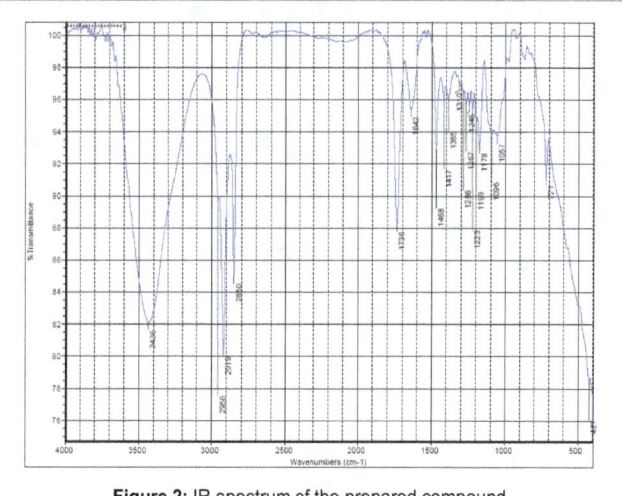

Figure 2: IR spectrum of the prepared compound.

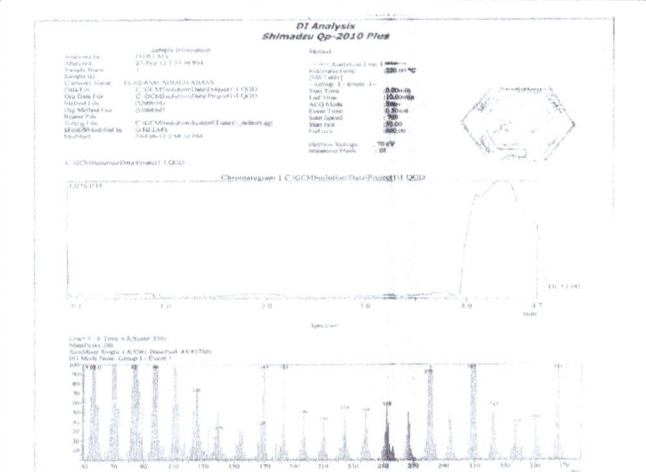

Figure 3: Mass spectrum of the prepared compound.

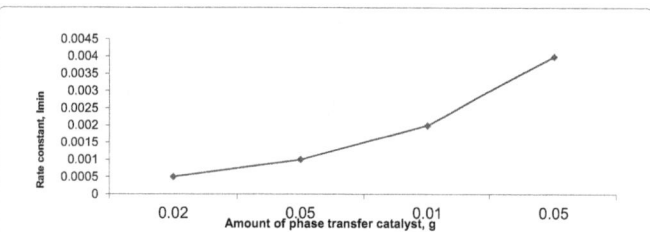

Figure 4: Relationship between the esterification rate and concentration of the phase transfer catalyst.

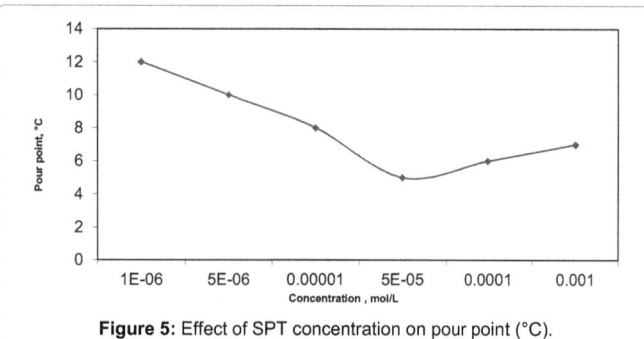

Figure 5: Effect of SPT concentration on pour point (°C).

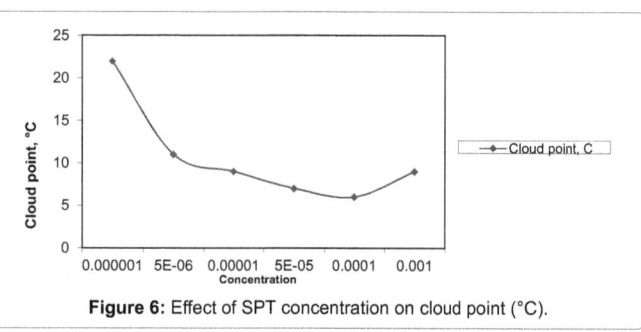

Figure 6: Effect of SPT concentration on cloud point (°C).

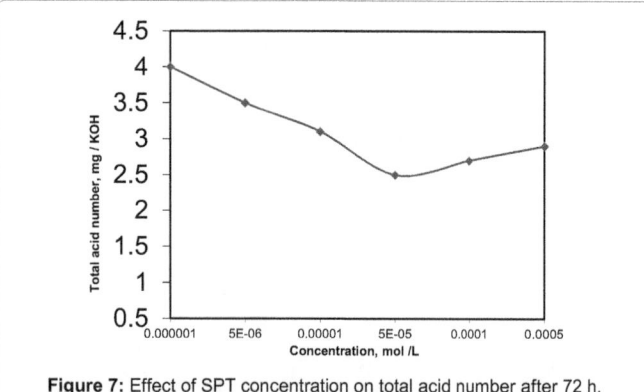

Figure 7: Effect of SPT concentration on total acid number after 72 h.

show that the viscosity and total acid number increase slightly over time during oxidation. These data confirm that the addition of SPT retards hydrocarbon degradation and the increasing of the oil viscosity with time. Thus, SPT increases the oil's stability toward oxidation and extends its lifetime. These results can be clearly observed in Figure 7. From this diagram, the addition of the additive to the oil retards the increase in the total acid number through oxidation. The best results are obtained at 5×10^{-5} mol/L (the CMC of the additive).

To explain these results, the author suggests a concept of revised micelle formation during hydrocarbon oxidation. The formation of micelles and their aggregation may take part in the oxidation process by inhibiting the chain propagation of the alkoxy free radicals R–O˙. The author believes that the micelles have cores in which the alkoxy free radicals are trapped. This means that the radicals that form oxidatively lose their ability to attack other species. Moreover, increasing the concentration of the additive does not change the degree of oxidation stability. It can be suggested that, the possibility of micelle to aggregates or destruct of micelle lead to free molecule of this additive tend to adsorbed at oil interface. This mechanism implies that before an additive is added into an oil, the critical value of micellization must be determined. This work agrees with early research by the author [11].

To elaborate this concept, further work will be needed to calculate the aggregation number of the micelles and their geometry at the oil interface. Moreover, the effects of non-ionic polymers on the degree of stability of the micelles, which can have a predominant role in oxidation processes, should be studied.

Conclusion

The mechanism of action for sorbitan palmitate as a multifunctional additive for the modification of the pour and cloud points and viscosity improver for a base oil was examined. The efficiency of this additive depends on its critical micelle concentration. The micelle cores act as traps for hydrocarbon oxide radicals in which to terminate hydrocarbon oxidation chains. The micellar inhibition depends on the incorporation of hydroperoxide or other polar oxygen-containing molecules in the reversed micelle, as the results revealed the increased oxidation stability of the oil. This information may be useful in further efficiency improvements for antioxidant additives.

References

1. Shubkin RL (1993) Synthetic lubricants and high performance functional fluids. Marcel Dekker, New York, USA.

2. Brown M, Fotheringham JD, Hoyes TJ, Mortier RM, Orszulik ST, et al. (2010) Synthetic base fluids, in Chemistry and technology of lubricants. Mortier RM, Fox MF, Orszulik ST (eds). Springer, London.

3. Chibber VK, Chaudhary RB, Tyagi OS, Anand ON (2001) 2nd World Tribology Congress, Sep. 3-17, Austria. pp. 193–202.

4. Rosen MJ (1972) The relationship of structure to properties in surfactants. Journal of the American Oil Chemists' Society 49: 293-297.

5. Rosen MJ (1974) Relationship of structure to properties in surfactants: II. Efficiency in surface or interfacial tension reduction. Journal of the American Oil Chemists' Society 51: 461-465.

6. Marszall L (1988) Cloud point of mixed ionic-nonionic surfactant solutions in the presence of electrolytes. Langmuir 4: 90-93.

7. Gryglewicz S (2000) Synthesis of dicarboxylic and complex esters by transesterification. 12th International Colloquium, Germany.

8. Huang ID, Dauermann L (1969) Exploratory process study. Base-catalyzed reaction of organic chlorides with sodium acetate. Industrial and Engineering Chemistry Product Research and Development 8: 227-232.

report [11]. The onset of the effect is observed at a concentration below the critical micelle concentration (CMC) and reaches a maximum at the CMC. Further increases in the concentration of the additive lead to the reversal of its adsorption orientation, as confirmed by Omar and Khidr [12].

There is a good relationship between the oxidation stability of an oil and its viscosity. The oxidation behaviour of the oil was studied by oxidizing it at 120°C and determining the total acid number at different times, whereas the viscosity was determined at 100°C. The results

9. Hennis HE, Thompson LR, Long JP (1968) Esters from the reactions of alkyl halides and salts of carboxylic acids. Comprehensive study of amine catalysis. Industrial and Engineering Chemistry Product Research and Development 7: 96–101.

10. Liotta CL, Harrs HP, McDermott MC, Gonzalez T, Smith K (1974) Chemistry of "naked" anions II. Reactions of the 18-crown-6 complex of potassium acetate with organic substrates in aprotic organic solvents. Tetrahedron Letters 15: 2417-2420.

11. El-Mehbad N (2013) Development antioxidants synthesized by phase transfer catalysts for lubricating oil. Tech Connect World 2013 Proceedings: Biotech Conference and Expo, May 12-16, 2013, Washington, DC, USA.

12. Khidr TT, Omar AMA (2003) Anionic/non-ionic mixture of surfactants for pour point depression of gas oil. Egypt J Pet 12: 21-26.

29

Characterization of Dynamic Pressure Response in Vertical Two Phase Flow

Agbakwuru J, Ogunlana A, Oshagbemi O, Rahman MA* and Imtiaz S

Texas A&M University at Qatar, Doha, Qatar

Abstract

One of the problems encountered in drilling, especially in offshore environments is "kicks". Kick is a sudden pressure imbalance in the wellbore during drilling operation. When this imbalances in pressure occurs the reservoir pressure has the ability to push the reservoir fluid into the wellbore. This may create a catastrophic event such as blow-out of the drilling rig. Thus, prior detection of the kick situation is critical to prevent any such catastrophic event. Currently, a kick situation is predicted or detected observing the properties of returned drilling mud from the wellbore. This method is not reliable as well as time consuming. The objective of this study is to develop a tool that will enable the prediction and detection of kick situations in managed pressure drilling (MPD). To achieve this goal, a two-phase experiment is conducted in 7.62 cm and 5 m long vertical pipe section. Instead of periodic sampling for kick situations, the newly developed tool enables the continuous monitoring of kick situations.

Keywords: Kicks; Wavelet transform; Two phase flow; Signal processing; Managed pressure drilling

Introduction

Managed Pressure Drilling (MPD) is a new drilling technique, which is developed for reducing the various drilling problems like kick, drilling fluid circulation loss, wellbore instability and formation damage. These drilling problems generally grow up during the conventional drilling process Managed Pressure Drilling is used to precisely manage the wellbore pressure when drilling with a narrow window between 1.38×10^6 and 2.068×10^6 Pa of the pore pressure and fracture pressure [1,2]. It is very useful for mature field because it can be revisited with better well control. Managed Pressure Drilling also reduces the Non Productive Time (NPT), which is time spent without drilling operations. It is believed that about 40% of the drilling problems happen as a result of pressure issues, lost return or kicks and stuck pipes. So MPD helps to reduce the drilling cost and stop the pressure related drilling hazards.

According to International Association of Drilling Contractors (IADC), Managed Pressure Drilling is an adaptive drilling process used to control precisely the wellbore pressure profile throughout the wellbore. The objectives are to ascertain the downhole pressure environment limits and to manage the annular hydraulic pressure profile accordingly. MPD is intended to avoid continuous influx of formation fluids to the surface. MPD categories contain three variations namely: Constant Bottom Hole Pressure (CBHP), Pressurized Mud Cap Drilling (PMCD) and Dual Gradient Drilling (DG). The CBHP method of managed pressure drilling uses annular frictional pressure and choke pressure in addition to mud hydrostatic pressure to achieve precise wellbore pressure control. This is the most common type of MPD being used.

The aim of MPD is to drill as close the pore pressure as possible and thereby reduce the dynamic overbalance. A reduction in dynamic overbalance often help to increase the rate of penetration (ROP), decrease surge and swab effect, reduce influx, and enhance well control (lost circulation, kicks). Lowering dynamic overbalance reduces the differential pressure in the well. As differential pressure is lowered, the force needed to break rock is lowered increasing ROP [3].

As defined, MPD controls the Bottom Hole Pressure (BHP) by applying surface back – pressure. This means that it can be chosen to drill with either a static underbalanced mud or a static balanced mud. Drilling with underbalanced mud yields a lager operating window, meaning if something unexpected happens in the well the operating margins are larger. However, if dynamic overbalance is not maintained at all times the sudden overbalance may cause severe damage to the unprotected formation. Government regulations are also strict regarding unbalanced drilling.

Conventional Methods of Detecting Kicks

A kick is an unwanted influx of gas or liquid into the wellbore, as a result of higher fluid pressure in the formation than in the wellbore. Conventionally, drilling mud of higher density than that of the formation fluid is used to prevent kicks. There are two basic conventional approaches of detecting kick – using pit gain (volumetric comparison), and using flow-out versus flow-in (rate comparison).

Pit gain (Volumetric comparison)

Pit gain is the variance in the amount of drilling fluid pumped into the well and the volume of drilling fluid pumped out of the well [4]. Under a no-kick situation i.e., stable situation, the two volumes should be equal i.e., zero gain. Whereas in a kick situation the amount pumped out is higher than the amount pumped in i.e., positive gain.

Flow-out vs. Flow-in (Rate comparison)

The flow-out rate is a measurement of the rate of fluid return from the well [4]. It is typically achieved by placing sensors in the flow line coming out of the wellbore. While flow-in rate is a measurement of the rate of fluid pumped into the well i.e., pump rate X volume per pump stroke. In a stable well, both flow rates should be approximately the same. An unexplained increased in the flow-out rate could be an indicator of a kick situation. This approach is normally used when the rig is not pumping fluid [4-6].

*Corresponding author: Rahman MA, Texas A&M University at Qatar, Doha, Qatar, E-mail: marahman@mun.ca

Challenges with the conventional method

Although the conventional methods have shown some benefits overtime, there are still some challenges with them. Some of the key challenges are:

1. In both approaches, it takes some time for the return fluid to get back to surface (pit or flow-out sensor) in order to detect a kick situation. This time delay could be very costly, especially in deep water environment. Hence, early detection is critical to mitigating the impact of kicks.

2. Certain kinds of operations can make it impossible to use pit gain as a kick indicator. For example, this happens when return flow from the well goes overboard instead of into a pit. Rig personnel generally cannot measure the volume of flow overboard, so they cannot make volume-in/volume-out comparison during such operations.

3. There is also the issue of false alarms. For example, even with the pump closed, there could still fluid out flow due to thermal expansion of the drilling fluid, rig heave or ballooning [4]. If not properly diagnose, this could be misinterpreted as a kick situation, which will subsequently lead to unnecessary costly non-productive time (i.e., time spent without drilling operations).

4. Due to the volume of the pit tanks, the sensitivity of an in-/out-flux is very small [7]. Hence, an in pit volume can be in the range of 0.5 – 1.0 m³ before the drilling crew responds [8].

Kick detection approach

In managed pressure drilling, the annulus is sealed while drilling. This closed loop system provides the additional advantage of making it easier to detect net flow rate and pressure anomalies within the wellbore. A detailed discussion of this advantage is documented by Grayson and Gans [9]. Reitsma has proposed a solution to detect kick by monitoring the time-trends of stand pipe pressure and annular discharge pressure in MPD [10]. Hauge described the modeling of kick detection and prevent using Ordinary Differential Equation (ODE) and Partial Differential Equation (PDE) models. Sonic measurement of the annular fluid, described by Hage and Avest has also been used to determine the phase shift induced by gas in the wellbore [7,11]. The proposed solution utilizes pressure anomalies in wellbore to detect kicks. By monitoring the time-trends of the down-hole pressure (with the aid of pressure sensors as shown in Error! Reference source not found), unwanted event such as kick can be detected. Downhole pressure measurements in time domain can be converted to frequency domain using Fast Fourier Transform (FFT) to detect unwanted signals. The proposed solution addresses the following key challenges faced by the conventional methods

Earlier detection of kick: By locating the pressure sensor(s) close to the source of the kick, the delay experienced with conventional methods is reduced. There is no need to wait for the flow to get to the surface.

Reduced false alarm: This method will eliminate false alarms due to out-flow drilling fluid expansion. This approach is more sensitive to kick situation as compared to the conventional methods. Hence, reduction is risk and subsequently improvement in safety.

Applications of proposed solution

The proposed solution can be applied in:

1. Managed Pressure Drilling (MPD) to detect pressure anomalies.

Pressure 1	Pressure 2	Pressure 3	Pressure 4	Pressure 5	3 Fluid Flow
54.320457	50.19384	14.580179	5.586773	11.627607	544.83972
54.320457	50.19384	14.580179	5.586773	11.246124	541.0752
54.320457	50.19384	14.924051	5.94955	11.627607	550.486501
54.320457	50.562089	15.207923	5.586773	11.627607	544.83972
54.320457	50.19384	14.236307	5.586773	11.627607	544.83972
54.686498	50.19384	14.580179	5.223995	11.627607	544.83972
54.320457	50.19384	14.580179	5.94955	11.246124	546.721981
54.686498	50.562089	14.580179	5.94955	11.627607	546.721981
54.320457	50.19384	14.580179	5.94955	11.627607	542.95746
54.320457	50.19384	14.924051	5.94955	11.246124	546.721981
54.320457	50.19384	15.955008	5.586773	12.00909	546.721981
54.320457	49.82559	15.267923	5.586773	12.00909	546.721981
54.320457	50.19384	14.024051	5.04055	12.390573	544.83972
54.320457	50.19384	13.20469	5.94955	11.627607	542.95746
54.320457	49.82559	14.580179	5.586773	11.246124	548.604241
57.248783	49.82559	14.580179	5.586773	12.00909	546.721981
54.320457	50.19384	14.924051	5.223995	11.627607	537.31068
54.320457	50.19384	15.267923	4.49844	11.627607	544.83972
54.320457	50.19384	14.580179	5.586773	11.246124	546.721981
54.320457	49.82559	14.580179	5.586773	11.627607	550.486501
54.320457	49.457341	14.580179	5.94955	11.627607	546.721981
54.320457	50.19384	14.580179	5.94955	10.483157	544.83972
54.320457	49.82559	14.580179	5.94955	11.627607	546.721981
54.320457	50.19384	15.267923	3.047331	11.246124	546.721981
54.320457	49.82559	14.580179	5.586773	10.86464	548.604241
54.320457	50.19384	15.267923	5.94955	12.00909	542.95746
53.054416	50.19384	14.024051	5.223995	11.246124	544.83972
54.320457	50.19384	14.580179	5.223995	12.00909	544.83972
54.686498	49.457341	14.580179	5.94955	11.627607	550.486501

Table 1: Experimental result showing the constant liquid rate (540 L/min) without air.

2. Conventional drilling where the operational mode does not allow the use of pit-gain or flow-out/flow-in methods to detect kicks.

3. Flow assurance to detect pressure anomalies. This study will focus on the application in MPD for detecting gas kick.

Laboratory Experiments

This section deals with the laboratory set-up and experiments performed. The experiment is to obtain good pressure readings by observing how two fluids (water and air) interact with each other to create abnormalities that might lead to kick situations in simulated wellbore.

Experimental set-up

The multiphase loop is consisted of two major sections, the horizontal and the vertical parts of the loop. Table 1 show the schematic diagram of the loop.

We focused majored only on the vertical section of the flow test section on our set of experiments. Installed along the loop, are pressure sensors, temperature sensors, air flow valve, liquid flow valve, air flow meters and liquid flow meter. Table 2 shows an image of the multiphase vertical flow loop with temperature and pressure sensors inserted in the pipe.

Design of experiments

Different flow phenomena and fluid interaction in the simulated wellbore have been studied. We obtained pressure reading by varying the flow rates of both liquid and air in two phase vertical pipe. We also

Date and Time	Pressure 1	Pressure 2	Pressure 3	Pressure 4	Pressure 5	3 Fluid Flow
18-07-2014, 11:58	72.320457	53.508082	12.516946	20.460648	26.123967	697.3028
18-07-2014, 11:58	71.890416	53.508082	12.173074	20.097871	25.742484	701.06732
18-07-2014, 11:58	72.622498	53.139833	12.516946	20.823426	25.742484	702.94958
18-07-2014, 11:58	71.890416	53.139833	12.516946	20.460648	25.742484	701.06732
18-07-2014, 11:58	71.890416	53.139833	12.516946	20.823426	25.742484	697.3028
18-07-2014, 11:58	71.890416	53.139833	12.860818	20.823426	26.123967	697.3028
18-07-2014, 11:58	71.890416	53.139833	12.516946	20.823426	26.886933	699.18506
18-07-2014, 11:58	72.320457	54.01283	12.20469	20.460648	26.50545	701.06732
18-07-2014, 11:58	71.890416	53.139833	12.204469	20.823426	26.123967	695.42054
18-07-2014, 11:58	71.890416	53.139833	13.204469	20.823426	26.50545	699.18506
18-07-2014, 11:58	72.320457	53.139833	13.204469	21.186203	26.123967	701.06732
18-07-2014, 11:58	71.890416	53.139833	12.516946	20.460648	26.886933	701.06732
18-07-2014, 11:58	72.320457	53.139833	12.860818	20.460648	26.123967	701.06732
18-07-2014, 11:58	71.890416	53.139833	12.173074	21.186203	26.886933	701.06732
18-07-2014, 11:58	72.320457	52.103335	12.860818	20.097871	25.742484	701.06732
18-07-2014, 11:58	71.890416	53.139833	13.892435	20.460648	26.123967	699.18506
18-07-2014, 11:58	72.320457	53.139833	13.2.469	21.186203	26.886933	701.06732
18-07-2014, 11:58	71.890416	53.508082	12.860818	20.823426	24.216551	695.42054
18-07-2014, 11:58	72.320457	53.139833	13.20469	20.460648	25.742484	697.3028
18-07-2014, 11:58	72.320457	53.139833	12.516946	20.460648	26.50545	701.06732
18-07-2014, 11:58	71.890416	53.139833	12.173074	20.460648	26.123967	699.18506
18-07-2014, 11:58	72.320457	53.508082	12.860818	20.823426	26.50545	701.06732
18-07-2014, 11:58	72.320457	53.508082	12.516946	20.823426	26.886933	699.18506
18-07-2014, 11:58	72.320457	53.139833	12.516946	20.460648	26.50545	701.06732
18-07-2014, 11:58	72.320457	53.508082	12.860818	21.186203	26.886933	701.06732
18-07-2014, 11:58	72.320457	53.139833	12.516946	20.460648	26.886933	701.06732
18-07-2014, 11:58	71.890416	53.139833	12.860818	21.186203	26.50545	701.06732
18-07-2014, 11:58	72.320457	53.508082	12.860818	20.460648	26.886933	693.538279
18-07-2014, 11:58	72.320457	53.139833	12.516946	20.460648	26.886933	697.3028

Table 2: Showing constant liquid (693 L/min) flow rate with a sudden injection of short air.

observed the various flow patterns in the vertical test section of the pipe. Our experimental procedures are summarized as follows:

1. The flow was started with liquid first, and then pressure measured as air was gradually injected.

2. Different flow rates were obtained by adjusting both the air and liquid flow valves (producing bubble or slug flow patterns).

3. Pressure measurements are taken on the simulated wellbore.

4. The pressure-time result is saved for further analysis.

5. Keeping air flow rate constant and varying the liquid rates. (max liquid rate achievable was 718 L/min and the min rate was 540 L/min).

6. Keeping liquid rate constant and sudden release of air for a very short period (1 sec) into the system (the busty effect).

7. Tables 1 and 2 show an experimental data at 540 L/min without air and at 693 L/min with a sudden injection of air at a very short time (1 sec).

8. We also monitored the behaviour pattern of pressure from pressure probe 2 (#2 as shown in Figure 1) and pressure probe 4 (#4).

9. Our main area of focus is on the pressure #2 and #4 of the test section of the vertical pipe.

10. From the pressure difference between pressure #2 and #4, we predicted kick situation.

Result Analysis

This section deals the analysis performed for this study. It presents the underlying principles and performance of the kick detection tool used.

Underlying principle and detection algorithm

The foundation or main underlying principles used in the kick detection tool is wavelet transformation. A wavelet is a function of zero average [12]. Its amplitude starts at zero, increases, and then decreases back to zero. There are various type of wavelets based on the algorithm used in generating the wavelets [13]. The most common types are Daubechy, Coiflet, Meyer, Symmlet. Wavelet applications can be found in several applications such as image recognition applications, seismograph, blood-pressure, heart-rate and ECG analyses, music editing, finger print verification, DNA analysis, speech recognition, and several other signal processing applications [14] (Figure 2).

In summary wavelet transformation is the use of wavelets to remove or filter noise (unwanted signals) in order identify a pattern or target [14].

Kick detection algorithm

The algorithm used in detecting kicks is summarized in the flow steps below:

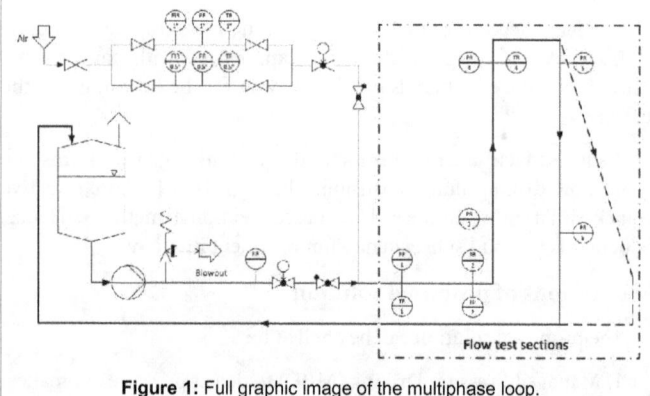

Figure 1: Full graphic image of the multiphase loop.

Step 1: Measure the pressure difference between a reference location and the point of interest in the wellbore (higher than the reference location).

Step 2: Appropriate wavelet transform level is calculated based on this pressure difference.

Step 3: Pressure response from the interest location is passed through the wavelet scale tool.

Step 4: Spikes (Kicks) are detected by the tool. The current version of the wavelet tool developed is a pure nominal scale that detects a kick or non-kick situation. Currently, the severity/magnitude of kick not calculated by the tool.

   i.   Flat line = Non-kick situation.

   ii.   Spike = Kick situation.

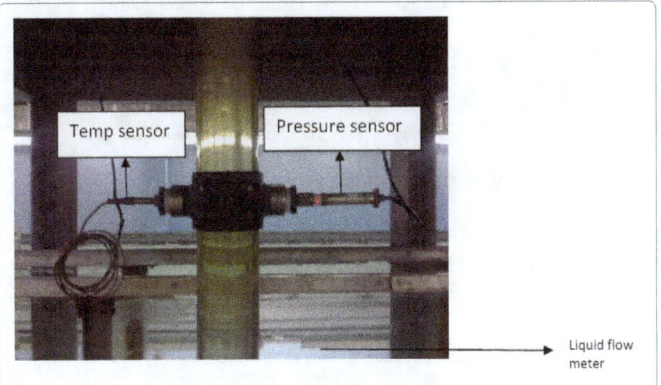

Figure 2: Temperature and pressure sensors.

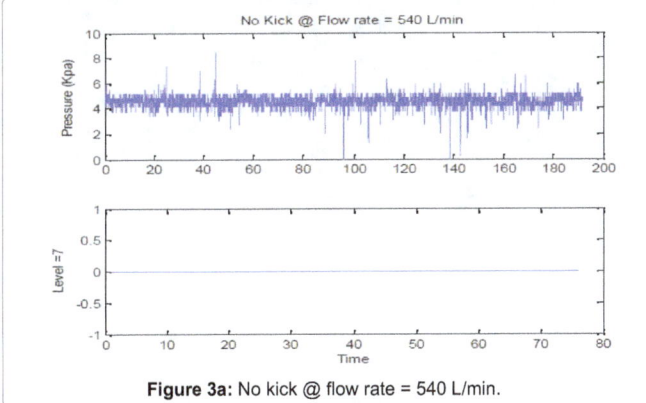

Figure 3a: No kick @ flow rate = 540 L/min.

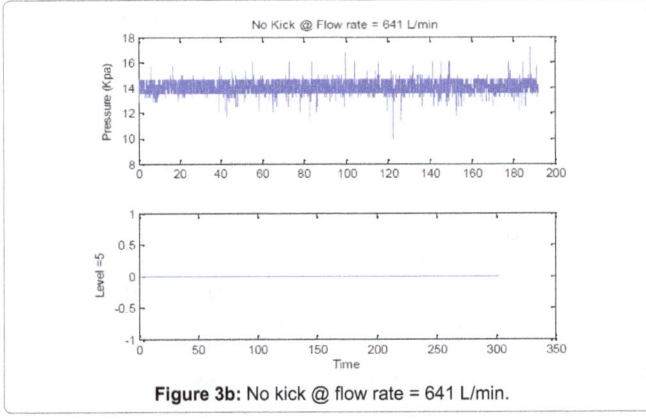

Figure 3b: No kick @ flow rate = 641 L/min.

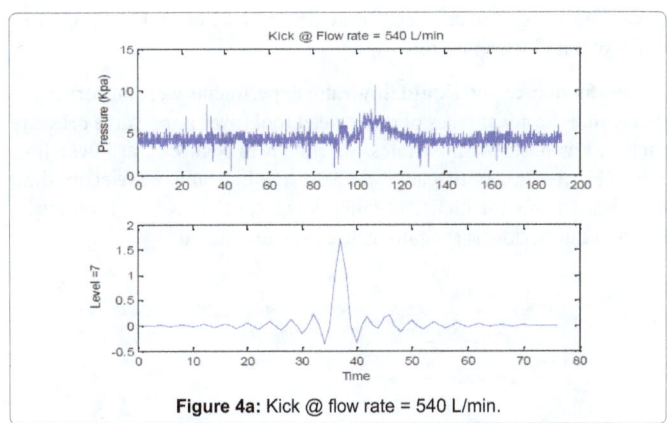

Figure 4a: Kick @ flow rate = 540 L/min.

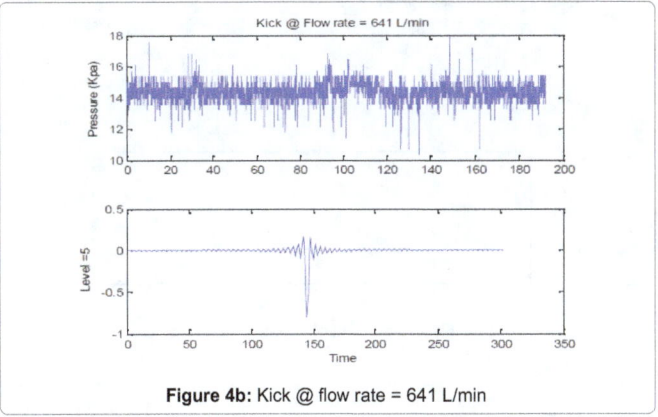

Figure 4b: Kick @ flow rate = 641 L/min

Results

This section presents the result and conclusion of the analysis performed in this study. The result is based on flow patterns of 540 L/min to 718 L/min and air burst durations of 0, 1, and 2 seconds. The reference point is taken as #P2 probe, while the point of interest is #P4 probe. Results from the wavelet tool are compared with results from pure gauge pressures at the location of interest.

No kick situation: Figures 3a, 3b is the response without kicks at the lower and upper ends of the achievable flow rates in the lab. The top charts are the gauge pressure response, while the bottom charts are the wavelet scale response. Observing the top charts exclusively could be misinterpreted for kick. By using the tool, no spike is observed, the line is flat, hence no kick

Bursty early kick situation: Figures 4a, 4b is the response with kicks at different flow rates. The top charts are the gauge pressure response, while the bottom charts are the wavelet scale response. By using the tool, a spike was clearly visible to indicate a potential kick situation.

Reliability test: Three runs (repeated measurements) where performed on sample size of 16 flow scenarios. Reliability tests were performed in SPSS and MedCalc using both coefficient of variation from duplicate measures test as well as Cronbach's Alpha test. As shown in Figures 5a, 5b, no variation was observed between duplicate runs and the Chronbach's alphas score is 1. For research purposes Chronbach's alpha should be more than 0.7 [15].

Validity test: The validity test performed using Kappa Agreement test. Twenty four flow scenarios were measured using the wavelet tool against actual physical observation (goal standard). As shown in Figures 6a, 6b, the Kappa score is 0.909 (90.09%) indicating very good scale. The wavelet scale successfully detected 15 of 16 kick situations and 8 of 8 no kick situations.

Performance and liquid flow rate: Experiments were performed to determine the advantages of the wavelet tool (over pure gauge pressure method) at different flow rates. As shown in Figure 7, at lower flow rates (Figure 7a), both gauge pressure method and wavelet method can clearly show the kick. At higher flow rates (Figure 7b), the wavelet method out performs the pure gauge pressure method.

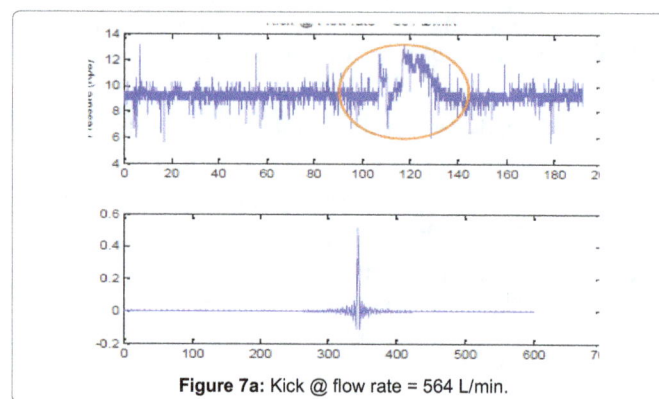

Figure 7a: Kick @ flow rate = 564 L/min.

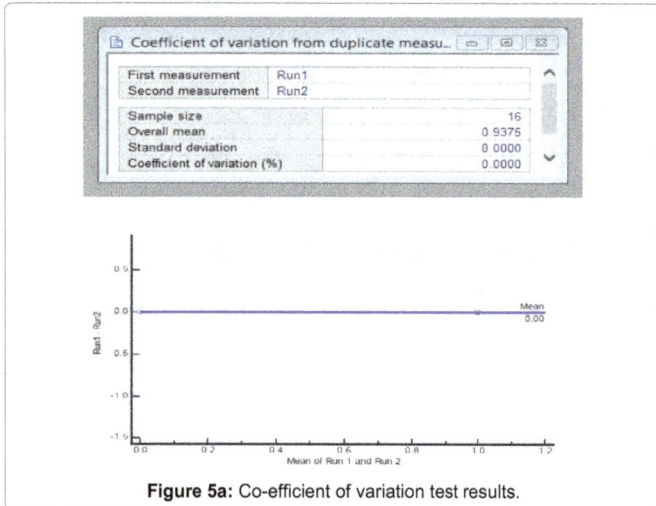

Figure 5a: Co-efficient of variation test results.

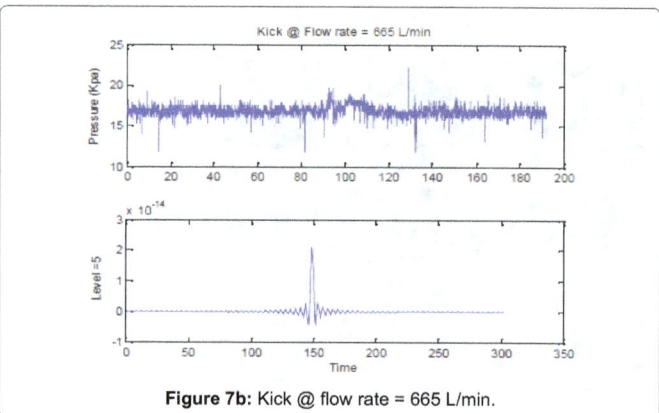

Figure 7b: Kick @ flow rate = 665 L/min.

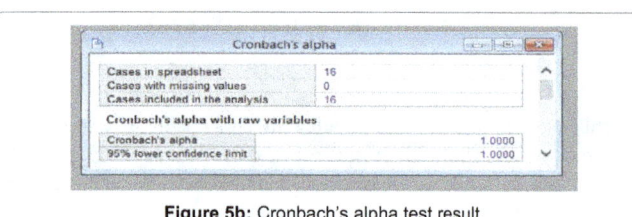

Figure 5b: Cronbach's alpha test result.

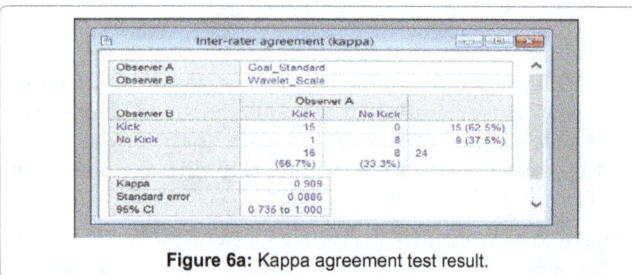

Figure 6a: Kappa agreement test result.

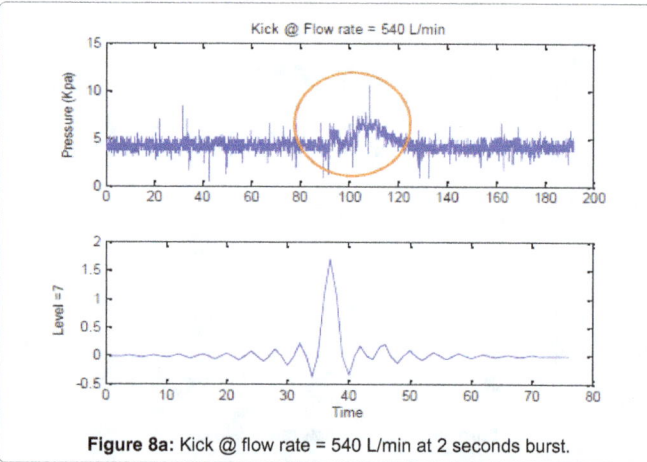

Figure 8a: Kick @ flow rate = 540 L/min at 2 seconds burst.

Performance and burst duration: Experiments were performed to determine the advantages of the wavelet tool (over pure gauge pressure method) at different air burst durations. To bursts durations – 1 second and 2 seconds were used for the analysis. At longer burst durations (Figure 8a), both pure pressure and wavelet methods can clearly show the kick. At shorter burst durations (Figure 8b), the wavelet method out performs the pure gauge pressure method.

Performance and burst rate: Experiments were performed to see how the tool reacts with burst rate. At low burst rates (Figure 9a) the tool is more accurate at detecting the number of bursts. At higher burst rated (Figure 9b), the performance diminishes. This is in agreement with the design of the tool. The tool is designed to detect early kick which is typically characterized with low rate bursts. High rated bursts signify late kick situations.

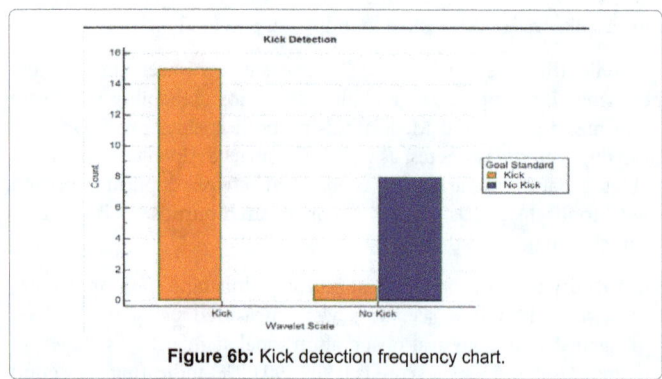

Figure 6b: Kick detection frequency chart.

Conclusion and Future Work

From the results and analysis present above, it shows that the wavelet method appears to be more efficient than pure gauge pressure gauge method in detecting bursty early kicks situations. Its benefits over the pure gauge method is more pronounced with higher liquid flow rates, with shorter kick burst duration, and as well as with shorter kick burst durations. This type of scenario is similar to a potential kick environment in deep see drilling i.e., higher drilling mud flow rates with higher drilling depths, and lots of fractures zones that are potential sources of bursty kicks. This current version of the tool does not calculate the severity/magnitude of the kick. For future work, the returned wavelet coefficients could be utilized in developing an algorithm to determine the kick severity. In conclusion, wavelet transformation is a viable tool in early detection of kicks.

Acknowledgements

We would like to thank the laboratory technicians, Matt, Craig and Trevor, and also Tobias Brueckner, an exchange student from Germany, for his assistance in the laboratory and the exchange of knowledge.

References

1. Miguel BD, Carlos AF, Rodolfo CF, Cesar AP (2012) Theoretical and experimental study of signal processing techniques for measuring hermatic compressor speed through pressure and current signals. International Compressor Engineering Conference.

2. Malloy KP (2008) Minerals management service in a probabilistic approach of managed pressure drilling in offshore applications.

3. Naduri S, Medley GH, Schubert JJ (2009) MPD: Beyond narrow pressure windows. IADC/SPE Managed Pressure Drilling and Underbalanced Operations Conference and Exhibition. San Antonio, Texas, USA.

4. Graham Bob, Reilly WK, Beinecke F, Boesch DF, Garcia TD, et al. (2011) Deep water: The Gulf Oil disaster and the future of offshore drilling. Report to the President, National Commission on the BP Deepwater Horizontal Oil Spill and Offshore Drilling.

5. Proakis JG, Manolakis DG (1996) Digital Signal Processing. Prentice-Hall.

6. Manolakis DG, Proakis JG (1996) Digital Signal Processing. Prentice-Hall.

7. Hauge E (2012) Automatic kick detection and handling in managed pressure drilling systems. Norwegian University of Science and Technology.

8. Anfinsen BT, Rommetveit R (1992) Sensitivity of early kick detection parameters in full-scale gas kick experiment with oil and water based drilling muds. IADC/SPE Drilling Conference. New Orleans, LA, USA.

9. Grayson B, Gans AH (2012) Closed loop circulating systems enhance well control and efficiency with precise wellbore monitoring and management capabilities. SPE/IADC Managed Pressure Drilling and Underbalanced Operations Conference and Exhibition. Milan, Italy.

10. Reitsma D (2010) A simplified and highly effective method to identify influx and losses during managed pressure drilling without the use of a coriolis flow meter. SPE/IADC Managed Pressure Drilling and Underbalanced Operations Conference and Exhibition. Kuala Lumpur, Malaysia.

11. Hage JI, Avest D (1994) Borehole acoustics applied to kick detection. Journal of Petroleum Science and Engineering.

12. Mallat SG (1999) A wavelet tour of signal processing. New York: Academic Press.

13. Misiti M, Misiti Y, Oppenheim G, Poggi JM (2013) MATLAB Wavelet Toolbox Documentation, The Mathworks Inc., Massachusetts.

14. Sifuzzaman M, Islam MR, Ali MZ (2009) Application of wavelet transform and its advantages Compared to fourier transform. Journal of Physical Sciences 13: 121-134.

15. Bland JM, Altman DG (1997) Statistical notes: Cronbach's alpha. BMJ 314-572.

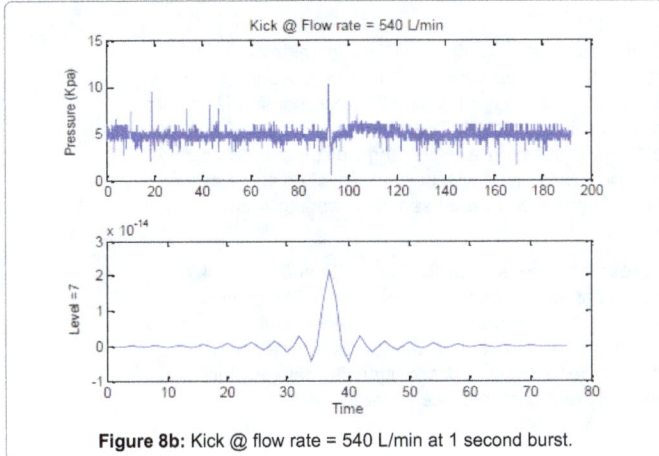

Figure 8b: Kick @ flow rate = 540 L/min at 1 second burst.

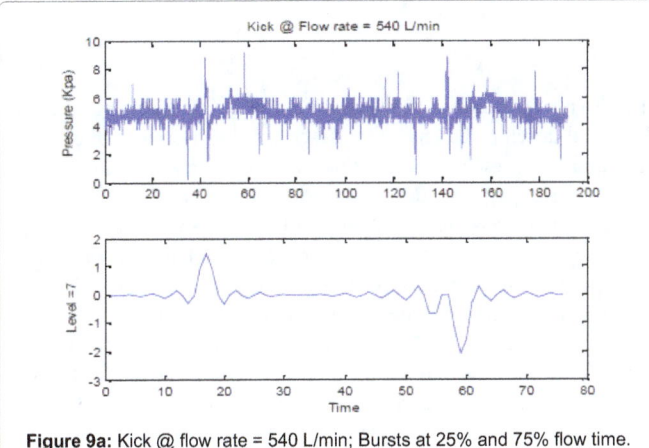

Figure 9a: Kick @ flow rate = 540 L/min; Bursts at 25% and 75% flow time.

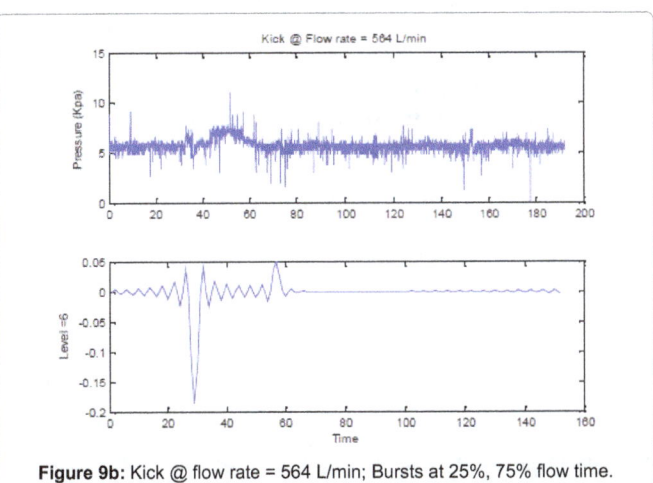

Figure 9b: Kick @ flow rate = 564 L/min; Bursts at 25%, 75% flow time.

30

A Current Viscosity of Different Egyptian Crude Oils: Measurements and Modeling Over a Certain Range of Temperature and Pressure

Shaimaa Saeed[1]*, Tarek M Aboul-Fotouh[2] and Ibrahim Ashour[1,3]

[1]*Chemical Engineering Department, Faculty of Engineering, Minia University, Egypt*
[2]*Mining and Petroleum Engineering Department, Faculty of Engineering, Al-Azhar University, Egypt*
[3]*Environmental Engineering, Zewail city of Science and Technology, 6 of October City, Egypt*

Abstract

Viscosity is an important characteristic of crude oil. The viscosity has significant effect for all events in the pipeline transportation of crude oil. The changing in viscosity of crude oil is depending on the temperature, pressure and chemical composition of the crude. In the determination process of crude oil viscosity operations, it is usual to use only temperature dependence of viscosity while the pressure is neglected. Therefore, the essential goal of this study is to obtain a model that can successfully calculate the viscosity in a wide range of temperatures and pressures. Moreover, at this study a mathematical model of changing the current viscosity of Egyptian crude oil with changing of temperature and pressure is determined. Different API gravities of 6 oil samples ranging from 13.4 to 40.4 are tested.

The current viscosity measured at the temperature and pressure ranges from 20 to 140°C and from 14.7 to 132.3 psi. These measurements are determined by using the quartz process viscometer apparatus. The comparison between the experimental data and the calculated values is indicated that the proposed model successfully calculate the experimental data with an average absolute percentage relative error of less than 3 and correlation coefficients of 0.991. It is noted that it's possible to predict a correlation for the dead crude oil viscosity using temperature and pressure change, because the pressure is important for the viscosity and cannot be neglected.

Keywords: Current viscosity; Viscometer apparatus; Crude oils; Mathematical; Correlation

Introduction

The viscosity is important for numerical simulations to determine the economics of the Enhanced Oil Recovery (EOR) projects and the success or failure of a given EOR schemes. The viscosity of crude oil depends on many factors; one of them is the source of chemical composition [1]. So, developing a model of viscosity to include different regions of the world seems to be a very impossible task [2]. Authors in that field are depending on viscosity measurements as interval readings, but this paper is depending on the current readings of viscosity to simulate the reality. So the aim of resent research is the determination of current Egyptian crude oil viscosity as a function of temperatures and pressures and creating mathematical model for this current viscosity with parameters directly indicate compositions which are more valuable and simple to use. In general, the viscosity is defined as the internal resistance of the fluid to flow. The crude oil viscosity is an important physical property that controls and influences the flow of oil through porous media and pipes [3]. Evaluation of viscosity of crude oil is an important for the design of various operations in the crude oil production field. Therefore, the viscosity of crude oil, pressure and temperature dependent, must be evaluated for both reservoir engineering and operation design.

The changing in viscosity with temperature and pressure change is usually predicted empirically. Despite of the importance of viscosity in engineering design, our understanding of such property is inferior to that of equilibrium properties. There are many difficulties in calculating viscosity measurements, especially for olive oil, which is a very important property that should be precisely evaluated for the reservoir simulation. Measuring the viscosity of dead oil is easier using empirical correlations at different temperatures other than the reservoir temperatures. These dead oil measurements are used in calculating live oil viscosity [4]. The difficult and high cost of viscosity measurements

at reservoir conditions are the main reason for the weakness of such data. Beal developed a chart that described the viscosities of 655 dead oil samples, representing 492 oil fields around the world and covering viscosities ranging from 0.8 to 155 cP, °API gravities from 10.1 to 52.5 and temperatures from 38°C to 105°C. In addition, Kartoatmodjo and Schmidt [5] represented an empirical correlation to predict the viscosity of dead oil with 3588 data points from 661 dead oil samples that covered gravities ranging from 14.4 to 58.9 API, viscosities ranging from 0.5 to 682 cP, and temperatures ranging from 75°F to 320°F. Labedi [6] also presented the dead oil viscosity in the range of 0.66 to 4.79 cP and °API ranging from 32.2 to 48.0 as a function of API gravity and temperature in the range of 38 to 152°C using 91 data points.

Several correlations for predicting dead oil viscosity are available in the literature review and some of them are discussed in this study, the Beggs and Robinson [7] represented a model for temperatures ranging from 21 to 146°C and El sharkawy and Alikhan [8] also represented a model based on crude oil samples from the Middle East for temperatures ranging from 38°C to 150°C. Naseri et al. [9] provided model for temperatures ranging from 40 to 146°C. Many authors decided that, Beal and Beggs and Robinson equations are

***Corresponding author:** Shaimaa Saeed, Chemical Engineering Department, Faculty of Engineering, Minia University, Egypt, E-mail: engshaimaa24@gmail.com

more accurate than previous efforts which might have been true for this viscosity range. However, large errors observed when this model applied outside of these temperatures, viscosity and API ranges. Darko Knežević et al. [10] developed a mathematical model of new correlation for the lubricating oil dynamic viscosity as a function of temperature and pressure. In this work viscosity of Egyptian crude oils with API ranges from 13.4 to 40.4 are measured experimentally over wide range of temperature and pressures and a mathematical model has been developed.

Experimental Study

The dynamic current viscosity, μ of dead crude oil for 6 samples in a temperature range from 20 to 140°C and pressure range from 14.7 to 132.3 psi is determined using a quartz process viscometer apparatus. The quartz process viscometer consists of two components, the electronic device for control and evaluation and the measuring head. The core of the quartz process viscometer is an oscillating quartz sensor that is subjected to dampening by the viscous properties of the surrounding liquid at the measuring head. The type of the sensor is SiO_2 of cylindrical shape with small dimensions. The quartz viscometer's microprocessor contains powerful extrapolation algorithm of the temperature dependent density of the liquid, resulting of a mathematical and physical analysis of the system [11]. Optionally the viscometer can be supplied with a high pressure sample container, which can be used to measure the viscosity of pressurized samples up to 100 bar (1450 psi) at a maximum temperature of 150°C (300°F) as shown in Figure 1. Moreover, the steps for how to use this apparatus are illustrated in Figure 2. These steps are expressed as the following:

1. The sample is filled into the pressure cell (1)

2. Inside the pressure cell a crude oil level of 100 ml is marked (10)

3. The formation of a vortex due to a stirrer with (200 r/m) found so recommended to fill in some more fluid (~103 ml).

4. The sensor head is screwed into the closure head from the below using a suitable copper gasket (3).

5. Before closing the unit, please place the O-ring seal (2) in the notch on the inner side.

6. A location bolt (4) shows the correct orientation of the flange lid; please make sure that the pin enters the corresponding bore (5).

7. After closing the unit, use the 8 screws (6) to close and fix the pressure cell.

8. To improve the temperature regulation, a customized insulation jacket (11) can be supplied.

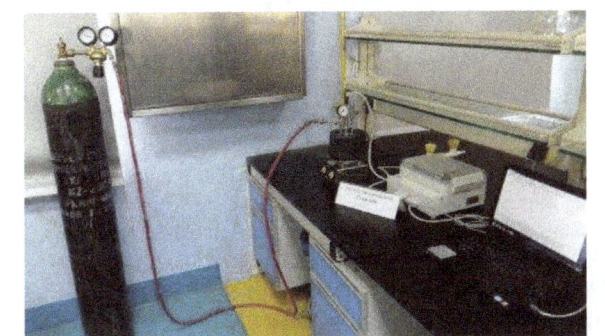

Figure 1: The process viscometer apparatus in the laboratory.

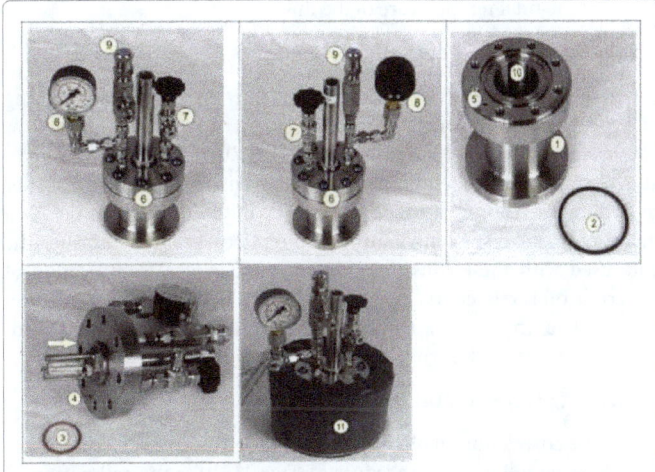

Figure 2: Steps of measurements in the process viscometer apparatus.

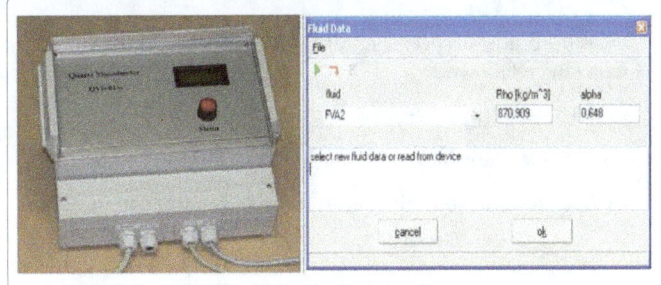

Figure 3: The control unit of the process viscometer apparatus.

9. The needle valve (7) is used to pressurize the unit. The pressure inside the unit is displayed on the pressure gauge (8).

10. For safety reasons an overpressure valve (9) is part of the unit. For the control unit:

At the lower border of the main window is the status line, which displays the settings of "the current measurement" (Figure 3):

• Given density.

• Given coefficient of thermal expansion

• Name of the fluid (if known).

• Number of measurements (or start temperature or measurement duration).

• Stop temperature.

• Temperature (or time) step.

According to the density extrapolation of the samples the process viscometer need to introduce thermal expansion factor (α) which calculated from the following equation (1):

$$\rho t = \rho 15 - (tf - t15) \tag{1}$$

Where:

ρt is the density of the fluid at final temperature in Kg/m^3,

$\rho 15$ is the density of the fluid at 15.5°C,

α is thermal expansion coefficient,

tf is the final temperature of the fluid,

t15 is the temperature of the fluid at 15.5°C.

Every crude oil gravity has certain (α) introduced in the beginning of process measurements. The measurements based on time step as you prefer (ex.: every 15 sec). Also magnetic stirrer is used in samples at 200 revolutions per minute (r/m) (current measurements). After recording the reading of runs from the apparatus, the Figures between the viscosities versus temperatures are plotted. Table 1 describes the data used with these runs. In addition, the chemical composition of the crude oil is related to (specific gravity and Watson characterization factor). And The Watson factor is related to the average boiling point and API gravity of the crude oil through the following relation [12]:

$Kw = (131.5 + API141.5)\ (T_b)1/3$

In the correlations which utilizing the Watson characterization factor, we follow the procedure of Twu [13] for the estimation of dead oil viscosity; which can be briefed as follows:

$\mu od = \gamma oT * vT$

γ_{oT} is the crude oil specific gravity at temperature T and is given by the following relation

$\gamma_{oT} = 0.999012(\gamma_{o60} VCF_T)$

$\gamma o\ 60$ is the crude oil specific gravity at 60 °F.

VCF_T is the crude oil volume correction factor at 60 °F:

$VCF_T = e^{\ [-\alpha 60\ \Delta(1+0.8\times 60\ \Delta T)\ ]}$

$\propto_{60}$ is the thermal expansion coefficient at 60 °F:

$$\alpha_{60} = \frac{K_0 + K_1 \gamma_{060}}{\gamma_{060}^2}$$

In the quartz process viscometer apparatus, the thermal expansion factor and the density in g/cm³ (specific gravity) of the crude oil at 60°F are essential input parameters for each run. Hence, the apparatus automatically apply this rule in its calculations for the current viscosity in experimental data. So, the composition is taking into consideration in the 5 recent model and the calculated parameters (constants) of the recent model are directly indicated the composition.

Results and Discussion

The new model

The experimental data of the dynamic current dead oil viscosity of 6 samples with different API values is measured with in the temperature range and pressure range of (20°C to 150°C), (14.7 to 132.3 Psi). The ASTM shows that dead oil viscosity is classified according to its standard API at 15.5°C which is the first parameter for any model, and the second parameter is the value of the measured temperature [14,15]. According to literature review, most of the previous models are based

on, sometimes two parameters, the API and temperature to calculate the dead viscosity. In most cases, API parameter has no physical meaning [16] Therefore, another parameter is used with temperature, such as pressure. The objective is to create a model in the following formats to calculate the dead oil viscosity.

$\mu_{od} \propto (T,)$ (2)

Where: μ_{od} is the dead oil viscosity in cP,

T is the temperature in °C,

P is the pressure in psi.

Before creating the new current viscosity model, understanding the relationship. Between the input and output variables is important. Specifically, identifying which parameters are insignificant and can be eliminated from the final model and the parameters that are highly correlated with the output. Accordingly, the current dead oil viscosity (μod) is considered to be a function of temperature (T) and pressure (P). As shown in plotting the data yields Figure 4. It illustrates how viscosity changes according to temperature and pressure changes. As illustrated in Figure 4 the viscosity decreasing by increasing temperature and when measuring the viscosity at different pressures are observed that the value of viscosity increased by increasing the pressure at certain API. More addition, due to the samples of crude oils is prepared from different Egyptian company's fields so there is a difference in composition. Figure 5 illustrates the relationship between current viscosity and frequency (the frequency is a parameter given by the quartz viscometer apparatus) for every run at (0.3, 6.9 bar) and 200 r/m. Additionally, it shows decreasing in current viscosity with increasing the frequency of the fluid. Moreover, Reynolds Number is calculated at temperature intervals (50°C, 100°C, 150°C) from the following relation to know the type of flow [17].

$N_{Re} = D\ N\ \rho\ /\mu$ (3)

Where: D is the inside diameter of the crude cylinder in the quartz process viscometer.

N = The agitator speed (rph).

ρ = The density of the crude given by the apparatus.

μ= The viscosity of the crude given by the apparatus

The NRe >10,000 the flow is turbulent.

The validity of the new viscosity model

After testing many previous viscosity equations for all the experimental viscosity data for this study as a function of pressure and temperature, the results show that the following equation is set to be presented the data in this work and give us correlation coefficients (R2) up to 0.99 as follow:

$$\mu_{0d}(P, T) = ae^{\left[\frac{b}{(T-c)}\right]} e^{\left[P/(a_1 + a_2 T)\right]}$$ (4)

Where: μ_{od} is the dynamic viscosity in cp, T is the temperature in °C and P is the pressure in psi. (a, b, c, a1, a2) are constants (parameters directly indicate composition). The model can predict the current dead oil viscosity with an average percentage absolute relative error (AARE) of 3% and correlation coefficient (R2) of 0.991. The results are evaluated and tested using different techniques. One of the main challenges in this study is that most of the existing models in the literature are limited to certain ranges of temperature, API value. But this study represents the current viscosity at different ranges of temperatures and pressures with

Variable	Range
API	13.4 to 40.4
Temperature (°C)	20 to140
Pressure (psi)	14.7 to 132.3
Density (g/cm³)	808.8 to 975.6
Thermal expansion factor (α)	0.65 to 0.7
Revolution / minute (r/m)	200

Table 1: Description of data used with samples.

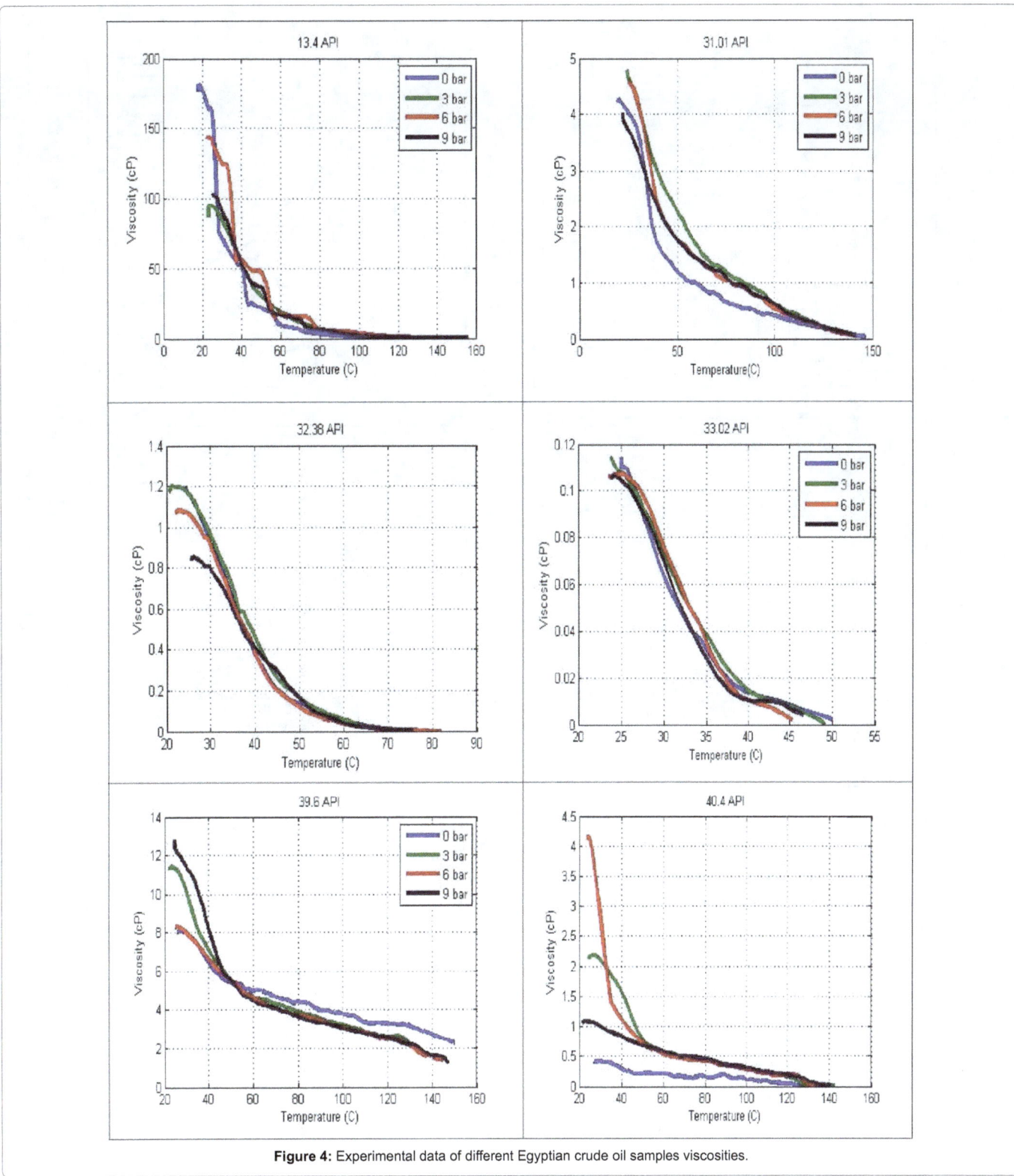

Figure 4: Experimental data of different Egyptian crude oil samples viscosities.

constant parameters which directly indicate composition.

Table 2 indicates the values of (a, b, c, a1, a2) related to each run of the data set. To check the ability of the new current viscosity model to indicate all experimental data, cross plots are used. The values are in a good agreement, with the average percentage absolute relative error

and standard division comparing with previous models as indicated in Table 3. Table 4 indicates the evaluation of the new model for prediction of the current dead oil viscosity. In addition, the absolute percentage average relative error (AARE%) and the standard percentage deviation (SD%) tests are performed between the calculated and the measured

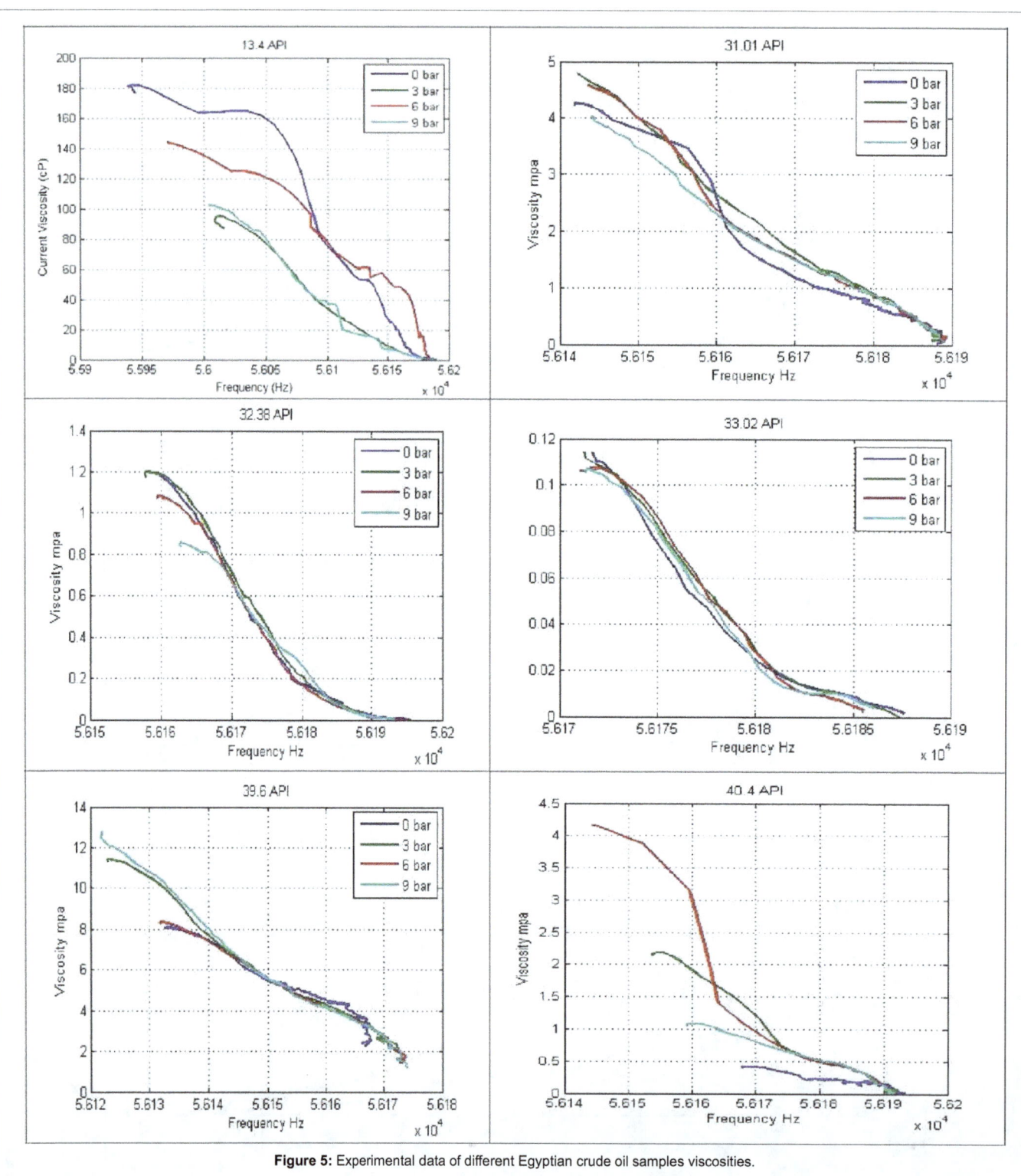

Figure 5: Experimental data of different Egyptian crude oil samples viscosities.

results by the following expressions:

$$AARE\% = \left| \frac{(measured - calculated)}{measured} \right| * 100 \qquad (5)$$

$$SD = \frac{1}{1-N} \sqrt{\sum_{i=1}^{N} \left(\left| \frac{measured - calculated}{measured} \right| - AARE \right) 2} \qquad (6)$$

Where: i is the sample number and N is the total number of samples.

The new model shows the lowest average percentage absolute

Parameters (directly indicate composition)	Values of parameters at different pressures (Psi)				°API
	14.7	44.1	88.2	132.2	
a	1.007	0.000212	6.475	2.717	
b	-20.38	4.405	102.7	115.7	13.4
c	-1.897	-0.6139	-9.704	-5.272	
a1	1.469	-10.28	-2241	-4769	
a2	0.04793	1.491	239	483.6	
a	0.9112	0.9423	0.0293	1.006	
b	0.142	3.759	0.0085	-4349	
c	-376.3	-1105	26.11	0.7489	39.6
a1	5.301	11.57	14.35	0.4612	
a2	0.0538	0.2569	0.0488	0.24	
a	0.3132	0.0043	0.0017	0.00603	
b	65.56	-0.00346	-0.001	302.1	
c	-5.811	26.1	25.45	-139.7	31.01
a1	-7.439	5.511	9.61	24.52	
a2	5.435	0.0324	0.0583	0.1663	
a	0.00017	0.0044	511.6	4.585	
b	0.7921	26.34	161.8	1395	
c	0.6613	57.57	63.64	239.4	33.02
a1	0.8194	9.282	-17.72	-10.79	
a2	0.0576	0.0796	-0.121	2.401	
a	1.08E-06	0.00323	0.0704	3.34E-06	
b	-0.0124	-2.86	125.6	0.00297	
c	26.97	19.82	-6.978	21.67	40.4
a1	1.078	4.876	625	9.934	
a2	0.002261	0.05478	22.28	0.0183	
a	0.05061	0.07867	0.0368	1.21E-07	
b	-34.77	-89.97	-86.09	-28.9	
c	4.174	2.507	8.988	13.2	32.38
a1	-0.1303	-0.208	-2.118	5.831	
a2	0.1268	0.2811	0.4903	0.05881	

Table 2: The values of modeling parameters at different pressure and API.

References	Average absolute relative error % (AARE %)	Standard deviation% (SD %)
Beal	31.6	37.3
Beggs and Robinson	21.2	28.0
Glaso	27.4	31.9
Labedi	29.7	42.6
Kartoatmodjo and Shmidt	33.1	37.25
Ibrahim Ashour	19.2	25.8
Osama Alomair	8	203.7

Table 3: Average absolute relative error and standard deviation of previous model.

relative error and standard percentage deviations relative to the others. Figure 6 shows the behavior of a new model according to the experimental and calculated values. The new model provides the best prediction without any scattering between the experimental data and the calculated one, thus the new model presents high value of correlation coefficient. Figure 6 consists of plots for 6 samples at (13.4, 31.01, 32.38, 33.02, 39.6, 40.4) API.

Conclusion

Although, crude oils of different compositions can have the same

°API gravity, significant errors may be introduced when the viscosities of crude oils are estimated from general viscosity trends and the API gravity. The new current dead oil viscosity model is a function of temperature T and pressure P with constant parameters that directly indicate composition which are more valuable and simple to use. Impossible to find any similarity between this study and previous one because all measurements are current viscosity, the crude oil is in continues motion in the apparatus at 200 revolutions per minute (r/m). The temperature is increased automatically up to 150°C and the reading is taken automatic with time step as the user prefer. The pressure range used is chosen to unique the pressure inside the pipe line which the fluid (crude oil) is transferred across. So, this process is simulated to the reality. The validity of the agreement between the experimental current dead oil viscosity data and the predicted values indicates that the new model successfully represents the experimental data with an average percentage absolute relative error less than 3% and correlation coefficients ranging from 0.99 to 0.94 at different temperatures and pressures. From the statistical analysis, the new model is present as one of the best models by comparing it with other models published in the literature. The new model shows it is easy to use different temperatures

API	Pressure (psi)	Average Absolute Relative Error (AARE%)	(Standard Deviation) SD %	R2
	14.7	0.056	0.00452	0.9739
13.4	44.1	0.144	0.937	0.9186
	88.2	0.1756	0.474	0.8867
	132.3	0.1037	0.856	0.918
	14.7	0.0792	0.0023	0.886
31.01	44.1	0.03	0.0388	0.9862
	88.2	0.0078	0.00569	0.9881
	132.3	0.0199	0.0302	0.9798
	14.7	0.0772	0.3135	0.9138
32.38	44.1	0.116	0.4587	0.8327
	88.2	0.1377	0.177	0.9007
	132.3	0.0826	0.5554	0.991
	14.7	0.0557	0.725	0.9784
33.02	44.1	0.3047	4.36	0.9285
	88.2	0.0787	1.127	0.9665
	132.3	0.1876	2.502	0.9207
	14.7	8.12E-05	0.00019	0.9787
39.6	44.1	4.95E-05	1.95E-05	0.9897
	88.2	0.00068	0.000652	0.9879
	132.3	0.00632	0.00998	0.9844
	14.7	0.00451	0.00134	0.9304
40.4	44.1	0.00804	0.00763	0.9876
	88.2	0.0413	0.0213	0.9744
	132.3	0.0315	0.0917	0.9522

Table 4: Accuracy of Egyptian crude oil model for estimating viscosity of recent study.

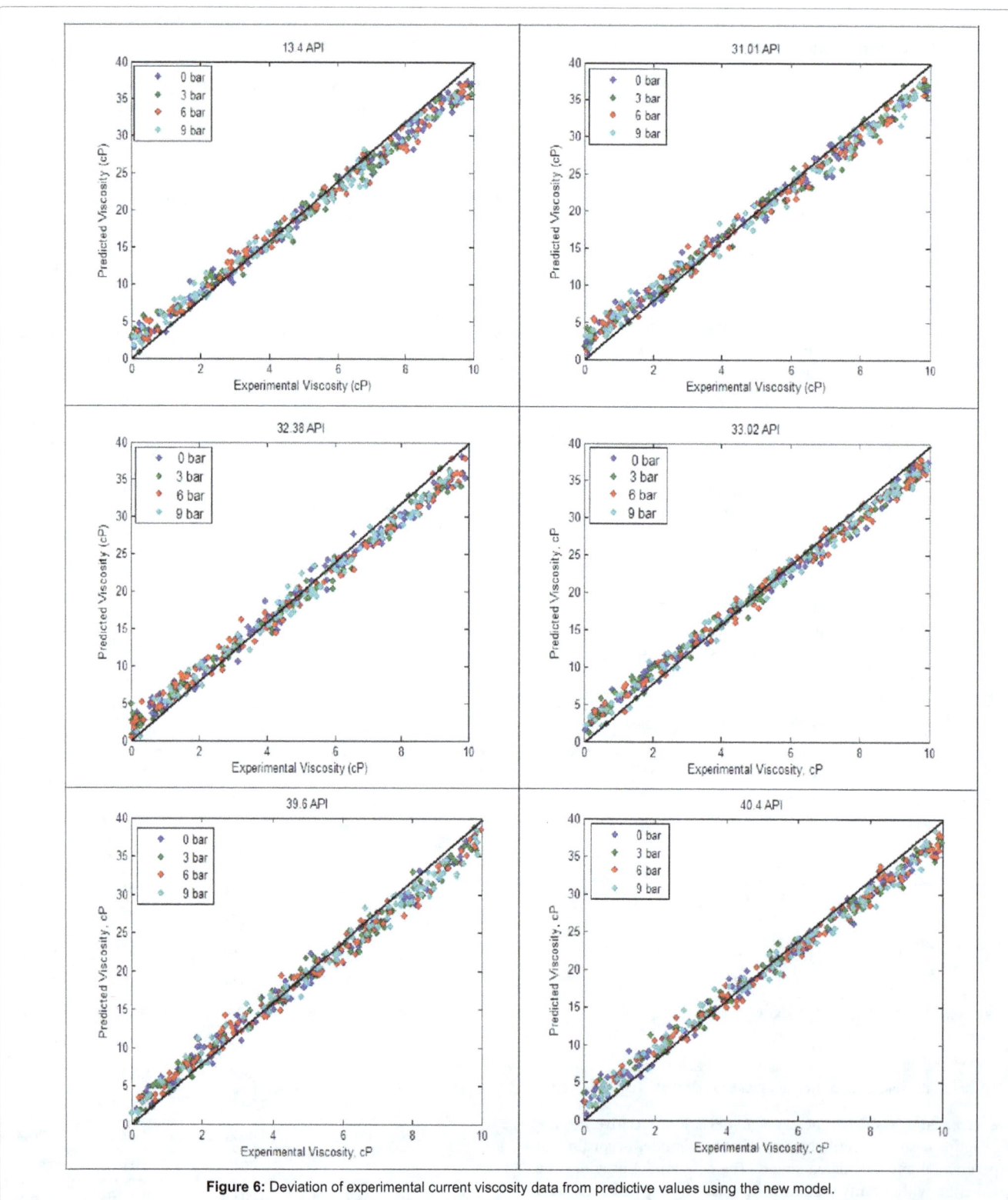

Figure 6: Deviation of experimental current viscosity data from predictive values using the new model.

and pressures to predict current viscosity, provide best accuracy over a wide range of oil gravities, and could be used to predict better outcomes in future works.

References

1. Ahrabi F, Ashcroft SJ, Shearn RB (1987) High pressure volumetric phase composition and viscosity data for a North Sea crude oil and NGL mixtures. Chem Eng Res 67:329–334

2. Ashour I, Al-Rawahi N, Vakili-Nezhaad G, Fatemi A (2012) A new correlation for prediction of viscosities of Omani-Fahud-Field Crude Oils.

3. Sattarin M, Modarresi H, Bayat M, Teymori M (2007) New viscosity correlations for EAD crude oils. Pet Coal 49: 33-39.

4. Beal C (1946) Viscosity of air, water, natural gas, crude oil and its associated gases at oil field temperature and pressures. Trans. AIME 165: 114-127.

5. Kartoatmodjo F, Schmidt Z (1994) Large data bank improves crude physical property correlation. Oil Gas J. 4: 51–55.

6. Labedi R (1992) New viscosity correlations for dead crude oils J Pet Sci Eng. 8: 221-234.

7. Beggs HD, Robinson JR (1975) Estimating the viscosity of crude oil systems. JPT 9: 1140-1141.

8. Elsharkawy AM, Alikhan AA (1999) Models for predicting the viscosity of Middle East crude oils. Fuel 78: 891-903.

9. Naseri A (2005) A correlation approach for prediction of crude oil viscosities. S.A.: J. of Pet. Sci. Eng. 47: 163-174.

10. Knežević D, Savić V (2006) Mathematical modeling of changing of dynamic viscosity, as a function of temperature and pressure, of mineral oils for hydraulic systems UDC 532.12: 665.6. FACTA UNIVERSITATIS 4: 27-34.

11. F5 Technologie GmbH Im Büchenorte 3 D-31515 Wunstorf, Germany.

12. Hassan Naji S (2011) The dead oil viscosity correlations a C-sharp simulation approach JKAU: Eng. Sci. 22: 61-87.

13. Twu CH (1985) Internally consistent correlation for predicting liquid viscosities of petroleum fractions. Ind. Eng. Chem. Process Des. Dev. 34: 1287-1293.

14. Alomair O (2015) Heavy oil viscosity and density prediction at normal and elevated temperatures. J of Pet Exploration and Production Technology 6: 253-263.

15. Mehrotra AK (1991) Generalized one parameter viscosity equation for light and medium hydrocarbon. Ind Eng Chem Res 30: 1367-1372.

16. Standing MB (1947) A pressure-volume-temperature correlation for mixtures of California oils and gases. Drill Prod Pract, American Petroleum Institute, New York, pp. 275-284.

17. John Pietranski F (2012) Mechanical agitator power requirements for liquid batches. (PDH online Course K103, 2 PDH).

31

Application of a Novel Ultrasonic Technology to Improve Oil Recovery with an Environmental Viewpoint

Hesam Arabzadeh[1] and Mahmood Amani[2]*

[1]*Department of Chemical Engineering, Ataturk University, Erzurum, Turkey*
[2]*Petroleum Engineering Program, Texas A&M University at Qatar, Doha, Qatar*

Abstract

It is proven by recent studies that sonication has a positive influence over the oil flow within the porous media. Accordingly, the researchers in this paper evaluated the influence of sonication over the oil recovery by means of free fall gravity drainage. Furthermore, the influence of sonication on the oil permeability was assessed in three samples that had different bead size in average. By use of the Hagroot backward method and Matlab simulation, the optimal petrophysical situation for sonication was determined. The authors concluded that sonication positively affects the oil recovery for the non-asphaltenic samples, while it has a reverse effect on the asphaltenic samples because of increasing the viscosity in long-term. Furthermore, it was witnessed that gravity drainage was heightened by increase of beads' size in the non-asphaltenic sample. Accordingly, this mechanism can be useful in oil recovery by means of gravity drainage, specifically in fracture reservoirs.

Keywords: Improved oil recovery; Ultrasound; Gravity drainage; Hagroot method; viscosity; Matlab simulation; Relative permeability

Introduction

Oil extraction from the reservoirs is primarily initiated by the pressure gradient of the reservoir that brings about the fluid flow to the production wells and afterward to the surface [1]. When the reservoir pressure is not sufficient, down-hole pumps are implemented or gas lift methods are used to get the oil to the surface. Sole reliance on the characteristics of the rocks in the subsurface, the oil properties, and the extraction method, will contribute to recovering only 10 to 15 percent of the total oil in the reservoir in what is referred to as the primary recovery stage [2]. The methods used in the second stage- namely secondary recovery- includes injection of fluid into the reservoir in order to displace the fluid produced in the well, so that the pressure of the reservoir is maintained. These methods include water flooding and brine and gas reinjection, among which water flooding is the most commonly used. Up to 40 percent of the total oil in the reservoir can be extracted in this stage [3].

There is a variety of factors that negatively affect the oil extraction process in the primary and secondary recovery. These factors include capillary forces, high mobility ration, and finally heterogeneity of reservoir rocks. So far, up to 55 percent of the existing oil in the reservoir can be extracted. Hereby, the third stage in oil extraction appears, that is tertiary recovery [4]. This stage is commonly called Enhanced Oil Recovery (EOR). The common techniques used in the EOR are thermal recovery, gas injection, and chemical injection [5], and the goals of implementing the EOR methods are decreasing the oil-water interfacial tension, lowering capillary pressure, and finally, lowering the oil-water mobility ratio by means of increasing the viscosity of water [6].

However, researchers are looking for new techniques to reduce the related risks of the EOR process. One of these techniques is ultrasonic wave radiation which supports EOR while prevents imposing damages to the producing formation. The studies on the implementation of ultrasonic waves and their manner of influence started in 1950 with a study on the correlation between the water level and the stimulation caused by earthquakes [7]. In fact, the seismic waves emitted by the earthquake were identified as the causes of the increase in fluid's level and its pressure. The same result was achieved for the wells around

highways and railways. As a result, the researchers came to conclusion that the behavior of the sound waves and the phenomenon of simulation by the sound waves must be understood better in order to safely use these waves. Since the 1970s, a great number of studies on ultrasonic waves have been conducted [8-11].

Implementation of ultrasound waves in EOR methods has a number of advantages including easy and quick application, protection of wellbore formation against damages, low operational costs and high profitability, high compatibility with other EOR methods, and finally having a wide range of applications [12,13].

The Enhanced Oil Recovery methods can be utilized in the following conditions [14]:

• When the drilling mud has been flowing for a long time, and the well is damaged due to overuse of the drilling mud.

• When injection of water and acid into the well is in vain, and does not increase the recovery rate.

• When the wells show a high potential for production, but the production rate is low.

• When the wells are producing heavy oil and paraffin.

• When the well is damaged by salt and sediments.

Literature Review

Application of electromagnetic methods or wave treatment requires the utilization of various physical fields instead of a substance to play

---

*Corresponding author: Mahmood Amani, Petroleum Engineering Program, Texas A&M University at Qatar, Doha, Qatar, E-mail: mahmood.amani@qatar.tamu.edu

the role of an agent. Since the other methods of enhancing oil recovery, including chemical flooding, polymer flooding, gas flooding, and steam flooding, have a number of disadvantages like being costly, requiring too many surface apparatus, being environmental-unfriendly, and having technical constraints, the advantage of acoustic treatment lies in the need for lower resources, lower energy, and lower expenses than the other methods of IOR currently being used [15]. The results of a number of studies revealed that acoustic treatment of the oil well, especially within the ultrasonic range, is among the best methods of boosting oil extraction from the wells [16-18]. The advancements in the efficient ultrasound generators, choice of appropriate wells, and finally physically modeling of the processes by means of mathematics has improved the performance of this method in the recent decades [19].

Studies of oil recovery by using vibration first began in the 1950s, at the time the researchers witnessed an increase in oil extraction after the cultural noises and occurrence of earthquakes. The first scholars who posed the possibility of using sound waves to increase oil production were Duhon and Campbell [20] who tested water flooding on the well cores while radiating ultrasonic waves with frequencies between 1 and 5.5 Megahertz. In their study, the water was injected to the center of the sandstone core of the well, while the ultrasound-generating probe was placed in the center of the core too. The receiver probe was located in the bottom of the core. The authors concluded that the usage of ultrasound waves enhanced the oil extraction and made the fluid displacement better. They also observed a negative correlation among the frequency, cavitation, and recovery- i.e., an increase in frequency of the waves brings about a decrease in cavitation as well as oil recovery.

The two studies of Chen, Fairbanks and Chen [21,22] showed that the usage of ultrasonic waves in porous media is capable of boosting the percolation rate, while the heat made by ultrasound waves plays a minor role in this increase. These researchers radiated ultrasound waves with the frequency of 20 Kilohertz and the power of 10 W/cm² to the oil and water that was flowing in the porous sandstone and the capillary. Moreover, the results of a research conducted by Cherskiy showed a notable boost in the permeability of the cores which were saturated with water while there was acoustic radiation [23]. The other two studies that approved the results of the previous studies were conducted by Neretin and Yudin and Pogosyan [24,25]. Neretin and Yudin [24] reported that displacement of oil by water in the well's loose sands increases while ultrasound waves are radiated. Pogosyan [25] concluded that ultrasound waves boost the gravitational separation of water and kerosene.

There was a shift of focus in 1990, and in the next three decades, the scholars paid attention on the simulation of elastic waves as well as seismic methods. In this regard, the systematic review of Beresnev and Johnson [26] collected and analyzed the studies on the methods that took advantage of elastic wave simulation in oil extraction- both ultrasound and seismic waves. This systematic review came to conclusion that both elastic waves and seismic waves are able to increase the rate of permeability as well as production within the porous media.

In another study, Beresnev [26] tried to analyze the mobility of the oil droplets within the porous media- through a capillary-oriented mechanism- afterward the ultrasound waves are radiated. Figure 1 displays a schematic view of the trapped oil. The scale of this research was a pore scale of an oil droplet which was not wet. The droplet was entrapped at the pore's opening in a channel. As shown in the Figure below, the right meniscus is smaller than the left one. The result will be the formation of an imbalanced capillary pressure (ΔP) inside the droplet. This imbalance stands against the pressure from outside, and

Figure 1: Trapped oil under ΔP (pressure gradient. R_{right} and R_{left} represent the droplet radii. Cap. Press stands for capillary.

increases as the droplet travels toward the pore's throat. The external pressure gradient is parallel with the internal pressure gradient (ΔP_S). As a result, (ΔP_S) must increase so that the droplet passes the pore's throat. Ultrasound waves are able to generate this extra pressure (P_U). In a mathematical way, the droplet passes through the pore's throat if we have $\Delta P_S + P_U > \Delta P$. It is obvious that the higher the power of the ultrasound wave, the higher the generated force will be.

Mobilization of the oil after radiation of ultrasonic waves was further analyzed in the study of Amro [27]. These researchers came to conclusion that ultrasound waves are able to improve the oil mobility, while changing its permeability. Moreover, they compared the oil extraction after ultrasound treatment in residual oil saturation and original oil in well. The results revealed that residual oil saturation brought about more oil recovery.

In a recent study by Keshavarzi [28], the influence of ultrasonic waves on free gravity drainage of the oil was analyzed. The researchers in this study utilized a glass bead pack porous medium for conducting the free fall gravity drainage tests under the conditions of ultrasound radiation and non-radiation. Moreover, the permeability of the wetting phase was obtained in this research. The results of this study revealed that radiation of ultrasound waves notably boosts the recovery factor in the process of free gravity drainage. Furthermore, the permeability of the wetting phase as well as the non-wetting phase increased by radiation of ultrasound waves.

Research Methodology

As mentioned in the previous section, Keshavarzi [28], conducted a study on the role of ultrasonic wave on two-phase relative permeability in a free gravity drainage process. This recent study analyzed the relative permeability of merely one sample (one glass bead pack). As a result, the researchers in the present study decided to conduct a series of similar experiments on three samples to confirm or refute the findings of this study. Consequently, the same methodology as the Keshavarz's was used in this research:

The test setup included a Plexiglas cylinder whose length was 30 centimeters. The diameter of the cylinder was 4 cm, while its inner diameter was 3 cm. Moreover, a graduated cylinder was used for measuring the amount of the extracted liquid. The ultrasonic generator had an average effective output of 80 Watts, and a frequency of 22 KHz. Figure 2 depicts a schematic view of the apparatus [28]. There were three samples used in this research with an average bead size of 80-100 µm, 170-200 µm, and 240-270 µm. In order to calculate the medium porosity, the volume of the entering fluid to the cylinder was measured; and, the calculated amount for medium porosity was 40 ± 0.3% for the first sample, 52 ± 0.48% for the second sample, and finally 63 ± 0.52% for the third sample. By means of fluid injection into the cell with a certain pressure gradient and calculation of the flow rate, the bead pack's absolute permeability was measured to be 30 ± 1 Darcy for the first sample, 38 ± 1.4 Darcy for the second sample, and finally 43 ± 1.2 Darcy for the third sample.

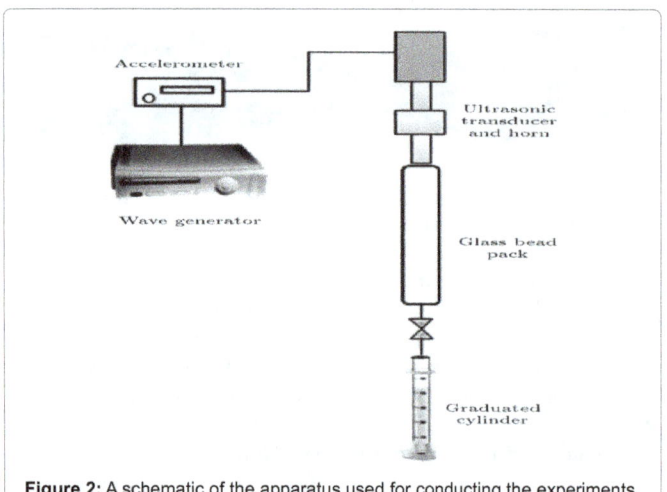

Figure 2: A schematic of the apparatus used for conducting the experiments.

Fluid	Density $\left(\frac{kg}{m^3}\right)$	Viscosity (cp)	Asp. content
Crude oil A	920	5.39	0.23%
Crude oil B	984	74	4.2%

Table 1: Physical characteristics of the used wetting fluids in this study (in a standard condition).

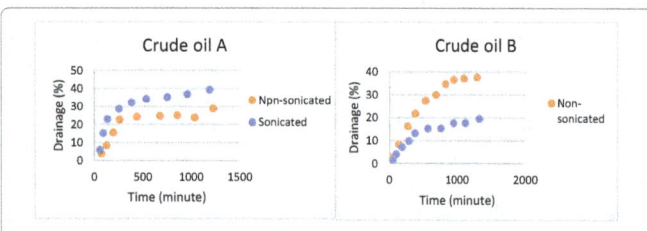

Figure 3: The effects of ultrasound radiation on ultimate drainage recovery of the sample with an average bead size of 80-100 μm.

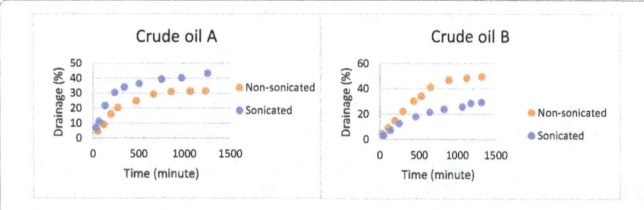

Figure 4: The effects of ultrasound radiation on ultimate drainage recovery of the sample with an average bead size of 170-200 μm.

The researchers used air in the non-wetting phase, and such materials as crude oil A and crude oil B (asphaltenic) in the wetting phase. Table 1 displays the characteristics of the wetting fluids used in the study.

The researchers decanted the glass beads into the Plexiglas cylinder and assembled the apparatus before conducting each experiment. In order to avoid mistakes in calculations, the researchers used first-hand dry glass beads in every experiment. The cell was completely saturated with the liquid sample after vacuuming the cell to a favorable pressure. Then, the researchers positioned the cell in a vertical standing, while its faces were open. After that, the liquid produced against the time was determined. The temperature in the experiment environment was the ambient laboratory of 25°C, and the experiments were conducted under the radiation and non-radiation of ultrasonic waves [28].

In order to measure the relative permeability of the samples, the authors took advantage of Hagroot [29] backward method. According to this method, the value of normalized oil production, after the time of breakthrough, is measured by the following formula:

$$Np = 1 - \left(1 - \frac{1}{n}\right)\left(\frac{1}{nk_{r0}^0 t_D}\right)^{\frac{1}{n-1}} \qquad (1)$$

In the above equation, n is the exponent of Corey equation and k_{r0}^0 is the Corey constant.

The curves of relative permeability in the wetting phase were measured by comparing the recovery data with time. Afterward, these results along with the production data were utilized for drawing the relative permeability curves in the non-wetting phase. This process included the history matching of the wetting phase data with the production data [28].

Results

Figure 3 displays the results achieved from the free fall gravity drainage tests. The relative error in calculation of oil recovery rate was measured to be 3 to 6 percent for all three samples.

As can be seen in Figure 3, it can be mentioned that in general, ultrasonic radiation boosted the rate of oil recovery, the same as drainage recovery for the first sample with an average bead size of 80-100 μm. The crude oil A showed an increase of 38 percent from the initial recovery of 29% to 40% after ultrasound radiation. However, the crude oil B sample reflected a reduction of ultimate recovery rate after sonication. This sample showed a reduction of 50% in its ultimate oil recovery rate after sonication, and its recovery rate decreased from 38% before sonication to 19% after sonication. This reduction was related to the asphaltenic nature of the sample. Najafi [30] explained the reason of this phenomenon in their study on Quantifying the role of ultrasonic wave radiation on kinetics of asphaltene aggregation in a toluene–pentane mixture. The researchers concluded that disintegration of asphaltene micelle structures of the oil and its dissolution in the fluid leads to an increment in the oil viscosity after being exposed to ultrasound waves for extended periods for time. Finally, an increase in the viscosity of a liquid means a reduction in the drainage recovery.

The effects of radiating ultrasonic waves on the sample with average bead size of 170-200 μm are reflected in Figure 4.

In general, it was observed that ultrasound waves' radiation to crude oil A increased the ultimate rate of drainage recovery. However, this increase was almost 2% higher than the increase in the sample with the average bead size of 80-100 μm. As observed in Figure 4, the ultimate recovery rate of Crude oil A reflected an increase of 40 %, that is from the drainage recovery rate of 32% to drainage recovery rate of 44.8% in its final recovery. Similar to the previous sample, there was a decrease in the ultimate recovery rate of the crude oil B sample from 49% to 29.4%, that is a general decrease of 40%.

Figure 5 depicts the influence of sonication on ultimate drainage recovery of the sample with the average bead size of 240-270 μm. There was a congruity between the achieved results for this sample and the results achieved for the two previous samples- i.e., an increase in the ultimate recovery rate of crude oil A, and a decrease in the final recovery rate of the crude oil B. In general, it was observed that the increase in the recovery rate of the material in this sample was approximately 5% higher than the second sample, and 7% higher than the first sample. The acquired results for this sample showed an increase of 45% in the

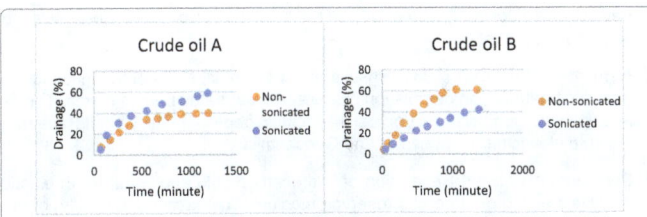

Figure 5: The effects of ultrasound radiation on ultimate drainage recovery of the sample with an average bead size of 240-270 μm.

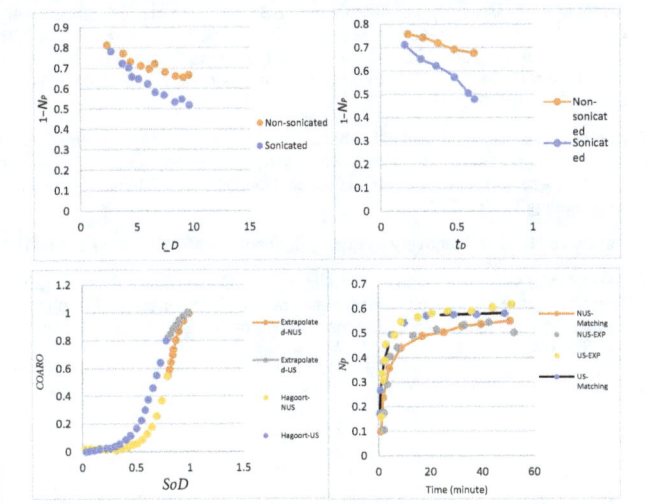

Figure 6: Steps of measuring the constants of Corey type equation for relative permeability of crude oil A-wetted sample.

final recovery rate of crude oil sample A from 40% to 58%. The decrease in the final recovery rate of the crude oil B-wetted sample was 30%, i.e., from primary 62% to 43.4% after radiation.

Due to the fact that Hagroot method cannot be used for samples which have alterations in their physical characteristics, this method was not applicable for crude oil B sample with asphaltenic nature. Consequently, the relative permeability of wetting as well as non-wetting phases was calculated for the samples wetted with crude oil A. As an example, the procedure of relative permeability calculation for crude oil A-wetted sample with the average bead size of 170-200 μm is depicted in Figure 6a. The breakthrough takes place in $t_{D=0.18}$ for non-radiation of ultrasound waves, and in $t_{D=0.16}$ for ultrasound radiation. As observed, breakthrough under ultrasound radiation condition took place earlier. For evaluation of n, the exponential data after the time of breakthrough must be utilized. As a result, a crossing line is drawn through the data (Figure 6b). By using the following equation, the value of n was calculated to be 5.80 for non-radiation of ultrasound waves, and 3.97 for the radiation of ultrasound waves.

$$\frac{d\ln(1-Np)}{d\ln t_D} = -\left(\frac{1}{n-1}\right) \quad (2)$$

In order to calculate the amount of k_{r0}^0 through Eq. 1, the value of N_p acquired from drawing a straight line at $t_D=1$ was used, and the value of k_{r0}^0 was calculated to be 1.79 for non-radiation of ultrasound condition and 1.87 for radiation of ultrasound waves. Figure 6c displays the results achieved for the relative permeability of crude oil A-wetted sample by use of backward Hagroot method. At the end, it was observed that ultrasound radiation reduced the value of n for crude oil A, while increased the value of $CO\,A_{r0}^0$ for this material.

With regard to the calculation of relative permeability in the non-wetting phase, the history matching of the results acquired from Eq. 3 and the experimental data of production was done (Figure 6d).

$$\frac{dNp}{dt} = \frac{k_g\left(-\dfrac{pc}{H} + \Delta\rho g(1-Np)\right)}{H\phi\mu_g\left[Np(1-M)+\right]M} \quad (3)$$

The results revealed a boost in the relative permeability of all the three samples in the non-wetting phase. Figure 6 displays the relative permeability of the crude oil A-wetted samples in both phases. As can be seen in the Figure 6, radiation of ultrasound waves increased the relative permeability of the gas and oil. Moreover, the relative permeability of gas end point improved, while there was a decrease in the amount of critical gas saturation.

Conclusion

The researchers in this study analyzed the influence of ultrasound radiation on the free fall gravity drainage and relative permeability of three different samples of sand pack beads with different sizes, and with different wetting and non-wetting conditions. Following up the methods used in the study of Keshavarzi [28], it was concluded that ultrasound waves are able to improve the recovery in the gravity drainage process, which is the main recovery parameter in the fracture reservoirs, for crude oil A (non-asphaltenic), while these waves increase the viscosity of the oil in asphaltenic samples, and, as a result, reduce its recovery in long exposure time. Furthermore, it was concluded that the increase in the size of the beads in the samples led to an increase of oil recovery in non-asphaltenic samples by the help of ultrasound waves; and, decreased the amount of gravity drainage in the asphaltenic samples.

References

1. Bavière M (1991) Basic concepts in enhanced oil recovery processes. Springer, Berlin, Germany.

2. Amit R (1986) Petroleum reservoir exploitation: Switching from primary to secondary recovery. Operations Research 34: 534-549.

3. Nielson JP (1989) U.S. Patent No. 4,856,587. Washington, DC: U.S. Patent and Trademark Office.

4. Vinegar HJ, DeRouffignac EP, Glandt CA, Mikus T, Beckemeier MA (1994) U.S. Patent No. 5,297,626. Washington, DC: U.S. Patent and Trademark Office.

5. Bilak R (2006) US Patent No. 7,069,990. Washington, DC: U.S. Patent and Trademark Office.

6. Thomas S (2008) Enhanced oil recovery - An overview. Oil & Gas Science and Technology-Revue de l'IFP 63: 9-19.

7. Griffing V (1950) Theoretical explanation of the chemical effects of ultrasonics. J Chem Phys 18: 997-998.

8. Zakiewicz B (1981) U.S. Patent No. 4,305,463. Washington DC: U.S. Patent and Trademark Office.

9. Muslimov RH (2005) Modern methods of enhanced oil recovery: The design, optimization, and performance evaluation. AN RT, Kazan.

10. Mullakaev MS (2012) Ultrasonic intensification of technological processes of production and refinement of oil, cleaning of oil-contaminated water and soil. Moscow 22: 12.

11. Abramov VO, Mullakaev MS, Abramova AV, Esipov IB, Mason TJ (2013) Ultrasonic technology for enhanced oil recovery from failing oil wells and the equipment for its implemention. Ultrason Sonochem 20: 1289-1295.

12. Mohammadian E, Shirazi MA, Idris AK (2011) Enhancing oil recovery through application of ultrasonic assisted waterflooding. In SPE Asia Pacific Oil and Gas Conference and Exhibition. Society of Petroleum Engineers.

13. Abramova A, Abramov V, Bayazitov V, Gerasin A, Pashin D (2014) Ultrasonic technology for enhanced oil recovery. Engineering 6: 177-184.

14. Speight JG (2013) Enhanced recovery methods for heavy oil and tar sands. Elsevier. Enhanced recovery methods for heavy oil and tar sands.

15. Mullakaev MS, Abramov VO, Abramova AV (2015) Development of ultrasonic equipment and technology for well stimulation and enhanced oil recovery. J Pet Sci Eng 125: 201-208.

16. Watkins GC, Sharp KC (1985) Enhanced oil recovery: Retrospect and prospect. J Can Pet Technol.

17. Kobayashi T, Kobayashi T, Fujii N (2000) Effect of ultrasound on enhanced permeability during membrane water treatment. Jpn J Appl Phys 39: 2980-2981.

18. Caicedo S (2009) Feasibility study of ultrasound for oil well stimulation based on wave-properties considerations. SPE Prod Oper 24:81-86.

19. Mullakaev MS, Abramov OV, Abramov VO (2008) Development and study of operating efficiency of technological ultrasonic installations. Chem Pet Eng 44: 433-440.

20. Duhon RD, Campbell JM (1965) The effect of ultrasonic energy on flow through porous media. In: Second Annual Eastern Regional Meeting of SPE/AIME, Charleston, WV, Virginia.

21. Chen WI (1969) Influence of ultrasonic energy upon the rate of flow of liquids through porous media. Chemical Engineering. PhD thesis, West Virginia University, USA.

22. Fairbanks HV, Chen WI (1971) Ultrasonic acceleration of liquid flow through porous media Chem Eng Prog Symp Ser 67: 108.

23. Cherskiy NV (1977) The effect of ultrasound on permeability of rocks to water

24. Neretin VD, Yudin VA (1981) Results of experimental study of the influence of acoustic treatment on percolation processes in saturated porous media. Topics in nonlinear geophysics (Voprosi nelineinoy geofisiki): All Union Research Institute of Nuclear Geophysics and Geochemistry.

25. Pogosyan AB (1989) Separation of hydrocarbon fluid and water in an elastic wavefield acting on a porous reservoir medium. Transactions (Doklady) of the USSR Academy of Sciences, Earth Science Sections 307: 575- 577.

26. Beresnev IA, Vigil D, Li W (2005) Elastic waves push organic fluids from reservoir rock. Geophys Res Lett 32:13303.

27. Amro MM, Al-Mobarky A, Al-Homadhi ES (2007) Improved oil recovery by application of ultrasound waves to waterflooding. In: SPE 105370, presented at the 15th SPE Middle East Oil and Gas Show and Conference held in Bahrain International Exhibition Centre, Kingdom of Bahrain, 11–14th March 2: 1015-1022.

28. Keshavarzi B, Karimi R, Najafi I, Ghazanfari MH, Ghotbi C (2014) Investigating the role of ultrasonic wave on two-phase relative permeability in a free gravity drainage process. Scientia Iranica Transaction C Chemistry Chemical Engineering 21: 763-771.

29. Hagoort J (1980) Oil recovery by gravity drainage. Old SPE Journal 3: 139-150.

30. Najafi I, Mousavi SMR, Ghazanfari MH, Ghotbi C, Ramazani A, et al. (2011) Quantifying the role of ultrasonic wave radiation on kinetics of asphaltene aggregation in a toluene–pentane mixture. Petroleum Science and Technology 29: 966-974.

transactions (Doklady) of the U.S.S.R. Academy of Sciences. Earth science sections 232: 201-204.

32

Rigorous Modeling of Solution Gas – Oil Ratios for a Wide Ranges of Reservoir Fluid Properties

Arash Kamari[1], Sohrab Zendehboudi[2], James J Sheng[3]*, Amir H Mohammadi[1,4]* and Deresh Ramjugernath[1]

[1]*Thermodynamics Research Unit, School of Chemical Engineering, University of Kwa-Zulu-Natal, Howard College Campus, King George V Avenue, Durban 4041, South Africa*
[2]*Department of Chemical Engineering, Massachusetts Institute of Technology (MIT), Cambridge, MA 02139, USA*
[3]*Department of Petroleum Engineering, Texas Tech University, TX, USA*
[4]*Institute of Research in Chemical Engineering and Petroleum (IRGCP), Paris Cedex, France*

Abstract

The reservoir fluid properties, the solution gas–oil ratio (GOR), are of great importance in various aspects of petroleum engineering. Therefore, a rapid means for estimating such parameters is much sort after. In this study, the linear interaction and general optimization method is applied in the development of a precise and reliable model for estimating the solution GOR. In order to develop a model that would be comprehensive, a reliable and extensive databank comprising of more than 1000 datasets collected from various geographical locations, including Asia, Mediterranean Basin, North America, Africa, and Middle East was compiled. Furthermore, the model developed was benchmarked against widely-used empirical methods in order to evaluate the performance of method proposed in predicting solution GOR data. The results show that the model proposed in this study outperforms the empirical methods to which it was compared. This study also investigated the influence of the reservoir fluid properties on the estimated solution GOR for the newly-developed model. Results show that bubble point pressure and gas gravity have the largest and the smallest influences on the predicted solution GOR, respectively. Finally, the Leverage approach was applied to determine the applicability domain for the proposed method via the detection of outlier data points. It was determined that only 26 data points, out of more than 1000 data, are identified as outlier data points.

Keywords: Reservoir fluid properties; Solution gas–oil ratio; GOR; API; Outlier data

Abbreviations: AARD: Average Absolute Relative Deviation; APRE: Average Percent Relative Error; API: Oil API Gravity; EOR: Enhanced Oil Recovery; LINGO: Linear Interactive And General Optimizer; OFVF: Oil Formation Volume Factor; Pb: Bubble Point Pressure, Psi; Rs: Solution Gas-Oil Ratio, SCF/STB; RMSE: Root Mean Square Error; SCF: Standard Cubic Feet; STB: Stock Tank Barrel; TR: Reservoir Temperature, °F; Γg: Gas Specific Gravity.

Introduction

The properties of reservoir fluids [1] are normally determined from bottom-hole and/or surface recombined samples. The fluid properties are required for a large number of reservoir engineering calculations, which include, selection of the most important enhanced oil recovery (EOR) method for a reservoir candidate, estimation of hydrocarbon reserves, performance prediction, calculations related to the production operation, production optimization, well-testing studies, fluid flow through porous media, etc [1-10]. In other words, reservoir fluid properties such as bubble point pressure, oil formation volume factor (OFVF), and solution GOR are key parameters in petroleum engineering calculations and are obtained through laboratory measurements, theoretical methods, and/or empirically derived correlations. Generally, petroleum engineers seek a rapid way to obtain these parameters, taking into account both economic and technical issues. The determination of reservoir fluid properties using laboratory experiments are not simple and can be time-consuming and expensive [11-13]. The reservoir fluid properties, in the absence of experimental measurements, must be determined through empirical methods.

Over the years, various empirical methods have been reported for the determination of reservoir fluid properties related to oil samples from different geographical locations worldwide. To this end, in one of the first attempts, Elam [14] in 1957 proposed a correlation for the

estimation of saturation pressure as a function of temperature, gas specific gravity, oil gravity and solution GOR using as a basis of 231 data points for Texas crude oil. One year later, Lasater [15] presented a bubble point-pressure correlation for black oil data taken from Canada, western and mid-continental USA and South America. His model was developed using 158 samples of 137 various crude oils. He reported an average error of 3.8% for his model. He also observed that the existence of CO_2 in crude oil samples results in an increment in the saturation pressure. Vasquez and Beggs [16] proposed some empirically derived methods for the estimation of reservoir fluid properties using a universal databank collected from various regions of the world. Moreover, they separated the experimentally obtained data into two classes. The first group contained oils with gravities less than 30°API. The second group contained oils with gravities more than 30°API. In contrast with Lasater's results [15], they found that CO_2 content decreases the saturation pressure.

In 1983, Ostermann [17] developed two correlations for the estimation of saturation pressure of crude oil samples taken from

***Corresponding authors:** Sheng J, Department of Petroleum Engineering, Texas Tech University, TX, USA, E-mail: james.sheng@ttu.edu

Amir H Mohammadi, Institute of Research in Chemical Engineering and Petroleum (IRGCP), Paris Cedex, France,
E-mail: mohammadi@ukzn.ac.za; amir-hossein.mohammadi@mines-paristech.fr

different regions in Alaska based on a limited number of data points. Al-Marhoun [9] developed an empirical correlation applying data gathered from the Middle East region. In 1990, Rollins et al. [18] proposed an empirically derived method to calculate the stock-tank gas–oil ratio as a function of oil API gravity, separator pressure and temperature, and gas gravity. In the same year, Sutton and Farshad [19] reviewed several PVT correlations and compared the accuracy of each model for several PVT parameters for application in the Gulf of Mexico. In their study, Glaso's correlations [20] provided acceptable results for calculation of saturation pressure, solution GOR, and OFVF. They reported that Vazquez and Begg's correlations [16] had higher accuracy for solution GOR for more than 1400 SCF/STB and saturation pressures more than 7000 psi. In 1992, Dokla and Osman [21] studied 51 crude oil samples from UAE and developed a new empirical methods for OFVF, saturation pressure and solution GOR. They reported that PVT correlations should be derived using local data sets because universal correlations are not always accurate enough. Moreover, Omar and Todd [22] developed models for OFVF and saturation pressure on the basis of Standing's correlations [23] using 93 PVT datasets from Malaysian oil reservoirs. Their models showed better accuracy for Malaysian oil samples. Furthermore, Petrosky and Farshad [24] proposed some empirically derived methods for the determination of reservoir fluid properties using data collected from the Gulf of Mexico. They showed that the empirical methods proposed outperformed other methods developed for the Gulf of Mexico, involving those reported by Standing [23], Vasquez and Beggs [16], Glaso [20], and Al-Marhoun [9]. Elsharkawy et al. [25] also compared different correlations to characterize Kuwaiti crude oils using a limited number of oil samples in this year.

Ghetto et al. [26] proposed some empirical methods for the calculation of saturation pressure, solution GOR, OFVF, oil compressibility, and oil viscosity for heavy and extra-heavy oils. The data used in developing the correlations came from reservoir fluid samples extracted from the Mediterranean Basin, Africa, and the Persian Gulf. In 1998, Khairy et al. [27] developed some empirical methods for the estimation of saturation pressure and bubble point OFVF. They compared their model with nine published correlations. In 1999, Velarde and McCain [28] developed a set of empirical methods for calculating solution GOR and OFVF, and modified OFVF using 195 laboratory tests. In 2007, Mazandarani and Asghari [29] tuned Al-Marhoun 's correlation [9] for Iranian field data to obtain a modified correlation using about fifty fluid samples collected from different Iranian oil fields. In 2008, Taghaz et al. [30] tested the accuracy of PVT correlations to determine the solution GOR of Libyan oils using about 1600 data points from different oil fields in the Sirte basin. They concluded that no correlation is suitable for Libyan oils. In 2012, Shafiie et al. [31] optimized Standing [23] and McCain correlations for solution GOR and OFVF, based on Iranian crude oil samples, and developed a new model using Genetic Algorithms. Very recently, Arabloo et al. [11] developed simple and accurate empirical methods for the prediction of saturation pressure and OFVF using a large databank compiled from various geographical locations. Here, it is worth mentioning that smart techniques have previously been implemented for the estimation of reservoir fluid properties and petroleum engineering problems in addition to empirical methods [32-40].

In this study, the LINGO (Linear Interactive and General Optimizer) [41] methodology is implemented to propose an efficient, precise, and rapid-to-use model for the determination of solution GOR. To achieve a comprehensive model covering properties for all regions, a reliable databank was collected which comprises an extensive range of reservoir conditions, as well as PVT properties. The widely-used empirically derived correlations were used for comparison to benchmark the performance of the model proposed in this study. To this end, an error analysis was conducted graphically and statistically. The influence of the reservoir fluid properties on the solution GOR values calculated were also studied. Finally, applicability domain of the proposed method was determined through the detection of outlier data points using Leverage approach.

Solution Gas–Oil Ratio

As mentioned above, the solution GOR plays a key role in PVT analysis related to petroleum engineering calculations. As a consequence, solution GOR affects the OFVF, the viscosity compressibility of oil, and it is also needed to determine the in-situ total reservoir fluid rates. As a definition, solution GOR is the amount gas dissolved in oil with regards pressure. Here, it should be noted that reservoirs containing light oils have more dissolved gas than a reservoir with heavy oils. With an increase in pressure, solution GOR increases approximately linearly until the attainment of bubble point/saturation pressure (P_b); after which it is a constant and the oil is supposed to be under-saturated (Figure 1). Figure 1 is a typical illustration of the trend of solution GOR versus pressure.

As a result, most of empirically derived methods reported in the open literature for the determination of reservoir fluid properties have been developed on the basis of data related to a specific region and limited PVT studies [42]. This drawback can decrease the precision of these methods in predicting reservoir fluid properties at a particular solution gas–oil ratio. This means that the aforementioned empirical correlations may lead to significant deviation when they are utilized for the estimation of reservoir fluid properties for other geographical locations. For that reason, it is of important to collect a comprehensive databank covering a wide range of reservoir fluid properties for all regions in the world. Therefore, a reliable and comprehensive databank [9,17,21,22,26,43-47] comprising more than 1000 data series collected from various geographical locations including Asia, Mediterranean Basin, North America, Africa, and Middle East was compiled in this study. The databank collected includes reservoir fluid properties, viz. solution GOR (R_s, SCF/BBL), bubble point pressure (P_b, psi), reservoir temperature (T_R, °F), and gas gravity (γ_g), as well as oil gravity (API). A statistical description of the properties, including maximum, minimum, and average values is summarized in Table 1.

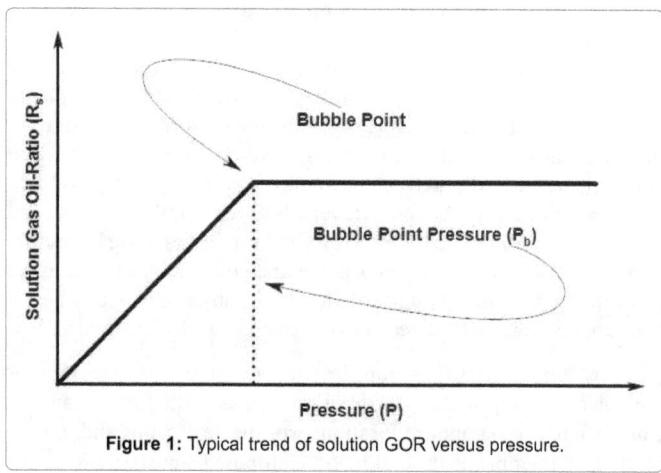

Figure 1: Typical trend of solution GOR versus pressure.

Property	Unit	Minimum	Average	Maximum	Role
R_s	SCF/STB	7.08	515.32	3298.66	Output
γ_g	–	0.52	1.00	3.44	Input
T_R	°F	54.9	173.30	360.93	Input
API	–	6.00	33.41	56.8	Input
P_b	psi	58.01	1755.58	7127.01	Input

Table 1: Statistical analysis of reservoir fluid properties used for the estimation of solution gas–oil ratio.

Proposing the New Model

The key aim of the present study is to propose a comprehensive, accurate and reliable model for the determination of solution GOR using data collected from various crudes worldwide. To this end, the LINGO methodology [41] is used for pursuing our objective in this study. Basically, the technique is an interactive linear and discrete tuning tool. As a result, the methodology has been utilized in mathematics, science, and industry, and employed to solve computer problems mathematically [48-50]. Furthermore, quadratic programming, as well as linear and nonlinear, and integer programming are the most important problems solved by LINGO software [11]. Additionally, the LINGO methodology can solve the root and algebraic equations/problems linearly and nonlinearly. It should be mentioned that the LINGO software includes a number of common mathematical functions within the programming language to use by operators for finding/solving their programming problem [51]. In this study, the LINGO methodology is used to develop a reliable model to determine the solution GOR as a function of reservoir fluid properties, including bubble point/saturation pressure, reservoir temperature, and gas gravity, as well as oil gravity (API) as follows:

$$R_s = f(\gamma_g, T_R, API, P_b) \tag{1}$$

In the development of the model, the databank collected was separated into two sets of data, viz. the training and test sets. Approximately 80% of entire the databank was used in the development of model (training set), and the rest (20%) was assigned to the test set for checking the model developed and evaluating its accuracy, performance, and capability. To measure the accuracy of the model developed, the average absolute relative deviation (AARD) is selected as an objective function. Finally, a simple form of equation with four easy functions including ×,+,−,/ was obtained as follows:

$$RS = A + B - 15.849 \tag{2}$$

$$A = 0.14624\,P_b - 0.14624\,API + \frac{802.44}{P_b} + \frac{(2.727 P_b - APIT_R + 2715.5)^2}{(API - 995.53)^2} \tag{3}$$

$$B = (\,0.0064332\,P_b + 0.0064332\,API_g\,) \times (API\,\gamma_g - 14.811) \tag{4}$$

where P_b denotes the bubble point pressure (psi), API stands for oil API gravity, T_R denotes the reservoir temperature (°F), and γ_g is gas specific gravity.

Variables Relevancy Analysis

To show the degree of dependency of the reservoir fluid properties selected as input variables (*i.e.* saturation pressure, reservoir temperature, and gas specific gravity as well as oil API gravity) on the solution GORs estimated by Eq. (2), a sensitivity analysis was performed. Hence, the relevancy factor (r) [52] is utilized in this study for measuring the degree of effect of each reservoir fluid property applied in Eq. (2) for the determination of solution GOR. Regarding the relevancy factor approach, an input variable has a higher influence on

the output parameter if the calculated absolute value of r between the input and output variables is greater than the r values for other input variables. Consequently, the positive or negative influence of input variables (saturation pressure, reservoir temperature, and gas specific gravity as well as oil API gravity) on the solution GORs is however not determined by an absolute value of r. Consequently, following equation is used to calculate the r values through the relevancy analysis [40]:

$$r(\mathrm{Inp}_k, \mu_g) = \frac{\sum_{i=1}^{n}(\mathrm{Inp}_{k,i} - \overline{\mathrm{Inp}_k})(\mu_i - \overline{\mu})}{\sqrt{\sum_{i=1}^{n}(\mathrm{Inp}_{k,i} - \overline{\mathrm{Inp}_k})^2 \sum_{i=1}^{n}(\mu_i - \overline{\mu})^2}} \tag{5}$$

where $\mathrm{Inp}_{k,i}$ stands for i^{th} value of the kth input variables and $\overline{\mathrm{Inp}_k}$ denotes the average value of the kth input variables (*i.e.* bubble point pressure, reservoir temperature, and gas gravity as well as oil gravity), μ_i indicates the ith value of the solution gas–oil ratios determined by Eq. (2), and μ is the average value of the solution gas–oil ratios determined by Eq. (2).

Results and Discussion

Performance evaluation of the new model

A graphical and statistical error analysis was conducted to evaluate the performance of the method over a wide range of the reservoir fluid properties, and to compare the results obtained using the model against the most widely used empirically derived correlations. Hence, AARD, root mean square error (RMSE), and average relative percent error (ARPE) were considered as statistical error parameters, and a parity diagram or scatter plot, as well as a relative distribution error curve are used as two graphical illustrations to evaluate the performance of the method proposed to estimate the solution GORs. The results of error analysis are summarized in Table 2. The results in the table indicate that the model proposed has acceptable accuracy with respect to the large number of data points and the wide range of reservoir fluid properties employed in its development. The AARD value reported for the model is 19.83%. The value indicates that the method output values are in agreement with corresponding experimental records of the solution GOR. Furthermore, the calculated APRE and RMSE values are 1.73% and 203.05, respectively.

To assess the performance of the proposed model graphically, a scatter diagram, as well as a relative distribution error plot of values estimated using Eq. (2) were plotted. Figure 2 shows a comparison of values estimated by the model developed in this study versus experimental values of solution GOR on a parity diagram. It is clear

Method	AARD, %	APRE, %	RMSE
Glaso [20]	79.25	32.37	468.33
Petrosky and Farshad [24]	62.70	-48.16	217.38
Kartoatmodjo and Schmidt [54]	57.80	-48.15	395.93
Standing [23]	47.84	-38.61	312.88
Farshad et al. [2]	43.07	-28.87	267.08
Vazquez and Beggs [16]	42.29	-31.99	389.08
Al-Marhoun [9]	42.01	-21.28	348.17
Dindoruk and Christman [56]	36.89	-13.78	254.27
Macary and El-Batanony [53]	36.54	1.40	238.87
Al-Shammasi [55]	32.95	-16.72	242.81
Baniasadi et al. [42]	23.15	2.29	197.39
This present study	19.83	1.73	203.05

Table 2: Error analysis performed for the proposed model and comparable methods investigated in this study.

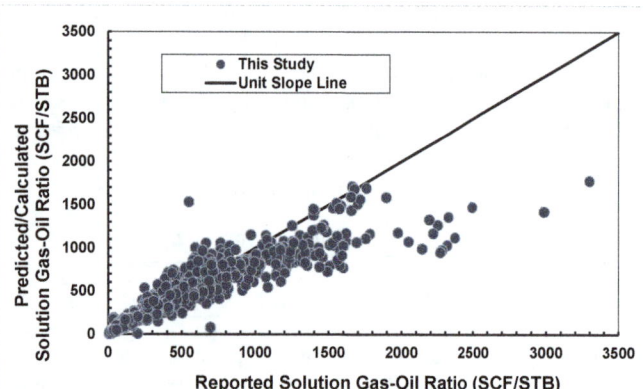

Figure 2: Scatter diagram of the predicted solution gas-oil ratio values versus the experimental records.

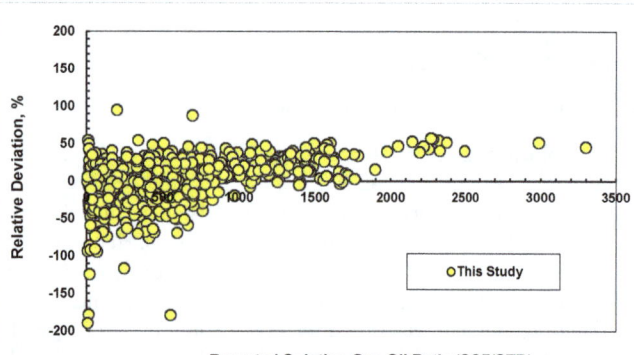

Figure 3: Relative error distribution plot of the predicted solution gas-oil ratio values versus the experimental records.

from the figure that the estimated GOR values approximately match the experimental values, resulting in data clustered around the parity line. This shows the capability of the model proposed in this study in predicting more than 1000 data values for solution GOR. Another graphical comparison is shown in Figure 3, which illustrates the calculated average relative percent error between the model and experimental data for solution GOR. As can be seen in the figure, the relative errors for the estimated data are clustered around the zero line. This demonstrates that there is acceptable agreement between the model predictions and experimental data for solution GOR. Figure 4 shows the plot of the predicted values against experimental data for the solution GOR with respect to the sorted data index.

A comprehensive comparison analysis was undertaken between the model developed in this study and widely-used empirically derived methods including the Farshad et al. method [2], Macary and El-Batanony method [53], Petrosky and Farshad method [24], Vazquez and Beggs method [16], Al-Marhoun method [9], Kartoatmodjo and Schmidt method [54], Al-Shammasi method [55], Standing method [23], Glaso [20], Baniasadi et al. method [42], and Dindoruk and Christman method [56] in order to evaluate the performance of the method in predicting solution GOR data. Table 2 reports the statistical results obtained for the comparisons undertaken. The table shows that the model developed in this study performs the best model for the calculation of solution GOR. Figure 5 illustrates graphically the calculated AARD for the model developed, as well as all comparative methods investigated in the study. From Table 2 and Figure 5, it can be concluded that the methods of Baniasadi et al. [42], Al-Shammasi

[55], Macary and El-Batanony [53], Dindoruk and Christman [56], and Al-Marhoun [9] are, after the method proposed in this study, the most accurate for the calculation of solution GOR with AARD values of 23.15, 32.95, 36.54, 36.89, and 42.01%, respectively. Table 3 lists some random data points selected from the databank, and Table 4 summarizes the estimated values for the data points presented in Table 3 using method developed and empirical methods discussed above. Table 4 also confirms superior performance of the model developed in this study over the empirical methods to which it was compared.

Influence of the reservoir fluid properties on solution GOR

As pointed out earlier, reservoirs containing light oils have more dissolved gases than reservoir with heavy oils. Therefore, it would be interesting to determine the accuracy of the model developed for various ranges of oil API gravity. To this end, the capability of the model presented in this study for estimating the solution GOR was observed across the spectrum of the light to heavy oils. The solution GORs estimated by the proposed model was partitioned into four classes of oil API gravities, viz. 6-15, 15-25, 25-35, and 35-56.8°. The results of analysis in terms of the calculated AARD values is shown in Figure 6. As can be seen in the figure, the model errors for the estimation of solution GORs of light oils is less than that for heavy oils. In other words, it can be concluded that the model developed in this study is more applicable for crudes with higher values of oil API gravity. To further investigate the influence of the reservoir fluid properties, including saturation pressure, reservoir temperature, gas specific gravity, and oil API gravity, a sensitivity analysis was performed in this study using the relevancy factor approach. Figure 7 shows the results of sensitivity analysis. The figure indicates that bubble point pressure and

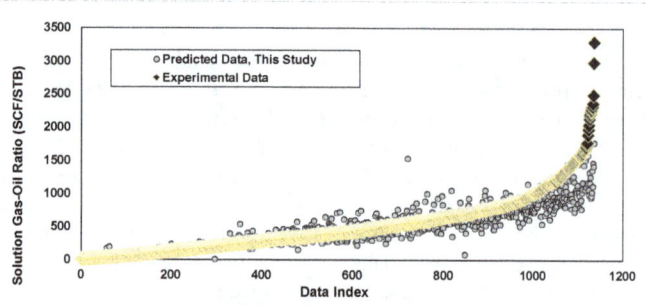

Figure 4: Fitting curve, as sorted data index, for the predicted solution gas-oil ratio values versus the experimental records.

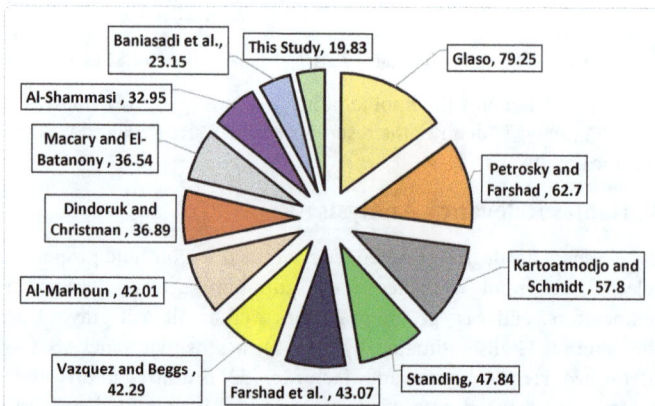

Figure 5: Graphical comparison of the developed model against the comparative methods studied in terms of statistical error parameter of AARD.

Data Index	P_b	γ_g	API	T_R	R_{si}
1	2082.77	0.756	7.5	153.5	208.7
2	2076.97	0.815	10.5	152.6	260.0
3	554.99	0.68	12	74.9	52.4
4	599.98	0.74	14.8	82.8	68.0
5	825.28	1.411	19.4	172.4	177.8
6	3199.97	0.75	20.9	110.5	556.2
7	285	0.74	23	114.8	32.5
8	430.04	1.04	25	99.4	95.5
9	1499.98	0.64	27	107.7	239.9
10	909.97	0.67	30	88.25	171.8
11	3057	0.778	32	175	679.0
12	400.01	0.8	34	71.6	76.6
13	2775.01	0.823	35.7	140.5	689.4
14	1340	0.8	36.3	87.8	313.4
15	1415	1.2468	37.2	248	486.0
16	5760.99	0.924	40.1	302	1760.6
17	1153.05	0.85	40.4	105.5	299.6
18	2221	0.693	45.3	238	547.0
19	1386.97	0.763	46.5	116.03	367.6
20	1962.07	0.78	52.5	138.25	636.7
21	1170.47	0.649	56.8	140.7	300.9

Table 3: Records of some data points existing in the databank compiled in this study.

gas specific gravity has the largest and smallest influences, respectively, on the solution GOR values predicted by the model.

Detection of outlier solution GOR data points

The detection of outlier data points that exist in a databank used to develop a predictive model is important to know in order to determine the applicability domain of the model developed. To this end, the Leverage methodology [57-59] is utilized in this study to identify outlier data points in the solution GOR databank that was compiled. Detailed

information on the Leverage methodology in terms of mathematical equations, as well as a step-by-step procedure is reported elsewhere [57-59]. The Williams diagram is sketched to show graphically the applicability domain of the proposed method. The existence of a majority of solution GOR data in the domain $0 \leq H \leq 0.1428$ and $-3 \leq$ Standardized Residuals $\leq$ demonstrates that the method is statistically valid. The data points which are located in the domain range $-3 \leq$ Standardized Residuals $\leq$ are recognized as valid solution GOR data, and data which are outside the range are considered as outliers. The results show that only 26 data points in the solution GOR databank (among more than 1000 data points) were identified as outlier data points (Figure 8).

Conclusion

The linear interaction and general optimization method, as a modeling approach, was applied in the development of an accurate and reliable model for calculating solution GOR data. A comprehensive databank comprising more than 1000 data samples collected from various geographical locations was compiled and used to develop a comprehensively applicable model. The performance of the model developed was compared to some widely-used empirical methods. The influence of the reservoir fluid properties on the estimated solution GOR data was also investigated. Finally, applicability domain of the proposed method was determined through the detection of outlier data points using the Leverage approach. It is found that only 26 data points (among more than 1000 data values) are identified as outlier data points. The results obtained indicate that the model proposed in this study outperforms all comparable models studied with an AARD value of 19.83%. The sensitivity analysis conducted in this study indicates that bubble point pressure and gas gravity have the largest and smallest influences, respectively, on the predicted solution GOR data. Furthermore, the model proposed in this study has greater applicability for the estimation of solution GORs for reservoirs containing light oils.

Data Index	This Study	ARD%	Baniasadi	ARD%	D& Christman	ARD%	Al-Marhoun	ARD%	Macary	ARD%	Al-Shamasi	ARD%
1	219.0	4.9	102.8	50.8	291.9	39.9	194.7	6.7	288.9	38.4	199.0	4.6
2	250.2	3.8	150.2	42.2	323.3	24.4	260.2	0.1	310.8	19.5	236.7	9.0
3	52.4	0.0	39.5	28.5	78.5	42.3	35.6	35.5	64.2	42.3	43.6	3.4
4	66.1	2.9	57.7	15.2	92.7	36.3	50.8	25.4	75.5	36.0	59.4	6.9
5	174.6	1.8	36.5	6.5	91.1	133.6	40.2	3.0	40.0	2.5	28.5	27.0
6	555.0	0.2	38.9	22.5	53.7	69.0	20.2	36.4	44.3	70.9	27.4	5.9
7	30.3	6.6	170.0	17.6	167.8	18.7	142.0	31.2	161.5	4.1	170.4	1.2
8	80.2	16.1	669.0	12.1	631.0	17.1	856.4	12.5	1107.4	78.3	900.9	45.0
9	240.4	0.2	241.8	0.8	184.0	23.3	163.4	31.9	233.7	19.4	264.7	35.2
10	152.3	11.3	167.3	2.6	123.2	28.3	106.1	38.2	132.3	5.6	163.9	16.9
11	658.8	3.0	776.1	10.2	729.1	3.5	996.8	41.5	1094.5	90.4	1064.8	85.2
12	75.8	1.1	324.2	5.2	257.2	24.7	364.2	6.5	265.4	4.9	366.0	31.2
13	678.0	1.6	749.8	23.6	653.9	7.8	665.6	9.7	909.5	83.7	967.1	95.3
14	311.7	0.5	331.2	5.7	250.8	20.0	344.2	9.8	252.9	1.1	362.2	41.6
15	490.7	1.0	366.7	3.8	404.7	6.2	591.2	55.2	200.3	47.4	356.3	6.5
16	1694.3	3.8	473.4	0.4	376.5	20.2	595.1	26.2	369.2	4.1	546.3	41.9
17	299.5	0.0	616.6	22.1	492.0	37.8	510.4	35.5	535.7	32.3	692.3	12.5
18	547.4	0.1	633.3	0.1	461.5	27.2	440.4	30.5	493.2	22.2	702.5	10.8
19	371.2	1.0	426.7	16.1	265.6	27.8	376.6	2.5	290.3	3.2	487.6	62.5
20	601.4	5.5	690.7	8.5	478.9	24.8	699.1	9.8	526.9	1.4	889.1	71.1
21	323.9	7.6	400.1	33.0	177.6	41.0	230.1	23.5	240.0	20.2	497.8	65.4

Table 4: A point-to-point comparison between the results obtained with the proposed model and comparative methods for the experimental records reported in Table 3.

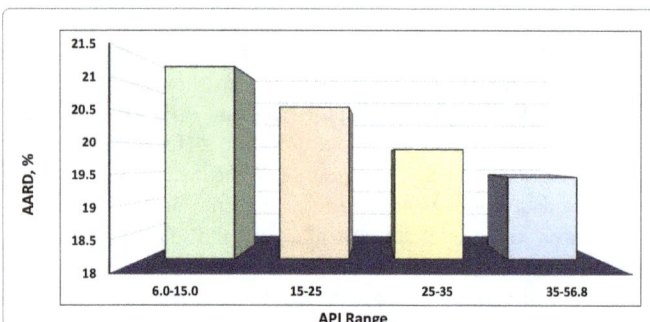

Figure 6: Accuracy of the model developed in this study in different API ranges.

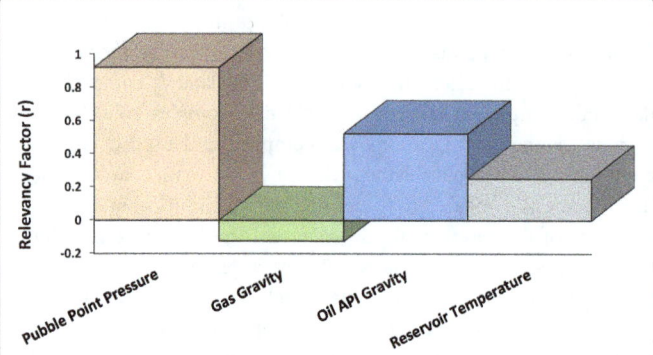

Figure 7: Degree of importance for each input parameter for the prediction of solution gas-oil ratio.

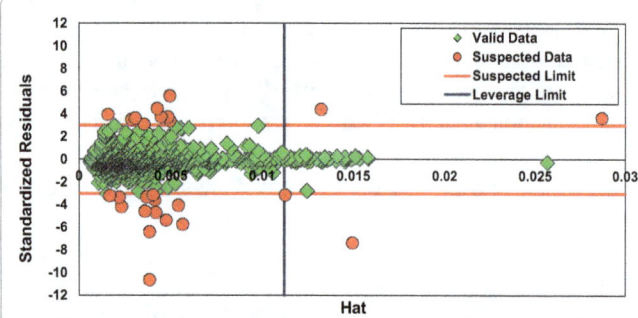

Figure 8: Graphical plot of the leverage analysis for the recognition of outlier data points.

References

1. Elsharkawy AM, Alikhan AA (1997) Correlations for predicting solution gas/oil ratio, oil formation volume factor and under-saturated oil compressibility. J Pet Sci Eng 17: 291-302.

2. Frashad F, LeBlanc J, Garber J, Osorio J (1996) Empirical PVT correlations for Colombian crude oils, in: SPE Latin America/Caribbean petroleum engineering conference. Society of petroleum engineers, Trinidad, Spain.

3. Alizadeh N, Mighani S, Hashemi kiasari H, Hemmati-Sarapardeh A, Kamari A (2013) Application of fast-SAGD in naturally rractured heavy oil reservoirs: A case study. Middle East oil & Gas show (MEOS) 18th Conference, Barhain.

4. Kamari A, Mohammadi AH (2014) Screening of enhanced oil recovery methods. Nova Science Publishers.

5. Kamari A, Nikookar M, Hemmati-Sarapardeh A, Sahranavard L, Mohammadi AH (2014) Screening of Potential Application of EOR Processes in a naturally fractured oil reservoir. Enhanced Oil Recovery: Methods, Economic benefits and impacts on the environment. Nova Science Publishers, USA.

6. Kamari A, Nikookar M, Sahranavard L, Mohammadi AH (2014) Efficient

7. Kamari A, Hemmati-Sarapardeh A, Mohammadi AH, Hashemi-Kiasari H, Mohagheghian E (2015) On the evaluation of Fast-SAGD process in naturally fractured heavy oil reservoir. Fuel 143: 155-164.

8. Hemmati-Sarapardeh A, Khishvand M, Naseri A, Mohammadi AH (2013) Towards reservoir oil viscosity correlation. Chemical Engineering Science 90: 53-68.

9. Al-Marhoun MA (1988) PVT correlations for Middle East crude oils. J Pet Technol 40: 650-666.

10. Al-Marhoun MA (2004) Evaluation of empirically derived PVT properties for Middle East crude oils. Journal of Petroleum Science and Engineering 42: 209-221.

11. Arabloo MA, Amooie A, Hemmati-Sarapardeh A, Ghazanfari MH, Mohammadi AH (2014) Application of constrained multi-variable search methods for prediction of PVT properties of crude oil systems. Fluid Phase Equilibria 363: 121-130.

12. Asoodeh M, Bagheripour P (2012) Estimation of bubble point pressure from PVT data using a power-law committee with intelligent systems. Journal of Petroleum Science and Engineering 90: 1-11.

13. Elsharkawy P (AM) An empirical model for estimating the saturation pressures of crude oils. J Pet Sci Eng 38: 57-77.

14. Elam FM (1957) Prediction of bubble point pressures and formation volume factors from field data. University of Texas, Austin, Texas.

15. Lasater J (1958) Bubble point pressure correlation. J Pet Technol 10: 65-67.

16. Vazquez M, Beggs HD (1980) Correlations for fluid physical property prediction. J Pet Technol 32: 968-970.

17. Ostermann R, Ehlig-Economides C, Owolabi O (1983) Correlations for the reservoir fluid properties of Alaskan crudes. University of Alaska.

18. Rollins JB, McCain Jr, Creeger TJ (1990) Estimation of solution GOR of black oils. J Pet Tech 42: 92-94.

19. Sutton RP, Farshad F (1990) Evaluation of empirically derived PVT properties for Gulf of Mexico crude oils. SPE Reservoir Engineering 5: 79-86.

20. Glaso O (1980) Generalized pressure-volume-temperature correlations. J Pet Tech 32: 785-795.

21. Dokla ME, Osman ME (1992) Correlation of PVT properties for UAE crudes. SPE formation evaluation 7: 41-46.

22. Omar M, Todd A (1993) Development of new modified black oil correlations for Malaysian crudes, in: SPE Asia Pacific oil and gas conference. Society of Petroleum Engineers.

23. Standing M (1947) A pressure-volume-temperature correlation for mixtures of California oils and gases. Drilling and Production Practice.

24. Petrosky G, Farshad F (1993) Pressure-volume-temperature correlations for Gulf of Mexico crude oils, in: SPE Annual Technical Conference and Exhibition. Society of Petroleum Engineers.

25. Elsharkawy AM, Elgibaly AA, Alikhan AA (1995) Assessment of the PVT correlations for predicting the properties of Kuwaiti crude oils. J Pet Sci Eng 13: 219-232.

26. Ghetto GD, Paone F, Villa M (1994) Reliability analysis on PVT correlations, In: European Petroleum Conference. Society of Petroleum Engineers.

27. Khairy M, El-Tayeb M, Hamdallah M (1998) PVT correlations developed for Egyptian crudes. Oil and Gas Journal.

28. Velarde J, Blasingame T, McCain W (1999) Correlation of black oil properties at pressures below bubble-point pressure–a new approach. J Can Petrol Tech 38: 1-6.

29. Mazandarani MT, Asghari SM (2007) Correlations for predicting solution gas-oil ratio, bubble-point pressure and oil formation volume factor at bubble-point of Iran Crude Oils, In: European Congress of Chemical Engineering, Copenhagen.

30. Taghaz A, Eltaeb N, Alakhdar S (2008) Comparison study of published PVT correlations and its application to estimate reservoir fluid properties for Libyan oil reservoirs, in: Tenth Mediterranean Petroleum Conference and Exhibition. Libya.

31. Shafiie O, Moghadasi J, Shahbazian M, Zargani F (2012) Optimization of formation volume factor and solution gas-oil ratio correlations for southern Iranian oilfields using genetic algorithm. Journal of American Science 8: 19-25.

32. Kamari A, Arabloo M, Shokrollahi A, Gharagheizi F, Mohammadi AH (2015) Rapid method to estimate the minimum miscibility pressure (MMP) in live reservoir oil systems during CO_2 flooding. Fuel 153: 310-319.

33. Kamari A, Bahadori A, Mohammadi AH, Zendehboudi S (2014) Evaluating the unloading gradient pressure in continuous gas-lift systems during petroleum production operations. J Pet Sci Eng 32: 2961-2968.

34. Kamari A, Bahadori A, Mohammadi AH, Zendehboudi S (2015) New tools predict monoethylene glycol injection rate for natural gas hydrate inhibition. J Loss Prev Process Ind 33: 222-231.

35. Kamari A, Gharagheizi F, Bahadori A, Mohammadi AH (2014) Determination of the equilibrated calcium carbonate (calcite) scaling in aqueous phase using a reliable approach. J Taiwan Inst Chem Eng 45: 1307-1313.

36. Kamari A, Mohammadi A, Bahadori A, Zendehboudi S (2014) A reliable model for estimating the wax deposition rate during crude oil production and processing. J of Pet Sci Tech 32: 2837-2844.

37. Kamari A, Mohammadi AH, Bahadori A, Zendehboudi S (2014) Prediction of Air Specific Heat Ratios at Elevated Pressures Using a Novel Modeling Approach. J of Chem Eng Tech 37: 2047-2055.

38. Zendehboudi S, Shafiei A, Bahadori A, James LA, Elkamel A, et al. (2013) Asphaltene precipitation and deposition in oil reservoirs–Technical aspects, experimental and hybrid neural network predictive tools. Chem Eng Res Design.

39. Esfahani S, Baselizadeh S, Hemmati-Sarapardeh A (2015) On determination of natural gas density: Least square support vector machine modeling approach. J Nat Gas Sci Eng 22: 348-358.

40. Hosseinzadeh M, Hemmati-Sarapardeh A (2014) Towards a predictive model for estimating viscosity of ternary mixtures containing ionic liquids. Journal of Molecular Liquids 200: 340-348.

41. LINGO-Softwate (2011) Optimization Modeling Software for Linear, Nonlinear, and Integer Programming, LINDO Systems, LINGO Version 11, in, 2011.

42. Baniasadi H, Kamari A, Heidararabi S, Mohammadi AH, Hemmati-Sarapardeh A (2015) Rapid method for the determination of solution gas-oil ratios of petroleum reservoir fluids. J Nat Gas Sci Eng 24: 1-10.

43. Abdul-Majeed GH, Salman NH, Scarth B (1988) An empirical correlation for oil FVF prediction. J of Can Pet Technol.

44. Bello O, Reinicke K, Patil P (2008) Comparison of the performance of empirical models used for the prediction of the PVT properties of crude oils of the Niger delta. J of Pet Sci Technol 26: 593-609.

45. Mahmood MA, Al-Marhoun MA (1996) Evaluation of empirically derived PVT properties for Pakistani crude oils. J Pet Sci Eng 16: 275-290.

46. Moghadam JN, Salahshoor K, Kharrat K (2011) Introducing a new method for predicting PVT properties of Iranian crude oils by applying artificial neural networks. J of Pet Sci Technol 29: 1066-1079.

47. Obomanu D, Okpobiri G (1987) Correlating the PVT properties of Nigerian crudes. J Energy Resour Technol 109: 214-217.

48. Chuang YF, Lee HT, Lai YC (2012) Item-associated cluster assignment model on storage allocation problems. Computers & Industrial Engineering 63: 1171-1177.

49. Carvalho M, Lozano MA, Serra LM, Wohlgemuth V (2012) Modeling simple trigeneration systems for the distribution of environmental loads. Environmental Modelling & Software 30: 71-80.

50. Vidal D, Blobel J, Pérez Y, Thormann M, Pons M (2007) Structure-based discovery of new small molecule inhibitors of low molecular weight protein tyrosine phosphatase. Eur J Med Chem 42: 1102-1108.

51. Z. Miao (2011) Discussion of optimize method of fire alarm dispatching based on operation research principle. Procedia Engineering 11: 689-694.

52. Chen G, Fu K, Liang Z, Sema T, Li C, et al. (2014) The genetic algorithm based back propagation neural network for MMP prediction in CO 2-EOR process. Fuel 126: 202-212.

53. Macary S, El-Batanoney M (1993) Derivation of PVT correlations for the Gulf of Suez crude oils. Sekiyu Gakkai Shi 36: 472-478.

54. Kartoatmodjo T, Schmidt Z (1994) Large data bank improves crude physical property correlations. Oil & Gas Journal.

55. Al-Shammasi A (1999) Bubble point pressure and oil formation volume factor correlations, in: SPE Middle East Oil Show & Conference, pp: 241-256.

56. Dindoruk B, Christman PG (2004) PVT properties and viscosity correlations for Gulf of Mexico oils. SPE Reservoir Evaluation & Engineering 7: 427-437.

57. Mohammadi AH, Gharagheizi F, Eslamimanesh A, Richon D (2012) Evaluation of experimental data for wax and diamondoids solubility in gaseous systems. Chemical Engineering Science.

58. Gharagheizi F, Eslamimanesh A, Ilani-Kashkouli P, Mohammadi AH, Richon D (2012) QSPR molecular approach for representation/prediction of very large vapor pressure dataset. Chemical Engineering Science, 76: 99-107.

59. Rousseeuw PJ, Leroy AM (2003) Robust regression and outlier detection. Wiley.

33

Comparative Study on Oil Recovery Enhancement by WAG Injection Technique in a Fractured Oil Reservoir in the Southwest of Iran

Hajnajafi Reza[1], Amid Arman[2] and Hajnajafi Ghazal[3]

[1]Department of Petroleum Engineering, University of New South Wales, Sydney, Australia
[2]Department of Environmental Management HSE, Tehran North branch of Islamic Azad University, Tehran, Iran
[3]Department of Geology, University of Science and Research of Islamic azad university, Tehran, Iran

Abstract

Water alternating gas injection (WAG) and simultaneous water and gas injection (SWAG) are injection techniques used to enhance oil recovery. Gas injection is today regarded as the second method for increasing oil recovery. While very little oil remains in the areas swept by gas, the amount of volumetric displacement by injected gas has always drawn much concern. This is due to high mobility ratio of the gas injected to reservoir oil. To solve this problem, WAG and SWAG injection techniques are used to control fluid flow profile. High microscopic displacement efficiency of gas (in whole scale) and high volumetric displacement efficiency of water (in microscopic scale) considerably increases oil recovery in the front space of water. In this paper, we investigate various WAG injection methods using Eclipse and compare the results to determine the best injection scenario for the reservoir under study.

Keywords: Oil recovery enhancement; Fractured reservoirs; Simulation; SWAG injection technique; WAG injection technique

Introduction

SWAG is a modern technique used to recover oil. In this method, a mass of water and gas is simultaneously injected into the reservoir. The most important advantage of this technique over WAG is better control of gas mobility thanks to injection of a specified quantity of water. This method was first executed at Seeligson field in the southwest of Texas in 1963. SWAG injection technique considerably reduces costs by: A) eliminating the separation between water and gas during injection; B) eliminating changes in operation equipment; and C) reducing the costs incurred by GOR [1]. Simulation studies indicate that SWAG injection with 1:1 ratio increases oil-in-place production by 5% compared with 4.5% increase offered by WAG injection method [1]. In the next paragraphs, we review the history of WAG and SWAG processes.

Surgnchev conducted a comparative study on WAG, FAWAG and SWAG injection methods using a three-dimensional model [2]. Permeability of the model was 220 md in the upper part and 2-20 md in the lower part. In this model, vertical and horizontal wells were designed for production and injection, as illustrated in Figure 1. Simulation results indicated an oil-in-place recovery of 51% after injection, with oil being mostly recovered from the small section in upper part of the model. In SSWAG model, gas was injected to the upper part with thickness of 100 ft and water was injected to the lower part with thickness of 40 ft. Moreover, WAG and SSWAG were compared as a third method for increasing oil recovery. According to the results, recovery efficiency increased from 26.6% to 31.1% [2]. Esmaiel studied water-to-gas ratio in WAG injection process and concluded that the increase of water-to-gas ratio will increase oil recovery and water cut rate. So, an optimal ratio must be selected to reduce water cut rate [3].

Meshal algharaib conducted a study on the improved SWAG process parameters. They proposed a model in which gas was injected in the lower part and water was injected in the upper part of the reservoir. In the present study, we investigate WAG and SWAG injection methods. Simulation results indicated that these methods offer better sweeping and recovery efficiency and are more cost effective. This study is limited to determining optimal quantity of water and gas [4]. Mirkalaei studied WAG injection processes to determine optimal injection rate [5].

Jiang studied WAG ratio and reached the following conclusion: determining WAG ratio is one of the important design parameters and considerably affects operational and economic conditions of the project. Optimal WAG ratio is affected by the type of stone wettability. High WAG ratio has the highest impact on oil efficiency in water-friendly reservoirs and reduces the amount of residual oil [6]. Ghaderi conducted a study on the impact of CO_2 miscible injection and WAG injection methods in oil reservoirs with compressed formation [7]. They found that injection of water masses is necessary for reducing fingering impact. Fingering occurs quickly during CO_2 injection due to low viscosity of CO_2. To delay this, the amount of injected water should be more than CO_2. This is especially important in fractured reservoirs. In normal reservoirs and those with very low permeability, however, low viscosity of CO_2 may help the process.

In this study, we investigate various WAG injection methods and compare the results in order to determine the best injection scenario for the reservoir under study.

Literature Review

WAG injection technique was first executed in 1957 in Alberta and produced successful results [8]. Thanks to the advantages of this technique over separated injection methods (e.g. control of relative mobility of displaced and displacing phases, prevention of early fingering in oil production wells, the capability to produce oil from unswept areas, creation of controllable and sustainable progression, and the capability to use operational tools), WAG injection technique

---

*Corresponding author: Hajnajafi Reza, Department of Petroleum Engineering, University of New South Wales, Sydney, Australia, E-mail: rezahajnajafi@gmail.com

has become widespread in different points of the world, such as the US, Canada, North Sea, Russia, Turkey and Venezuela, particularly in the past two decades. Researchers have recently studied different aspects of WAG injection method in order to determine the changes in reservoir conditions during injection period.

Cobanoglu designed different scenarios for injection quantity, cycle, and the number of produced and injected wells using ECLIPSE100 Simulator and studied and compared immiscible gas injection and WAG injection methods in Baty Kozluca field in Turkey. According to the results, immiscible gas injection technique considerably increased the field efficiency due to inappropriate mobility ratio. He reported that WAG immiscible injection offers more efficiency compared with immiscible gas injection [9].

Hustod and Klov studied WAG injection method and compared it with water-gas injection method in layers with different permeability levels in the North Sea. They found that water and gas fingering in high-permeability layers and immobility and bad sweeping process in low-permeability layers reduce injection efficiency in these methods. According to their results, WAG injection method prevents gas movement in high-permeability layers, develops a three-phase area in the reservoir, and stabilizes the progression. So, they concluded that WAG injection method offers more efficiency compared to water-gas injection methods [10].

Shi and coworkers studied WAG injection method in Kuparuk field in the north of Alaska for a period of 20 years using field results. They first used gas injection method to increase oil recovery in the field. Due to early fingering and GOR increase, however, they employed WAG injection technique which increased oil production by 120 MMSTB during injection period [11].

Instefjord studied WAG injection method in Gullfaks field for a period of 10 years. According to his results, WAG injection technique increased oil production by 2 MMSTB during injection period. He reported that the execution of WAG injection technique increased displacement efficiency and reduced the percentage of produced water [11]. Trnerr and coworkers, Quale and coworkers, Siri, Skauge and Aarra and Snorer and Quraini have studied WAG injection technique in Seeligsou, Stephansen, Joffer Viking, Snorer and West Sak fields respectively. All of these studies have demonstrated the advantages of WAG injection method over other methods for recovering oil [12]. In the past decade, almost 40% of gas injection projects throughout the world, such as Canada, Russia, Turkey and Norway, have been executed by WAG injection technique. 80% of these projects have produced successful results [13].

Research Method

We investigated various water-gas injection methods using simulations made by Eclipse Simulator and compared the results to determine the best injection scenario for the reservoir. In doing so, we studied oil recovery enhancement using various water-gas injection methods at an under-saturated fractured reservoir in the southwest of Iran. Phase behavior of reservoir fluid was studied by PVTi module of Eclipse 100 (immiscible). After matching the history of reservoir production and pressure data, we simulated various water-gas injection methods based on such parameters as injection quantity and oil production quantity.

Details of matrix and reservoir fracture

Oil in place is 3.7 MMMbbl, water and oil contact level is 11400 ft,

and reservoir peak is 9200 ft below the level of free water. Average porosity of fractures is approximately 0.002% and average permeability of fractures is 1200 md. Average porosity of matrix is 0.1 and average horizontal permeability of the matrix is 1 md for all layers.

Details of reservoir oil

Initial pressure of the reservoir in the depth of 11200 ft is 5600 psi and reservoir temperature is 214 F. Bubble point pressure of the reservoir is 2400 psi and the ratio of dissolved gas to oil is 570 scf/stb. Density and API of reservoir oil are 54 Ib/ft^3 and 43 respectively. For reservoir water in the pressure of 5600 psi, volumetric coefficient of formation water is 1.3bbl/stb and water viscosity is 0.66 cp.

Model description

We simulated dynamic model of the reservoir using Eclipse 100 (immiscible) in three-dimensional and three-phase manner by fully-explicit equation solving method. For the purpose of simulation, we divided reservoir A into many blocks. The number of blocks in X, Y and Z is 40, 120 and 50 respectively. The first 18 blocks in part Z belong to matrix and the other 32 blocks belong to fractures. Since the reservoir was fractured, we modeled it with binary porosity.

Optimization of injection wells

There are six production wells in the reservoir. We consider four of them as production wells and the other two as injection wells. We investigated different scenarios regarding the situation and number of injection wells. The intensity of injection flow in all scenarios was set on $2*10^6$ scf/day. The scenario with an injection well in the reservoir peak and a well in lower part of the reservoir was elected as the best scenario for injection. This scenario offers the advantage of late gas penetration into production wells because reservoir slope in this location is fairly good.

Injection method

In water and gas injection, Buckley and Leverett's theory is equally applicable. In vertical gas and oil flow, however, overlooking the impact of gravity is impossible. So, a variety of equations should be used for minor f$_g$ gas flow, depending on whether injection is performed in oil area (flow is assumed to be horizontal) or in gas cap (flow is assumed to be vertical). Therefore, *Diffr* oil is between time *j* (reservoir pressure - P$_j$) and j+1 (reservoir pressure - P$_j$+1), assuming that there is no water penetration. From the equation we can obtain reinjection quantity to maintain full pressure. When pressure remains constant, the numerator of equation is 0 and since recovery is not 0, the denominator of equation should also be 0 and the equation is indefinite and vague. So, I is obtained. In both injection operations in gas cap and oil area, the first step is to prepare f$_j$ (S$_g$) curve with the assumption of viscosity in saturation [14]. To estimate appropriate minor movement of gas during injection and achieve a harmonious injection progression, the following equation was used [15]. Using cross-injection, an appropriate injection quantity can be applied.

Data Analysis

We simulated the reservoir in three modes: depletion in normal state, production through gas injection, and production through water injection. In water and gas injection simulation, we defined and executed many scenarios, as shown in Table 1. We investigated water and gas injection scenarios in two distinct parts, selected the best scenario in each group, and then compared them. To adjust the injected water, which was supplied from sea water, to reservoir and its fluid, and

Symbol	Description
WI1	$4_p/1_{in}$
WI2	$4_p/1_{in}$
WI3	$4_p/1_{in}$
WI4	$2_{in}3_{pro}$
WI5	$5_{wat}/4_p/1_{in}$
GI1	$4_p/1_{in}$
GI2	$4_p/1_{in}$
GI3	$4_p/1_{in}$
GI4	$2_{in}3_{pro}$
GI5	$4_{inj}1_{pro}$
WAG1	1:1
WAG2	1:2
WAG3	1:3
WAG4	2:3
WAG5	3:4
WAG cycle1	4 month water, 8 months gas
WAG cycle2	6 months water, 6 months gas
WAG cycle3	8 months water, 4 months gas
WAG cycle4	12 months water, 12 months gas
SWAG1	1:1
SWAG2	1:2
SWAG3	1:3
SWAG4	2:3
SWAG5	3:4
M	Ratio of initial gas cap volume to initial oil area volume (for injection to oil area without gas cap m=0)
I	Produced part of gas which has been re-injected.
R	Average GOR between j and j+1.
$f_g = 1/\left(1+(k_{ro}/k_{rg})(\mu_g/\mu_0)\right)$	Gas injection in oil area
$Diffr = \left((1-r_j)Diff\left(B_{o/B_G}-R_S\right)-(1+m)B_o Diff\left(1/B_G\right)\right)/\left(B_{o/B_G}-R_S+\overline{R}\left(1-I\right)\right)$	
$I = (B_o + R_C * Bg/(R_S+R_C)B_g)$	
$v_c = \left(\rho_o - \rho_G\right)\left[K.g.\sin\alpha\right]/\left(\mu_o - \mu_g\right)$	Minor movement of gas
$q = v_c.A$	Appropriate quantity for injection

Table 1: Research symbols.

to prevent the corrosion, we added some supplements to it. The gas used in this study is methane.

Gas injection

To achieve the best daily injection quantity in gas injection scenarios, we studied gas injection operation in several fields in which gas injection method had been used to increase recovery, and selected an initial injection quantity based on characteristics and geological information of the reservoir, pressure gradient and depth of the reservoir in injection layer, gradient level of the fracture, and the potentiality of surface installations. To determine the best injection quantity, we frequently changed the quantity and executed the scenarios with each quantity. This way we assessed the sensitivity of reservoir to the increased or decreased quantity and selected the best injection quantity. The best quantity for gas injection was 8 million cubic foot per day for each injection well. For the purpose of better comparison, Figures 1 and 2 illustrate the results obtained from all gas injection scenarios with

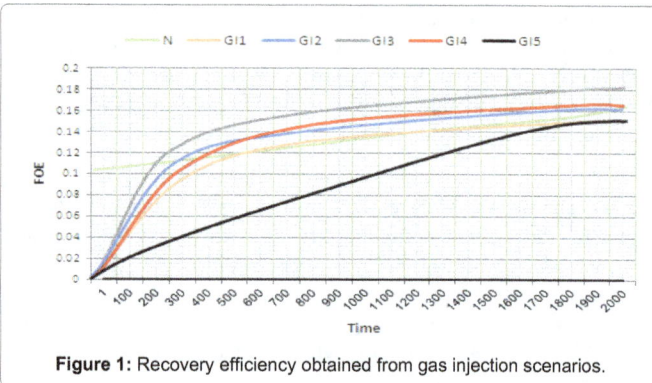

Figure 1: Recovery efficiency obtained from gas injection scenarios.

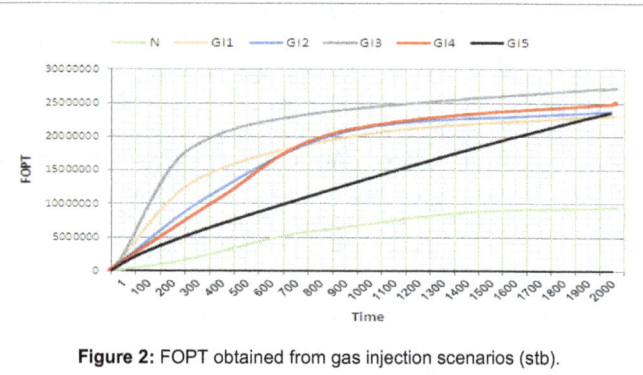

Figure 2: FOPT obtained from gas injection scenarios (stb).

injection rate of 8 million cubic foot per day for each well in different patterns for the intended parameters.

According to the results obtained from gas injection scenarios, the third scenario is the best scenario and offers the best recovery efficiency. In this scenario, injection is made through one well and production is made through four wells defined in the model. Moreover, injection is made in both gas area and the reservoir.

Water injection

All water injection scenarios were designed like gas injection scenarios, with the only different that injection layers were defined in lower parts of the reservoir, mainly in aquifer. To achieve the best daily injection quantity in water injection scenarios, we studied water injection operation in several fields in which water injection method had been used to increase recovery [13]. First, we selected an initial injection quantity based on characteristics and geological information of the reservoir, pressure gradient and depth of the reservoir in injection layer, gradient level of the fracture, and the potentiality of surface installations. To determine the best injection quantity, we frequently changed injection quantity and executed the scenarios with each quantity. This way we assessed the sensitivity of reservoir to quantity changes and selected the best injection quantity (8000 barrels per day). So, we injected 8000 barrels of water to the reservoir each day. Figures 3 and 4 illustrate the results obtained from execution and simulation of eight water injection scenarios with injection rate of 8000 STB/Day.

According to the results obtained from water injection scenarios, the fifth scenario is the best scenario and offers the best recovery efficiency. In this scenario, four water injection and one oil production well have been defined. Water is injected through injection wells (Figure 5).

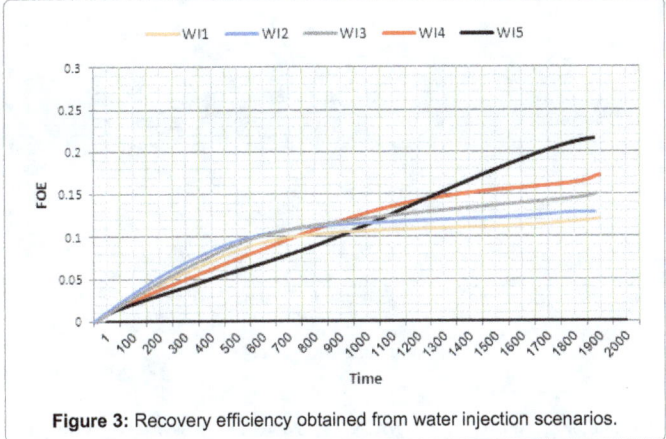

Figure 3: Recovery efficiency obtained from water injection scenarios.

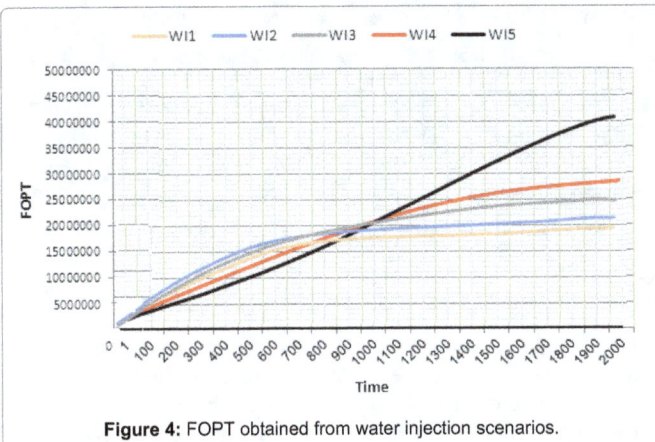

Figure 4: FOPT obtained from water injection scenarios.

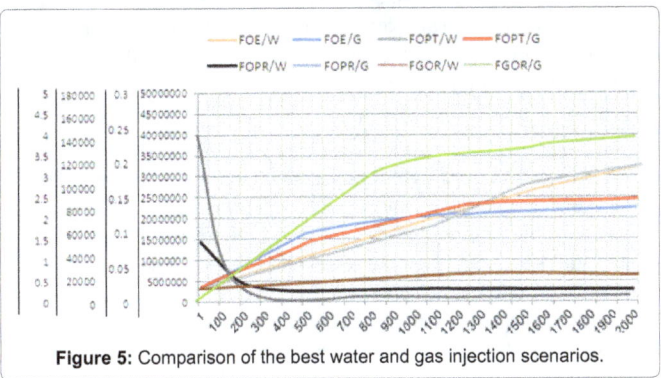

Figure 5: Comparison of the best water and gas injection scenarios.

Comparison of water and gas injection

We compared the results obtained from the best water and gas injection scenarios in order to determine which injection method offers the best recovery efficiency, field oil production rate and daily production rate, represents lowest gas-to-oil ratio and water cut, maintains reservoir pressure, and is merely based on simulation results without consideration of injection scheme. As mentioned earlier, the best gas injection scenario is the injection from one well and production through four wells with injection rate of 8 million cubic foot per day, and the best water injection scenario is the injection from four wells and production through one well with injection rate of 8,000 barrels per day.

The comparison between the results obtained from the best water

injection scenario and the best gas injection scenario indicated that the former produces better results, so water injection scenario was determined as the best scenario.

WAG

To investigate the impact of this parameter, we set fluid injection volume to be 0.3 time empty space volume of the model. Figures 6 and 7 contain the results. WAG refers to the ratio of injected water to injected gas. If this ratio exceeds its optimal value, water cut increases. If it is less than its optimal value, gas-to-oil ratio (GOR) increases. In SWAG, only the ratio of gas to water is important, with time ratio (cycle) being of no importance because injection is simultaneous. WAG ratio is controlled by gas ability to wet reservoir rock. Immiscible WAG increases the volume of swept oil and improves sweeping efficiency. It also reduces costs through reducing the volume of gas injection to reservoir. Determination and use of optimal WAG ratio is an important parameter in design and greatly affects operational and economic conditions of the project. Further, WAG ratio may be increased following the increase of optimal gas production rate. As you can see, injection ratio of 1:1 is the best mode among different injection ratios for the model. Moreover, the more water-to-gas ratio, the earlier gas breakthrough occurs and the more water cut will be [16,17].

To compare the impact of injection cycles, we evaluated four different modes. The results are contained in Figures 8 and 9. In WAG process, each injection cycle is divided into two sub-cycles. In the first sub-cycle, water is injected from injection wells. In the second sub-cycle, gas is injected from injection wells. Generally it is better to inject water to reservoir in the first sub-cycle. If gas is injected in the first sub-cycle, the injected gas quickly reaches the production well because of its high mobility and gas middle-break occurs. If water is injected in the first sub-cycle and gas is injected in the second sub-cycle, water prevents the

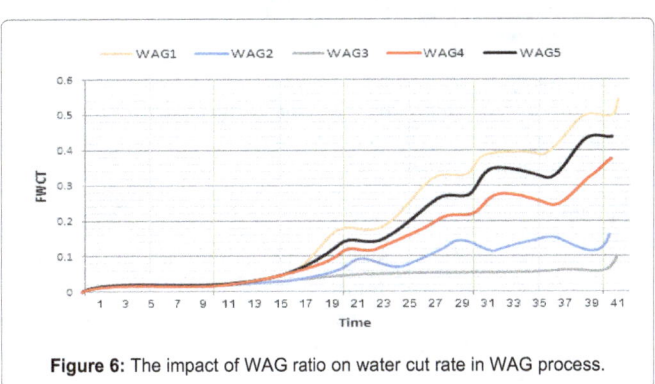

Figure 6: The impact of WAG ratio on water cut rate in WAG process.

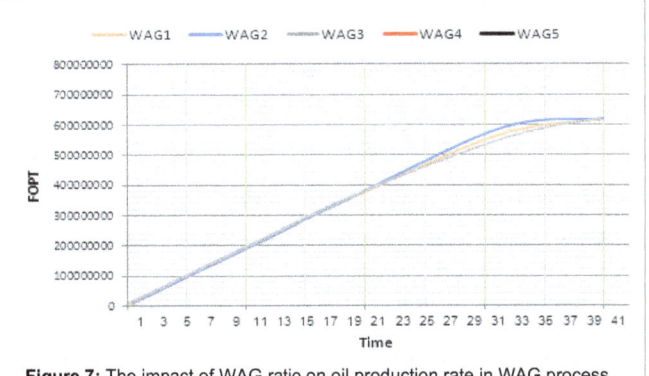

Figure 7: The impact of WAG ratio on oil production rate in WAG process.

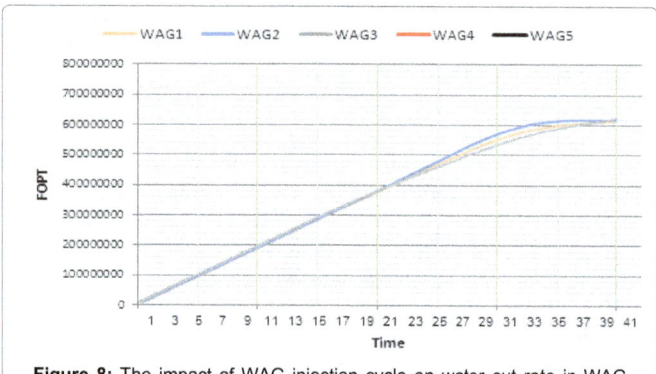

Figure 8: The impact of WAG injection cycle on water cut rate in WAG process.

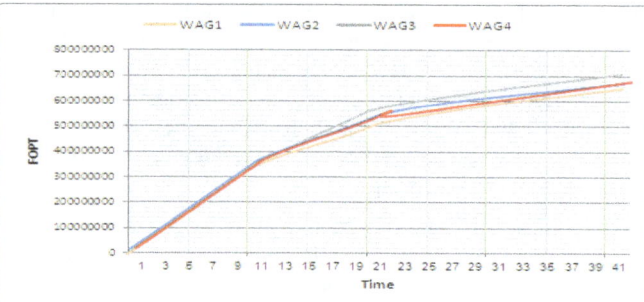

Figure 9: The impact of WAG injection cycle on oil production rate in WAG process.

quick movement of gas and the occurrence of gas breakthrough. This is especially important in cases where reservoir has a high permeability. As you can see, the injection cycle consisting of 12 months of water and 12 months of gas is the optimal cycle for the model.

SWAG process

To investigate the impact of this parameter, we set fluid injection volume on 0.3 time empty space volume of the model. The results are contained in Figure 10. As you can see, injection ratio of 1:3 is the best mode for the model.

Final comparison

We compared WAG and SWAG scenarios to determine the optimal one. This comparison was made for injection ratio of 1:1 and injection por volume of 0.3. The results indicated that SWAG injection scenario was more efficient than WAG injection scenario. In this step, we compared different injection techniques to determine the optimal method. Tables 2 and 3 represent efficiency rates and FOPT of the methods in question. As you can see, SSWAG offers the highest efficiency. So, it could be said that displacement efficiency in this method is better than in other injection methods because of the type of injection and the enhanced efficiency of gas and water injection in the movement and displacement of reservoir oil toward production well. As you can see, SSWAG offers more FOPT compared with other injection methods.

Comparison of WAG and IWAG methods

Since WAG injection includes different water and gas injection methods, it is necessary to compare it with IWAG in order to determine an optimal method for increasing oil recovery and enhancing production efficiency. To make a proper comparison between the above methods, we not only investigated WAG injection method but also designed and studied a variety of water and gas injection techniques. From each method, we

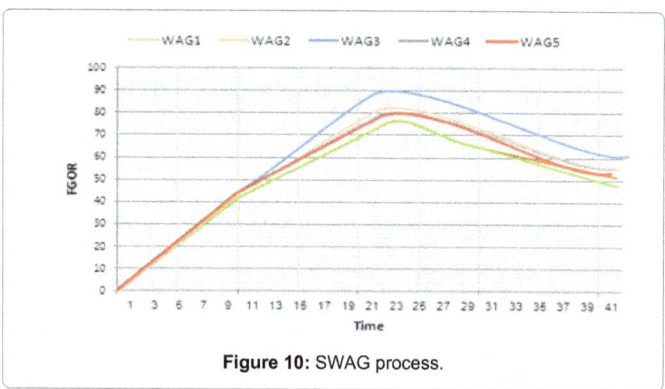

Figure 10: SWAG process.

Parameter	IWAG	HWAG	SWAG	SSWAG	WAG$_{aw}$f
Efficiency	73000	76000	77000	82000	70000
FOPT	7.7%	7.9%	8%	8.5%	7.3%

Table 2: Comparison of WAG methods.

Parameter	Natural	Gas I	Water	WAG
Daily production	4.2%	5.2%	6.2%	8%
FOPT	4200	5200	6200	8000
Saturation percentage	21.7	21.5	21.4	20.9

Table 3: Comparison of WAG and IWAG injection methods.

selected a scenario with the highest production rate. Then we compared the scenarios with highest production efficiency in each method.

We compared daily production rates in oil recovery enhancement methods. As you can see, WAG injection method offers more production rate compared with other methods. It also offers better efficiency rate and FOPT. Further, the least saturation percentage of residual oil belongs to WAG. So, this method offers better macroscopic and microscopic efficiency compared with other methods. Therefore, WAG injection method enhances final efficiency and increase oil recovery and production.

Conclusion

We studied various water and gas injection methods using Eclipse simulator and compared the results to determine the best injection scenario for the reservoir. In doing so, we investigated oil recovery rate in an under-saturated fractured reservoir in the southwest of Iran using various water and gas injection methods. Phase behavior of reservoir fluid was studied by PVTi module of Eclipse 100 (immiscible). After adjusting the history of reservoir production and pressure data, we simulated various water and gas methods based on such parameters as injection quantity and oil production rate. The following represents the summary of results:

1. The best gas injection scheme is four production wells in the sides and one injection well in the middle of the project. The best water injection scheme is four injection wells and one production well.

2. In the fractured reservoir, water injection offers higher FOE than gas injection.

3. Water injection scenario also offers acceptable results in FOPT, FPR, FGOR and FOPR. In general, water injection offers better displacement efficiency compared to gas injection.

4. Due to high permeability of reservoir rock, gas quickly reaches the production wells (gas breakthrough) and GOR increases. This

explains the lower FOPT in gas injection scenarios compared to water injection.

5. Another reason why gas injection scenario offers less oil recovery efficiency lies in field cross-slope which causes the gas to go toward higher layers and the oil of lower layers to be not swept.

6. SWAG scenario offers more water cut compared with WAG.

7. The efficiency of SWAG and WAG scenarios is almost equal.

8. SSWAG offers more efficiency compared with other WAG injection methods. This indicates that this method offers more sweeping efficiency, both microscopic and macroscopic, compared to other injection methods.

9. WAG method offers more efficiency rate and FOPT compared with natural production methods as well as water and gas injection techniques. So, this method was introduced as optimal recovery enhancement method in the field under study.

10. WAG method offers less residual oil saturation percentage. So, sweeping efficiency (microscopic and macroscopic) of this method is better than other injection methods.

11. Among various WAG injection methods in 4-point and 5-point schemes, SSWAG in 4-point scheme offers the highest efficiency and FOPT.

12. 4-point scheme offers better efficiency and less residual oil saturation percentage compared with 5-point scheme. This indicates that the increased number of production wells does not enhance efficiency and merely increases production speed. Therefore, 5-point scheme increases project costs, particularly the costs of drilling production wells.

13. The efficiency of SWAG injection method is less than SSWAG. This indicates that SWAG in single-phase offers less displacement and production efficiency compared with SSWAG.

References

1. Berge LI, Stensen JA, Crapez B, Quale EA" (2002) SWAG Injectivity Behavior Based on Siri Field Data" paper SPE 75126, presentation at the SPE/DOE Improved Oil Recovery Symposium held in Tulsa, Oklahoma U.S.A.

2. Surguchev LM, Korbal R (1992) "Screeing of WAG injection Strategies for heterogeneous Reservoirs", SPE 25075, Presented at the European Conference held in Cannes, France.

3. Esmaiel TE, Fallah S, Kruijsdijk CV (2005) "Determination of WAG Ratios and Slug Sizes Under Uncertainty in a Smart Wells Environment", Delf University, SPE 93569, Presented at the 14th SPE Middle East Oil & Gas Show and Conference, Bahrain Exhibition Center, Bahrain.

4. Algharaib M, Gharbi R, Malallah A (2007) "Parametric Investigations of Modified SWAG injection Technique", SPE 105071, Presented at the 15th SPE Middle East Oil & Gas Show and Conference held in Bahrain International Exhibition Centre, Kingdom of Bahrain.

5. Mirkalaei SMM, Hosseini J, Masoudi R, Ataei A, Demiral B, et al. (2011) " Investigation of Different I-WAG schemes toward Optimization of Displacement Efficiency" SPE 144891, Enhanced Oil Recovery Conference held in Kuala Lumpur, Malaysia.

6. Jiang H, Nuryaningsih L, Adidharma H (2012) "The Study of Timing of Cyclic injections in Miscible CO2 WAG" SPE 153792, Western Regional Meeting held in Bakersfield, California, U.S.A.

7. Ghaderi SM, Clarkson CR, Chen S, Kaviani D (2012) "Evaluation of Recovery Performance of Miscible Displacement and WAG Process in Tight Oil Formations" SPE 152084, Vienna, Austria.

8. Quijada MG (2005) Optimization of a CO_2 flood design wasson field - west Texas, Master Of Science Thesis, Texas A&M University, USA.

9. Cobanoglu M (2001) "A Numerical study to evaluate the use of WAG as an EOR method for oil production improvement at B. Kozluca field", Turkey, PP, SPE 72127, presentation at the SPE Asia Pacific Improved Oil Recovery Conference held in Kuala Lumpur, Malaysia.

10. Klov M, Hustod N (2003) "Experimental investigation of various methods of tertiary gas injection", PP, SPE 80579, presented at the 2003 Society of Petroleum Engineers Annual Technical Conference and Exhibition, Houston USA.

11. Instefjord R, Todnem CA (2002) "10 Years of WAG injection in Lower Brent at the Gullfaks field", PP, SPE 78344, presentation at the SPE 13th European Petroleum Conference held in Aberdeen, Scotland, U.K.

12. Shi W, Corwith J (2008) "Kuparuk MWAG project after 20 years", SPE 113933, presentation at the SPE/DOE Improved Oil Recovery Symposium held in Tulsa, Oklahoma USA.

13. Rehman T (2006) A Techno economical evaluation of miscible flooding, Dalhousie university.

14. Boerrigter PM, Van de leemput BLEC, Johan Pieters, Krijin wit, Yama JGJ (1993) Fractured Reservoir Simulation Case studies ",Koninklijike/shell E&P laboratorium SPE/1993, Bahrain.

15. Daltaban TS, Noyola A (2002) "An investigation into the technical feasibility of gas injection into fractured CHUC reservoir in the gulf of Mexico", SPE 74357, Villahermosa, Mexico.

16. Rogers JD, Grigg RB (2000) "A literature analysis of the WAG injectivity abnormalities in the CO2 process" SPE 59329, presented at the 2000 review SPE/DOE Improved Oil Recovery symposium on held in Tulsa, Oklahoma.

17. Willhite GP (1986) "Water flooding," society of engineers, Richardson, TX.

34

A Numerical Study of Temperature Profile by Coupling Memory-Based Diffusivity Model With Energy Balance During Thermal Flooding

Amjed Hassan M and Enamul Hossain M*

Department of Petroleum Engineering, King Fahd University of petroleum & Minerals (KFUPM), Dhahran 31261, Saudi Arabia

Abstract

Accurate estimation of the temperature distribution within a reservoir undergoing a thermal recovery operation is a key factor in process design, reservoir management and production forecasting. The thermal and rheological properties of the reservoir rock and fluids play significant roles in the heat transfer between the formation matrix and flowing fluids.

The memory-based diffusivity equation is implemented as a momentum-balance to present continuous alteration of rock and fluid properties and to investigate the temperature propagation during thermal flooding process. This model is coupled with recently developed energy balance equation to investigate the different parameters that influence the temperature profile. Numerical solution of the coupled mathematical model is presented for the case of equal rock and fluid temperature. It is assumed that the rock attains the fluid temperature instantaneously, that is, the rock and fluid temperatures are assumed equal throughout the reservoir. Matlab 7.10 program is used to carry out the computation and provides temperature profiles.

Results show that, coupling the memory-based diffusivity model with energy balance leads to more reasonable temperature profiles during the thermal flooding. The distribution of reservoir temperature with respect to time and distance can be estimated by coupling the memory-based equation and the mathematical tool which were developed by Hossain et al., in addition, it can be concluded that, the fluid velocity, time and the rheological properties, have important effects on the temperature distributions throughout the reservoir. In future, the results of the numerical solution can be integrated with lab experiment results to predict performance of thermal flooding process and better understanding of reservoir management.

Keywords: Memory concept; Diffusivity equation; Rock; Fluid interaction; Porous media; Integro-differential equation; Heat transfer coefficient; Peclet number

Introduction

The heat transfers from region of higher temperature to lower-temperature region due to the temperature differences in all physical models. This transport process continues until the system attains a uniform temperature. Conduction, radiation, and convection are considered to be the main mechanisms of the heat flux. In any system, conduction, radiation play significant rule in heat transfer, while convective heat transfer is of utmost interest in porous media where fluid velocity is the major concern. In the hydrocarbon reservoirs, the temperature distribution is an important issue due to its utilization in detecting water or gas influx or type of fluid entering into the wellbore, therefore, it will enhance the better understanding of reservoir management and future prediction of the performance of reservoir. During thermal EOR operations, the highly complex characteristics of rock–fluid interaction play a vital role in heat transfer between the rock matrix and flowing hydrocarbons. Such heat transfer is the major factor governing the temperature profile within the reservoir. It is very important to understand the complex rheology of fluid flowing through formation rock. The most challenging issues during the development of models are the time dependency of the rock and fluid properties and the scale of time. Therefore, the notion of memory can be introduced to model the variations of rock and fluid properties with time. In addition, memory term can be used to analyze the rheological behavior of rock and fluid properties when the rock is not in thermal equilibrium with the fluid(s) [1].

Hossain et al. [2] demonstrated extensive studies of fluid memory concept on the available literature and models. They presented that

different fluid properties such as stress or density have been used to identify the fluid memory concept. Caputo [3] modified Darcy's law to consider the memory concept by introducing the order of the fractional derivative (α) and the pseudo-permeability ratio of medium with memory to fluid viscosity (η). These two parameters have been introduced instead of formation permeability and fluid viscosity in conventional Darcy's law to simulate the effects of changing permeability and viscosity over time. Zhang [4] defined memory concept as a function of time and space as the forward time events is dependent on previous time events.

In the application of porous media, Iaffaldano et al. [5] investigated the permeability variations with time in sand formations; they developed a memory model for flowing of fluids in porous formation, their model matches with the experimental results. They concluded that the flux rate variations observed during the experiments were compatible with the compaction of sand. This variation was due to the amount of fluid that went through the grains locally. As a result, there was a reduction of porosity.

*Corresponding author:** Enamul Hossain M, Department of Petroleum Engineering, King Fahd University of petroleum & Minerals (KFUPM), Dhahran 31261, Saudi Arabia, E-mail: amjed.moh06@gmail.com

Hossain [6,7] developed a new porous media diffusivity equation with the inclusion of rock and fluid memories by invoking time-dependent viscosity and permeability. The developed model is highly non-linear and has an option of considering the memory variables. The non-linear solution of this model shows the difference between the predication of the classical diffusivity equation and the proposed memory- based model. The results show that the proposed model could be used for a wide range of applications. Their work establishes the contribution of memory effect on the fluid flow through the reservoir formations.

Zavala-Sanchez et al. [8] conducted dynamic model to study the effective mixing and spreading dynamics at preasymptotic times in terms of effective average transport coefficients. They found that, the system remembers its initial state, which was defined as memory effects for the effective transport coefficients.

Giuseppe et al. [9] proposed a modified constitutive equation, the introduced the memory formalism on both the flux term and the pressure gradient variations. They modelled the memory effect by using the fractional order derivatives. In addition, they performed three laboratory experiments to validate their model, homogeneous and heterogeneous formations with different particle size were employed. Their model successfully provides an efficient and useful mean to quantify the memory effects.

Hossain et al. [10] studied the memory effect and pressure variations on the stress-strain relationship. They obtained the variation of shear stress as a function of strain rate for crude oil to investigate the effects of fluid memory. They found that, the fluid memory causes nonlinearity, which leads to chaotic behaviour of the stress-strain relationship. Their model can be used in rheological study and reservoir simulation; also it can be used to select the proper surfactant or foam for EOR processes.

During the enhanced oil recovery processes, the fluid flows through a porous medium and this will lead to alteration of rock and fluid properties during thermal recovery [11-14]. The formation permeability and porosity may vary locally with time due to several reasons such as chemical dissolution of the medium, swelling and flocculation, pore plugging and fine precipitations [5,9,15]. Most importantly, the alteration of rock and fluid properties controls the temperature distribution within the reservoir formation. Yoshioka et al. [16,17] found that, the fluid velocity and time have strong effects on the temperature profile. Therefore, it is important to investigate the effects of memory on the rheology of the rock–fluid system [18].

Hossain et al. [1] studied the role of memory concept to investigate the temperature profile during thermal EOR process; they used continuous time function to model the changing rock and fluid properties. They found that, the temperature profile for rock and fluid are sensitive to the fluid memory, rheological properties and time. Their studies provide a better understanding of heat transfer phenomena during thermal operations in porous media.

The traditional models are unable to handle the alteration of rock and fluid properties with time during thermal operations [19]. In 2008 Hossain showed that, continuous heat transfer within the fluid and rock matrix may result in changing the porosity, permeability, and rheology of the reservoir due to memory and temperature variation. However, previous works did not consider the thermal effects in terms of Peclet number and other heat transfer coefficients when the rock and fluid temperatures are considered equal (Ts = Tf). In addition, the continuous alteration of fluid and pore space properties may be greatly influenced by the fluid memory, especially in geothermal reservoirs [11].

Hossain et al. [1] developed model equations to study the effects of including the memory function in the fluid flow behaviour during hot water injection into a hydrocarbon reservoir. They investigated the major role of alteration of various rock and fluid properties during thermal operations in terms of Peclet number and they proposed three dimensionless numbers associated with heat transfer in porous media. Their study assumes that the rock attains the fluid's temperature instantaneously, which imply, the rock and fluid temperatures are equal throughout the system. In addition, the developed model equation describes how the fluid and rock properties are dependent on the continuous time function.

In this work, the temperature distributions within a reservoir undergoing a thermal flooding process are estimated by coupling the mathematical model developed by Hossain et al., [1] and the memory-based diffusivity equation. The fluid velocity has a significant role on the heat diffusion throughout the reservoir. The previous studies assumed constant fluid velocity over time, however in this work; the fluid velocity was calculated using numerical concepts based on the pressure distribution. First, the memory-based diffusivity equation was solved to obtain the reservoir pressures, then the pressure gradient was determined using finite difference concept, thereafter, the numerical integration was applied to find the distribution of fluid velocity. Finally, the temperature profiles were calculated based on the fluid velocity distribution. Results concluded that, the rock and fluid memories have significant roles on the temperature profiles and the rheological parameters of the rock and fluid. Moreover, the temperature distribution with respect to time and distance can be estimated by coupling the memory- based equation and the mathematical tool, also, the fluid velocity, time and the rheological properties, have important effects on the temperature distributions throughout the reservoir.

Mathematical Formulation

The model considers a porous medium of uniform cross sectional area and that is homogeneous along the x axis. Normal practice assumes fluid flow in porous media to be governed by Darcy's law. In this study, the constitutive equations with memory effects are rewritten by introducing the fractional order derivatives to account for the non-local aspects of fluid flow models [6-8].

$$\frac{1}{n}\frac{\partial \eta}{\partial x}\left[\frac{\int_0^t (t-\xi)^{-\alpha}\left(\frac{\partial^2 p}{\partial \xi \partial x}\right)d\xi}{\Gamma(1-\alpha)}\right] + c_f\frac{\partial p}{\partial x}\left[\frac{\int_0^t (t-\xi)^{-\alpha}\left(\frac{\partial^2 p}{\partial \xi \partial x}\right)d\xi}{\Gamma(1-\alpha)}\right] + \frac{\partial}{\partial x}\left[\frac{\int_0^t (t-\xi)^{-\alpha}\left(\frac{\partial^2 p}{\partial \xi \partial x}\right)d\xi}{\Gamma(1-\alpha)}\right] = \frac{\phi c_t}{\eta}\frac{\partial p}{\partial t}$$

The derivative of fractional order is defined as:

$$Z = \frac{\partial^\alpha}{\partial t^\alpha}\left(\frac{\partial p}{\partial x}\right) = \frac{\int_0^t (t-\xi)^{-\alpha}\left(\frac{\partial^2 p}{\partial \xi \partial x}\right)d\xi}{\Gamma(1-\alpha)}$$

With $0 \leq \alpha < 1$. In Eq. (1), $\Gamma(1-\alpha)$ is the Euler gamma function, using the definition of Z-value, Eq. (1) may be written as:

$$\frac{1}{n}\frac{\partial \eta}{\partial x}Z + c_f\frac{\partial p}{\partial x}Z + \frac{\partial Z}{\partial x} = \frac{\phi c_t}{\eta}\frac{\partial p}{\partial t}$$

Equation (3) is called as the diffusivity equation with memory because it is a combination of the motion equation and the continuity equation. The model is a non-linear and integro-differential form of memory for any axial flow of single phase fluid flow equation for a porous formation.

As a matter of fact, the equation is strictly non-linear because three parts of the equation are non-linear, the first part is non-linear due to dependence of η on pressure, the second part is non-linear since Z is a function of pressure, and the fourth part is a multiplication of porosity, total compressibility, η and pressure derivative with time, therefore, this term also is non-linear.

If there is no thermal equilibrium in the system, a 1D energy balance equations can be written to develop the model equation for describing the temperature propagation in porous media as [14,20], the final form can be written as follow:

$$N_{HA4}\frac{\partial T^*}{\partial t^*} + N_{PeL}\frac{L}{L_c}\frac{\partial T^*}{\partial x^*} - N_{HA3}\frac{\partial^2 T^*}{\partial x^{*2}} = 0$$

$$N_{HA3} = \frac{(k_s + k_f)}{k_e}; N_{PeL} = \frac{L_c\rho_f c_{pf} u_x}{k_e}, \text{ and}$$

$$N_{HA4} = \frac{Mk}{k_e\,\phi\mu c_t}$$

Computations on Eq. (4) are carried out for different rock/fluid parameters where the initial and boundary conditions are defined as:

$T_f^*(x, 0) = T_s(x, 0) = Ti$, in terms of dimensionless form: T_f^* (x, 0) = $T_s^*(x, 0) = 1$

$T_f^*(0, t) = T_s(0, t) = T_{st}$, in terms of dimensionless form: T_f^* (0, t) = $T_s^*(0, t) = T_{st}/T_i$

$T_f^*(L, t) = T_s(L, t) = Ti$, in terms of dimensionless form: T_f^* (L, t) = $T_s^*(L, t) = 1$

Numerical solution

To solve Eq. (4) numerically with the specified initial and boundary conditions, finite difference method with explicit scheme can be used, and then the following equation might be achieved.

$$T_i^{*n+1} = (1 - 2ha_7)T_i^{*n} - (a_6 - ha_7)T_{i+1}^{*n} + (a_6 + ha_7)T_{i-1}^{*n}$$

Where

$$h = \Delta t^* / (\Delta x^*)^2; a_6 = \frac{\Delta t^*}{\Delta x^*}\frac{N_{PeL}}{2N_{HA4}}\frac{L}{L_c}; a_7 = \frac{N_{HA3}}{N_{HA4}}$$

In Eq. (5) the local Peclet number can be defined as $(NPe)L = Lc \rho f$ $cpfum/ke$, where Lc represents a characteristic length such as the mean pore throat diameter of the porous medium. During the computation, Lc can be calculated using Winlad's correlation [21], $Lc = 2\pi rpt$ and log $rpt = 0.732 + 0.588$ log $k - 0.864$ logϕ, where k is in mD and rpt is in microns. Kolodzie and Pittman [21,22] proposed the preceding correlation as the best permeability estimator for sandstones. Other correlations for carbonate rocks are also proposed in the literature. During the computation of T*, the same procedure is used to calculate the Lc for all the figures [23].

To solve Eq. (5), the distribution of fluid velocity should be known, which can be obtained using numerical concepts based on the pressure distribution. First, the memory- based diffusivity equation is solved to obtain the reservoir pressures, then the pressure gradient can be

Fluid and rock properties	Fluid and rock properties
C_{pg} = 29.7263 [KJ/Kg-K]	Sg = 20% [vol/vol]
C_{po} = 2.0934 [KJ/Kg-K]	So = 60% [vol/vol]
C_{ps} = 0.8792 [KJ/Kg-K]	Sw = 20% [vol/vol]
C_{pw} = 4.1868 [KJ/Kg-K]	THw = 550 K
h_c = 280.87 [KJ/h-m²-K]	T_i = 300 K
k_g = 0.0143 [KJ/h-m-K]	ρg = 16.7121 [Kg/m³]
k_o = 1.3962 [KJ/h-m-K]	ρo = 800.923 [Kg/m³]
k_s = 9.346 [KJ/h-m-K]	ρs = 2675.08 [Kg/m³]
k_w = 3.7758 [KJ/h-m-K]	ρw = 1000.0 [Kg/m³]
k_i = 15 x 10⁻¹⁵ [m²]	Φ = 25% [m³/m³]
p_i = 48263299 [pa]	μf = 0.12 pa.s [Ns/m²]
c_f = 1.2473 x 10⁻⁹ [1/pa]	q_i = 17.5 m³/d [110 bbl/day]
c_s = 5.8015 x 10⁻¹⁰ [1/pa]	A = 300 m × 20 m = 6000 m²

Table 1: Fluid and rock property values for numerical computation.

determined using finite difference concept, thereafter, the numerical integration is applied to find the distribution of fluid velocity. Finally, the temperature profiles can be calculated based on the fluid velocity distribution [24,25].

Results and Discussion

Computations are carried out for a reservoir 1000 m long, where hot water is injected at a constant rate of 17.5 m³ of equivalent water volume per day. All assumed rock and fluid parameters are listed in Table 1. The time and distance steps are set at Δx* = 0.0167 and Δt* = 0.00001. Temperature variation is obtained for the case where the fluid and rock temperatures are equal.

Variation of reservoir pressure

Equation (3) was solved numerically using space-based derivative concept, initially Darcy diffusivity equation was assumed to be valid and the distribution of the permeability with time and space remain unchanged. In addition, the reservoir formation was considered to be homogeneous and isotropic in its initial phase. All the coefficients are computed at the average pressure of the grid cell 'i', at time step 'n' [13].

Figure 1 presents the pressure variation against distance from the wellbore towards the outer boundary of the reservoir. The pressure response increases towards the reservoir boundary after 1 Day, 6 months and 2 years respectively. Minimum formation pressure is found around the wellbore and it gradually increases up to its initial reservoir pressure near to the outer boundary. After 1 day of production, the pressure reaches the initial pressure at almost 20 m from the wellbore whereas it reaches at around 185 and 490 after 6 months and 2 years of the production respectively.

Figure 2 illustrates the variation of pressure with time for the distance of 5 m, 150 m and 500 m from the wellbore towards the boundary of the reservoir using the same Eq. (3). Due to the reservoir depletion, the reservoir pressure decreases as the time goes with more pressure drop away from the wellbore.

Variation of fluid velocity

The fluid velocity with memory effect can be determined based on the pressure distribution. The fluid velocity was calculated using numerical concepts as follow; the pressure distribution was estimated by solving the memory-based diffusivity equation, then finite difference concept was used to determine the pressure gradient, thereafter, the numerical integration was applied to find the distribution of fluid velocity.

A Numerical Study of Temperature Profile by Coupling Memory-Based Diffusivity Model With Energy...

223

Figure 3 illustrates the variations of fluid velocity against distance from the wellbore for 1, 5, 12 and 24 months. Velocities increase and reach its pick near to the wellbore, increasing the production time result in increasing the velocity profile. Velocity has a decreasing trend till reach zero after 100 m, 220 m, 350 m and 540 m after 1, 5, 12 and 24 months respectively. This is due to the fact that velocity has zero value because there is no pressure gradient at these locations. These characteristics clearly identify the extreme memory effect on the

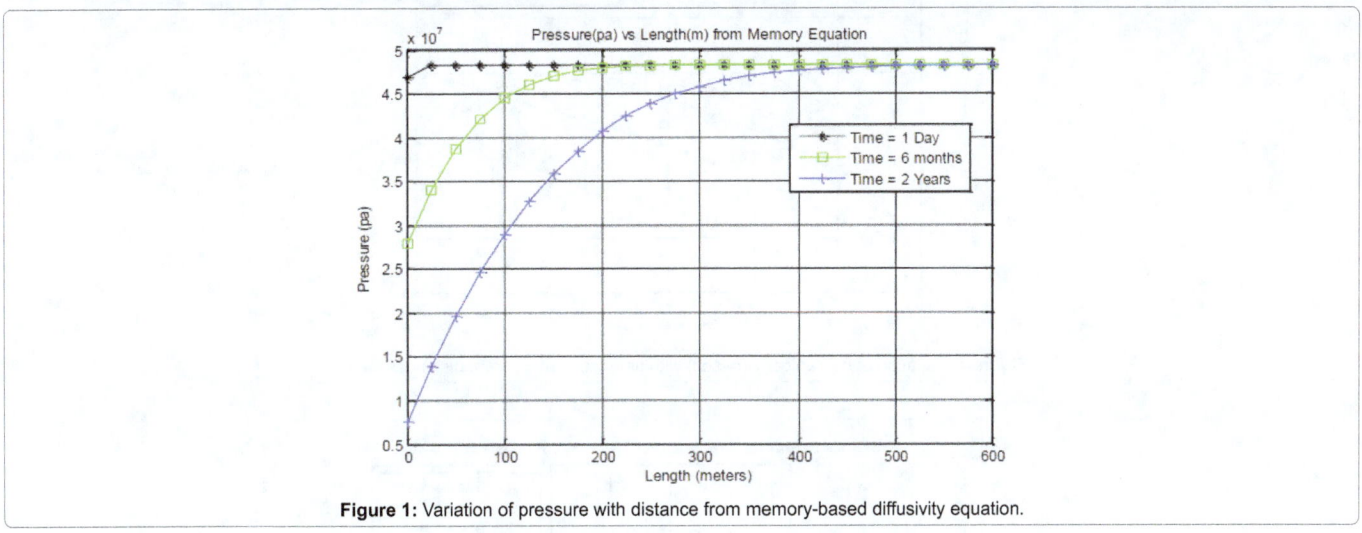

Figure 1: Variation of pressure with distance from memory-based diffusivity equation.

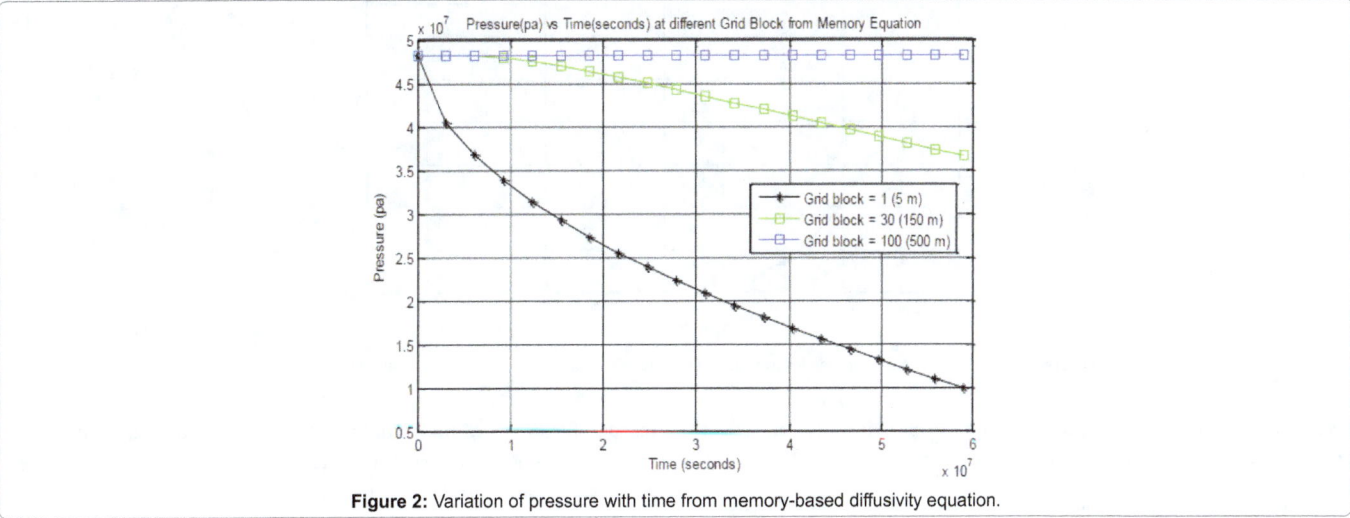

Figure 2: Variation of pressure with time from memory-based diffusivity equation.

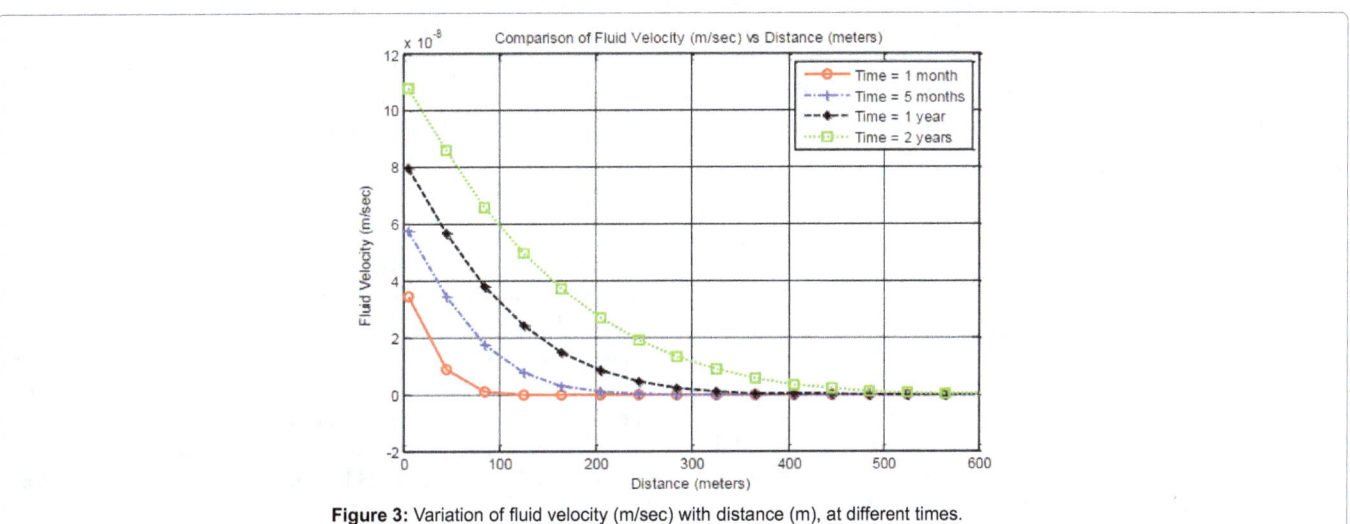

Figure 3: Variation of fluid velocity (m/sec) with distance (m), at different times.

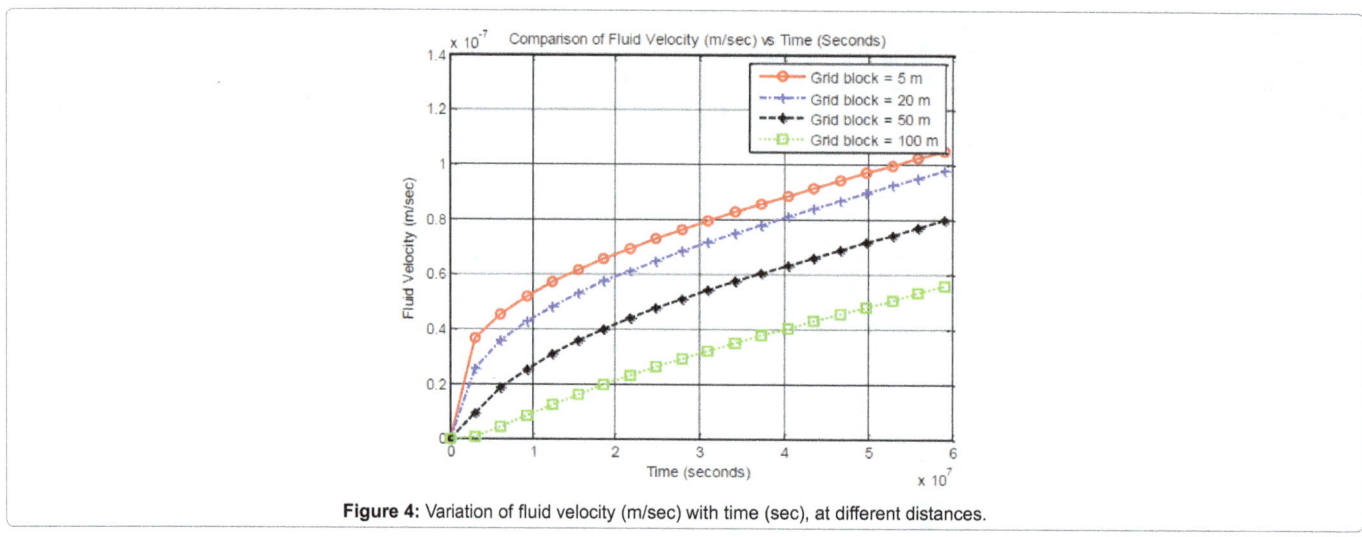

Figure 4: Variation of fluid velocity (m/sec) with time (sec), at different distances.

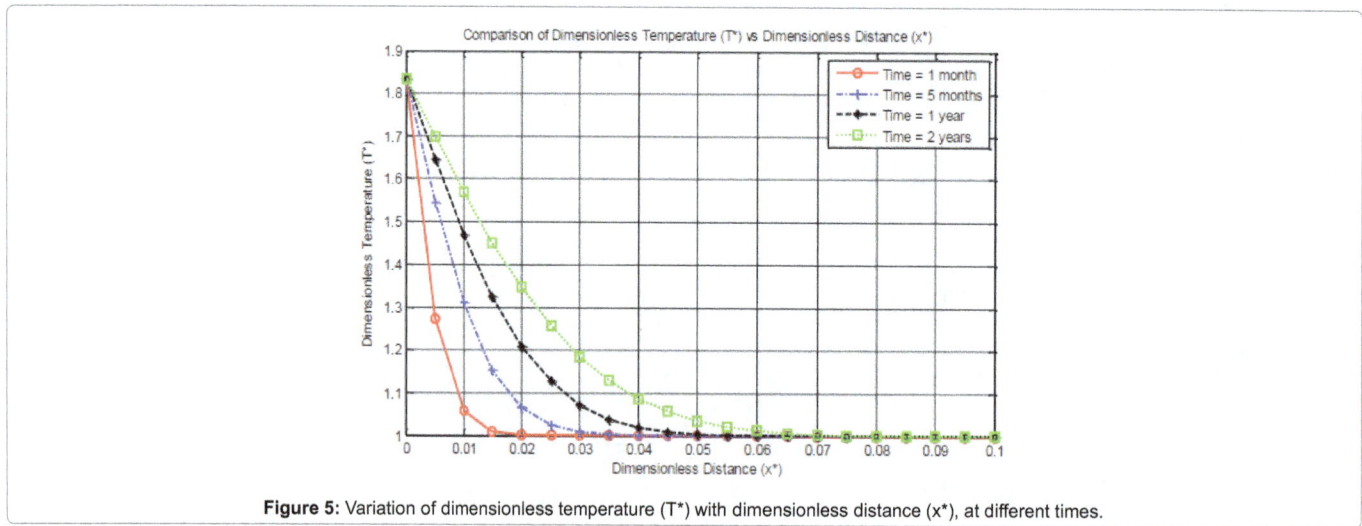

Figure 5: Variation of dimensionless temperature (T*) with dimensionless distance (x*), at different times.

reservoir performance around the wellbore and diminish its effect gradually toward the boundary of the reservoir.

Figure 4 depicts the changes of fluid velocity with memory effects over time for 5 m, 20 m, 50 m and 100 m away from the wellbore. In the area near to wellbore, fluid velocity rises very fast and requires significant time to stabilize. The same trend can be observed for 100 m away from wellbore, however there is less velocity compared to 5 m for a particular time.

Variation of reservoir temperature

Figure 5 shows the variation of dimensionless temperature against dimensionless distance. While Figure 6 illustrates the changing of reservoir temperature with respect to distance after 1 month, 5 months,1 year and 2 years. In both figures, the temperature response decreases towards the reservoir boundary after 1 month, 5 months, 1 year and 2 years respectively.

Since hot water is injected at temperature higher than reservoir temperature, so maximum reservoir temperature is found around the wellbore and it gradually decreases to its initial reservoir temperature near to the outer boundary. After 5 month of production, the

temperature reaches the initial reservoir temperature at almost 35 m from the wellbore whereas it reaches at around 55 and 70 after 1 year and 2 years of the production respectively.

Figure 7 shows the variation of dimensionless temperature against dimensionless time. While Figure 8 illustrates the changing of reservoir temperature with respect to time at 5 m, 20 m, 50 m and 100 m away from the wellbore. In both figures, the temperature response increases with time due to the heat transfer from hot water to reservoir fluid/rock, the rate of heat transfer depends on the thermal properties of reservoir rock and fluids; also the fluid velocity has a significant role on the heat diffusion throughout the reservoir.

Usually hot water is injected at temperature higher than reservoir temperature, so sufficient time is required to achieve stability conditions between the injected water and reservoir fluids/rock, i.e. the reservoir temperature approaches to temperature of injected fluid, however, the reservoir temperature will never be equal to the temperature of injected fluid as stated in the thermodynamic. As a result, after 2 years of continuous hot water injection, the reservoir temperature increases to 510 K, 400 K, 310 K, and 305 K at locations of 5 m, 20 m, 50 m, and 100 m respectively.

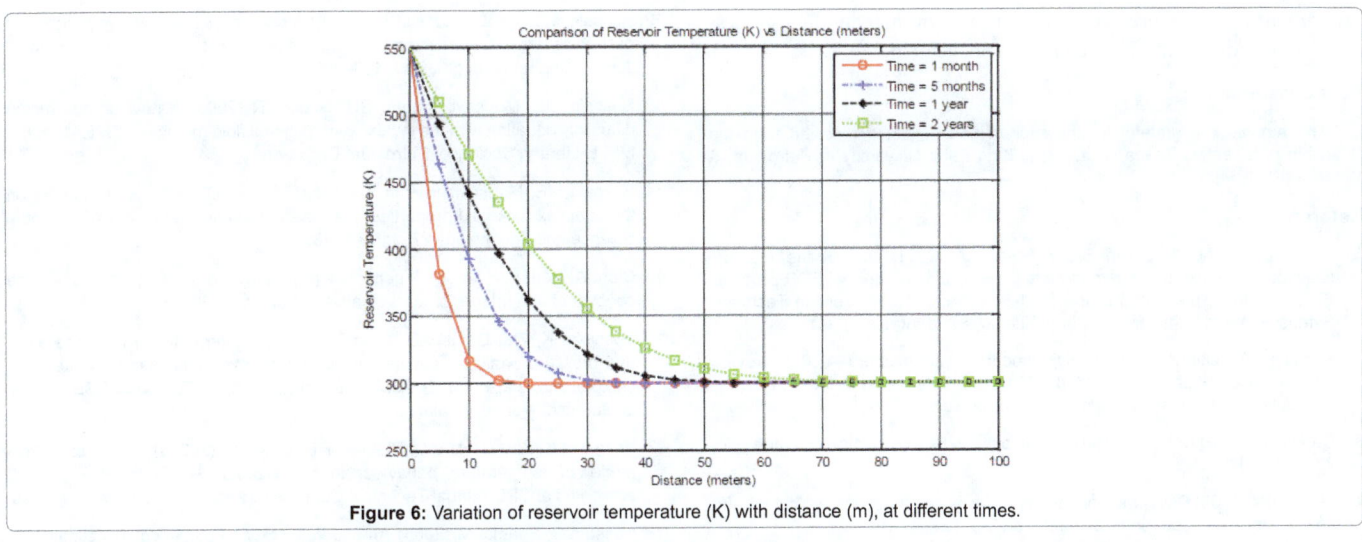

Figure 6: Variation of reservoir temperature (K) with distance (m), at different times.

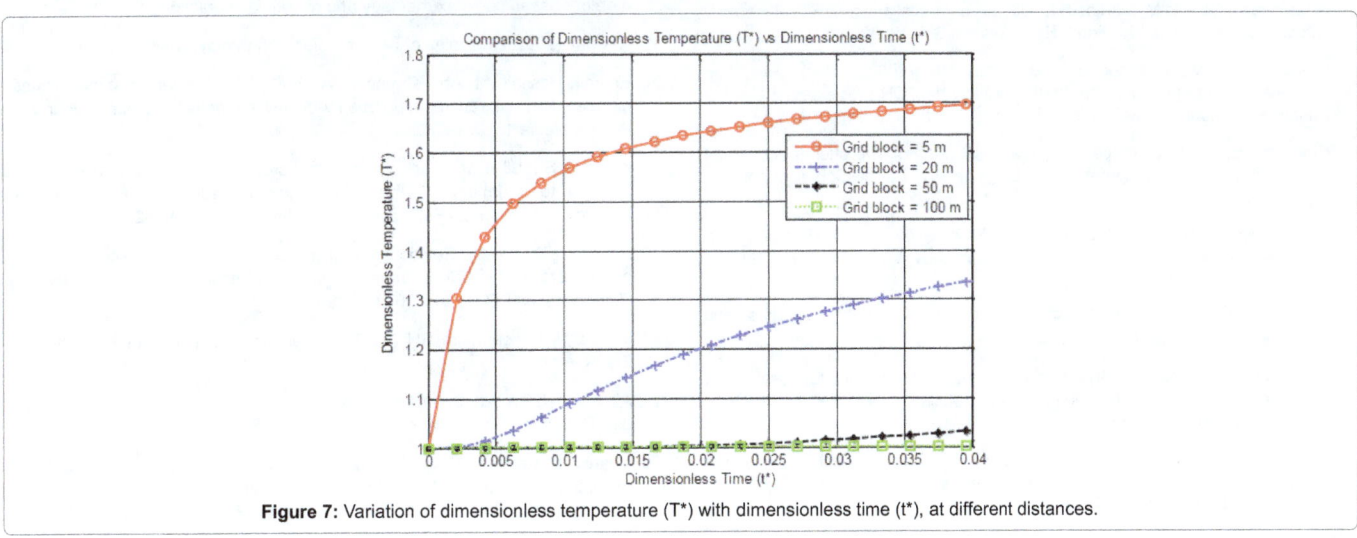

Figure 7: Variation of dimensionless temperature (T*) with dimensionless time (t*), at different distances.

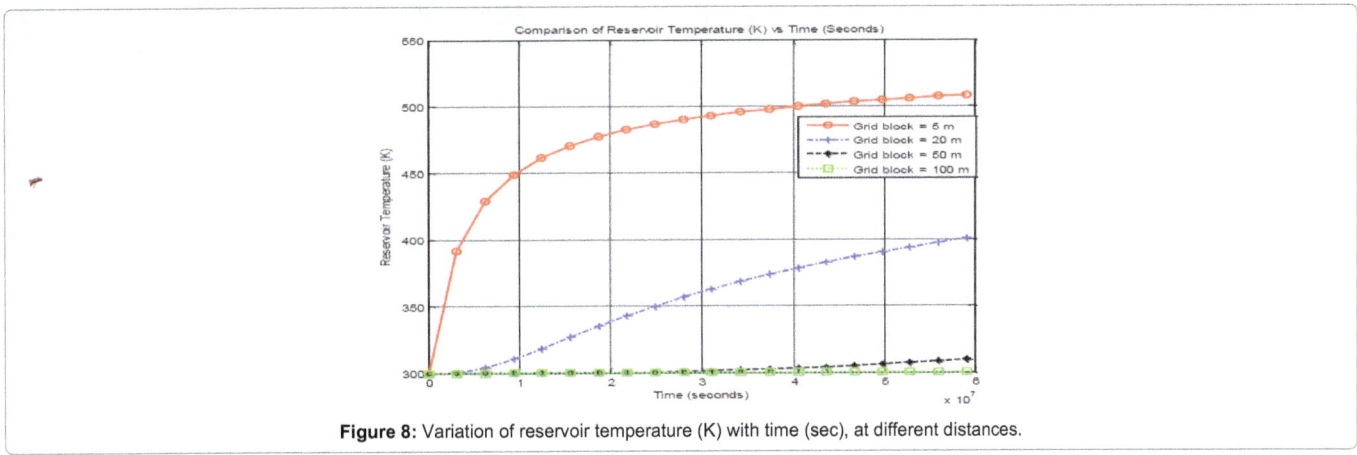

Figure 8: Variation of reservoir temperature (K) with time (sec), at different distances.

Conclusion

The temperature propagation due to thermal flooding process was investigated based on the implementation of memory concept, a numerical solution of the mathematical models which was developed by Hossain et al. is proposed. Results show that, the distribution of reservoir temperature with respect to time and distance can be estimated by coupling the memory-based equation and the mathematical tool which were developed by Hossain et al., in addition, it can be concluded that, the fluid velocity, time and the rheological properties, have important effects on the temperature distributions throughout the reservoir. In future, the results of the numerical solution can be integrated with lab experiment results to predict performance of thermal flooding process and better understanding of reservoir management.

Acknowledgement

The authors are grateful for the support and guidance received from the Deanship of Scientific Research (DSR) at King Fahd University of Petroleum & Minerals (KFUPM).

References

1. Hossain ME, Abu-Khamsin SA, Al-Helali AA (2011) Use of the memory concept to investigate the temperature profile during a Thermal EOR process. SPE – 149094, presented at the 2011 SPE Saudi Arabia Section Technical Symposium and Exhibition held in Al-Khobar, Saudi Arabia.

2. Hossain ME, Islam MR (2006) Fluid properties with memory–A critical review and some additions. CIE–00778, Proc. 36th International Conference on Computers and Industrial Engineering, Taipei, Taiwan.

3. Caputo M (1999) Diffusion of fluids in porous media with memory. Geothermics 23: 113-130.

4. Zhang HM (2003) Driver memory, traffic viscosity and a viscous vehicular traffic flow model. Transportation Research Part B 37: 27–41.

5. Iaffaldano G, Caputo M, Martino S (2006) Experimental and theoretical memory diffusion of water in sand. Hydrol. Earth System Sci. 10: 93-100.

6. Hossain ME, Mousavizadegan SH, Islam MR (2008b) Rock and fluid temperature changes during thermal operations in EOR processes. J Nat Sci Sustainable Technol 2: 347–378.

7. Hossain ME, Mousavizadegan SH, Islam MR (2008c) The effects of thermal alterations on formation permeability and porosity. Petrol. Sci. Technol. 26: 1282–1302.

8. Zavala-Sanchez V, Dentz M, Sanchez-Vila X (2009) Characterization of mixing and spreading in a bounded stratified medium. Adv. Water Resour. 32: 635–648.

9. Giuseppe ED (2009) Flux in porous media with memory: Models and experiments. Transport in Porous Media.

10. Hossain ME, Islam MR (2009) An advanced analysis technique for sustainable petroleum operations. VDM Verlag (ed.) Dr. Muller Aktiengesellschaft & Co. KG, Saarbracken, Germany.

11. Hossain ME, Mousavizadegan SH, Ketata C, Islam MR (2007a) A novel memory based stress-strain model for reservoir characterization. J Nat Sci Sustainable Technol. 1: 653–678.

12. Hossain ME, Liu L, Islam MR (2007b) Inclusion of the memory function in describing the flow of shear-thinning fluids in porous media. International Journal of Engineering (IJE) 3: 458– 477.

13. Hossain ME, Mousavizadegan SH, Islam MR (2008) A new porous media diffusivity equation with the inclusion of rock and fluid memories. SPE–114287-MS, E-Library, Society of Petroleum Engineers.

14. Hossain ME, Mousavizadegan SH, Islam MR (2009b) Effects of memory on the complex rock-fluid properties of a reservoir stress-strain model. Petroleum Science and Technology 27: 1109-1123.

15. Cloot A, Botha JF (2006) A generalised groundwater flow equation using the concept of non-integer order derivatives. Water SA 32: 1–7.

16. Yoshioka K, Zhu D, Hill AD, Lake LW (2005a) Interpretation of temperature and pressure profiles measured in multilateral wells equipped with intelligent completions, paper SPE-94097 presented at SPE Europec/EAGE annual conference, Madrid, Spain.

17. Yoshioka K, Zhu D, Hill AD, Dawkrajai P, Lake LW (2005b) A comprehensive model of temperature behaviour in a horizontal well. paper SPE-95656 presented at SPE Annual Technical Conference and Exhibition, Dallas, Texas.

18. Hossain S, Enamul M (2008) An experimental and numerical investigation of memory-based complex rheology and rock/fluid interactions.

19. Kaviany (2002) Principles of heat transfer. John Wiley, New York.

20. Enamul Hossain M, Abu-Khamsin SA (2012) Development of dimensionless numbers for heat transfer in porous media using a memory concept. Journal of Porous Media 15: 957-973.

21. Kolodzie S (1980) Analysis of pore throat size and use of the Waxman-Smits equation to determine OOIP in Spindle Field. Colorado Society of Petroleum Engineers, 55th Annual Fall Technical Conference Paper 9382.

22. Pittman ED (1992) Relationship of porosity and permeability to various parameters derived from mercury injection-Capillary Pressure Curve for Sandstone. AAPG Bulletin 76: 191-198.

23. Marx JW, Langenheim RH (1959) Reservoir heating by hot fluid injection. Trans. AIME.

24. Almehaideb RA (2003) Improved correlations for fluid properties of UAE crude oils. Petroleum Science and Technology 21: 1811–1831.

25. İscan AG, Kök MV, Bagci AS (2006) Estimation of permeability and rock mechanical properties of limestone reservoir rocks under stress conditions by strain gauge. Journal of Petroleum Science and Engineering 53: 13-24.

Permissions

The contributors of this book come from diverse backgrounds, making this book a truly international effort. This book will bring forth new frontiers with its revolutionizing research information and detailed analysis of the nascent developments around the world.

We would like to thank all the contributing authors for lending their expertise to make the book truly unique. They have played a crucial role in the development of this book. Without their invaluable contributions this book wouldn't have been possible. They have made vital efforts to compile up to date information on the varied aspects of this subject to make this book a valuable addition to the collection of many professionals and students.

This book was conceptualized with the vision of imparting up-to-date information and advanced data in this field. To ensure the same, a matchless editorial board was set up. Every individual on the board went through rigorous rounds of assessment to prove their worth. After which they invested a large part of their time researching and compiling the most relevant data for our readers.

The editorial board has been involved in producing this book since its inception. They have spent rigorous hours researching and exploring the diverse topics which have resulted in the successful publishing of this book. They have passed on their knowledge of decades through this book. To expedite this challenging task, the publisher supported the team at every step. A small team of assistant editors was also appointed to further simplify the editing procedure and attain best results for the readers.

Apart from the editorial board, the designing team has also invested a significant amount of their time in understanding the subject and creating the most relevant covers. They scrutinized every image to scout for the most suitable representation of the subject and create an appropriate cover for the book.

The publishing team has been an ardent support to the editorial, designing and production team. Their endless efforts to recruit the best for this project, has resulted in the accomplishment of this book. They are a veteran in the field of academics and their pool of knowledge is as vast as their experience in printing. Their expertise and guidance has proved useful at every step. Their uncompromising quality standards have made this book an exceptional effort. Their encouragement from time to time has been an inspiration for everyone.

The publisher and the editorial board hope that this book will prove to be a valuable piece of knowledge for researchers, students, practitioners and scholars across the globe.

List of Contributors

Chirwa EMN and Bezza FA
Water Utilisation and Environmental Engineering Division, Department of Chemical Engineering, University of Pretoria, Pretoria 0002, Republic of South Africa

Emina R, Obiadi II and Obiadi CM
Department of Geological Sciences, Nnamdi Azikiwe University, Awka, Nigeria

Abbas Mamudu, Olafuyi Olalekan and Giegb-efumwen Peter Uyi
University of Benin, Benin City, Nigeria

Nkemakolam Izuwa and Basil C Ogbunude
Department of Petroleum Engineering, Federal University of Technology, Owerri, Peace Wokoma, University of Port Harcourt, Nigeria

Marfo SA
World Bank African Centre of Excellence in Oilfield Chemicals Research, IPS, Uniport, Nigeria : Petroleum Engineering Department, University of Mines and Technology, Tarkwa, Ghana

Appah D
Department of Gas Engineering, Uniport, Nigeria

Joel OF
Centre for Petroleum Research and Training, IPS, Uniport, Nigeria

Ofori-Sarpong G
Minerals Engineering Department, University of Mines and Technology, Tarkwa, Ghana

As'ad AM, Yeneneh AM and Obanijesu EO
Department of Chemical Engineering, Curtin University, Bentley Campus, Perth, W.A. 6102, Australia

Sellami MH, Bellemharbet K and Djabbour N
Process Engineering Department, Laboratory of Process Engineering, Ouargla University 30000 Algeria

Loudiyi K
Renewable Energies Laboratory (REL) Al Akhawayne University, Ifrane, Morocco

Peprah Agyare Godwill and Jackson Waburoko
China University of Geosciences, Wuhan, Hubei, China

Aramesh Shahbazi and Behnam Rezaei Nasab
Faculty of Law and Political Sciences, Allameh Tabatabaei University, Tehran, Iran

Banapurmath NR, Sankalp Patil, Naveen G and Sanketh Tonannavar
B.V.B College of Engineering and Technology, Hubli, Karnataka, India

Chavan AS, Bansode SB and Tandale MS
Dr. Babasaheb Ambedkar Technological University, Lonere, Raigad, MS, India

Keerthi Kumar N
B.M.S. Institute of Technology and Management, Bangalore, Karnataka, India

Oje Obinna A
Department of Chemistry/Biochemistry/Molecular Biochemistry, Federal University Ndufu Alike Ikwo, Ebonyi State, Nigeria

Ubani Chibuike S and Onwurah INE
Department of Biochemistry, University of Nigeria, Nsukka, Enugu State, Nigeria

Mahmood Amani and Mohamed Almodaris
Texas A&M University, Qatar

Awolayo AN and Hemanta K Sharma
Department of Chemical and Petroleum Engineering, University of Calgary, 2500 University Drive, N.W. Calgary, Alberta, Canada

Souleyman A Issaka, Abdurahman H Nour and Rosli Mohd Yunus
Faculty of Chemical and Natural Resources Engineering, University Malaysia Pahang, Malaysia

Maryam Shirinkar and Masoud Zolanvar
North Branch of Islamic Azad University of Tehran, Iranian Offshore Oil Company, Iran

Elkady MF
Chemical and Petrochemical Engineering Department, Egypt-Japan University of Science and Technology, New Borg El-Arab City, Alexandria, Egypt
Advanced Technology and New Materials and Research Institute (ATNMRI), City of Scientific Research and Technological Applications, Alexandria, Egypt

Ahmed Zaatout and Ola Balbaa
Chemical Engineering Department, Faculty of Engineering, Alexandria University, Alexandria, Egypt

Mahmood Amani, Abdul Salam Abd, Abdulrahman Al-Hardan, Alireza Roustazadeh, Rommel Yrac
Texas A&M University, Qatar

Durkin SM and Moroziyuk OA
Department of Exploitation and Development of Oil and Gas Fields and Subsurface Hydromechanics, Candidate of Technical Sciences, Ukhta State Technical University, Ukhta, Russia

Ruzin LM and Polishvayko DV
Department of Exploitation and Development of Oil and Gas Fields and Subsurface Hydromechanics, Doctor of Technical Sciences, Ukhta State Technical University, Ukhta, Russia

Abzaletdinov GA
Department of Petroleum Engineering, Louisiana State University, Baton Rouge, USA

Muhammad Ali Theyab and Pedro Diaz
London South Bank University, 103 Borough Road, SE1 0AA, London, UK

Saeed Morsali
Faculty of Applied Sciences, Department of Environmental Science, Gazi University, Ankara, Turkey

Prince MJA
Department of Petroleum Engineering, AMET University, Chennai - 603112, Tamil Nadu, India

Arinze Emmanuel Emeka
Department of Civil Engineering, Michael Okpara University of Agriculture, Umudike, Abia State, Nigeria

Mahmood Amani, Idris M, Abdul Ghani M, Dela Rosa N, Carvero A and Yrac R
Texas A&M University at Qatar, Qatar

Tarek M Aboul Fotouh
Department of Mining and Petroleum Engineering, Faculty of Engineering, Al-Azhar University, Egypt
Department of Chemical Engineering, The British University in Egypt, Egypt

Omar A Mazen
Department of Chemical Engineering, The British University in Egypt, Egypt

Ibrahim Ashour
Department of Chemical Engineering, Faculty of Engineering, Minia University, Egypt
Zewail City University of Science and Technology, Sheikh Zayed 12588, 6th October, Egypt

Kuiqian Ma, Yanlai Li and Ting Sun
China National Offshore Oil Corporation (CNOOC) Ltd., Tianjin, China University of Petroleum-Beijing, P.R. China

Rahman MA, Jacqueline Stevens, Jared Pardy and Danika Wheeler
Faculty of Engineering and Applied Sciences, Memorial University of Newfoundland, St. John's, NL, Canada

Sulaiman WRW, Abbas AH, Jaafar MZ and Idris AK
Faculty of Chemical and Energy Engineering, University of Technology Malaysia, Malaysia

Noura El-Mehbad
Faculty of Science, Najran University, Saudi Arabia

Agbakwuru J, Ogunlana A, Oshagbemi O, Rahman MA and Imtiaz S
Texas A&M University at Qatar, Doha, Qatar

Shaimaa Saeed
Chemical Engineering Department, Faculty of Engineering, Minia University, Egypt

Tarek M Aboul-Fotouh
Mining and Petroleum Engineering Department, Faculty of Engineering, Al-Azhar University, Egypt

Hesam Arabzadeh
Department of Chemical Engineering, Ataturk University, Erzurum, Turkey

Mahmood Amani
Petroleum Engineering Program, Texas A&M University at Qatar, Doha, Qatar

Arash Kamari and Deresh Ramjugernath
Thermodynamics Research Unit, School of Chemical Engineering, University of Kwa-Zulu-Natal, Howard College Campus, King George V Avenue, Durban 4041, South Africa

Sohrab Zendehboudi
Department of Chemical Engineering, Massachusetts Institute of Technology (MIT), Cambridge, MA 02139, USA

James J Sheng
Department of Petroleum Engineering, Texas Tech University, TX, USA

Amir H Mohammadi
Thermodynamics Research Unit, School of Chemical Engineering, University of Kwa-Zulu-Natal, Howard College Campus, King George V Avenue, Durban 4041, South Africa
Institute of Research in Chemical Engineering and Petroleum (IRGCP), Paris Cedex, France

Hajnajafi Reza
Department of Petroleum Engineering, University of New South Wales, Sydney, Australia

Amid Arman
Department of Environmental Management HSE, Tehran North branch of Islamic Azad University, Tehran, Iran

Hajnajafi Ghazal
Department of Geology, University of Science and Research of Islamic azad university, Tehran, Iran

Amjed Hassan M and Enamul Hossain M
Department of Petroleum Engineering, King Fahd University of petroleum & Minerals (KFUPM), Dhahran 31261, Saudi Arabia

Index